SUPRAMOLECULAR PHOTOCHEMISTRY

ELLIS HORWOOD SERIES IN PHYSICAL CHEMISTRY

Series Editor: Professor T. J. KEMP, Department of Chemistry, University of Warwick

Atherton, N.	Electron Spin Resonance Spectroscopy
Ball, M.C. & Strachan, A.N.	Chemistry and Reactivity of Solids
Cullis, C.F. & Hirschler, M.	Combustion and Air Pollution: Chemistry and Toxicology
Davies, P.B. & Russell, D.K.	Laser Magnetic Resonance
Horvath, A.L.	Handbook of Aqueous Electrolyte Solutions
Jankowska, H., Kwiatkowski, A. & Choma, J.	Active Carbon
Jaycock, M.J. & Parfitt, G.D.	Chemistry of Interfaces
Jaycock, M.J. & Parfitt, G.D.	Chemistry of Colloids
Keller, C.	Radiochemistry
Ladd, M.F.C.	Symmetry in Molecules and Crystals
Mason, T.J. & Lorimer, P.	Sonochemistry: Theory, Applications and Uses of Ultrasound in Chemistry
Navratil, O., Hala, J., Kopune, R., Leseticky, L., Macasek, F. & Mikulai, V.	Nuclear Chemistry
Paryjczak, T.	Gas Chromatography in Adsorption and Catalysis
Quinchon, J. & Tranchant, J.	Nitrocelluloses
Sadlej, J.	Semi-empirical Methods in Quantum Chemistry
Snatzke, G., Ryback, G. & Slopes, P. M.	Optical Rotary Dispersion and Circular Dichroism
Southampton Electrochemistry Group	Instrumental Methods in Electrochemistry
Wan, J.K.S. & Depew, M.C.	Polarization and Magnetic Effects in Chemistry

BOOKS OF RELATED INTEREST

Balzani, V. & Scandola, F.	Supramolecular Photochemistry
Buxton, G. V. & Salmon, G.A.	Pulse Radiolysis and its Applications in Chemistry
Epton, R.	Chromatography of Synthetic and Biological Polymers: Vols 1 & 2
Harriman, A.	Inorganic Photochemistry
Hearle, J.	Polymers and their Properties: Vol. 1
Horspool, W. & Armesto, D.	Organic Photochemistry
Kennedy, J.F., *et al.*	Cellulose and its Derivatives
Kennedy, J.F., *et al.*	Wood and Cellulosics
Krestov, G.A.	Thermodynamics of Solvation: Solution and Dissolution; Ions and Solvents; Structure and Energetics
Lazár, M., *et al.*	Chemical Reactions of Natural and Synthetic Polymers
Milinchuk, V.K. & Tupikov, V.I.	Organic Radiation Chemistry Handbook
Nevell, T.J.	Cellulose Chemistry and its Applications
Štěpek, J. *et al.*	Polymers as Materials for Packaging
Švec, P. *et al.*	Styrene-based Plastics and their Modifications

SUPRAMOLECULAR PHOTOCHEMISTRY

VINCENZO BALZANI
'G. Ciamician' Department of Chemistry
University of Bologna, Italy

FRANCO SCANDOLA
Department of Chemistry
University of Ferrara, Italy

ELLIS HORWOOD
NEW YORK LONDON TORONTO SYDNEY TOKYO SINGAPORE

.4311620

CHEMISTRY

First published in 1991 by
ELLIS HORWOOD LIMITED
Market Cross House, Cooper Street,
Chichester, West Sussex, PO19 1EB, England

A division of
Simon & Schuster International Group
A Paramount Communications Company

© Ellis Horwood Limited, 1991

Typeset in Times by Ellis Horwood Limited
Printed and bound in Great Britain
by Bookcraft (Bath) Limited, Midsomer Norton, Avon

British Library Cataloguing in Publication Data

Balzani, Vincenzo *1936–*
Supramolecular photochemistry.
1. Photochemistry
I. Title II. Scandola, Franco
541.35
ISBN 0–13–877531–1

Library of Congress Cataloging-in-Publication Data

Balzani, Vincenzo, 1936–
Supramolecular photochemistry / Vincenzo Balzani, Franco Scandola.
p. cm. (Ellis Horwood series in physical chemistry)
Includes bibliographical references and index.
ISBN 0–13–877531–1
1. Photochemistry. 2. Macromolecules. I. Scandola, Franco, 1942– . II. Title. III. Series.
QD715.B34   1990
541.3'5–dc20
90–43938
CIP

Table of contents

Preface 11

Acknowledgements 13

List of abbreviations 15

1. Scope and limitations 19

2. Principles of molecular photochemistry 25
  2.1 Introduction 25
  2.2 The nature of light 25
  2.3 Potential energy surfaces 26
  2.4 Electronic states and electronic configurations 30
  2.5 Light absorption 34
  2.6 Unimolecular processes of excited states 37
    2.6.1 Vibrational relaxation 37
    2.6.2 Radiative deactivation 38
    2.6.3 Radiationless deactivation 38
    2.6.4 Chemical reaction 40
    2.6.5 Kinetic aspects 41
  2.7 Bimolecular processes of excited states 43
    2.7.1 General considerations 43
    2.7.2 Thermodynamic aspects 44
    2.7.3 Kinetic aspects 45

3. Supramolecular properties 51
  3.1 Introduction 51
  3.2 Supermolecules as localized multicomponent electronic systems 54
  3.3 Properties of molecular components 58
  3.4 Intercomponent processes 60

3.4.1 Radiative processes 61
3.4.2 Thermal electron transfer 63
3.4.3 Photoinduced electron transfer 65
3.4.4 Electron-transfer kinetics. 65
3.4.5 Electronic energy transfer 71
3.4.6 Energy *vs* electron transfer. 73

4. Control and tuning of excited-state properties of molecular components . . . 76
4.1 Introduction. 76
4.2 Perturbing spectroscopic levels. 77
4.3 Introducing new energy levels 80
4.4 Introducing nuclear constraints. 82
    4.4.1 Covalent links 83
    4.4.2 Host–guest interactions 85
4.5 Conclusions 86

5. Covalently-linked systems: photoinduced electron transfer 89
5.1 Introduction. 89
5.2 Kinetic considerations 94
5.3 Organic systems with steroid-type bridges. 98
5.4 Organic systems with norbornylogous bridges 102
5.5 Organic systems with aromatic bridges. 107
5.6 Porphyrin–quencher systems 110
    5.6.1 Porphyrin–quinone systems with trypticene bridges 110
    5.6.2 Porphyrin–quinone systems with bicyclooctane bridges. 113
    5.6.3 Porphyrin–quinone systems with polyene bridges 114
    5.6.4 Porphyrin–quinone systems with flexible bridges. 116
    5.6.5 Structural effects in porphyrin–quinone systems 119
    5.6.6 Capped porphyrin–quinone systems. 120
    5.6.7 Porphyrin–viologen systems 121
    5.6.8 Porphyrin–host covalently-linked systems 122
5.7 Systems based on metal complexes 123
    5.7.1 Chromophore–quencher complexes 125
    5.7.2 Binuclear metal complexes. 132
    5.7.3 Polynuclear metal complexes 140
5.8 Triad and tetrad systems featuring photoinduced charge
    separation 145
5.9 Macromolecular systems 150

6. Covalently-linked systems: electronic energy transfer 161
6.1 Introduction. 161
6.2 Supramolecular systems based on organic molecular components. . . . 161
    6.2.1 Systems with rigid or conformationally-restricted bridges 161
    6.2.2 Systems with flexible bridges. 167
6.3 Supramolecular systems based on metal complexes 170
    6.3.1 Introduction 170
    6.3.2 Complexes linked through bridges 170

6.3.2.1 Identical complexes. 170
6.3.2.2 Complexes of different metals 172
6.3.2.3 Complexes of different ligands. 173
6.3.3 Metal-containing fragments bridged by polyimine ligands 174
6.3.3.1 Introduction 174
6.3.3.2 Symmetrical homometallic binuclear complexes 174
6.3.3.3 Unsymmetrical homometallic binuclear complexes. 178
6.3.3.4 Heterometallic binuclear complexes 178
6.3.3.5 Complexes of higher nuclearity 180
6.3.4 Metal-containing fragments bridged by cyanide ligands 184
6.3.4.1 Oligonuclear ruthenium complexes. 184
6.3.4.2 Ruthenium–chromium complexes 186
6.3.4.3 Cobalt–chromium complexes 188
6.3.4.4 Chromium–chromium complexes. 189
6.3.4.5 Cobalt–cobalt complexes 190
6.3.5 Other systems 190

7. Structural changes in photoflexible systems. 197
7.1 Introduction. 197
7.2 Photoactive components 197
7.2.1 Olefin-type compounds 198
7.2.2 Azobenzene and related compounds 199
7.2.3 Spiropyrans and related compounds. 200
7.2.4 TICT compounds. 202
7.3 Supramolecular systems 204
7.3.1 Systems involving crown ethers 204
7.3.1.1 Crown ethers with an intramolecular bridge 204
7.3.1.2 Cylindrical and phane-type crown ethers 205
7.3.1.3 "Butterfly" and capped crown ethers 208
7.3.2 Systems involving cyclodextrins. 210
7.3.3 Systems involving metal complexes 211
7.3.4 Other systems 212
7.4 Micelles, membranes, liquid crystals, and polymers. 215
7.4.1 Photoeffects on micelles 215
7.4.2 Photocontrol of membrane functions 215
7.4.3 Photoresponsive liquid crystals 217
7.4.4 Photoresponsive polymers 217

8. Ion pairs 226
8.1 Introduction. 226
8.2 Electronic energy levels 227
8.3 Correlations between optical and thermal quantities 230
8.4 Photoinduced processes 231
8.4.1 Dynamic and static quenching. 233
8.4.2 Direct measurements of electron-transfer rates. 236
8.4.3 Photochemical reactions 237
8.4.3.1 Metal complexes 239

      8.4.3.2 Organic compounds 241
     8.4.4 Luminescence 243
   8.5 Control and tuning of excited state properties 243

9. Electron donor–acceptor complexes and exciplexes 248
   9.1 Introduction 248
   9.2 Absorption spectra 251
   9.3 Luminescence 254
   9.4 Photoinduced processes 257

10. Host–guest systems 267
   10.1 Introduction 267
   10.2 Crown ethers and related species 268
     10.2.1 Complexation of metal cations 268
     10.2.2 Adducts with metal complexes 274
     10.2.3 Other systems 278
   10.3 Aza macrocycles and related species 279
     10.3.1 Complexation of metal cations 280
     10.3.2 Adducts of metal complexes with polyammonium macrocyclic
        receptors 280
   10.4 Cyclophanes and related species 283
   10.5 Cyclodextrin inclusion compounds 288
     10.5.1 Introduction 288
     10.5.2 Photophysics 289
     10.5.3 Photochemistry 295
   10.6 Adducts of DNA 299
     10.6.1 Introduction 299
     10.6.2 Intercalation 299
     10.6.3 Spectroscopy and photophysics 301
     10.6.4 Photochemistry 304
     10.6.5 Conclusions 306

11. Other systems 319
   11.1 Caged metal ions 319
     11.1.1 Introduction 319
     11.1.2 Cobalt complexes 319
     11.1.3 Chromium complexes 324
     11.1.4 Lanthanide complexes 326
     11.1.5 Ruthenium complexes 330
     11.1.6 Conclusions 333
   11.2 Catenanes, rotaxanes, and related species 333
     11.2.1 Introduction 333
     11.2.2 Catenanes and rotaxanes 334
     11.2.3 Catenands and catenates 337
     11.2.4 Helicates 341
   11.3 Proton-transfer processes 342
     11.3.1 Introduction 342

11.3.2 Photoinduced tautomerization 345
11.3.3 Excited-state proton-transfer in metal complexes 347

12. Photochemical molecular devices 355
12.1 Introduction 355
12.2 Artificial *vs* natural devices 356
12.3 Machinery of photochemical molecular devices 356
12.4 PMDs based on photoinduced electron transfer 358
12.4.1 Conversion of light into chemical or electrical energy 359
12.4.2 Photoinduced electron collection 363
12.4.3 Remote electron-transfer photosensitization 365
12.4.4 Switching electric signals 366
12.5 PMDs based on electronic energy transfer 367
12.5.1 Spectral sensitization 367
12.5.2 Antenna effect 370
12.5.3 Remote photosensitization 371
12.5.4 Light-energy up-conversion 374
12.6 PMDs based on photoinduced structural changes 375
12.6.1 Switching electric signals 376
12.6.2 Switching receptor ability 377
12.6.3 Modification of cavity size 377
12.6.4 Activation of coreceptor catalysis 378
12.7 From PMDs to practical microdevices 380
12.7.1 Interfacing with the macroscopic world 380
12.7.2 Spectral sensitization of semiconductor electrodes 381
12.7.3 Sensitization of charge separation in Langmuir–Blodgett films . . . 382
12.7.4 (Supra)molecular shift register 383
12.8 Synthesis and self-organization 385

Author index 395

Subject index 422

Preface

Photochemistry is a natural phenomenon as old as the world, and also a modern branch of science, at the interface between light and matter and at the crossroads of chemistry, physics, and biology. Photochemistry is of paramount importance to life (photosynthesis, vision, phototaxis, etc.) as well as to technology (image reproduction, photocatalysis, photodegradation, etc.). In the last decade photochemistry has reached a remarkable level of experimental and theoretical efficacy. Hitherto most of the fundamental photochemical investigations have dealt with molecular species (*molecular photochemistry*).

In the same way as combination of atoms leads to molecules, combination of molecular components leads to *supramolecular* species (supermolecules). The current literature clearly shows that chemical research is rapidly moving from molecular to supramolecular species. There are at least four reasons for this trend: (1) the high degree of knowledge reached on molecular species; (2) the extraordinary progress made by synthetic methods; (3) the continuous search for new chemical functions; (4) the need to fill the gap which separates chemistry from biology. The award of the 1987 Nobel Prize in Chemistry to C. J. Pedersen, D. J. Cram, and J. M. Lehn "for their development and use of molecules with structure-specific interactions of high selectivity" has given new impetus to studies of supramolecular chemistry.

One of the most interesting aspects of the chemistry of supramolecular systems is their interaction with light and the great variety of processes that may ensue. This is the realm of **supramolecular photochemistry**. In the last few years supramolecular photochemistry has grown very rapidly along several independent directions. The unravelling of the structure of the reaction centre of bacterial photosynthesis (which led J. Deisenhofer, R. Huber, and H. Michel to the 1988 Nobel Prize in Chemistry) has highlighted the tight relationship between supramolecular organization and the achievement of specific light-induced functions, and has greatly stimulated research in the field of supramolecular photochemistry. The aim of this book is to provide a unifying view of the basic aspects and of the various applicative facets of this field. Photochemistry and supramolecular chemistry are, by their own nature, interdisciplinary areas, and this is even truer for supramolecular photochemistry. Decisive contributions to the solution of some topical problems, such as the conversion of sunlight into chemical energy and information and signal processing at the molecular level, are likely to come from progress in this discipline.

Acknowledgements

It is a privilege to express our sincere appreciation to R. Ballardini, F. Barigelletti, S. Campagna, L. De Cola, L. Flamigni, M.T. Gandolfi, M.T. Indelli, A. Juris, M. Maestri, S. Monti, and L. Prodi for their critical reading of various parts of the manuscript and for several suggestions. R. Ballardini, L. De Cola, M.T. Gandolfi, M. Maestri, and L. Prodi have also contributed in many ways to the preparation of the manuscript and of the figures, and C. Chiorboli has taken care of the author index. Support from the other members of our research groups is also acknowledged. We are indebted to V. Cacciari and G. Gubellini for the drawings, to N. Armaroli for gathering documentation, and to Giovanna Balzani and Tiziana Cremonini for their careful typing of the manuscript.

We would also like to thank our many colleagues who sent us reprints and preprints of their papers, and the Editors of various journals for their permission to reproduce some figures.

Last but not least, we would like to record here our debt of gratitude to our first teacher, Professor V. Carassiti, who introduced us into the wonderful world of photochemistry.

Bologna and Ferrara   Vincenzo Balzani
May 1990   Franco Scandola

List of abbreviations

(Includes abbreviations of ligands and other chemical compounds.)

A = acceptor (in Chapter 9 only)
BL = bridging ligand
BLM = bilayer membrane
C = connector
CLDA = covalently-linked donor–acceptor
cr = charge recombination
cs = charge separation
CT = charge transfer
D = donor (in Chapter 9 only)
EDA = electron donor–acceptor
E_{op} = optical electron-transfer transition
FC = Franck–Condon
FCWD = Franck–Condon weighted density of states
H = holder
HOMO = highest occupied molecular orbital
IP = ion pair
IPCT = ion-pair charge transfer
IT = intervalence transfer
L = ligand; connector (Chapters 3 and 5); luminophore (Chapter 12)
LB = Langmuir–Blodgett
LC = ligand-centred
LF = ligand field
LLCT = ligand-to-ligand charge transfer
LMCT = ligand-to-metal charge transfer
LUMO = lowest unoccupied molecular orbital
M(BL)CT = metal-to-(bridging ligand) charge transfer
MC = metal-centred
MLCT = metal-to-ligand charge transfer

MMCT=metal-to-metal charge transfer
MO=molecular orbital
OSCT=outer-sphere charge transfer
Pel=electron-transfer photosensitizer
Pen=energy-transfer photosensitizer
Pi=photoisomerizable compound
PMD=photochemical molecular device
Rel=electron relay
Ren=energy relay
Sel=electron store
SSCT=second-sphere charge transfer
TICT=twisted-intramolecular-charge-transfer
TRMC=time-resolved microwave conductivity
U=energy up-converter

Ligands and other chemical compounds

(The structure number or the number of the figure in which the compound is illustrated is given in parentheses.)

ANS=1-anilino-naphthalene-8-sulphonate
BiBzIm^{2-}= 2,2'-bibenzimidazolate ion (Fig. 6.3)
bidpq=2,2',3,3'-tetra-2-pyridyl-6,6'-biquinoxaline (Fig. 6.3)
biq=2,2'-biquinoline (Fig. 10.13)
bpa=1,2-bis(4-pyridyl)ethane
bpe=t-1,2-bis(4-pyridyl)ethylene (**5.26**)
bpm=2,2'-bipyrimidine (Fig. 6.3)
bpt$^-$=3,5-bis(pyridin-2-yl)-1,2,4-triazolate ion (Fig. 6.3)
bpy=2,2'-bipyridine (Fig. 10.13)
4,4'-bpy = 4,4'-bipyridine
bpy.bpy.bpy cryptand=(ligand of **11.5**)
CA=chloranil
CD=cyclodextrin (Fig. 10.8)
4-CNpy=4-cyanopyridine
2.2.1 cryptand=4,7,13,16,21-pentaoxa-1,10-diazabicyclo[8.8.5]tricosane
                (ligand of **11.4**)
CS=cis-stilbene (**10.35**)
cyclam=1,4,8,11-tetraazacyclotetradecane
dec-bpy=4,4'-dicarboxyethyl bipyridine
diamsar=diaminosarcophagine (ligand of **11.3** with Y=NH$_2$)
DIP=4,7-diphenylphenanthroline (Fig. 10.13)
DMABN=p-dimethylaminobenzonitrile (**7.7**)
DMF=dimethylformamide
DPA=diphenylamine
DPP=2,9-diphenyl-1,10-phenanthroline (for a derivative, see **11.18**)
2,3-dpp=2,3-bis(2-pyridyl)pyrazine (Fig. 6.3)
2,5-dpp=2,5-bis(2-pyridyl)pyrazine (Fig. 6.3)
dppe=1,2-bis(diphenylphosphino)ethylene (P–P in **5.29**)

dpq=2,3-di-2-pyridylquinoxaline (Fig. 6.3)
dpte=1,2-bis(phenylthio)ethane (**5.26**)
DPS=dimethylphenylsiloxy
DQT^{2+}=diquat (**10.18**)
EDTA=ethylenediaminetetraacetate
en=ethylenediamine
FN=fumaronitrile
HAT=1,4,5,8,9,12-hexaazatriphenylene (Fig. 6.3, Fig. 10.13)
HMB=hexamethylbenzene
Me_2bpy=4,4'-dimethyl-2,2'-bipyridine
mnt^{2-}=1,2-dicyano-1,2-ethylenedithiolate
MQ^+=monoquat (as a ligand, *see* **5.20**)
MTHF=2-methyltetrahydrofuran
MV^{2+}=methylviologen, or paraquat (**10.19**)
ppz=4',7'-phenanthrolino-5',6':5.6-pyrazine (Fig. 6.3)
PQT^{2+}=paraquat, or methylviologen (**10.19**)
PTZ=phenothiazine
pz=pyrazine
RB^{2-}=rose bengal
sar=sarcophagine (ligand of 11.3 with Y=H)
sep=sepulchrate: 1,3,6,8,10,13,16,19-octaazabicyclo-[6,6,6]-icosane
    (ligand of **11.1**)
TAP=1,4,5,8-tetraazaphenanthrene (Fig. 10.13)
TCNB=tetracyanobenzene
TCNE=tetracyanoethylene
tim=2,3,9,10-tetramethyl-1,4,8,11-tetra-azacyclotetradeca-1,3,8,10-tetraene
tppq=2,3,7,8-tetra-2-pyridylpyrazino[2,3-g]quinoxaline (Fig. 6.3)
tpy=2,2':6,2''-terpyridine
tribpy=polypyridine-type ligand (**6.10**)
TS=*trans*-stilbene (**10.34**)

1

Scope and limitations

In the last two decades knowledge at the molecular level has made extraordinary progress in several fields of chemistry. Rational synthetic methods have been designed to prepare **molecules** of any shape and size, that can barely be described by the IUPAC nomenclature rules and are therefore called by trivial names often derived from objects encountered in everyday life: barrels [1], baskets [2], belts [1], bowls [3], boxes [4,5], bridges [6], butterflies [7], cages [5,8,9], calixes [10], caps [11], carcerands [12], cavitands [13], clefts [14], collars [1], containers [12], corands [15], crowns [16], cryptands [17], cucurbituril [18], cyclophanes [19], cylinders [5], fences [20], footballene [21], gates [5], gondola [22], helicenes [23], hinges [24], katapinands [25], ladders [26], lanterns [27], lepidopterene [28], octopus [29], ovalene [30], pagodanes [31], podands [32], propellanes [33], scorpiands [34], sepulchrands [35], spacers [36], speleands [37], spherands [15], staffanes [38], stellanes [39], strips [9,40], torands [41], trinacrene [42], tweezers [43,44], vessels [13], wheel-and-axle [45] and wires [46]. It is now possible to say that almost any desired molecular structure can be prepared, provided sufficient resources are applied to the problem.

A natural consequence of this development, as well as an opportunity [47] to reach important goals, is the trend of today's chemistry to study more and more complex systems. Looking at the current chemical literature, one is impressed by the rapid growth of research in a field (better defined in section 3.1) that can be loosely called *supramolecular chemistry*†, that is the chemistry of systems (**supermolecules**†) made up of molecular components in the same way as molecules are made up of atoms [48]: adducts [49], cage-type compounds [5,12,35,50], catenanes [51,52], chromoionophores [53], chundles (channel + bundle) [5], clathrates [54], covalently-linked molecular components [55] (diads [6], triads [56,57], tetrads [58]), electron donor–acceptor complexes [59], helicates [60,61], host–guest systems [62–66], inclusion compounds [54], intercalates [67,68], ion pairs [69,70], knots [71], liquid crystals [72], molecular devices [5,73,74], polynuclear metal complexes [75–77], rotaxanes [78,79], sandwich-type systems [80], second-sphere coordination compounds [49,81,82], stacks [83,84], starburst dendrimers [85] and supercomplexes [5,49,81].

† Although the prefixes *supra* and *super* have slightly different meanings in Latin, chemists use them in an interchangeable way. The commonest terms in the literature are *supra*molecular and *super*molecule. We will conform ourselves to this use.

In the field of photochemistry, which is at the crossroads of chemistry, physics, and biology, the present state of the art can be summarized as follows. On one hand, the photochemical and photophysical processes of thousands of organic molecules [86–88], coordination compounds [89–91], and organometallic complexes [91,92] have been elucidated and suitable theoretical treatments are now available to rationalize the structural, energetic, and dynamic properties of the most important excited states of several families of molecules. On the other hand, the natural photochemical processes which occur in living organisms are revealing more and more their intrinsic complexities and are not yet completely understood [93–98]. It now seems important to realize that in between molecular photochemistry and photobiology there is an immense and yet scarcely explored territory: that of *supramolecular photochemistry* [99], where the knowledge accumulated from molecular photochemistry can be profitably used to make progress towards the understanding of photobiological processes and the design of artificial systems capable of performing useful light-induced functions.

The field of supramolecular photochemistry is very extended and, of course, difficult to fence. Our aim in this book is to illustrate the fundamental concepts of this field and to review and discuss some important results and possible applications.

In an attempt to reach such goals, this book has been organized as follows. In Chapter 2, the principles of molecular photochemistry and photophysics are recalled. Chapter 3 lays down the distinction between 'supermolecules' and 'large molecules' and points out that in a supermolecule there are properties related to each molecular component and properties related to intercomponent processes. Supramolecular photochemistry concerns both the control and tuning of the excited state properties of molecular components (Chapter 4) and the study of the photoinduced intercomponent processes. The latter topic, of course, is the main object of the book and it is dealt with in Chapters 5–11. A clearcut classification of the supramolecular systems and of their photochemical processes is often difficult. The approach taken to review the very large amount of available results and to give the reader a rational discussion of the most important issues is based on the way in which the molecular components are held together and/or interact with one another. Particular emphasis is given to photoinduced electron transfer (Chapter 5), electronic energy transfer (Chapter 6), and photoinduced structural changes (Chapter 7) in covalently-linked supramolecular species since such processes constitute the bases for the design of molecular devices of applicative interest and for the understanding of the most important photobiological processes. Photochemical processes in ion pairs (Chapter 8), electron donor–acceptor complexes (Chapter 9), host–guest systems (Chapter 10), and other supramolecular species (caged metal ions, catenanes, rotaxanes, etc., Chapter 11) are also discussed. Photochemical molecular devices and their possible applications are dealt with in Chapter 12. Each chapter is essentially self-sufficient, but extensive use is made of cross-references in an attempt to give a unified view of the field.

As it appears from the above outline, we have focused our attention on systems made of a *small number of discrete molecular components* held together by covalent bonds, electrostatic interactions, hydrogen bonds, or other intermolecular forces. Systems like polymers, semiconductors, micelles, films, derivatized electrodes, and liquid crystals are not discussed in detail. Photobiological processes are not treated explicitly, but the relationships between natural and artificial systems are discussed and the role played by model supramolecular structures to elucidate the mechanisms of natural processes is underlined.

The systems discussed in this book belong to different areas of chemistry. For example, host–guest species are usually considered a topic of organic chemistry, polynuclear metal complexes are almost exclusively dealt with in inorganic chemistry books and journals, and picosecond laser flash photolysis studies belong to the realm of physical chemistry. Our attempt to give a unified view of such different areas, which have traditionally been separated, is probably a risky enterprise. We are confident, however, that it is also a way to stimulate intuition which, after all, is the most important quality of chemists.

REFERENCES

[1] Stoddart, J. F. (1989) *J. Inclusion Phenom.* **7** 227.
[2] Dung, B. and Vögtle, F. (1988) *J. Inclusion Phenom.* **6** 429.
[3] Tucker, J. A., Knobler, C. B., Trueblood, K. N., and Cram D. J. (1989) *J. Am. Chem. Soc.* **111** 3688.
[4] Odell, B., Reddington, M. V., Slawin, A. M. Z., Spencer, N., Stoddart, J. F., and Williams, D. J. (1988) *Angew. Chem. Int. Ed. Engl.* **27** 1547.
[5] Lehn, J. M. (1988) *Angew. Chem. Int. Ed. Engl.* **27** 89.
[6] Oevering, H., Paddon-Row, M. N., Heppener, M., Oliver, A. M., Cotsaris, E., Verhoeven J. W., and Hush, N. S. (1987) *J. Am. Chem. Soc.* **109** 3258.
[7] El Khalifa, M., Pétillon, F. Y., Saillard, J. Y., and Talarmin, J. (1989) *Inorg. Chem.* **28** 3849.
[8] Grammenudi, S. and Vögtle, F. (1986) *Angew. Chem. Int. Ed. Engl.* **25** 1122.
[9] Kohnke, F. H., Mathias, J. P., and Stoddart, J. F. (1989) *Angew. Chem. Int. Ed. Engl.* **28** 1103.
[10] Gutsche C. D. (1989) *Calixarenes.* Royal Society of Chemistry, Cambridge.
[11] Ganesh, K. N. and Sanders, J. K. M. (1980) *J. Chem. Soc., Chem. Commun.* 1129.
[12] Cram, D. J., Karbach, S., Kim, Y. H., Baczynskyj, L., Marti, K., Sampson, R. M., and Kalleymeyn, G. W. (1988) *J. Am. Chem. Soc.* **110** 2554.
[13] Cram, D. J., Karbach, S., Kim, Y. E., Knobler, C. B., Maverick, E. F., Ericson, J. L., and Helgeson, R. C. (1988) *J. Am. Chem. Soc.* **110** 2229.
[14] Rebek, J., Jr. (1987) *Science* **235** 1478.
[15] Cram, D. J. (1986) *Angew. Chem. Int. Ed. Engl.* **25** 1039.
[16] Pedersen, C. J. (1988) *Science* **241** 536.
[17] Dietrich, B., Lehn, J. M., and Sauvage, J. P. (1968) *Tetrahed. Lett.* 2285.
[18] Mock, W. L. and Shih, N. Y. (1986) *J. Org. Chem.* **51** 4440.
[19] Diederich, F. (1988) *Angew. Chem. Int. Ed. Engl.* **27** 362.
[20] Collman, J. P. (1977) *Acc. Chem. Res.* **10** 265.
[21] Kroto, H. W. (1987) *Nature* **329** 529.
[22] Vinod, T. K. and Hart, H. (1988) *J. Am. Chem. Soc.* **110** 6574.
[23] Martin, R. H. (1974) *Angew. Chem. Int. Ed. Engl.* **13** 649.
[24] Hamilton, A. D. and Van Engen, D. (1987) *J. Am. Chem. Soc.* **109** 5035.
[25] Park, C. H. and Simmons, H. E. (1968) *J. Am. Chem. Soc.* **90** 2431.

[26] Barr, D., Clegg, W., Hodgson, S. M., Lamming, G. R., Mulvey, R. E., Scott, A. J., Snaith, R., and Wright, D. S. (1989) *Angew. Chem. Int. Ed. Engl.* **28** 1241.

[27] Bryson, N. J., Brenner, D., Lister-James, J., Jones, A. G., Dewan, J. C., and Davison, A. (1989) *Inorg. Chem.* **28** 3825.

[28] Ferguson, J., Robbins, R. J., and Wilson, G. J. (1986) *J. Phys. Chem.* **90** 4222.

[29] Hyatt, J. A. (1978) *J. Org. Chem.* **43** 1808.

[30] Giniger, R. and Amirav, A. (1986) *Chem. Phys. Lett.* **127** 387.

[31] Fessner, W. D., Sedelmeier, G., Spurr, P. R., Rihs, G., and Prinzbach, H. (1987) *J. Am. Chem. Soc.* **109** 4626.

[32] Weber, E. and Vögtle, F. (1981) *Topics Curr. Chem.* **98** 1.

[33] Mehta, G. and Subrahmanyam, D. (1989) *J. Chem. Soc., Chem. Commun.* 1365.

[34] Pallavicini, P. S., Perotti, A., Poggi, A., Seghi, B., and Fabbrizzi, L. (1987) *J. Am. Chem. Soc.* **109** 5139.

[35] Sargeson, A. M. (1979) *Chem. Brit.* **15** 23.

[36] Glaudemans, C. P. J., Lehnmann, J., and Scheuring, M. (1989) *Angew. Chem. Int. Ed. Engl.* **28** 1669.

[37] Canceill, J., Collet, A., Gabard, J., Kotzyba-Hilbert, F., and Lehn, J. M. (1982) *Helv. Chim. Acta* **65** 1894.

[38] Murthy, G. S., Hassenruck, K., Lynch, V. M., and Michl, J. (1989) *J. Am. Chem. Soc.* **111** 7262.

[39] Gleiter, R., Sigwart, C., and Kissler, B. (1989) *Angew. Chem. Int. Ed. Engl.* **28** 1525.

[40] Kenney, P. W. and Miller, L. L. (1988) *J. Chem. Soc., Chem. Commun.* 84.

[41] Bell, T. W., Firestone, A., and Ludwig, R. (1989) *J. Chem. Soc., Chem. Commun.* 1902.

[42] Ashton, P. R., Isaacs, N. S., Kohnke, F. H., Stagno D'Alcontres, G., and Stoddart, J. F. (1989) *Angew. Chem. Int. Ed. Engl.* **28** 1261.

[43] Leppkes, R. and Vögtle, F. (1981) *Angew. Chem. Int. Ed. Engl.* **20** 396.

[44] Zimmerman, S. C. and Wu, W. (1989) *J. Am. Chem. Soc.* **111** 8054.

[45] Toda, F., Ward, D. L., and Hart, H. (1981) *Tetrahed. Lett.* **22** 3865.

[46] Arrhenius, T. S., Blanchard-Desce, M., Dvolaitzky, M., Lehn, J. M., and Malthete, J. (1986) *Proc. Natl. Acad. Sci. USA* **83** 5355.

[47] Pimentel, G. C. (1985) *Opportunities in chemistry.* National Academy Press, Washington (DC).

[48] Stoddart, J. F. (1988) *Chem. Brit.* **24** 1203.

[49] Colquhoun, H. M., Stoddart, J. F., and Williams, D. J. (1986) *Angew. Chem. Int. Ed. Engl.* **25** 487.

[50] Belser, P., De Cola L., and von Zelewsky, A. (1988) *J. Chem. Soc., Chem. Commun.* 1057.

[51] Dietrich-Buchecker, C. O., Sauvage, J. P., and Kern, J. M. (1984) *J. Am. Chem. Soc.* **106** 3043.

[52] Ashton, P. R., Goodnow, T. T., Kaifer, A. E., Reddington, M. V., Slawin, A. M. Z., Spencer, N., Stoddart, J. F., Vicent, C., and Williams, D. J. (1989) *Angew. Chem. Int. Ed. Engl.* **28** 1396.

[53] Löhr, H. G. and Vögtle, F. (1985) *Acc. Chem. Res.* **18** 65.

[54] Weber, E. (ed.) (1987) *Molecular inclusion and molecular recognition — Clathrates I. Topics Curr. Chem.* **140.**

[55] Connolly, J. S. and Bolton, J. R. (1988). In Fox, M. A. and Chanon, M. (eds) *Photoinduced electron transfer.* Part D. Elsevier, p.103.

[56] Gust, D., Mathis, P., Moore, A. L., Liddell, P. A., Nemeth, G. A., Lehman, W. R., Moore, T. A., Bensasson, R. V., Land, E. J., and Chachaty, C. (1983) *Photochem. Photobiol.* **37S** S46.

[57] Wasielewski, M. R., Niemczyk, M. P., Svec, W. A., and Pewitt, E. B. (1985) *J. Am. Chem. Soc.* **107** 5562.

[58] Gust, D., Moore, T. A., Moore, A. L., Barrett, D., Harding, L. O., Makings, L. R., Liddell, P. A., De Schryver, F. C., Van der Auweraer, M., Bensasson, R. V., and Rougée, M. (1988) *J. Am. Chem. Soc.* **110** 321.

[59] Jones, G. II (1988). In Fox, M. A. and Chanon, M. (eds) *Photoinduced electron transfer.* Part A, Elsevier, p. 245.

[60] Lehn, J. M. and Rigault, A. (1988) *Angew. Chem. Int. Ed. Engl.* **27** 1095.

[61] Constable, E. C., Ward, M. D., and Tocher, D. A. (1990) *J. Am. Chem. Soc.* **112** 1256.

[62] Vögtle, F. (ed.) (1981) *Host–guest complex chemistry I. Topics Curr. Chem.* **98.**

[63] Vögtle, F. (ed.) (1982) *Host–guest complex chemistry II. Topics Curr. Chem.* **101.**

[64] Vögtle, F. and Weber, E. (eds) (1984) *Host–guest complex chemistry III. Topics Curr. Chem.* **121.**

[65] Saenger, W. (1980) *Angew. Chem. Int. Ed. Engl.* **19** 344.

[66] Breslow, R. (1982) *Science* **218** 532.

[67] Barton, J. K. (1986) *Science* **233** 727.

[68] Sherman, S. E. and Lippard, S. J. (1987) *Chem. Rev.* **87** 1153.

[69] Balzani, V. and Scandola, F. (1988). In Fox, M. A. and Chanon, M. (eds) *Photoinduced electron transfer.* Part D. Elsevier, p. 148.

[70] Vogler, A. and Kunkely, H. (1990) *Topics Curr. Chem.* **158** in press.

[71] Dietrich-Buchecker, C. O. and Sauvage, J. P. (1989) *Angew. Chem. Int. Ed. Engl.* **28** 189.

[72] Ringsdorf, H., Schlarb, B., and Venzmer, J. (1988) *Angew. Chem. Int. Ed. Engl.* **27** 113.

[73] Balzani, V., Moggi, L., and Scandola, F. (1987). In Balzani, V. (ed.) *Supramolecular photochemistry.* Reidel, p.1.

[74] Carter, F. L., Siatowski, R. E., and Wohltjen, H. (eds) (1988) *Molecular electronic devices.* North-Holland.

[75] (1983) *Prog. Inorg. Chem.* **30** (issue dedicated to H. Taube).

[76] Scandola, F., Indelli, M. T., Chiorboli, C., and Bignozzi, C. A. (1990) *Topics Curr. Chem.* **158** in press.

[77] Balzani, V. (1990) *J. Photochem. Photobiol. A: Chemistry* **51** 55.

[78] Ogino, H. and Ohata, K. (1984) *Inorg. Chem.* **23** 3312.

[79] Anelli, P. L., Ashton, P. R., Ballardini, R., Balzani, V., Gandolfi, M. T., Goodnow, T. T., Kaifer, A. E., Pietraszkiewicz, M., Prodi, L., Reddington, M. V., Slawin, A. M. Z., Spencer, N., Stoddart, J. F., Vicent, C., and Williams, D. J., *J. Am. Chem. Soc.* submitted.

[80] Davis, J. H., Jr., Sinn, E., and Grimes, R. N. (1989) *J. Am. Chem. Soc.* **111** 4784.

[81] Balzani, V., Sabbatini, N., and Scandola, F. (1986) *Chem. Rev.* **86** 319.

[82] Scandola, F. and Indelli, M. T. (1988) *Pure Appl. Chem.* **60** 973.

[83] Ortholand, J. Y., Slawin, A. M. Z., Spencer, N., Stoddart, J. F., and Williams, D. J. (1989) *Angew. Chem. Int. Ed. Engl.* **28** 1394.

[84] Alexander, J., Ehrenfreund, M., Fiedler, J., Huber, W., Räder, H.J., and Müllen, K. (1989) *Angew. Chem. Int. Ed. Engl.* **28** 1531.

[85] Tomalia, D. A., Naylor, A. M., and Goddard III, W. A. (1990) *Angew. Chem. Int. Ed. Engl.* **29** 138.

[86] Turro, N. J. (1978) *Modern molecular photochemistry*. Benjamin.

[87] Coxon, J. M. and Halton, B. (1986) *Organic photochemistry*. Cambridge University Press.

[88] Wayne, R. P. (1988) *Principles and applications of photochemistry*. Oxford University Press.

[89] Balzani, V. and Carassiti, V. (1970) *Photochemistry of coordination compounds*. Academic.

[90] Adamson, A. W. and Fleischauer, P. D. (eds) (1975) *Concepts of inorganic photochemistry*. Wiley.

[91] Ferraudi, G. J. (1988) *Elements of inorganic photochemistry*. Wiley.

[92] Geoffroy, G. L. and Wrighton, M. S. (1979) *Organometallic photochemistry*. Academic.

[93] Hader, D. P. and Tevini, M. (1987) *General photobiology*. Pergamon.

[94] Breton, J. and Vermeglio, H. (eds) (1988) *The photosynthetic bacterial reaction center. Structure and dynamics*. Plenum.

[95] Deisenhofer J. and Michel, H. (1989) *Angew. Chem. Int. Ed. Engl.* **28** 829.

[96] Huber, R. (1989) *Angew. Chem. Int. Ed. Engl.* **28** 848.

[97] Boxer, S. G., Goldstein, R. A., Lockhart, D. J., Middendorf, T. R., and Takiff, L. (1989) *J. Phys. Chem.* **93** 8280.

[98] Friesner, R. A. and Won, Y. (1989) *Photochem. Photobiol.* **50** 831.

[99] Balzani, V. (ed.) (1987) *Supramolecular photochemistry*. Reidel.

2

Principles of molecular photochemistry

2.1 INTRODUCTION

The aim of this chapter is that of providing a concise illustration of the fundamental concepts that will be used to discuss the photochemical and photophysical properties of supramolecular systems. For an extensive discussion of the topics treated in this chapter, the reader should refer to the available monographs on photochemistry [1]. It should be noted, however, that practically all the photochemistry books treat either organic or inorganic molecules, whereas a great number of supramolecular systems result from assembling of both organic and inorganic components. In fact, to enter the field of supramolecular photochemistry one needs to know not only general photochemical and photophysical concepts but also some important peculiarities of the excited states of organic and inorganic molecules as well as the most common terms and definitions used in both these fields of photochemistry.

2.2 THE NATURE OF LIGHT

Usually "light" is taken to mean electromagnetic radiation in the visible, near ultraviolet, and near infrared spectral range.

In the wave model, electromagnetic radiation is characterized by a wavelength, λ, a frequency, ν, and a velocity, c. The three quantities are related by the relationship $\lambda\nu = c$. The value of c is constant (2.998×10^8 m s^{-1} in vacuum), whereas λ (and ν) may cover a wide range of values. The SI units for λ and ν are the metre and the hertz, respectively. In some cases, the wavenumber $\bar{\nu}$ (defined as the number of waves per centimetre) is also used to characterize electromagnetic radiation. The electromagnetic spectrum encompasses a variety of types of radiation from γ-rays to radiowaves, distinguished by their wavelengths (or frequencies, or wavenumbers). In photochemistry we are concerned with the region ranging from 100 to 1000 nm (3×10^{15} to 3×10^{14} Hz, or 10^5 to 10^4 cm^{-1}).

In the quantum model a beam of radiation is regarded as a stream of *photons*, or *quanta*. A photon has no mass but it has a specific energy, E, related to the frequency of the radiation, ν, by (2.1),

$$E = h\nu \tag{2.1}$$

where h is Planck's constant (6.63×10^{-34} J s). From the above relationships it follows that the photon energy is 1.99×10^{-18} and 1.99×10^{-19} J, respectively, for light of 100 and 1000 nm. This picture of light as made up of individual photons is essential to photochemistry.

The interaction of light with molecular systems is generally an interaction between *one* molecule and *one* photon. It can be written in the very general form (2.2),

$$A + h\nu \rightarrow {}^{*}A \tag{2.2}$$

where A denotes the ground state molecule, $h\nu$ the absorbed photon, and *A the molecule in an electronically excited state. As the equation implies, the excited molecule *A is the molecule A with an extra energy $h\nu$.

To appreciate the size of the photon energy, we must compare it with the energies of the chemical bonds, which are normally expressed in kilojoules or kilocalories *per mole*. A mole is an Avogadro's number of molecules, i.e. 6.02×10^{23} molecules. We may extend the concept of mole to photons, defining an *einstein* as one mole of photons. When *one mole* of molecules absorbs *one einstein* of photons, this is equivalent to the one photon absorbed by one molecule, (2.2). The energy of one einstein of photons at 100 nm is 1198 kJ (286 kcal), and that of one einstein of photons at 1000 nm is 119.8 kJ (28.6 kcal). These energy values are of the same order of magnitude of those required to break chemical bonds (e.g., 190 kJ mol^{-1} for the Br–Br bond of Br_2; 416 kJ mol^{-1} for the C–H bond of CH_4). The energy which a molecule obtains when it absorbs a photon of light is therefore not at all negligible. Whether or not light absorption causes bond breaking will depend on the competition among various deactivating processes (section 2.6). In any case, because of the availability of such an extra amount of energy, *an excited molecule has to be considered as a new chemical species* which has its own chemical and physical properties, often quite different from the properties of the ground state molecule.

2.3 POTENTIAL ENERGY SURFACES

Central to any understanding of photochemistry is the concept that molecular systems possess well-defined *electronic* states: ordinary (thermal) chemistry is the chemistry of the ground electronic state, while photochemistry is the chemistry of (or originating from) electronic excited states. It is important to keep in mind that this common notion relies on a non-trivial conceptual separation between electronic and nuclear motions [2,3]. This separation, known as the Born–Oppenheimer (BO) approximation, has important implications in photochemistry.

The total energy of a molecule, represented by the Hamiltonian, H, consists of the potential and kinetic energies of both electrons and nuclei, as shown in (2.3),

$$H = V_e + V_N + V_{eN} + T_e + T_N \tag{2.3}$$

where V_e is the mutual potential energy of the electrons, V_N is the mutual potential energy of the nuclei, V_{eN} is the potential energy of the electrons with respect to the nuclei, and T_e and T_N are the kinetic energies of electrons and nuclei, respectively. The Schroedinger equation for a molecular system is given by (2.4),

$$H\Psi (q,Q) = E\Psi (q,Q) \tag{2.4}$$

where the $\Psi (q,Q)$ are wavefunctions of the true stationary states of the system, that depend on both electronic (q) and nuclear (Q) coordinates. Based on the huge difference in the masses of electrons and nuclei, however, the molecular system can be considered as made up of a fast subsystem (the electrons) and a slow one (the nuclei). The difference in the timescale of the two types of motions is the conceptual basis for the Born–Oppenheimer approximation.

Let us neglect T_N in (2.3) and define an electronic Hamiltonian (2.5).

$$H_e = V_e + V_N + V_{eN} + T_e \tag{2.5}$$

The electronic wavefunctions $\Psi_k(q,Q)$ will be the solutions of (2.6),

$$H_e\Psi_k(q,Q) = E_k\Psi_k(q,Q) \tag{2.6}$$

where k as an electronic quantum number. Conceptually, neglecting T_N in the Hamiltonian corresponds to solving the electronic problem at fixed values of the nuclear coordinates. For each geometry (Q), a set of electronic wavefunctions and energies (different k) can be obtained. For each k, the set of energy values obtained at different geometries defines a surface called the "adiabatic potential energy surface" of the kth electronic state. Fig. 2.1 shows schematically the adiabatic potential energy curves of the ground and first singlet excited states of a heteronuclear diatomic molecule (e.g., HCl). The change in the electronic wavefunction brought about by changes in the nuclear geometry can be appreciated by looking at the curves (dashed lines) for pure ionic and covalent valence-bond wavefunctions: in going from equilibrium geometry to stretched geometry, the ground state changes from highly ionic to highly covalent, while the opposite change occurs for the first excited state.

The problem of the motion of the nuclei is next solved by defining an effective Hamiltonian (2.7)

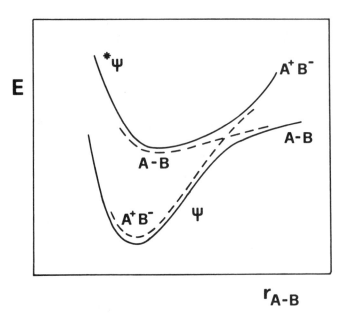

$$r_{A-B}$$

Fig. 2.1 — Adiabatic potential energy curves (full lines) of the ground and the first excited state for a heteronuclear diatomic molecule. Dashed lines represent the energies of pure ionic and covalent structures.

$$H_N = E_k(Q) + T_N \tag{2.7}$$

and nuclear eigenfunctions $X(Q)$ as the solutions of (2.8),

$$H_N X_{k,\upsilon}(Q) = E_{k,\upsilon} X_{k,\upsilon}(Q) \tag{2.8}$$

where υ is a nuclear (vibrational) quantum number. Conceptually, this corresponds to letting the slow nuclear subsystem move in a potential field that is determined by the fast electronic subsystem: the electronic wavefunction is considered to respond instantaneously to the changes in the nuclear coordinates.

It can be easily shown that a wavefunction of the form (2.9)

$$\Psi = \Psi_k(q,Q) \, \Psi_{k,\upsilon}(Q) \tag{2.9}$$

is a good solution of the complete Schroedinger equation (2.4) for the system with $E = E_{k,\upsilon}$, provided that $(\delta\Psi/\delta Q)(\delta X/\delta Q)$ and $\delta^2\Psi/\delta Q^2$ are small. This point is important in evaluating the accuracy of the BO approximation. It is evident that the approximation is expected to work well in situations (such as that near to equilibrium distances in Fig. 2.1) in which the electronic wavefunctions are slowly changing functions of the nuclear coordinates, but to be much less valid when (as in the region of the crossing dashed lines of Fig. 2.1) the electronic wavefunctions changes abruptly with nuclear motion. In the latter case, there may be some dynamic tendency for the states to preserve their electronic identity instead of following the changes predicted by the adiabatic BO approximation.

Since a nonlinear N-atomic molecule has $3N$-6 internal nuclear degrees of freedom, each electronic state is described by a potential hypersurface in a $3N$-5 dimensional space. The ground state hypersurface has usually a deep minimum corresponding to the stable geometry of the molecule (Fig. 2.2a). Two or more

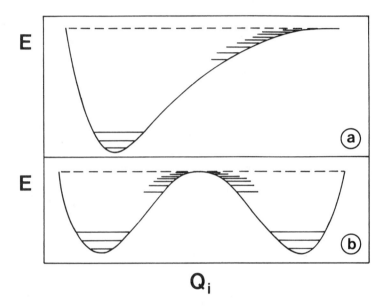

Fig. 2.2 — Potential energy curves. For more details, see text.

minima separated by energy barriers may occur, however, in systems for which isomers are possible (Fig. 2.2b). It is useful to distinguish two types of nuclear motions along the ground-state potential energy hypersurface: (i) small-amplitude motions around the minima; (ii) large-amplitude motions carrying the molecule to highly distorted geometries.

It is customary to describe the small-amplitude nuclear motions around the minima as a superposition of normal modes of vibration [3], i.e., harmonic oscillations along suitable symmetry-adapted combinations of bond length and bond angle changes. Large-amplitude nuclear motions, that occur in relatively flat regions of the potential energy hypersurface, can be viewed as strongly anharmonic vibrations with very closely spaced levels. In this case, however, a simple description of the system as moving smoothly on the potential energy hypersurface along classical trajectories is also appropriate for most purposes. Large amplitude nuclear motions of this type are required to achieve chemical change, i.e. to break bonds or to go from one isomer to the other. Classical models of nuclear motions are thus widely used in the field of chemical kinetics. It should also be noted that a description of chemical reactions in terms of potential energy surfaces is not limited to unimolecular processes. For a bimolecular reaction, the potential energy hypersurface of a "supermolecule" including the nuclei of both reactants can be considered, with reactants and products corresponding to different minima of this surface.

For electronically excited states, nuclear motions can be described by the same arguments used for the ground-state. Generally speaking, electronic excitation of a molecule results in some kind of bond weakening (*vide infra*). Two consequences of this fact should be stressed. First, excited-state minima tend to be less deep than the corresponding ground-state minima, so that small-amplitude vibrations in the excited state will generally have lower frequencies and smaller energy spacings than the corresponding vibrations in the ground state. Second, large-amplitude motions along relatively flat pathways leading to highly distorted geometries will be found more frequently in excited states than in the ground state. These differences are expected to become more pronounced as one goes to higher and higher electronically excited states.

2.4 ELECTRONIC STATES AND ELECTRONIC CONFIGURATIONS

Molecules are multielectron systems, for which the electronic Schroedinger equation (2.4) cannot be solved exactly. Approximate electronic wavefunctions of a molecular system can be conveniently written as products of one-electron wavefunctions, each consisting of an orbital and a spin part (2.10).

$$\Psi = \Phi S = \Pi_i \, \varphi_i s_i \qquad\qquad (2.10)$$

The φ_is are appropriate molecular orbitals (MO) and s_i is one of the two possible spin eigenfunctions, α or β. In writing the multielectron wavefunction Ψ, care must be taken to meet the prescriptions dictated by the Pauli principle. The orbital part of this product multielectron wavefunction defines the *electronic configuration* to which the electronic state belongs.

Formaldehyde is a simple organic molecule that can be used to exemplify the procedure to construct states from electronic configurations. The MO diagram of formaldehyde is shown schematically in Fig. 2.3. It consists of three low-lying σ-bonding orbitals, a π-bonding orbital of the CO group, an atomic nonbonding orbital n of oxygen, a π-antibonding orbital of the CO group, and three high-energy σ-antibonding orbitals. The lowest-energy electronic configuration is (neglecting the filled σ-bonding orbitals) $\pi^2 n^2$. Higher energy configurations can be obtained from the lowest one by promoting electrons to antibonding orbitals: $\pi^2 n \pi^*$, $\pi n^2 \pi^*$, and so on. In a very coarse zero-order description, the energy associated with a particular electronic configuration would be given by the sum of the energies of the occupied MOs.

In order to obtain a more realistic description of the energy states of the molecule, two features should be added to the simple configuration picture: (i) spin functions must be attached to the orbital functions describing the electronic configu-

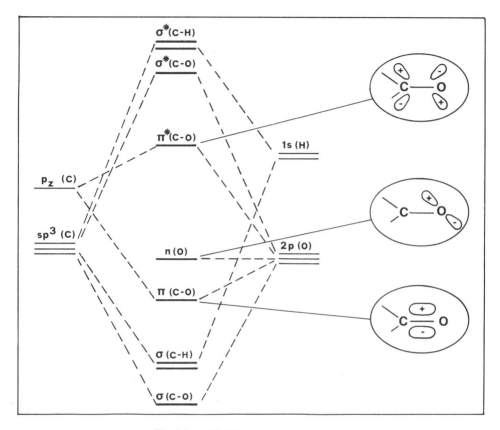

Fig. 2.3 — MO diagram for formaldehyde.

rations (2.10), and (ii) interelectronic repulsions must be taken into consideration. These two closely interlocked points have important qualitative consequences.

According to the Pauli principle, in fact, the electrons in a given electronic configuration repel each other more strongly when their spins are paired than when their spins are parallel, and this may lead to the splitting of an electronic configuration into several energy states [4].

In the case of formaldehyde, the inclusion of spin and interelectronic repulsion leads to the schematic state diagram shown in Fig. 2.4b, in which each excited electronic configuration (Fig. 2.4a) is split into a pair of triplet (lower) and singlet (higher) states. Notice that the splitting of the singlet-triplet pair of states arising from the $\pi\pi^*$ configuration is larger than that of the pair corresponding to the $n\pi^*$ configuration. This arises from the dependence of the interelectronic repulsion on the amount of spatial overlap between the MOs containing the two electrons, and this overlap is greater in the first than in the second case (see the MOs shapes in Fig. 2.3). In spectroscopy, the electronic states are designated by symbols that specify the symmetry of the wavefunction in the symmetry group of the molecule (e.g., A_1, A_2, etc. in the C_{2v} group of formaldehyde) and the spin multiplicity (number of unpaired

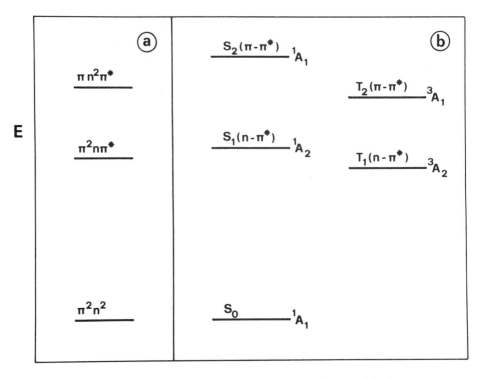

Fig. 2.4 — Configuration (a) and state (b) diagrams for formaldehyde.

electrons $+1$, as a left superscript). In organic photochemistry, on the other hand, it is customary to label the singlet and triplet states of molecules as S_n and T_n respectively, with $n = 0$ for the singlet ground state and $n = 1, 2$, etc. for the states arising from the various excited electron configurations (often indicated in parentheses). Both notations are shown for formaldehyde in Fig. 2.4b. State energy diagrams such as that of Fig. 2.4b are used for the description of the primary photophysical processes that follow excitation of a molecule (section 2.6) and are commonly called "Jablonski diagrams". The situation sketched above (i.e., singlet ground state, pairs of singlet and triplet states arising from each excited configuration, lowest excited state of multiplicity higher than the ground state) is quite general, being typical of systems, such as *organic molecules*, with low-symmetry closed-shell ground-state configuration.

For inorganic systems, the construction of electronic states via electronic configurations from the MO description follows the same general lines described above, although a number of differences in the Jablonski diagrams must be allowed for (*vide infra*). Typical inorganic molecules of photochemical interest are *coordination compounds* of transition metals. A schematic MO diagram for an octahedral transition metal complex is shown in Fig. 2.5. The various MOs can be conveniently classified according to their predominant atomic orbital contribution as : (i) strongly bonding, ligand-centred σ_L orbitals; (ii) bonding, ligand-centred π_L orbitals; (iii)

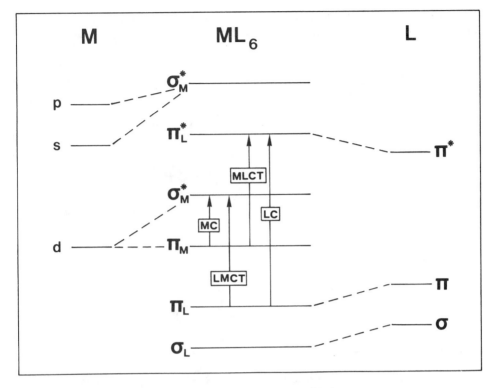

Fig. 2.5 — MO diagram for an octahedral transition metal complex. The arrows indicate the four types of transitions based on localized MO configurations: metal centred (MC); ligand-to-metal charge transfer (LMCT); metal-to-ligand charge transfer (MLCT); ligand centred (LC).

essentially nonbonding $\pi_M(t_{2g})$ metal-centred, predominantly d orbitals; (iv) anti-bonding $\sigma^*_M(e_g)$ metal-centred, predominantly d orbitals; (v) ligand-centred, anti-bonding π^*_L orbitals, and (vi) strongly antibonding, metal-centred σ^*_M orbitals. In the ground electronic configuration of an octahedral complex of a d^n metal ion, orbitals of types (i) and (ii) are completely filled, while n electrons occupy the orbitals of types (iii) and (iv). As usual, excited configurations can be obtained from the ground configuration by promoting one electron from occupied to vacant MOs. At relatively low energies, one expects to find electronic transitions of the following types (Fig. 2.5): *metal-centred* (MC, often indicated also as d–d) transitions from orbitals of type (iii) to orbitals of type (iv); *ligand-centred* (LC, often indicated also as π–π^*) transitions of type (ii) → (v); *ligand-to-metal charge transfer* (LMCT) transitions of type (ii) → (iii) or (ii) → (iv); *metal-to-ligand charge transfer* (MLCT, often indicated also as d–π^*) transitions of type (iii) → (v) or (iv) → (v). The relative energy ordering of these types of configurations in any particular complex depends on the nature of metal and ligands in more or less predictable ways.

The step from configurations to states is conceptually less simple for coordination compounds than for organic molecules, since one is often dealing with systems of high symmetry (i.e., with degenerate MOs) and open-shell ground configurations (i.e., with partially occupied HOMOs). The following example can be used to emphasize some of the peculiar intricacies of these systems. Fig. 2.6 shows schemati-

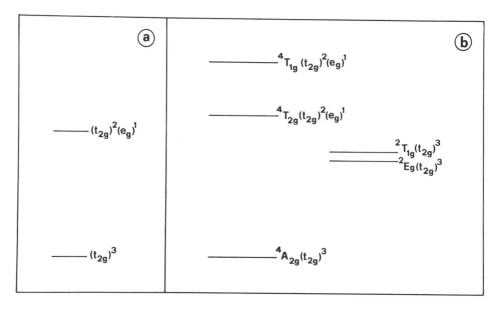

Fig. 2.6 — Configuration (a) and state (b) diagrams for an octahedral Cr(III) complex. Only the lower-lying excited states of each configuration are shown.

cally the lowest energy part of the electron configuration diagram (a) and state diagram (b) of an octahedral Cr(III) complex (d^3 electronic configuration). Points to be noticed in comparison with the "organic" Jablonski diagrams are: (i) spin multiplicities other than singlet and triplet can occur (e.g., doublet and quartet in the example given) although for each electronic configuration the state of highest multiplicity remains the lowest one; (ii) excited states can exist that belong to the same electronic configuration as the ground state (this implies that the ground state has the highest multiplicity); (iii) more than one pair of states of different multiplicity can arise from a single electron configuration. It is clear that, with respect to the organic case, coordination compounds tend to have more complex and specific Jablonski diagrams.

2.5 LIGHT ABSORPTION

A molecule can be promoted from the ground electronic state to an electronically excited state by the absorption of a quantum of light. The necessary condition is that the photon energy, $h\nu$, matches the energy gap between the ground and the excited state. This energy gap, for low-energy states of common organic and inorganic molecules, corresponds to light in the visible and near ultraviolet regions. When the above condition is satisfied, the probability of transition from the ground state Ψ_i to the excited state Ψ_f is proportional to the square of the so-called transition moment (2.11),

$$M_{if} = \langle \Psi_i | \mu | \Psi_f \rangle \qquad (2.11)$$

where μ is the dipole moment operator [5] defined as Σer_j, with e representing the electron charge and r_j representing the vector distance from the kth electron to the centre of positive charge of the molecule.

If the complete wavefunctions (2.9) and (2.10) are substituted into (2.11), the transition moment splits into the product of three terms as shown in (2.12).

$$M_{if} = \langle \Phi_i | \mu | \Phi_f \rangle \, \langle S_i | S_f \rangle \, \Sigma_n \langle X_{i,0} | X_{f,n} \rangle \qquad (2.12)$$

Notice that the dipole-moment operator is considered to be independent of nuclear coordinates (Condon approximation) and spin, and thus it only appears in the integral containing the orbital part of the electronic wavefunction. In (2.12), the summation in the third term is made over all the transitions between vibrational levels of the ground and excited states that contribute to the intensity of the electronic transition. For common molecular systems in the ground electronic state at room temperature, only transitions of the type $X_{i,0} \to X_{f,n}$ (thereafter referred to as $0 \to n$) need to be considered in most cases.

In practice, a very wide range of intensities is observed for electronic absorption bands of polyatomic molecules, with maximum molar absorption coefficients ranging from the lower limit of detection (0.1–0.01 M^{-1}cm^{-1}) up to values of the order of 5×10^5 M^{-1}cm^{-1}. Without carrying out any actual calculation of M_{if}, it is easy to identify typical cases in which the first or the second term in (2.12) are expected to vanish, thus leading to a zero predicted intensity of the electronic transition. The rules defining such cases are known as *selection rules* for light absorption. Because of the orthogonality of spin wavefunctions, the second term in (2.12) is expected to vanish whenever the initial and final states have different spin multiplicity, and the corresponding electronic transitions are called *spin-forbidden*. This spin selection rule is valid to the extent to which spin and orbital functions can be rigorously separated. Departures from this approximation can be dealt with in terms of a perturbation called *spin-orbit coupling*, by which states of different spin multiplicity can be mixed. This perturbation increases as the fourth power of the atomic number of the atoms involved. Thus spin-forbidden transitions of typical organic molecules (for example, $S_0 \to T_1$) are actually almost unobservable ($\varepsilon_{max} < 1$ M^{-1}cm^{-1}), whereas spin-forbidden transitions of metal complexes can reach quite sizeable intensities (for example, 10^2 M^{-1}cm^{-1} for $5d$ metal complexes).

It can be shown by the methods of group theory [3,6] that the first term in (2.12) vanishes if the symmetry properties of the initial and final orbital functions are not appropriate (that is, if the product $\Phi_i \Phi_f$ does not belong to the same irreducible representation of the point group of the molecule as μ). In these cases, the transition is said to be *symmetry-forbidden*. Examples of symmetry-forbidden transitions are the $n–\pi^*$ transitions of organic molecules and the MC transitions in centrosymmetric coordination compounds. In practice, such transitions have low but sizeable intensity ($\varepsilon_{max} \sim 10^2$ M^{-1}cm^{-1}) because of the poor separability of electronic and nuclear functions (breakdown of the Condon approximation). This again can be dealt with as a perturbation, called *vibronic coupling*. In this case, the transition acquires some intensity because vibrations of appropriate symmetry lead to admixture of the "forbidden" (zero-order) excited state by some "allowed" (zero-order) state.

The integrated band intensity only depends on the first two (electronic) terms of (2.12), since it can be shown that the summation in the third (nuclear) term always amounts to 1. This term, however, is important in determining the *shape* of the absorption band. It is often called the *Franck–Condon* term as it represents the quantum mechanical counterpart of the classical statement of the Franck–Condon principle (i.e., the nuclei cannot change either their position or their kinetic energy during the electronic transition). Let us consider the hypothetical situation sketched in Fig. 2.7a, in which the excited state has a potential energy surface identical in

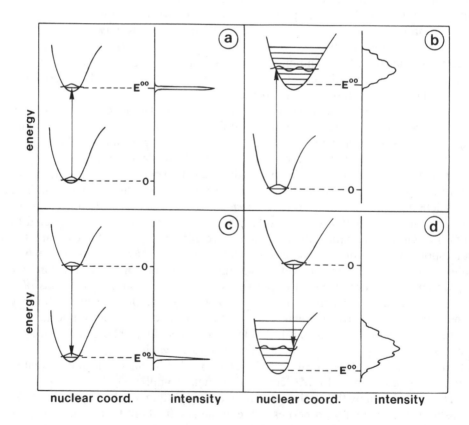

Fig. 2.7 — Relationship between excited state distortion and bandwidth of the absorption (a and b) and emission (c and d) bands.

shape and equilibrium geometry to that of the ground state. In this case, the ground- and excited-state nuclear wavefunctions are solutions of the same vibrational problem and constitute an orthonormal set. Thus $\langle X_{i,0}|X_{f,0}\rangle = 1$ and all other terms in the summation are zero. In other words, all the intensity of the electronic transition is concentrated in a sharp, line-shaped band corresponding to the 0–0 transition. The opposite case is shown in Fig. 2.7b, where the excited state is highly distorted (different equilibrium geometry and force constant) with respect to the ground state. In this case, the values of the vibrational overlap integrals for the various 0–n transitions should be evaluated to obtain the intensity distribution. It can be easily seen that the maximum vibrational overlap occurs with the excited-state vibrational level that intercepts the excited-state potential energy surface at the equilibrium geometry of the ground state (Fig. 2.7b). The overlap integrals for the other 0–n transitions decrease smoothly in going towards higher and lower n values. Thus, the intensity will be spread over a relatively broad, gaussian-shaped band centred around the "vertical" transition. In the intermediate case of a small degree of excited-state distortion, more or less symmetric bandshapes (with lower slope on the high-energy side) are expected.

The schematic pictures used to discuss the effect of the Franck–Condon factors on absorption bandshape (Fig. 2.7a,b) represent sections of the potential energy surfaces along a single coordinate. In an N-atomic molecule, the same type of arguments should be applied to $3N-6$ sections along different nuclear coordinates, and the actual bandshape should be regarded as a convolution of the various single-mode profiles. In many cases, the superposition of different vibrational progressions may lead to complete loss of the vibrational structure in the absorption band.

2.6 UNIMOLECULAR PROCESSES OF EXCITED STATES

2.6.1 Vibrational relaxation
Light absorption often generates the excited state in a high vibrational level because of the Franck–Condon principle and of excited-state distortion. The newborn electronically excited molecules can thus be regarded as "hot" species with respect to the surrounding ground-state molecules that have a Boltzmann equilibrium distribution largely centred on the zero vibrational level. The vibrationally excited molecules will tend to dissipate their excess vibrational energy (thermalize) by interaction (collisions) with surrounding molecules. This process is usually called *vibrational relaxation*.

For most practical systems (solid, liquid, and atmospheric pressure gaseous phase) vibrational relaxation occurs in the picosecond time scale. Since most of the interesting chemistry and physics that takes place in electronically excited states occurs on a much longer timescale (see below), *thermally equilibrated excited states* should be considered as the only relevant intermediates in photochemistry, regardless of the initial amount of vibrational excitation with which they may have been created.

The fact that vibrational relaxation is fast has important implications as far as the thermodynamic definition of electronically excited states is concerned. As thermally equilibrated species, electronically excited molecules have partition functions in the usual statistical sense and can be discussed in terms of thermodynamic state functions such as standard free energy, enthalpy, and entropy. The evaluation of such quantities for electronically excited states will be discussed in section 2.7.2.

2.6.2 Radiative deactivation

Electronically excited molecules can return to the ground state by emitting a quantum of light (2.13).

$$^*A \rightarrow A + h\nu' \tag{2.13}$$

This radiative transition is the reverse process of light absorption (2.2) and is often indicated as *spontaneous emission* (in order to distinguish it from the process known as stimulated emission, that is relevant to lasers). It can be shown that the probability of spontaneous emission depends on the third power of the frequency and is regulated by the same factors affecting that of light absorption. Therefore, a discussion parallel to that given in section 2.5 could be made here. In particular, the same spin and symmetry selection rules as discussed for absorption hold for radiative deactivation, with the same arguments applying to the degree of validity of the rules. In the language of photochemistry, it is customary to call *fluorescence* a spin-allowed emission (e.g., $S_1 \rightarrow S_0$ in an organic molecule) and *phosphorescence* a spin-forbidden emission (e.g., $T_1 \rightarrow S_0$ in an organic molecule). Typical values for the probability (per unit time) of strongly allowed radiative transitions is of the order of $10^9 \ s^{-1}$. For strongly spin-forbidden emissions (e.g., phosphorescence of organic molecules), probabilities as small as $1 \ s^{-1}$ can be obtained.

The role of the Franck–Condon factor in radiative deactivation is again that of determining the band shape. As shown in Fig. 2.7c, emission from an undistorted excited state will result in a sharp, line-shaped emission band at the 0–0 energy, whereas in a highly distorted case the emission band will be broad, gaussian-shaped, and centred at energies lower than the 0–0 energy (Fig. 2.7d). Since in a distorted case the absorption band maximum is at energies higher than the 0–0 energy, there must be a shift, called the *Stokes shift*, between the maxima in the absorption and emission spectra for the same transition. The magnitude of the Stokes shift is a simple, direct measure of the extent of distortion between the ground and the excited state. If the two states are considered as identical harmonic oscillators displaced in the nuclear space, the intensity distribution of the various 0–n transitions should be identical in absorption and emission, and a "mirror image" relationship (mirror at the 0–0 energy) should exist between absorption and emission spectra.

2.6.3 Radiationless deactivation

In radiative deactivation, energy conservation is provided by the emission of light (2.13). If an excited state is to be converted into the ground state (or a lower excited state) *without* emission of radiation, a two-step mechanism must operate: (i) isoenergetic conversion of the electronic energy of the upper state into vibrational energy of the lower state (radiationless transition); (ii) vibrational relaxation of the lower state. Since step (ii) is known to be very fast, step (i) will be the rate-determining process. The probability of the isoenergetic radiationless transition is given [7], according to perturbation theory, by Fermi's Golden Rule (2.14),

$$k_{nr} = (2\pi/h) \langle \Psi_i | H' | \Psi_f \rangle^2 \langle X_{i0} | X_{fn} \rangle^2 \qquad (2.14)$$

where H' is an appropriate perturbation (spin–orbit coupling or vibronic coupling) that promotes the transition. Radiationless transitions between states of different or equal spin multiplicity are called *intersystem crossing* and *internal conversion*, respectively. Because of the first term in (2.14), intersystem crossing has a much smaller probability than internal conversion.

Contrary to what happens with the related expression (2.12) for radiative transitions, the Franck–Condon term in (2.14) is made up of a single overlap integral for each vibrational mode, corresponding to the unique isoenergetic transition. In order to discuss the role of the Franck–Condon factor in radiationless transitions, it is worthwhile considering the two archetypal situations shown in Fig. 2.8.

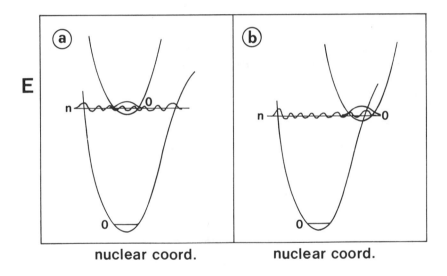

Fig. 2.8 — Limiting cases for potential energy surfaces involved in radiationless deactivations.

The situation of Fig. 2.8a is that of "nested" surfaces. It is easy to see that the vibrational overlap tends to be poor because of the oscillatory behaviour of the wavefunction of the high vibrational level of the ground state. In this situation, the probability of radiationless transition is small. For a given ground-state vibrational frequency, the probability decreases exponentially with the energy gap between the states (*energy gap law*), since the higher is the energy gap, the higher is the vibrational quantum number of the isoenergetic level of the ground state, and the smaller is the overlap. At a constant energy gap, the probability depends on the vibrational energy spacings of the ground state, since the smaller is the energy spacing, the higher is the vibrational quantum number of the isoenergetic level of the ground state, and the smaller is the overlap. Therefore, high-energy vibrations (e.g., C–H stretching in organic molecules) are more effective than low-frequency ones as energy-accepting modes [7]. In fact, deuteration (lowering the frequency) is used as a tool for reducing the rate of radiationless transitions.

The situation sketched in Fig. 2.8b is that of potential energy surfaces crossing in the vicinity of the excited-state minimum. It is seen that in this case there is always a relatively good vibrational overlap, independent of the vibrational quantum number of the ground-state level. In this situation, the probability of radiationless transition tends to be high, insensitive to the energy gap (as long as it does not alter the crossing situation), and independent of the vibrational frequency of the accepting mode.

In intermediate situations (minima nested but with crossing points not too far from the excited-state minimum) it may be more convenient for the molecule to go through the crossing point, because of a more favourable Franck–Condon factor, despite the substantial activation energy required. In these cases, the rates of radiationless transitions may become very sensitive to temperature.

In principle, radiationless transitions can occur between excited states as well as between an excited state and the ground state. Generally speaking, electronically excited states are relatively closely spaced with respect to the large energy separation occurring between the lowest excited state and the ground state. Moreover, high excited states are more likely to have distorted geometries than the lowest excited state. Both these facts favour radiationless transitions among excited states (small energy gaps and frequent crossing situations) with respect to those between the lowest state and the ground state (large energy gap and often nested surfaces). This schematic picture is at the basis of the experimental observation that (i) internal conversion within each excited state manifold of any given multiplicity (for example, $S_n \rightarrow \rightarrow S_1$ and $T_n \rightarrow \rightarrow T_1$ in organic molecules) is exceedingly fast (usually in the picosecond region), (ii) intersystem crossing between the lowest excited states of any multiplicity (for example, $S_1 \rightarrow T_1$ in organic molecules) may be fast to moderately fast (sub-nanosecond to nanosecond region), and (iii) radiationless deactivations from the lowest excited states of any multiplicity to the ground state are much slower (in organic molecules, microseconds to nanoseconds for $S_1 \rightarrow S_0$ and seconds to milliseconds for $T_1 \rightarrow S_0$). This difference in rates is the basis of the famous Kasha rule (*vide infra*).

2.6.4   Chemical reaction

The unimolecular excited-state processes described in sections 2.6.2 and 2.6.3 do not cause any chemical change in the light-absorbing molecule and can thus be classified as *photophysical* processes. Excited states can, however, achieve deactivation to ground-state species also by a variety of *chemical* processes. As energy-rich species, excited states are, as a whole, expected to be more reactive than the corresponding ground states. It would be misleading, however, to consider this enhanced reactivity from a quantitative viewpoint only. In fact, the excess energy of electronically excited states is most often associated with an electronic structure quite different from that of the ground state. This may result in a reactivity that is qualitatively quite different. In this respect, it is worthwhile to think of electronically excited molecules as completely new chemical species, with their own qualitative and quantitative reactivity.

A point which should be stressed is that excited-state reactions must be very fast on the conventional chemical time-scale, since they have to compete with the photophysical deactivation processes. In practice, excited-state reactions must be almost activationless processes. Therefore, the key to understanding excited-state reactivity is the identification of low-energy channels along the excited-state surface leading, perhaps *via* some surface crossing, to the potential energy minima of the ground-state products. Progress in this direction has been made by the extensive use of correlation diagrams to identify low-energy reaction pathways [1e,8].

It is outside the scope of this introductory chapter to discuss in any detail the types of reactions given by the various types of excited states. A number of textbooks [1] on organic and inorganic photochemistry cover this subject exhaustively.

2.6.5   Kinetic aspects

Generally speaking, the three unimolecular processes described above compete for deactivation of any excited state of a molecule (Fig. 2.9a). Therefore their individual

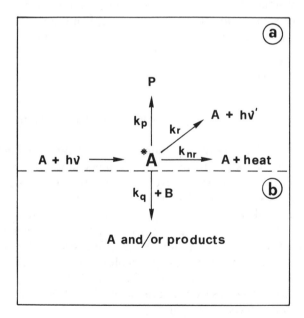

Fig. 2.9 — Unimolecular (a) and bimolecular (b) processes for the deactivation of an excited state. k_r, k_{nr}, and k_p are unimolecular rate constants for radiative decay, radiationless deactivation, and chemical reaction, and k_q is a bimolecular (quenching) rate constant.

specific rates *and* the kinetics of their competition in each excited state are of utmost importance in determining the actual behaviour of the excited molecule.

The probabilities of the various unimolecular processes discussed in the previous sections have the dimensions of a first-order rate constant. Thus, in the absence of other processes, an excited state $^*A$ will decay according to an overall first-order kinetics, with a *lifetime*, $\tau(^*A)$, given by (2.15).

$$\tau(^*A) = \frac{1}{k_r + k_{nr} + k_p} = \frac{1}{\Sigma_j \, k_j} \tag{2.15}$$

For each process of the $^*A$ excited state, an *efficiency* $\eta_i(^*A)$ can be defined as in (2.16).

$$\eta_i(^*A) = \frac{k_i}{\Sigma_j \, k_j} = k_i \, \tau(^*A) \tag{2.16}$$

The *quantum yield* Φ_i of a given process originating from $^*A$ is defined as the ratio between the number of molecules undergoing the process per unit time and the number of photons absorbed per unit time. If the excited state $^*A$ is directly reached by light absorption from the ground state, the quantum yield coincides numerically with the efficiency of the process. If the excited state $^*A$ is populated following one or more nonradiative steps from other excited states, then the value of the quantum yield is given by equation (2.17),

$$\Phi_i = \eta_i(^*A) \, \Pi_n \eta_n \tag{2.17}$$

where the η_n terms represent the efficiencies of the various steps involved in the population of $^*A$.

Kasha [9] has expressed in a rule the outcome of the competition between excited state unimolecular processes of a typical organic molecule: "The emitting level of a given multiplicity is the lowest excited level of that multiplicity". Stated in the above kinetic terms, Kasha's rule implies that: (i) the efficiency of internal conversion between singlet excited states of organic molecules down to S_1 is unity (i.e., the efficiencies of radiative and reactive processes of S_2, S_3, etc. are negligible); (ii) the lowest singlet state usually lives long enough so that intersystem crossing to T_1 and/or emission or chemical reaction can have appreciable efficiencies; (iii) the values of the quantum yields of processes of S_1 coincide with their efficiencies; (iv) the values of the quantum yields of processes from T_1 are the product of their efficiencies and the efficiency of $S_1 \rightarrow T_1$ intersystem crossing.

Kasha's rule has been extended by Crosby [10] to typical transition metal complexes: "In the absence of photochemistry from upper excited states, emission from a transition metal complex with an unfilled *d*-shell will occur from the lowest electronic excited state in the molecule or from those states that can achieve a significant Boltzmann population relative to the lowest electronic excited state". Crosby's rule can be obtained from the above arguments with the additional condition that the $S_1 \rightarrow T_1$ intersystem crossing in inorganic complexes has unitary efficiency, because of the heavy atom (metal) induced spin–orbit coupling.

Though in principle the actual behaviour of an excited molecule is the complex output of a cumbersome consecutive/competitive kinetic system, the exceedingly high rate of internal conversion between excited states is seen to simplify considerably the problem. In many instances, a careful evaluation of the factors affecting the kinetics of unimolecular processes of the lowest excited states gives the possibility of rationalizing and, to some extent, predicting the photochemical behaviour.

2.7 BIMOLECULAR PROCESSES OF EXCITED STATES

2.7.1 General considerations

As will clearly appear in Chapter 3, all photoprocesses dealt with in this book, including intercomponent processes (section 3.4), can be considered as "intramolecular" processes within *supermolecules*. It should also be noted, however, that such processes bear fundamental similarities with analogous bimolecular processes. For this reason, a short description of bimolecular processes of excited states is included in this introductory chapter.

As we have seen in section 2.6, each intramolecular decay step of an excited molecule is characterized by its own rate constant and each excited state is characterized by its *lifetime*, given by (2.15). It should be considered, however, that the excited state *A can also disappear because of interaction with other molecules (quenching processes, Fig. 2.9b). This may happen when the lifetime of the excited state is long enough for the excited molecule to encounter a suitable species B. Simple kinetic arguments show that in solution only those excited states that live longer than about 10^{-9} s may have a chance of being involved in encounters with other solute molecules. Usually, only the lowest excited state of any multiplicity satisfies this requirement (*vide supra*).

The most important bimolecular events are *energy-transfer* processes and *electron-transfer* chemical reactions, the latter involving either oxidation or reduction of the excited state:

$$*A + B \rightarrow A \ + *B \quad \textit{energy transfer} \qquad (2.18)$$

$$*A + B \rightarrow A^+ + \ B^- \quad \textit{oxidative electron transfer} \qquad (2.19)$$

$$*A + B \rightarrow A^- + \ B^+ \quad \textit{reductive electron transfer} \qquad (2.20)$$

Bimolecular energy- and electron-transfer processes are interesting for two reasons: (i) they can be used to quench an electronically excited state, i.e., to prevent its luminescence and/or intramolecular reactivity; (ii) they can be used to sensitize other species, for example to cause chemical changes of, or luminescence from, species that do not absorb light.

2.7.2   Thermodynamic aspects

In condensed phases, vibrational relaxation is a very fast process (10^{-12}–10^{-13} s) so that the electronically excited states involved in bimolecular processes are thermally equilibrated species (section 2.6.1). This means that these reactions can be dealt with in the same way as any other chemical reaction, i.e. by using thermodynamic and kinetic arguments.

For a thermodynamic treatment of reactions involving excited states [11], we need to define the standard excited state free energy, $G^0(*A)$. The readily available quantity for an excited state is its zero–zero energy $E^{00}(*A,A)$, i.e. the energy difference between the ground and excited states of a molecule, both taken at their zero vibrational levels (Fig. 2.7). Passing to thermodynamic quantities, the free energy difference between the excited and ground state molecules is given by (2.21).

$$\Delta G(*A,A) = \Delta H(*A,A) - T\,\Delta S(*A,A) \qquad (2.21)$$

In the condensed phase at 1 atm, $\Delta H \simeq \Delta E$, where ΔE is the internal (spectroscopic) energy. At 0 K, $\Delta E = NE^{00}$. This is also approximately true at room temperature if the vibrational partition functions of the two states are not very different. As far as the entropy term is concerned, it can receive three different contributions due to: (i) a change in dipole moment with consequent change in solvation; (ii) changes in the internal degrees of freedom; (iii) changes in orbital and spin degeneracy. This last contribution is the only one which can be straightforwardly calculated, but unfortunately it is also the least important in most cases. For a change in multiplicity from singlet to triplet it amounts to 0.03 eV at 25°C, which means that it can usually be neglected if one considers the experimental uncertainties which affect the other quantities involved in these calculations. The entropy contribution due to changes in dipole moment can be calculated if the change in dipole moment in going from the ground to the excited state is known. Finally, the contribution of changes of internal degrees of freedom is difficult to evaluate because the vibrational partition functions of the excited states are usually unknown.

Changes in size, shape and solvation of an excited state with respect to the ground state cause a shift (Stokes shift) between absorption and emission (section 2.6.2). When the Stokes shift is small (often a necessary condition to have a sufficiently long-lived excited state), the changes in shape, size, and solvation are also small and the entropy term in (2.21) may be neglected. In such a case, the standard free energy difference between the ground and excited states can be approximated as $\Delta G^0(*A,A) \simeq NE^{00}$ and the free energy changes of energy and electron transfer reactions can readily be obtained. An energy transfer process (2.18) will be thermodynamically allowed when $E^{00}(*A,A) > E^{00}(*B,B)$. As far as the electron-transfer processes (2.19) and (2.20) are concerned, within the approximation described above, the redox potentials for the excited state couples may be calculated from the standard potentials of the ground state couples and the one-electron potential corresponding to the zero–zero spectroscopic energy (i.e., the E^{00} value in eV):

$$E^0(A^+/*A) = E^0(A^+/A) - E^{00} \tag{2.22}$$

$$E^0(*A/A^-) = E^0(A/A^-) + E^{00} \tag{2.23}$$

The free energy change of the redox processes can then be readily calculated from the redox potentials, as is usually done for "normal" (i.e. ground state) redox reactions.

It should be noted that, as shown quantitatively by (2.22) and (2.23), an *excited state is both a stronger reductant and a stronger oxidant than the ground state* because of its extra energy content. Whether or not the excited state is a powerful oxidant and/or reductant depends, of course, on the redox potentials of the ground state.

2.7.3 Kinetic aspects

Leaving aside for the moment a detailed treatment of energy- and electron-transfer mechanisms, we will briefly recall here some fundamental kinetic aspects [12].

For processes requiring diffusion and formation of encounters, we can use the Stern–Volmer model which assumes statistical mixing of *A and B. The simplest case is that of a species *A which decays via some intramolecular paths and, in fluid solution, can encounter a quencher B (Fig. 2.9b). The excited state lifetimes in the absence (τ_0) and in the presence (τ) of the quencher B are given by (2.24) and (2.25), where k_q is the bimolecular constant of the quenching process.

$$\tau_0 = \frac{1}{k_p + k_r + k_{nr}} \tag{2.24}$$

$$\tau = \frac{1}{k_p + k_r + k_{nr} + k_q [B]} \tag{2.25}$$

Dividing (2.24) by (2.25), one obtains the well-known Stern–Volmer equation (2.26)

$$\tau_0/\tau = 1 + k_q \tau_0 [B] \tag{2.26}$$

that can be used to obtain k_q when τ_0 is known.

The rate constant k_q of the bimolecular quenching process is, of course, controlled by several factors. In order to elucidate these factors, a detailed reaction mechanism must be considered. Since both electron transfer and exchange energy transfer are collisional processes, the same kinetic formalism may be used in both cases [13]. For a reductive excited-state electron-transfer process (2.20), the reaction rate can be discussed on the basis of the mechanism shown in the scheme of Fig. 2.10, where k_d, k_{-d}, k'_d, and k'_{-d} are rate constants for formation and dissociation of the outer-sphere encounter complex, k_e and k_{-e} are unimolecular rate constants for the electron transfer step involving the excited state, and $k_{e(g)}$ and $k_{-e(g)}$ are the

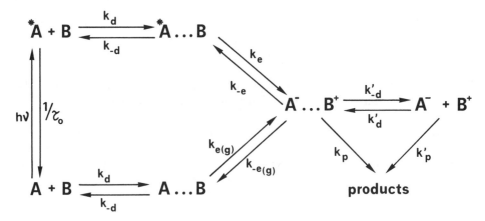

Fig. 2.10 — Kinetic mechanism for electron transfer reactions.

corresponding rate constants for the ground-state electron-transfer step. A simple steady state treatment [13,14] shows that the experimental rate constant of (2.20) can be expressed as a function of the rate constants of the various steps by (2.27),

$$k_{exp} = \frac{k_d}{1 + \dfrac{k_{-d}}{k_e} + \dfrac{k_{-d}\, k_{-e}}{k_x\, k_e}} \tag{2.27}$$

where k_x may often be replaced by k'_{-d} (for more details, see [13]). In a classical approach, k_{-e}/k_e is given by $\exp(-\Delta G^0/RT)$, where ΔG^0 is the standard free energy change of the electron-transfer step. An analogous expression holds for bimolecular energy transfer.

The key step of the process is, of course, the unimolecular electron- (or energy-) transfer step (k_e) and the essence of the problem is the fact that the equilibrium nuclear configuration of a species changes when it gains or loses an electron (or electronic energy). In fluid solution this configuration change may involve changes in bond lengths and angles and, at least in the case of electron transfer, solvent repolarization [15–19]. Since electronic motions are much faster than nuclear motions (Franck–Condon principle), in a classical approach an adjustment of the nuclear configuration prior to electron (or energy) transfer is required. This gives rise to an activation barrier, ΔG^+, as shown in Fig. 2.11. Once a suitable nuclear configuration has been reached, whether or not electron (or energy) transfer occurs is a matter of electronic factors (adiabaticity problem). In the classical approach [13,17], the unimolecular rate constant is given by (2.28),

$$k_e = \kappa \nu_N \exp(-\Delta G^+/RT) \tag{2.28}$$

where κ is the electronic transmission coefficient, ν_N is an effective frequency for

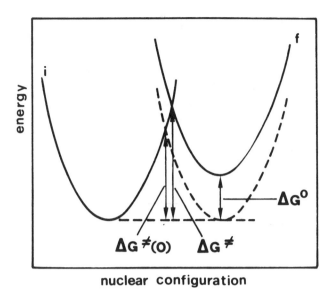

Fig. 2.11 — Energy surfaces of the initial and final states for an electron transfer reaction. For (2.20), i is $^*A \ldots B$ and f is $A^- \ldots B^+$.

nuclear motion, and $\Delta G^{\ddagger}$ is the free activation energy. The electronic transmission coefficient (i.e., the reaction probability at the crossing point) is usually unity for bimolecular electron-transfer processes [15,17,18], whereas it can be much lower for energy transfer [19,20]. (A detailed discussion of the factors which govern the electronic transmission coefficient is given in Chapters 5 and 6.) The free energy of activation, $\Delta G^{\ddagger}$, may be expressed by a free energy relationship like the classical Marcus quadratic equation (2.29) [15],

$$\Delta G^{\ddagger} = \Delta G^{\ddagger}(0) \ \{1 + [\Delta G^0 / 4 \ \Delta G^{\ddagger}(0)]\}^2 \tag{2.29}$$

where $\Delta G^{\ddagger}(0)$ is the so-called intrinsic nuclear barrier (Fig. 2.11). For a homogeneous series of reactions [21], such as those between the same oxidant $^*A$ and a series of structurally related reductants B_1, B_2, B_3, ... (or vice versa) that have variable redox potential but the same size, shape, electronic structure and electric charge, one can assume that throughout the series the reaction parameters k_d, k_{-d}, k'_{-d} in (2.27), κ and ν_N in (2.28) and $\Delta G^{\ddagger}(0)$ in (2.29) are constant. Under these assumptions, k_{exp} in (2.27) is only a function of ΔG^0, and a log k_{exp} vs ΔG^0 plot is a bell-shaped curve involving (i) a "normal" region for endoergonic and slightly exoergonic reactions, where log k_{exp} increases with increasing driving force, and (ii) an "inverted" region, for strongly exoergonic reactions, in which log k_{exp} decreases with increasing driving force.

The investigation of the inverted region is often quite difficult because the top of the bell-shaped curve is cut by diffusion (2.27). For energy-transfer processes difficulties may also arise because of the presence of parallel channels in the normal region related to higher excited states of the acceptor. For this and other (not yet fully understood) reasons [22], the inverted behaviour often escaped experimental detection in bimolecular processes [21], leading to the use of empirical asymptotic free-energy relationships [13,14]. However, following some pioneering work in rigid matrices [23], conclusive evidence for the predicted inverted behaviour has now been reached for both bimolecular [22,24–26] and unimolecular (intercomponent) [27,28] electron transfer processes.

As shown in Fig. 2.12, high intrinsic barriers $\Delta G^{\ddagger}(0)$ cause a decrease of the rate

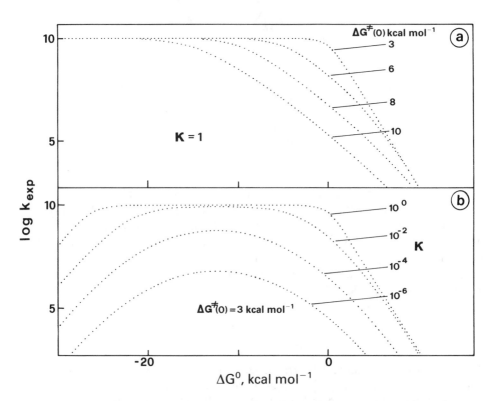

Fig. 2.12 — Effects of intrinsic barrier (a) and electronic transmission coefficient (b) on log k_{exp} vs ΔG^0 plots. The values of intrinsic barrier and electronic transmission coefficient are shown in the figure. Values used in the calculations for the other parameters are: $v_N = 6.2 \times 10^{12}$ s^{-1}; $k_d = 1 \times 10^{10}$ M^{-1}s^{-1}; $k_{-d} = k'_{-d} = 1 \times 10^{10}$ s^{-1}.

constant in the nearly isoergonic region, while low values of the transmission coefficient κ cause a decrease in the maximum rate constant value (nonadiabatic behaviour). Free energy relationships have been extensively used to establish whether a series of quenching reactions takes place via energy transfer, reductive electron transfer, or oxidative electron transfer. Such free energy relationships have also been used to estimate, by fitting procedures, plausible values for $\Delta G^{\neq}(0)$ and κ for homogeneous series of reactions [29].

REFERENCES

[1] For reference textbooks on photochemistry, see: (a) Calvert, J. G. and Pitts, J. N., Jr. (1966) *Photochemistry*. Wiley; (b) Balzani, V. and Carassiti, V. (1970) *Photochemistry of coordination compounds*. Academic; (c) Geoffroy, G. L. and Wrighton, M. S. (1973) *Organometallic photochemistry*. Academic; (d) Adamson, A. W. and Fleischauer, P. D. (eds) (1975) *Concepts of inorganic photochemistry*. Wiley; (e) Turro, N. J. (1978) *Modern molecular photochemistry*. Benjamin; (f) Barltrop, J. A. and Coyle, J. D. (1978) *Principles of photochemistry*. Wiley; (g) Ferraudi, G. J. (1988) *Elements of inorganic photochemistry*. Wiley; (h) Wayne, R. P. (1988) *Principles and applications of photochemistry*. Oxford University Press.

[2] Eyring, H., Walter, J., and Kimball, G. E. (1944) *Quantum chemistry*. Wiley.

[3] Herzberg, G. (1966) *Electronic spectra of polyatomic molecules*. van Nostrand Reinhold.

[4] Kauzmann, W. (1957) *Quantum chemistry*. Academic.

[5] Herzberg, G. (1950) *Spectra of diatomic molecules*. van Nostrand Reinhold.

[6] Cotton, F. A. (1971) *Chemical applications of group theory*. Interscience.

[7] Henry, B. R. and Siebrand, W. (1973). In Birks, J. B. (ed.) *Organic molecular photophysics*. Vol. 1. Wiley, p. 153.

[8] Vanquickenborne, L.G. (1982) *Comments Inorg. Chem.* **2** 23.

[9] Kasha, M. (1950) *Faraday Soc. Discussion* **9** 14.

[10] Crosby, G. A. (1983) *J. Chem. Educ.* **60** 791.

[11] Grabowski, Z. R. and Grabowska, A. (1976) *Z. Physik. Chem. Neue Folge.* **101** 197.

[12] Balzani, V., Moggi, L., Manfrin, M. F., Bolletta, F., and Laurence, G. S. (1975) *Coord. Chem. Rev.* **15** 321.

[13] Balzani, V., Bolletta, F., and Scandola, F. (1980) *J. Am. Chem. Soc.* **102** 2152.

[14] Rehm, D. and Weller, A. (1970) *Isr. J. Chem.* **8** 259.

[15] Marcus, R. A. (1964) *Annu. Rev. Phys. Chem.* **15** 155.

[16] Hush, N. S. (1968) *Electrochim. Acta* **13** 1005.

[17] Sutin, N. (1983) *Prog. Inorg. Chem.* **30** 441.

[18] Marcus, R. A. and Sutin, N. (1985) *Biochim. Biophys. Acta* **811** 265.

[19] Scandola, F. and Balzani, V. (1983) *J. Chem. Educ.* **60** 814.

[20] Endicott, J. F. (1988) *Acc. Chem. Res.* **21** 59.

[21] Balzani, V. and Scandola, F. (1983). In Graetzel, M. (ed.) *Energy resources through photochemistry and catalysis*. Academic, p. 1.

[22] Mataga, N. (1989). In Norris J. R., Jr. and Meisel D. (eds) *Photochemical energy conversion*. Elsevier, p. 32.

[23] Miller, J. R., Beitz, J. V., and Huddleston, R. K. (1984) *J. Am. Chem. Soc.* **106** 5057.

[24] Gould, I. R., Ege, D., Mattes, S. L., and Farid, S. (1987) *J. Am. Chem. Soc.* **109** 3794.

[25] Gould, I. R., Moser, J. E., Armitage, B., and Farid, S. (1989) *J. Am. Chem. Soc.* **111** 1917.
[26] Mataga, N., Kanda, Y., and Okado, T. (1986) *J. Phys. Chem.* **90** 3880.
[27] Closs, G. L. and Miller, J. R. (1988) *Science* **240** 440.
[28] Wasielewski, M. R. and Niemczyk, M. P. (1984) *J. Am. Chem. Soc.* **106** 5043.
[29] Balzani, V. and Scandola, F. (1986) *Inorg. Chem.* **25** 4457, and references therein.

3

Supramolecular properties

3.1 INTRODUCTION

Any general definition of *supermolecule* is necessarily arbitrary and the word may have different meanings depending on the area to which it is applied [1]. Conceptually, the feature that distinguishes a supermolecule from a "large molecule" is the possibility to split the supermolecule into individual molecular subunits (*components*) capable, as they are or with minor modifications, of a separate existence. The subunits are therefore characterized by a set of intrinsic properties (section 3.3) that can in principle be derived from a study of the isolated subunits or of suitable model compounds. Many of the intrinsic properties of each component are expected to be maintained in the supramolecular structure, with relatively minor changes that can be ascribed to the mutual perturbation between the subunits. However, the properties of a supermolecule will not generally be a simple superposition of those of the component units. In fact, it is possible that processes involving two or more components take place in a supermolecule, such as (i) intercomponent transfer processes (for example, electron or energy transfer) or (ii) cooperative effects (for example, complexation of other species by two or more components). These processes may cause the disappearance of some intrinsic property of the components and/or the appearance of completely new properties, characteristic of the supermolecule as such (section 3.4). The study of the new distinctive photochemical and photophysical properties of the supermolecule constitutes the object of *supramolecular photochemistry*.

The above definition of a supermolecule is clear-cut when the components are individual molecules held together by intermolecular forces (for example, host–guest systems, ion pairs, electron donor–acceptor complexes) [1a,c]. Such a definition can be extended to systems containing covalently-linked components which satisfy specific conditions. For example, for the systems depicted in Fig. 3.1a [2,3], the classification as *supermolecules* is straightforward, since molecular subunits with well-defined intrinsic properties can be easily identified (Fig. 3.1b). Among these,

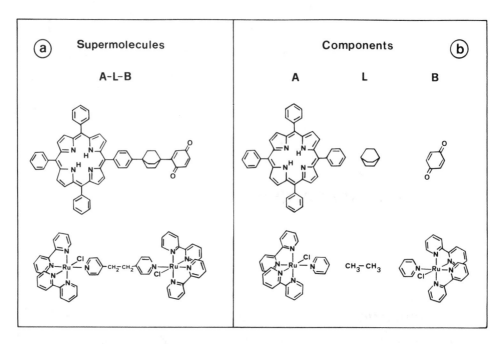

Fig. 3.1 — Covalently linked supermolecules (a) and their molecular components (b).

the porphyrin and quinone moieties in one case, and the ruthenium complex moieties in the other case are certainly the most important components of the supermolecule. In fact, these components have distinct spectroscopic, photophysical, photochemical, and redox properties that determine to a large extent the behaviour of the supermolecule. Fragments of this type can be conveniently called *active components* of the supermolecule. The bicyclooctane fragment and the $-CH_2-CH_2-$ bridge, on the other hand, lack intrinsic photochemically-interesting properties and have mainly a structural role: that of linking together, at a given distance and in a given geometry, the two active components. Fragments of this type can be conveniently called *connectors*. This does not imply, of course, that connectors necessarily play a totally passive role as components of a supermolecule (sections 3.2 and 3.4).

At the other extreme, the systems shown in Fig. 3.2 are clearly *large molecules* that upon fragmentation would completely lose their chemical identity. There are systems, however, for which a clear-cut decision is difficult. For example, in the case of stilbene (Fig. 3.3a) and the so-called Creutz–Taube ion [4] (Fig. 3.3b), individual molecular fragments or models thereof could still be identified, but their properties as free molecules would be remarkably different from those found in the whole system. Formally, such systems could still be viewed as supermolecules in which, however, the mutual perturbation of the fragments is so strong as to make this approach relatively useless. It should be noted that, in this loose sense, classical transition-metal coordination compounds could also be considered as super-

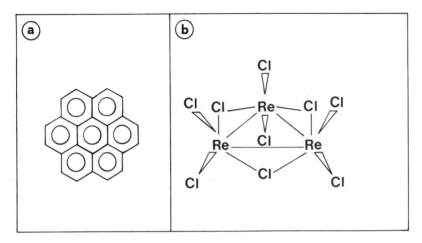

Fig. 3.2 — "Large molecules": (a) coronene; (b) Re_3Cl_9.

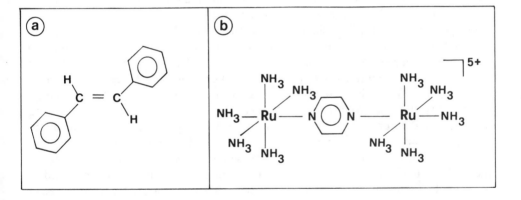

Fig. 3.3 — Borderline systems between the supermolecule and large-molecule cases.

molecules, since they are composed of metal ions and ligands that are to some extent capable of independent existence (the name "complexes" comes right from this characteristic). Here too, however, the mutual perturbation of the subunits (ligand field, metal-to-ligand back bonding) is generally so strong as to make a molecular approach more realistic than the supramolecular one.

From the above examples, it is clear that the heart of the "supermolecule" *vs* "large molecule" problem lies in the degree of interaction between the electronic subsystems of the component units. Provided that this interaction is small with respect to other relevant energy parameters, any multicomponent system approaches the ideal concept of a supermolecule.

3.2 SUPERMOLECULES AS LOCALIZED MULTICOMPONENT ELECTRONIC SYSTEMS

A class of compounds for which the argument concerning the extent of electronic interaction between their components has been developed in considerable detail is that of the so-called *mixed-valence* polynuclear transition-metal complexes [5]. Let us take as an example a binuclear complex such as **3.1**, where L stands for a neutral, symmetrical ligand [6]. In a valence-localized description, that is in terms of integral

$$[Ru(NH_3)_5-L-Ru(NH_3)_5]^{5+}$$

3.1

oxidation states of the metal centres, the overall charge corresponds to a Ru(II)–Ru(III) complex. In a fully delocalized description, on the other hand, a Ru(II$\frac{1}{2}$)–Ru(II$\frac{1}{2}$) complex would result. The factors determining the localized or delocalized nature of the complex can be easily appreciated following the approach originally developed by Hush [7]. Consider the two valence-localized "electronic isomers" Ru(II)–Ru(III) and Ru(III)–Ru(II). A specific equilibrium geometry corresponds to each of these species, in terms of both *inner* (e.g., Ru–NH$_3$ distances at both centres) and *outer* (e.g., orientation of solvent molecules around both centers) nuclear degrees of freedom. This is depicted in Fig. 3.4a using parabolic potential energy curves for the two electronic isomers and a generalized nuclear coordinate involving both inner and outer nuclear displacements. Fig. 3.4a emphasizes the fact that at the equilibrium geometry of each electronic isomer the other isomer can be considered as an electronically excited state. The energy separation between these two states at the equilibrium geometry is usually called the reorganizational energy and is indicated by λ (this quantity is related to to the intrinsic barrier $\Delta G^{\neq}(0)$ introduced in section 2.7.3, $\Delta G^{\neq}(0) = \lambda/4$). At the crossing point both electronic isomers have the same energy and geometry. This is the nuclear configuration where there are no Franck–Condon restrictions to electron exchange between the two centres. Its energy is in this model one quarter of the reorganizational energy.

If for some reason (for example, very long centre-to-centre distance, insulating character of L) the electronic interaction between the Ru(II) and Ru(III) centres, H_{AB}, is absolutely negligible, the curves in Fig. 3.4a adequately represent the system at any geometry along the nuclear coordinate. The system is expected to exhibit properties which are a perfect superposition of the properties of isolated Ru(NH$_3$)$_5$L^{3+} and Ru(NH$_3$)$_5$L^{2+}. Furthermore, even if the system acquires sufficient activation energy to reach the intersection region, the probability of electron exchange is negligible. In the field of mixed-valence chemistry, this is usually called Class I behaviour [8]. An example approaching this type of behaviour within the complexes of type **3.1** is obtained when L is the bridge **3.2** [6].

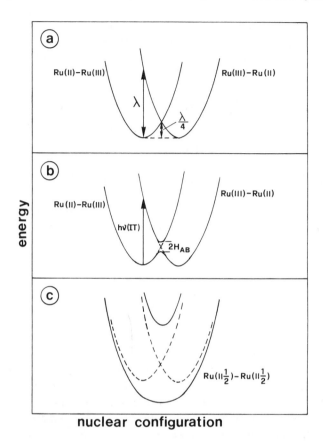

Fig. 3.4 — Potential energy curves for mixed-valence compounds with negligible (a), weak (b), and strong (c) electronic coupling. In (b) and (c), the dashed curves represent zero-order states.

$$N\!\!-\!\!\bigcirc\!\!-\!\!CH_2\!\!-\!\!CH_2\!\!-\!\!\bigcirc\!\!-\!\!N$$

3.2

In most cases, however, some electronic interaction is likely to occur between the Ru(II) and Ru(III) centres, either as a consequence of direct orbital overlap or via some through-bridge mechanism (section 3.4). In such cases, the curves in Fig. 3.4a are only zero-order representations. The electronic interaction has almost no effect on the zero-order curves in the vicinity of the equilibrium geometries, where the difference in energy between the electronic isomers is much larger than H_{AB}, but causes mixing of the zero-order states (avoided crossing) in the vicinity of the crossing point (Fig. 3.4b). Systems of this type can still be considered as valence-localized, and will still exhibit the properties of the isolated $Ru(NH_3)_5L^{3+}$ and $Ru(NH_3)_5L^{2+}$ components. However, new properties promoted by the Ru(II)–Ru(III) interaction can also be observed (section 3.4), such as, for example, optical intervalence transfer (IT) transitions with $h\nu = \lambda$(3.1),

$$[Ru^{II}(NH_3)_5\text{--}L\text{--}Ru^{III}(NH_3)_5]^{5+} \xrightarrow{h\upsilon} [Ru^{III}(NH_3)_5 - L - Ru^{II}(NH_3)_5]^{5+} \qquad (3.1)$$

or thermally activated electron-transfer processes (3.2)

$$[Ru^{II}(NH_3)_5\text{--}L\text{--}Ru^{III}(NH_3)_5]^{5+} \rightarrow [Ru^{III}(NH_3)_5\text{--}L\text{--}Ru^{II}(NH_3)_5]^{5+} \qquad (3.2)$$

interconverting the two electronic isomers. The barrier to thermal electron transfer is only negligibly smaller than calculated on the basis of the zero-order curves ($\lambda/4$). This type of behaviour is usually called Class II [8]. An example of Class II behaviour within complexes of type **3.1** is obtained when L is the bridge **3.3** [6].

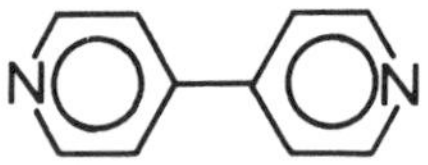

3.3

If strong electronic coupling is provided by the bridging ligand, the zero-order levels can be substantially perturbed even in the vicinity of their equilibrium geometries. In the limit of very large electronic coupling, when $H_{AB} \simeq \lambda$, the true first-order curves will show a single minimum at an intermediate geometry (Fig. 3.4c). In this case, the binuclear complex is better considered a fully delocalized $Ru(II\frac{1}{2})$–$Ru(II\frac{1}{2})$ species, with properties that are mostly unrelated to those of the hypothetical $Ru(NH_3)_5L^{3+}$ or $Ru(NH_3)_5L^{2+}$ components. This case is commonly indicated as Class III [8]. An example of Class III behaviour within complexes of type **3.1** is obtained when L is **3.4** [6].

$$N\equiv C-C\equiv N$$

3.4

The above classification of mixed-valence compounds has been illustrated using symmetric redox systems, that is systems made of identical subunits in which there is no net driving force for intramolecular electron transfer. The arguments concerning the degree of electron delocalization are, however, general and can be easily extended to systems which exhibit redox asymmetry. Clearly, mixed-valence Class I and II compounds belong to our operational definition of "supermolecule", while

Class III systems approach the "large molecule" limit. The discussion of the mixed-valence complexes emphasizes that a supermolecule should be amenable to a description in terms of *localized electronic configurations*. This requires that the degree of electronic coupling between the molecular components is small. It is important to recognize that, in this context, "small" is not intended in an absolute sense, but with respect to the energy of vibrational trapping of the electron on each molecular component.

Let us consider a generic supermolecule A–L–B where A and B are active components and L is a connector. So far we have discussed the problem of localization in a supermolecule with respect to oxidation states, that is, with respect to electronic configurations that differ for one *electron* being localized on different components and that can be interconverted by means of an *intercomponent electron transfer process* (3.3).

$$A–L–B \rightarrow A^+–L–B^- \tag{3.3}$$

An important situation that can be discussed along the same general lines is that of an *electronically excited* supermolecule (for example, *[A–L–B]). In such a system, it is possible to have electronic configurations (*A–L–B and A–L–*B) that differ in the excitation being localized on different molecular components. These configurations may be interconverted by an *intercomponent energy-transfer process* (3.4).

$$*A–L–B \rightarrow A–L–*B \tag{3.4}$$

Electron and energy transfer are quite similar processes that can be dealt with by using very similar kinetic models [9,10]. The most important differences between the two processes are as follows: (i) vibrational trapping of an electron always receives a substantial contribution from solvent repolarization (outer-sphere reorganization energy), whereas that of electronic energy mainly depends on intramolecular degrees of freedom (inner-sphere contribution), and (ii) the electronic coupling term H_{AB} is a simple one-electron matrix element in the electron-transfer case, but is a two-electron exchange or coulombic integral in the energy-transfer case. Although due to these differences the statement cannot be general, it is reasonable to say that most supermolecules that behave in a localized way with respect to electrons will do so for electronic energy as well. The supramolecular systems dealt with in this book are generally systems in which a localized description of *both* electronic excitation and oxidation states is appropriate.

The relationship between these two aspects of the problem becomes evident if one realizes that the "reactants" and "products" of energy- and electron-transfer processes are but different electronic states of the supermolecule. From such a viewpoint, all the intercomponent processes can be treated as radiationless or radiative transitions between such states. This picture is useful in emphasizing the conceptual relationship between intramolecular electron and energy transfer within a supermolecule. It should be stressed that electron and energy transfer can be considered as distinct processes only to the extent to which oxidation states and electronic excitation are actually localized in the system investigated. General aspects of electron and energy transfer between components of a supermolecule are discussed in section 3.4.

3.3 PROPERTIES OF MOLECULAR COMPONENTS

As discussed in the previous section, our operational definition relies on the notion that supramolecular systems can be described in terms of localized electronic configurations. As a consequence, in a supramolecular structure the molecular components have well defined *individual* properties. In this section we will simply underline a few points concerning relevant properties of molecular components and the perturbations that they may experience when included into a supramolecular structure.

What properties are relevant in characterizing an active component of a super-molecule is a matter that largely depends on the role the component plays in the photophysical or photochemical bahaviour of the supermolecule. As will be shown in more detail later on, active components may participate in various elementary acts including light absorption, light emission, intercomponent electron or energy transfer, isomerization, complexation of ions or molecular species, protonation–deprotonation (the relationship between these elementary acts and the concept of "function" is particularly developed in Chapter 12). Therefore, quite a large number of properties are necessary to characterize an active molecular component. The most important ones, related to the photophysical and redox behaviour, can be illustrated on the basis of Fig. 3.5.

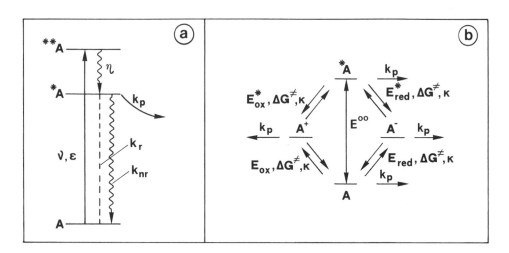

Fig. 3.5 — Photophysical and photochemical processes (a) and ground and excited-state redox processes (b) of a molecular component. In (b), the symbols $\Delta G^{\neq}$, κ, and k_p are not specifically labelled for the sake of simplicity.

In Fig. 3.5a the photophysical behaviour of a component A is schematically represented in terms of light absorption to give a first excited state **A, conversion to a lowest "active" state *A, and decay to ground-state species *via* radiative (k_r), radiationless (k_{nr}), and reactive (k_p) pathways. Quantities relevant to the definition of the photophysical properties of the component are: the frequency (v) and molar absorption coefficient (ε) of the absorption band, the efficiency of formation of the active excited state (η), the lifetime of the active excited state ($\tau = 1/(k_r + k_{nr} + k_p)$), the quantum yields of emission ($\Phi_r = k_r\tau$) and photoreaction ($\Phi_p = k_p\tau$).

In Fig. 3.5b the ground- and excited-state redox behaviour of a component A is schematized. The redox behaviour of the component is thermodynamically defined by the potentials for oxidation and reduction (E^{ox} and E^{red}) of the ground and excited state, and the energy (E^{00}) of the excited state (which is correlated with the redox potentials as indicated by equations (2.22) and (2.23)). From the kinetic point of view, the important parameters are the activation free energies (ΔG^+) and transmission coefficients (κ) of the corresponding self-exchange [11] reactions. As regards the redox behaviour of a component, another important aspect is that of the chemical stability (inversely related to the rate constant for chemical reaction, k_p) of the oxidized and reduced forms.

Of course, many other properties can be useful to characterize a component and to interpret its behaviour in a supramolecular system. Among these are the acid-base properties of the ground and excited states, the tendency of the excited state to isomerize, and the coordinative ability towards other species in the ground and excited states. No general discussion of the above properties will be given in this section. In subsequent sections relevant properties will be selected and discussed, depending on the role of specific components in specific supramolecular structures.

Connectors are not active components of a supramolecular structure, inasmuch as they do not have low-lying excited states and low energy levels. Their general role is that of keeping the active components of a supermolecule together. Therefore their main properties are *structural* in nature, for example end-to-end distance, degree of rigidity, conformational behaviour. It is important to realize, however, that connectors also have a more subtle but nevertheless important role: that of connecting the active components in an *electronic* sense. We shall see in the next and in later sections that a large part of the interesting photochemical and photophysical properties of supermolecules is determined by the occurrence of intercomponent electron- or energy-transfer processes. Except for the case of coulombic energy transfer (section 3.4.5 and Chapter 6), a certain degree of electronic interaction (orbital overlap) between the active components is required for these transfer processes to occur. At a few Ångstroms separation, direct orbital overlap is quite inefficient so that through-space interaction between components becomes negligible. Therefore, most of the electronic interaction between covalently-linked active components of a supermolecule is likely to be mediated by the connectors with a through-bond mechanism. It is thus evident that a very important intrinsic property of connectors is their ability to put the attached active components into electronic communication. By analogy with an electrical wiring scheme, we could view this property of the connectors as their "conductivity", although it may be difficult to find a single parameter describing this property in a quantitative way. The factors that affect the "conductivity" of a connector will be discussed to some extent in Chapters 5 and 6.

In principle, the properties of the molecular subunits of a supramolecular structure can be obtained from a study of the isolated components or of some suitable model molecules. It has already been pointed out (section 3.1), however, that in several cases the identification of real molecules that constitute suitable models for molecular components of a supermolecule is not a trivial problem. Strictly speaking, in a covalent A–L–B supermolecule in which A and B are active components and L is a connector, A and B would be radicals or coordinatively unsaturated species that can never exist as such. In some cases, for example when the connector is bound to the components *via* carbon–carbon bonds as in the supermolecules of Fig. 3.1, this is not a problem since molecular species such as AH and BH or AR and BR (R = alkyl group) are indeed almost identical in properties to A and B in the supermolecule. In other cases, however, the connector may interact more deeply with the electronic subsystem of the active components, for example, in the systems of Fig. 3.3b. In such a case, compounds that include the connector such as A–L and B–L should be used to approach the properties of A and B in the supermolecule. This, however, may be insufficient in some instances, as the properties of the A–L model can change further upon linking this fragment to component B. An example of such a situation is given by the series of binuclear complexes of the $Ru(bpy)_2(CN)$–CN–M type shown in Table 3.1 [12–15], in which spectroscopic energies ($Ru \rightarrow bpy$ MLCT transitions)

Table 3.1 — Absorption maxima and oxidation potentials of supramolecular systems of the type $Ru(bpy)_2(CN)$–CN–M [12–15][a]

M	$\lambda_{max}(nm)$[b]	$E_{1/2}(V$ *vs* SCE)[c]
—	428	+ 0.85
$Pt(dien)^{2+}$	416	+ 1.03[d]
$Ru(NH_3)_5^{2+}$	413	
$Ru(NH_3)_5^{3+}$	403	+ 1.07
$Ru(NH_3)_4py^{2+}$	417	
$Ru(NH_3)_5py^{3+}$	400	+ 1.12
$Ru(bpy)_2CN^{2+}$	400	+ 1.31

[a] Aqueous solution, unless otherwise noted; [b] $Ru \rightarrow bpy$ charge-transfer band; [c] oxidation potential of the Ru^{2+} ion of the $Ru(bpy)_2(CN)$–CN moiety; [d] DMF solution.

and redox potentials ($Ru(II) \rightarrow Ru(III)$ oxidation) of the $Ru(bpy)_2(CN)$–CN unit are substantially perturbed by the presence of various metal complex moieties, M. These changes can be rationalized in terms of the decrease in effective charge at the Ru centre due to the increased π-acceptor and decreased σ-donor ability of the bridging cyanide upon coordination to the second metal moiety. Intercomponent perturbations of this type should be expected, of course, only when the connector, as in the cyanide case, is a highly "conducting" one. These effects can be used in favourable cases to control and tune the ground- or excited-state properties of specific components (section 4.2).

3.4 INTERCOMPONENT PROCESSES

It has been stated in the foregoing sections that, except for the rather hypothetical case of totally noninteracting components, the behaviour of a supermolecule will differ from the superposition of those of the single components because of the occurrence of processes involving two (or more) components (*intercomponent processes*). Leaving to subsequent chapters detailed analyses and practical examples, we will simply mention here the principal types of intercomponent processes and discuss some of their general, characteristic features.

3.4.1 Radiative processes

The absorption spectrum of a supramolecular system can differ substantially from the sum of the spectra of the molecular components. Apart from those changes (usually shifts) that can be dealt with in terms of perturbations of the spectra of the single components (see previous section), some totally new bands can be present in the spectrum of the supermolecule. These bands correspond to the *optical electron-transfer* process shown in (3.5) and Fig. 3.6,

$$A.B \xrightarrow{h\nu} A^+.B^- \tag{3.5}$$

where A and B represent active components and the dot stands for any type of linking interaction (covalent bond, electrostatic interaction, intermolecular forces). Such a process should be clearly distinguished [16] from photoinduced electron

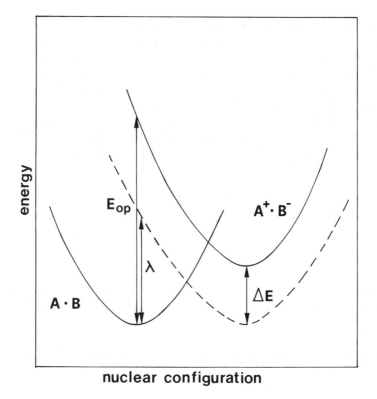

Fig. 3.6 — Optical electron transfer.

transfer, that corresponds to thermal (radiationless) electron transfer following electronic excitation of a single component (section 3.4.3). Optical electron-transfer transitions are well-known in the fields of mixed-valence compounds (where they are called intervalence transfer, IT, transitions, section 3.2), ion-pair systems (ion-pair charge transfer, IPCT, section 8.2), electron donor–acceptor complexes (charge transfer, CT, section 9.2).

The energy of an optical electron-transfer transition, E_{op}, is correlated according to the Hush theory [6,7] to the energy gradient between the minima of the A.B and $A^+.B^-$ curves, ΔE, and to the reorganizational energy (3.6).

$$E_{op} = \Delta E + \lambda \tag{3.6}$$

Contrary to what happens for the symmetric cases ($\Delta E = 0$) discussed in section 3.2, in an asymmetric case λ is a virtual quantity, that is, the vertical energy of a hypothetical system with the same degree of distortion but with no energy gradient (Fig. 3.6). Detailed expressions for the calculation of the reorganizational energy λ in terms of the internal distortions of the A and B components and solvent repolarization are available within the framework of the Hush theory [6,7], although in most practical cases their application may be hampered by the unavailability of the relevant molecular parameters. On the assumption that differences in entropy terms are negligible, the energy gradient ΔE can be estimated from the standard redox potentials of the components (or of reasonable models thereof). In doing so, a correction must be applied for the mutual influence of the two components in the supermolecule. For localized supermolecular systems (see section 3.2), the correction consists simply of the difference between the electrostatic work terms of the reactant and product states (3.7–3.9). In (3.8) and (3.9), Z is the electric charge of the components, r is the intercomponent distance, and ε is the effective dielectric constant of the medium. The energy of the optical electron-transfer transition can thus be used to obtain an experimental estimate of the reorganizational energy.

$$\Delta E = E^0(A^+/A) - E^0(B/B^-) + w(A^+.B^-) - w(A.B) \tag{3.7}$$

$$w(A^+.B^-) = \frac{Z(A^+)Z(B^-)e^2}{\varepsilon r} \tag{3.8}$$

$$w(A.B) = \frac{Z(A)Z(B)e^2}{\varepsilon r} \tag{3.9}$$

The optical electron-transfer band is expected to be gaussian-shaped, with a halfwidth that is directly related to the reorganizational energy by (3.10)[6,7].

$$\Delta \bar{v}_{1/2}(cm^{-1}) = 48.06(E_{op} - \Delta E)^{1/2} \tag{3.10}$$

The intensity of the optical electron-transfer band can be correlated, according to Hush [6,7] to the magnitude of the electronic coupling matrix element H_{AB}. The relationship is given by (3.11),

$$\varepsilon_{max} = \frac{2380r^2}{E_{op}\Delta\bar{v}_{1/2}}H_{AB}^2 \tag{3.11}$$

where ε_{max} is the maximum molar absorption coefficient, the energies and halfwidths are in cm^{-1} and the intercomponent distance, r, is in Ångstroms. We have seen in section 3.2 that the magnitude of the intercomponent electronic coupling relative to the reorganizational energy is crucial in determining the localized or delocalized nature of the supermolecule. The above equations show that important information about this point can be obtained, in principle, from a study of the optical electron-transfer spectrum of a supermolecule.

Whether or not electron-transfer transitions are actually observed in the spectrum of a supermolecule depends on several factors. Except for systems exhibiting exceptionally large interactions (that would probably not even warrant a localized description), the intensities of these bands are expected to be low with respect to those of the fragments. For example, with $H_{AB} = 100\,cm^{-1}$ and common values for the other parameters ($r = 10\,Å$, $E_{op} = 15000\,cm^{-1}$, $\Delta \bar{v}_{1/2} = 5000\,cm^{-1}$) equation (3.11) yields $\varepsilon_{max} = 30\,M^{-1}cm^{-1}$. Therefore, in order for such a band to be observable, it is necessary that it occurs in a region of the spectrum free from absorption by the components, usually the visible or near infrared. This in turn depends on the energetics (3.6), and requires that A and/or B are relatively easy to oxidize or reduce. It will be shown in Chapters 5 and 6 that in some cases electronic couplings as small as a few cm^{-1} may be sufficient to induce *radiationless* intercomponent processes such as electron or energy transfer. It is evident that in such cases, although the electronic coupling is an important parameter, information about its magnitude cannot be obtained from the spectra, as the optical electron-transfer transitions have negligible intensity.

In the situation shown in Fig. 3.6, no emission corresponding to the optical electron-transfer absorption process is expected. In fact, the excited state $A^+.B^-$ populated by the "vertical" absorption process becomes, after relaxation to its equilibrium nuclear geometry, the ground state of the system at that geometry. Cases are conceivable, however, in which, because of a greater ΔE (relative to λ), the relaxed state $A^+.B^-$ remains an excited state. In such cases, an emission analogue of the absorption process is possible (Fig. 3.7). Emission from electron-transfer states in some supramolecular systems have indeed been observed (Chapters 5 and 9). The field of twisted-intramolecular-charge-transfer (TICT) states of organic molecules (section 7.1.4) presents numerous examples of such type of emission, although in other respects the analogy between TICT systems and supermolecules should be considered with caution.

3.4.2 Thermal electron transfer

In a supermolecule A.B represented by an energy diagram such as that of Fig. 3.6, thermal electron transfer to give $A^+.B^-$ is thermodynamically unfavourable. It is possible, on the other hand, to have spontaneous thermal electron transfer in the reverse sense (3.12).

$$A^+.B^- \rightarrow A.B \tag{3.12}$$

This type of process is often called *back electron-transfer* or *charge recombination*. It is experimentally-accessible provided that some means is available to generate the thermodynamically unfavoured form and use it as a reactant. This is obviously difficult to do by ordinary chemical or electrochemical means, but is feasible under favourable conditions with photochemistry or radiation chemistry.

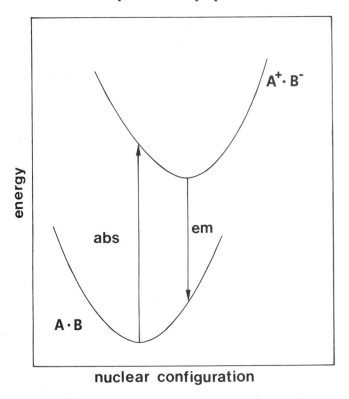

Fig. 3.7 — Absorption and emission electron-transfer processes in a supramolecular system.

Absorption of light can lead to the formation of the reactant species $A^+.B^-$ in two ways: following an optical electron-transfer process (as discussed in section 3.4.1), or following photoinduced electron transfer (a process that will be shortly presented in section 3.4.3 and discussed in more detail in Chapter 5). By using pulsed excitation, the back electron-transfer reaction (3.12) can be followed provided that the generating event (light pulse and, in the photoinduced mode, also the subsequent thermal process) is much faster than the electron-transfer process to be studied. Examples of charge recombination processes following photoinduced electron transfer will be presented in Chapters 5, 8, 9, and 10.

Radiation chemistry can be used as a means to obtain the reactant $A^+.B^-$ because of the nondiscriminating character of the strongly reducing or oxidizing radicals. Consider (for the sake of homogeneity with the previous notation) a supermolecule $A^+.B$. If this species is subjected to reduction by, for example, solvated electrons, the products $A.B$ and $A^+.B^-$ will be initially formed in a nearly statistical yield, regardless of the thermodynamic preference for reduction at site A^+. If the solvated electrons are delivered as a short pulse, in favourable cases it will be possible to follow the thermal electron-transfer reaction (3.12) while the statistical mixture of primary products relaxes towards its thermodynamic equilibrium (section 5.3). In this case the condition for the study of the process is that not only the radiation pulse but also the primary reduction process are much faster than the reaction to be studied. Since the primary reduction is a bimolecular diffusion-controlled process, the time resolution achievable in these experiments is not simply a matter of instrumentation. As shown in Chapter 5, the measured rate constants of electron-transfer processes in supramolecular systems span an extremely wide range, and the interpretation of these large rate variations in terms of molecular properties is a challenging problem. A classical kinetic model that can be used to discuss rate constants of thermal electron transfer is presented in section 3.4.4.

3.4.3 Photoinduced electron transfer

This process differs from optical electron transfer (section 3.4.1) in that in this case light absorption populates an excited state localized on a single component, and electron transfer occurs as a subsequent radiationless process (3.13).

$$\text{A.B} \xrightarrow{h\nu} {}^*\text{A.B} \xrightarrow{k_{el}} \text{A}^+\text{.B}^- \tag{3.13}$$

The relationship between this process, optical electron transfer, and charge recombination is shown in terms of potential energy curves in Fig. 3.8, where the component-localized excited state *A.B is assumed to be essentially undistorted along the reaction coordinate of electron transfer. The photoinduced electron-transfer process can be characterized by a rate constant, k_{el}, and by an efficiency, η_{el}. The rate constant can be directly measured in flash photolysis by the risetime of the $\text{A}^+\text{.B}^-$ absorption. Alternatively, k_{el} can be deduced from the lifetime of *A.B, τ, provided that the lifetime of free *A (or of some reasonable model thereof), τ_0, is known $(k_{el} = 1/\tau - 1/\tau_0)$. The efficiency can similarly be obtained by comparing the quantum yield of emission, Φ, with that of free *A (or of some reasonable model thereof), Φ_0, as $\eta_{el} = \Phi_0 - \Phi$. In these photoinduced intramolecular electron-transfer processes, the excited state of the A component is "quenched" by the B component in a way reminiscent of a bimolecular electron-transfer quenching process (section 2.7). The obvious difference is that in the intramolecular case diffusional processes are absent. This makes the intramolecular processes free from some kinetic complications present in corresponding bimolecular processes, such as diffusion-controlled plateaus (section 2.7.3). On the other hand, the lack of diffusional separation of the products may make the intramolecular processes more difficult to detect than the corresponding bimolecular one, due to fast charge recombination.

The kinetics of the step k_{el} in (3.13) can be analysed in terms of the general models for unimolecular electron-transfer processes (section 3.4.4), with the obvious specification that the thermodynamic driving force for the process must be obtained from the appropriate ground- and excited-state redox potentials (section 2.7.2), with correction for electrostatic work terms (3.7).

Several examples of photoinduced electron transfer in supramolecular systems will be discussed in Chapters 5, 8, 9, and 10.

3.4.4 Electron-transfer kinetics

Electron transfer is a very simple, weak-interaction chemical process in which no bond breaking or forming is involved. This permits a simple description of the whole

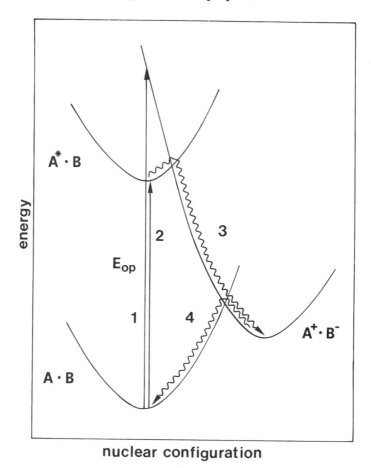

Fig. 3.8 — Relationship between optical (1), photoinduced (2 + 3), and thermal (4) electron-transfer processes in a supramolecular system.

reaction coordinate in terms of the known properties of reactants and products. The classical model of electron-transfer reactions is that developed by Marcus, Hush, and Sutin [11,17–20] (hereafter referred to as the Marcus model). Although more elaborate quantum mechanical models of electron-transfer processes have been subsequently developed [21], the Marcus model is still widely used, as it combines a relatively simple formalism with a remarkable amount of physical insight and predictive power. Some features of this model have been briefly mentioned in section 2.7.3 with regard to the unimolecular step of bimolecular electron-transfer processes.

Consider a general unimolecular electron transfer process (3.14)

$$A.B \xrightarrow{k_{el}} A^+.B^- .\tag{3.14}$$

According to the Marcus model, the rate constant of a unimolecular electron-transfer process can be expressed, in a way reminiscent of conventional transition-state theory, as [11]

$$k_{el} = \nu_N \kappa \exp(-\Delta G^{\neq}/RT) . \qquad (3.15)$$

The meaning of the various terms in (3.15) can be conveniently discussed in terms of the energy profiles of Fig. 3.9. As usual, the two curves represent the potential

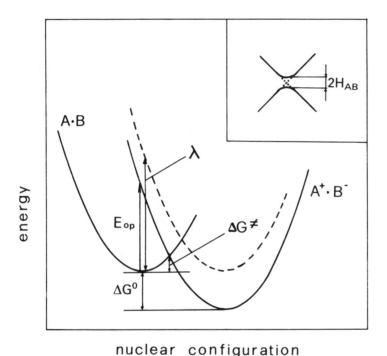

nuclear configuration

Fig. 3.9 — Energy profiles and kinetic parameters for a unimolecular electron-transfer reaction.

energies of reactants and products as a function of a reaction coordinate made up of an appropriate combination of displacements in the nuclear coordinates of the system. These coordinates are of two types: (i) inner, that is, internal coordinates (bond lengths and angles) of the reacting molecules; (ii) outer, that is, coordinates specifying the arrangement of the solvent surrounding reactants and products. The vertical displacement of the minima is related to the energetics of the reaction, while the horizontal displacement represents the different equilibrium solvation shells and different equilibrium molecular geometries of reactants and products. In order for the Franck–Condon principle to be obeyed, a distortion in the outer and inner nuclear coordinates of the reactants leading to a geometry where reactants and products are isoenergetic (crossing point of the two surfaces) is required prior to electron transfer. The transition state and the reaction coordinate of the electron transfer reaction (Fig. 3.9) correspond to the lowest energy pathway available in the multidimensional nuclear space of the system.

Thus, the activation free energy $\Delta G^{\pm}$ in (3.15) corresponds (converting from energies to free energies) to the energy difference between the crossing point and the reactant minimum in Fig. 3.9. In view of its physical meaning, the activation term in (3.15) can also be called the *nuclear* term of the rate expression. The term κ in (3.15) is the transmission coefficient of the reaction, that is the probability that the reactants, on reaching the geometry of the crossing point, convert into products. Owing to its physical origin (*vide infra*), κ can also be called the *electronic* factor of the rate constant. In (3.15), ν_N is the nuclear frequency factor of the reaction, which sets the maximum possible value for the rate constant. It can be expressed [11] as a weighted mean of the frequencies of the various nuclear modes involved in the reaction coordinate. As such, it tends to be dominated by the high-frequency inner modes (typical values, $4.5 \times 10^{13}\,\mathrm{s}^{-1}$ for C–C stretching of aromatic molecules and $(0.9\text{–}1.5) \times 10^{13}\,\mathrm{s}^{-1}$ for metal-ligand stretching in coordination compounds).

It is evident from Fig. 3.9 that the activation free energy is determined by the combined effects of the degree of distortion between products and reactants (horizontal displacement of the two curves) and the driving force of the reaction (vertical displacement of the two curves). Marcus theory expresses this combined dependence in terms of a parabolic free-energy relationship [17] (3.16).

$$\Delta G^{\pm} = (\lambda/4)\left(1 + \frac{\Delta G^0}{\lambda}\right)^2 \qquad (3.16)$$

In (3.16), ΔG^0 is the standard free energy change of the reaction. The parameter λ (already mentioned in sections 3.2 and 3.4.1) is the so-called *reorganizational energy*, corresponding to the vertical separation, at equilibrium geometry, between reactant and product curves for a hypothetical isoergonic reaction with the same nuclear distortions (dashed product curve in Fig. 3.9). The actual vertical energy difference corresponds to the energy of the optical electron-transfer transition discussed in section 3.4.1. The reorganizational energy can be split into the sum of two independent contributions (3.17) corresponding to reorganization of "inner" (bond lengths and angles within A and B) and "outer" (solvent orientation around the reacting pair) nuclear modes.

$$\lambda = \lambda_i + \lambda_0 \qquad (3.17)$$

These contributions can be calculated from expressions of Marcus theory [11,17–20], provided that the appropriate parameters are known. The outer reorganizational energy is given by a simple expression (3.18)

$$\lambda_0 = e^2 \left(\frac{1}{\varepsilon_{op}} - \frac{1}{\varepsilon_s}\right)\left(\frac{1}{2r_A} + \frac{1}{2r_B} - \frac{1}{r_{AB}}\right) \qquad (3.18)$$

when the reactants are considered as spheres in a dielectric continuum. In (3.18), ε_{op} and ε_s are the optical and static dielectric constants of the solvent, r_A and r_B are the radii of the reactants, and r_{AB} is the interreactant centre-to-centre distance ($r_{AB} > (r_A + r_B)$). Qualitatively, the outer part of the reorganizational energy increases with increasing solvent polarity and distance A–B. The inner part depends on the degree of geometrical distortion occurring in the A/A$^+$ and B/B$^-$ couples, and thus on the degree of delocalization and bonding character of the transferred electron. Common electron-transfer reactions involving delocalized aromatic molecules as reactants are expected to have small associated inner reorganizational energies. With metal coordination compounds, the magnitude of the inner contribution depends strongly on the antibonding or nonbonding character of the orbitals (for example, e_g^* or t_{2g} in octahedral complexes) involved in the electron-transfer process. Without resorting to calculations, the reorganizational energy of a "cross" reaction such as (3.14) can be obtained as the arithmetic mean of the experimental activation energies of the corresponding "self-exchange" reactions of the couples A/A^+ and B/B^- [17].

A point that has attracted a considerable deal of experimental and theoretical attention is the behaviour predicted by the Marcus free energy relationship (3.16) in the highly exoergonic ΔG^0 region. According to (3.16), $\Delta G^{\pm}$ equals $\lambda/4$ at $\Delta G^0 = 0$, goes to 0 at $\Delta G^0 = -\lambda$, and increases again for more negative ΔG^0 values. This can be easily visualized by considering how the crossing point shifts when the exoergonicity of the reaction increases (Fig. 3.10): from the right branch (Fig. 3.10a, "normal"

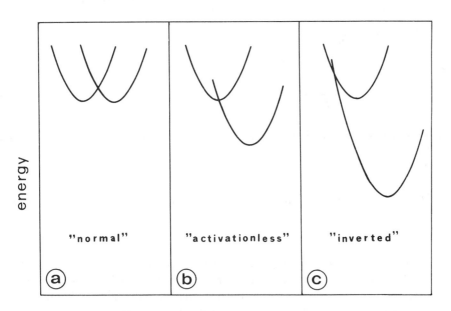

nuclear configuration

Fig. 3.10 — Potential energy curves for reactant and product states of an electron-transfer process in the three archetypal free energy ranges of the Marcus model.

activated process) through the minimum (Fig. 3.10b, activationless process) to the left branch (Fig. 3.10c, "inverted" activated process) of the reactant curve. Therefore, for moderately exoergonic reactions the driving force is expected to help the reaction kinetics, but for strongly exoergonic reactions the driving force is predicted to act against it. The ΔG^0 region in which this intuitively odd effect is expected ($\Delta G^0 < -\lambda$) is usually indicated as the *Marcus inverted region*. This feature is also predicted by quantum mechanical models of electron-transfer processes [21]. After a long period of experimental search and many unsuccessful attempts [22,23], the prediction of the inverted region is now supported by definite experimental evidence [24,25].

The transmission coefficient κ in equation (3.15) is related to the detailed shape of the potential energy curves in the intersection region. Strictly speaking, the reactant and product potential energy curves of Fig. 3.9 correspond to zero-order wavefunctions of the system. If there were no electronic interaction between these zero-order states, no mechanism for transition from reactants to products would be available. Actually, in most practical systems a small but finite electronic interaction occurs between A and B in the reactant pair, and a perturbation hamiltonian H_{AB} coupling the initial (A.B) and final (A$^+$.B$^-$) states of the system should be considered. This electronic coupling mixes the zero-order states in the intersection region, leading to the first-order avoided-crossing surfaces shown in the inset of Fig. 3.9. A quantitative expression for the transmission coefficient can be obtained [11] within the framework of the Landau–Zener treatment of avoided crossings. The relevant result is shown by (3.19) and (3.20)

$$\kappa = \frac{2[1 - \exp(-v_{el}/2v_N)]}{2 - \exp(v_{el}/2v_N)} \tag{3.19}$$

$$v_{el} = \frac{2H_{AB}^2}{h} \left(\frac{\pi^3}{\lambda RT}\right)^{1/2} \tag{3.20}$$

Two limiting cases can be identified [11] on the basis of equations (3.15), (3.19), and (3.20):

(i) If the electronic interaction H_{AB} is very small, $v_{el} \ll v_N$, $\kappa = (v_{el}/v_N) \ll 1$, and k_{el} is given by (3.21).

$$k_{el} = v_{el}\exp(-\Delta G^{\neq}/RT) \tag{3.21}$$

This is called the *nonadiabatic* limit of electron-transfer reactions, in which the rate-determining step is the electron transfer at the transition-state geometry. The unimolecular rate constant is much smaller than the nuclear frequency and is very sensitive to factors that may influence the degree of electronic interaction between the reactants (for example, centre-to-centre distance, steric hindrance of substituents, orientational factors, nature of interposed groups or medium, etc.).

(ii) If H_{AB} is sufficiently high that $v_{el} \gg v_N$, $\kappa = 1$ and k_{el} is given by (3.22).

$$k_{el} = v_N\exp(-\Delta G^{\neq}/RT) \tag{3.22}$$

This is called the *adiabatic* limit of electron transfer reactions, in which the rate-determining step is the nuclear motion that leads to the transition-state geometry. The unimolecular-reaction rate constant may approach (for small $\Delta G^{\neq}$) the nuclear frequency factor, and the reaction is insensitive to factors that may influence the degree of electronic interaction between the reactants.

The value of H_{AB} depends on the overlap between the electronic wavefunctions of the donor and acceptor groups, that should decrease exponentially with increasing donor–acceptor distance. The calculation of H_{AB} in real systems is generally a difficult theoretical problem (see section 5.2). In favourable cases, on the other hand, it is possible to estimate the magnitude of H_{AB} from spectroscopic data (section 3.4.1). It should be noticed that the amount of electronic interaction required to promote electron transfer is very small in a common chemical sense. In fact, it can be easily verified by substituting reasonable numbers for the parameters in equations (3.15), (3.19), and (3.20) that, for an activationless reaction, H_{AB} values of a few wavenumbers are sufficient to give rates in the sub-nanosecond time scale, and a few hundred wavenumbers may be sufficient to reach the limiting adiabatic regime.

The Marcus model, as outlined above, refers to a pair of A and B molecular reactants at fixed distance (for example, the reactants in the encounter or "precursor" complex of a bimolecular reaction). The model can be applied without any additional comment to describe intramolecular electron transfer in a supramolecular system in which the components are held together by electrostatic interaction or weak intermolecular forces. The extension of the model to a supramolecular system in which the active components A and B are covalently linked through a connector L (3.3), on the other hand, requires some comment. As compared with the analogous intermolecular reaction at the same centre-to-centre distance, the reorganizational energy (and thus the nuclear part of the rate constant) is not expected to be drastically altered by the presence of the connector. On the contrary, relatively important effects of the connector are expected to occur on the electronic part of the rate constant. In fact, depending on its length and electronic structure, the connector can induce a more or less important degree of delocalization between the active components, thus increasing H_{AB} with respect to the corresponding intermolecular value at the same centre-to-centre distance. The role of the connector in enhancing the electronic coupling between the active components in a supramolecular system can be described in terms of "superexchange" [26] (section 5.2). This through-bond mechanism can be viewed [27–30] in terms of configuration interaction between the initial (A–L–B) and final (A$^+$–L–B$^-$) zero-order states of the electron-transfer process and high-energy charge-transfer states involving the bridging ligand, such as A$^+$–L$^-$–B and A–L$^+$–B$^-$. The relevance of the magnitude of H_{AB} to the problem of localization vs delocalization in supramolecular systems has been pointed out in section 3.2.

3.4.5 Electronic energy transfer

The photophysical behaviour of a supermolecule A.B may differ from that of the individual components because of the occurrence of intercomponent electronic energy transfer (3.23).

$$\text{A.B} \overset{h\nu}{\rightarrow} {}^*\text{A.B} \overset{k_{en}}{\rightarrow} \text{A.}{}^*\text{B} \qquad (3.23)$$

When such a process is efficient the excited-state properties of the light-absorbing component (A) are quenched and the excited-state properties (for example, emission) of the other component (B) can be observed. These two phenomena can be described as intercomponent energy-transfer *quenching* and *sensitization*, respectively.

The energy-transfer process (second step of equation (3.23)) must obey energy conservation (that is, *B must be equal or lower in energy than *A), and requires some kind of electronic interaction between the donor and the acceptor. Following standard arguments [31], the electronic interaction between two molecular species can be split into two additive terms, a *coulombic* term and an *exchange* term. The two terms have different dependences on various parameters of the system (spin of ground and excited states, donor–acceptor distance, etc.) and each of them can become predominant depending on the specific system and experimental situation. This leads to the identification of two main energy-transfer mechanisms.

The coulombic (also called "resonance" or "Förster-type") mechanism [31,32] is a long-range mechanism that does not require physical contact between donor and acceptor. It can be shown that the most important term within the coulombic interaction is the dipole–dipole term, that obeys the same selection rules as the corresponding electric dipole transitions of the two partners ($*A \rightarrow A$ and $B \rightarrow *B$). Therefore, coulombic energy transfer is expected to be efficient in systems in which the radiative transitions connecting the ground and the excited state of each partner have high oscillator strength. Thus, the typical example of an efficient coulombic mechanism is that of singlet–singlet energy transfer (3.24) between large aromatic molecules, a process used by nature in the "antenna" part of the photosynthetic apparatus [33].

$$*A(S_1).B(S_0) \rightarrow A(S_0).*B(S_1) \tag{3.24}$$

In metal complexes (as seen in section 2.3.5) the only excited state of appreciable lifetime is generally the lowest, spin-forbidden excited state [34], so that coulombic energy transfer is not expected to be frequent in these compounds.

The exchange (also called "Dexter-type") mechanism [31,32] is a short-range mechanism that requires orbital overlap, and therefore physical contact, between donor and acceptor. The exchange interaction can be visualized as the simultaneous exchange of two electrons between the donor and the acceptor *via* LUMOs (from A to B) and HOMOs (from B to A). The spin selection rules for this type of mechanism arise from the need to obey spin conservation in the reacting pair as a whole. This allows the exchange mechanism to be operative in many cases in which the excited states involved are spin-forbidden in the usual spectroscopic sense. Thus, the typical example of an efficient exchange mechanism in organic photochemistry is that of triplet–triplet energy transfer (3.25), a process often used for selective quenching of triplets or photosensitization of triplet reactions [32].

$$*A(T_1).B(S_0) \rightarrow A(S_0).*B(T_1) \tag{3.25}$$

Exchange energy transfer from the lowest spin-forbidden excited state is also the rule for metal complexes [35].

The above mechanistic considerations are made for the general case of a pair of molecules *A.B and extend straightforwardly to supramolecular systems in which the components are held together by electrostatic interaction or intermolecular forces. When the donor and the acceptor are active components of a covalently-linked supramolecular structure, for example, *A–L–B, the presence of the connector is not expected to change substantially the situation as far as the coulombic mechanism (at the same centre-to-centre distance) is concerned. The situation may be different, on the other hand, for an exchange mechanism. It has been pointed out in sections 3.2 and 3.4.4 that the presence of a chemical link can increase the electronic-coupling matrix element for electron transfer in a binuclear complex relative to that of an analogous bimolecular reaction. Although the matrix elements involved in the two types of processes are somewhat different [36], the concept that electron delocalization *via* the bridging ligand can increase the interaction between components can be extended from electron to exchange energy transfer. Examples of coulombic and exchange energy transfer in supramolecular systems will be discussed in Chapter 6.

Intramolecular exchange energy-transfer processes bear several analogies with the bimolecular energy-transfer processes discussed briefly in section 2.7. A difference between bimolecular and intramolecular energy transfer should be pointed out. In bimolecular energy transfer involving diffusion (that can be described by a kinetic scheme analogous to that shown in Fig. 2.10), back energy transfer within the successor complex (analogous to k_{-e}) is usually neglected, even when the driving force of the forward process is very small. This is because the products are rapidly taken out of contact (usually on a nanosecond time scale) by the diffusive step k_{-d}. In a supramolecular system, such diffusion may be much slower or even impossible, and the possibility of back energy transfer and excited-state equilibration should be generally considered for systems involving small excited-state energy differences.

3.4.6 Energy *vs* electron transfer

An interesting general problem is the relationship between electron- and energy-transfer processes [9,10,35,36]. In this context, an important question is that concerning the relative rates of electron and exchange energy transfer in an ideal system in which both processes are thermodynamically allowed. No general answer to this question is available, but a few points can be stressed. To discuss this question, a kinetic model of energy transfer [9,35] can be used that considers electronic and nuclear factors in the rate constant of energy transfer, much in the same way as the Marcus model does for electron transfer. Generally speaking, energy transfer tends to have smaller reorganizational barriers than electron transfer, due to the much smaller solvent repolarization required for the former process. As far as the electronic term is concerned, the two-electron *vs* one-electron nature of the interaction intuitively implies more severe overlap requirements for energy than for electron transfer. In an elegant study on covalently-linked organic donor–acceptor systems, Closs, Miller, and coworkers [10] have shown that rates of energy transfer decrease with increasing bridge length much faster than those of the corresponding electron-transfer processes.

Experimental examples of energy- and photoinduced electron-transfer processes that illustrate the above points are discussed in Chapters 5 and 6.

REFERENCES

[1] (a) Lehn, J. M. (1985) *Science* **227** 849; (b) Balzani, V., Moggi, L., and Scandola, F. (1987). In Balzani, V. (ed.) *Supramolecular photochemistry.* Reidel, p. 1; (c) Lehn, J. M. (1988) *Angew. Chem. Int. Ed. Engl.* **27** 89; (d) Ringsdorf, H., Schlarb, B., and Venzmer, J. (1988) *Angew. Chem. Int. Ed. Engl.* **27** 113; (e) Vögtle, F. (1989) *Supramolekulare chemie.* Teubner, Stuttgart; (f) Tomalia, D. A., Naylor, A. M., and Goddard III, W. A. (1990) *Angew. Chem. Int. Ed. Engl.* **29** 138.

[2] Bolton, J. R., Ho, T-F., Liauw, S., Siemiarczuk, A., Wan, C. S. K., and Weedon, A. C. (1985) *J. Chem. Soc., Chem. Commun.* 559.

[3] Callahan, R. W., Brown, G. M., and Meyer, T. J. (1974) *J. Am. Chem. Soc.* **96** 7829.

[4] Creutz, C. and Taube, H. (1969) *J. Am. Chem. Soc.* **91** 3988.

[5] Brown, D. B. (ed.) (1980) *Mixed valence compounds.* Reidel.

[6] Creutz, C. (1983) *Prog. Inorg. Chem.* **30** 1.

[7] Hush, N. S. (1967) *Prog. Inorg. Chem.* **8** 391.

[8] Robin, M. B. and Day, P. (1967) *Adv. Inorg. Chem. Radiochem.* **10** 247.

[9] Balzani, V., Bolletta, F., and Scandola, F. (1980) *J. Am. Chem. Soc.* **102** 2152.

[10] Closs, G. L., Johnson, M. D., Miller, J. R., and Piotrowiak, P. (1989) *J. Am. Chem. Soc.* **111** 3751.

[11] Sutin, N. (1983) *Prog. Inorg. Chem.* **30** 441.

[12] Bignozzi, C. A. and Scandola, F. (1984) *Inorg. Chem.* **23** 1540.

[13] Bignozzi, C. A., Roffia, S., and Scandola, F. (1985) *J. Am. Chem. Soc.* **107** 1644.

[14] Bignozzi, C. A., Paradisi, C., Roffia, S., and Scandola, F. (1988) *Inorg. Chem.* **27** 408.

[15] Bignozzi, C. A., Roffia, S., Chiorboli, C., Davila, J., Indelli, M. T., and Scandola, F. (1989) *Inorg. Chem.* **28** 4350.

[16] Chanon, M., Hawley, M. D., and Fox, M. A. (1988). In Fox, M. A. and Chanon, M. (eds) *Photoinduced electron transfer.* Part A. Elsevier, p. 1.

[17] Marcus, R. A. (1964) *Annu. Rev. Phys. Chem.* **15** 155.

[18] Hush, N. S. (1968) *Electrochim. Acta* **13** 1005.

[19] Sutin, N. (1979). In Eichorn, G. L. (ed.) *Inorganic biochemistry.* Elsevier, p. 611.

[20] Marcus, R. A. and Sutin, N. (1985) *Biochim. Biophys. Acta* **811** 265.

[21] Ulstrup, J. (1979) *Charge transfer in condensed media.* Springer-Verlag.

[22] Rehm, D. and Weller, A. (1970) *Isr. J. Chem.* **8** 259.

[23] Indelli, M. T., Ballardini, R., and Scandola, F. (1984) *J. Phys. Chem.* **88** 2547, and references therein.

[24] Closs, G. L. and Miller, J. R. (1988) *Science* **240** 440.

[25] Gould, I. R., Moser, J. E., Armitage, B., and Farid, S. (1989) *J. Am. Chem. Soc.* **111** 1917.

[26] McConnell, H. M. (1961) *J. Chem. Phys.* **35** 508.

[27] Wasielewski, M. R. (1988). In Fox, M. A. and Chanon, M. (eds) *Photoinduced electron transfer.* Part A. Elsevier, p. 161, and references therein.

[28] Miller, J. R. (1987) *Nouv. J. Chim.* **11** 83.

[29] Mayoh, B. and Day, P. (1972) *J. Am. Chem. Soc.* **94** 2885.

[30] Richardson, D. E. and Taube, H. (1983) *J. Am. Chem. Soc.* **105** 40.

[31] Lamola, A. A. (1969). In Lamola, A. A. and Turro, N. J. (eds) *Energy transfer and organic photochemistry*. Interscience, p. 17.

[32] Turro, N. J. (1978) *Modern molecular photochemistry*. Benjamin.

[33] Witt, H. (1987) *Nouv. J. Chim.* **11** 91.

[34] Crosby, G. A. (1983) *J. Chem. Educ.* **60** 791.

[35] Scandola, F. and Balzani, V. (1983) *J. Chem. Educ.* **60** 814.

[36] Closs, G. L., Piotrowiak, P., MacInnis, J. M., and Fleming, G. R. (1988) *J. Am. Chem. Soc.* **110** 2652.

4

Control and tuning of excited state properties of molecular components

4.1 INTRODUCTION

In several problems of basic and applicative interest there is a need to modify the excited state behaviour of a molecule. This can often be done by assembling the molecule with one or more perturbing components. An important aim of supramolecular photochemistry is the elucidation of the mechanisms which allow control and tuning of the excited state properties when a molecule is incorporated into a supramolecular structure. In this chapter we describe briefly the types of perturbation that can be devised to play this role. More details and discussions of specific examples will be given in later chapters.

The photochemical and photophysical behaviour of a molecule depends on the relative efficiencies of luminescence, radiationless decay, and chemical reaction, and on the excited-state lifetime which controls the interactions with other species. As shown by (2.15) and (2.16), these quantities are determined by the rate constants of the various decay processes. In turn, such rate constants are controlled by symmetry and spin properties and by the relative positions and shapes of the potential energy curves that describe the various excited states [1].

For a general discussion of the ways in which the excited-state behaviour can be controlled and tuned by supramolecular interactions, we will make use of Fig. 4.1 [2] which shows schematically the potential energy curves of the ground state A and the lowest spin-forbidden excited state *A of a molecule. Light excitation of the ground state leads to upper, spin-allowed excited states (not shown in Fig. 4.1) whose radiationless decay is supposed to populate *A. Such an excited state can undergo radiative decay (with quantum yield Φ_r), chemical reaction (with quantum yield Φ_p), and radiationless decay. Radiative decay usually, but not always (section 7.1.4), takes place at nuclear coordinates not too different from those involved in light absorption. By contrast, radiationless decay often involves severe changes in the

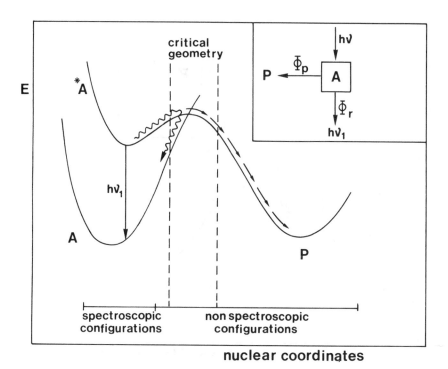

Fig. 4.1 — Potential energy curves for the ground state and the lowest spin-forbidden excited state of a luminescent and reactive molecule.

nuclear coordinates, and chemical reaction, of course, always implies a large nuclear rearrangement (for example, the breaking of a bond).

When a molecule is assembled with other components in a supramolecular structure, the following changes may occur in the energy level diagram [2,3]: (i) the spectroscopic levels of the original molecule may be perturbed; (ii) new energy levels may appear; (iii) changes may occur in the shape of the potential energy curves at nonspectroscopic nuclear configurations.

In the following sections we discuss some general aspects of these three types of perturbation and report some illustrative examples.

4.2 PERTURBING SPECTROSCOPIC LEVELS

The molecule to be perturbed is assembled with another component, T, that does not introduce new energy levels but only affects, to some extent, the energy levels of the original molecule (Fig. 4.2). No new bands appear in the absorption spectrum, and only small shifts of the original bands are observed. The photochemical and photophysical behaviour, however, may be substantially modified, with important consequences for the lifetime of the excited state and the efficiencies of the excited-state processes. In some cases, symmetry and/or spin perturbations can also be

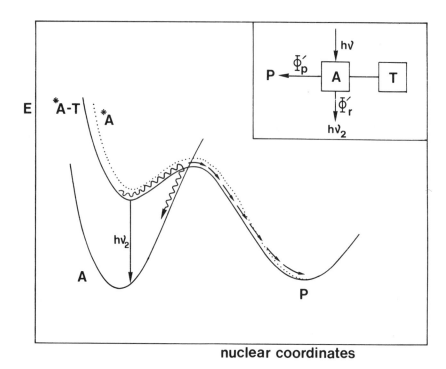

nuclear coordinates

Fig. 4.2 — Perturbation by a component T which does not introduce new energy levels. The dotted line represents the excited-state potential energy curve of the original molecule (cf. Fig. 4.1).

introduced, with noticeable consequences on excited-state decay. By the use of appropriate families of electronic perturbers, it may be possible to change gradually (i.e., to tune) most of the photochemical and photophysical properties of a molecule. For this reason, a perturber of this type is called a *tuner*, T.

As a consequence of the effect of a tuner on the energy of *A, there is a shift in the luminescence spectrum and a change in the thermodynamic ability of *A to participate in energy-transfer processes. The redox potentials of A are also affected and the same may happen to the redox potentials of *A (which depend on the energy of *A and on the redox potentials of A, section 2.7.2), with a consequent change in its thermodynamic ability to participate in electron-transfer processes.

The change in the energy of *A may also have dramatic effects on the decay processes since (i) the rate of radiative decay increases with increasing energy of *A and (ii) the rate of nonradiative decay to the ground state may decrease (activation-less decay to the ground state) or increase (activated surface crossing) with increasing energy of *A (Chapter 2). As a consequence, there are also changes in the quantum yields of luminescence and photoreaction (Fig. 4.2, inset).

Several examples of this type of perturbation have been reported in the literature. A very simple one is that concerning the complex $Ru(bpy)_2(CN)_2$ **4.1**. In DMF

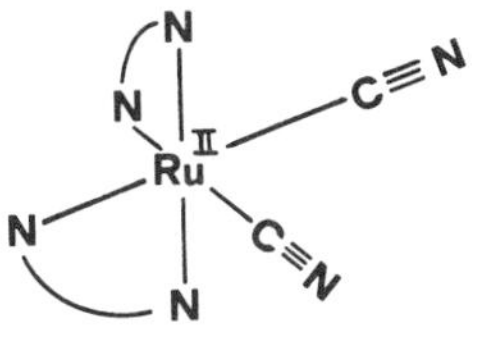

4.1 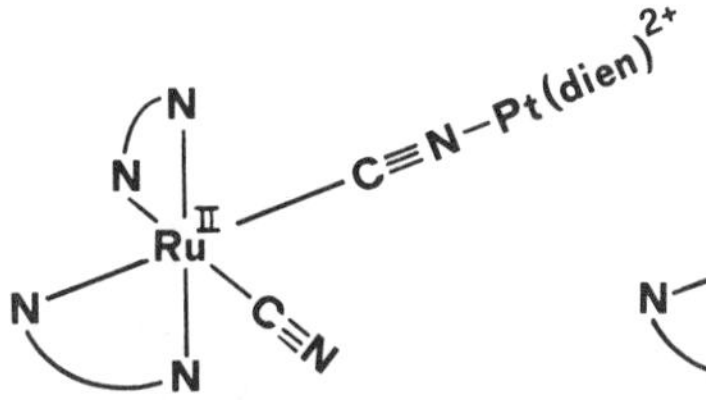   **4.2** 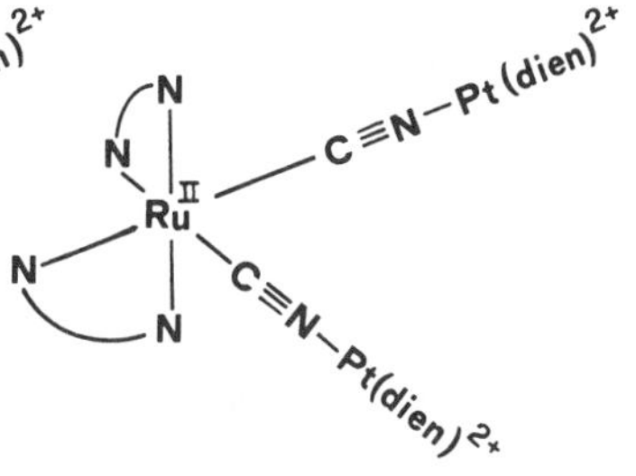   **4.3**

solution this molecule exhibits its lowest energy absorption maximum at 505 nm, attributed to a singlet MLCT (Ru→bpy) excited state, and a relatively long-lived (205 ns) luminescence with a maximum at 680 nm, attributed to a triplet MLCT (Ru→bpy) excited state [4–6]. The cyanide ligands of this complex can behave as nitrile ligands towards metal ions and metal complex moieties, giving rise to a variety of di- or trinuclear complexes. This happens, for example, with the $Pt(dien)^{2+}$ moiety, as shown by **4.2** and **4.3** [6]. $Pt(dien)^{2+}$ does not introduce low-lying excited states in the adduct since (i) its own (MC) excited states lie at very high energy, and (ii) its redox properties do not allow the occurrence of low-energy intercomponent charge-transfer transitions. Thus, the luminescence of $Ru(bpy)_2(CN)_2$ is not quenched on adduct formation. However, $Pt(dien)^{2+}$ has an electron-withdrawing effect on the Ru ion, with important consequences on both its absorption spectrum and excited-state properties (Table 4.1). It should be noticed that while the gradual

Table 4.1 — Properties of $Ru(bpy)_2(CN)_2$ (**4.1**) and of its 1:1 (**4.2**) and 1:2 (**4.3**) adducts

	4.1	**4.2**	**4.3**
λ_{max}(abs), nm	505	460	426
λ_{max}(em), nm	680	630	580
τ, ns	205	630	90
$*E^{00}$, eV	2.05	2.19	2.31
$*E_{1/2}$(ox), V	−1.32	−1.16	−1.45
$*E_{1/2}$(red), V	+0.37	+0.57	+0.81

From [6]. Deaerated DMF solution, room temperature. The potentials are *vs* SCE.

blue shift of the absorption and emission bands in going from $Ru(bpy)_2(CN)_2$ to the 1:1 and the 1:2 adducts with $Pt(dien)^{2+}$ is readily explained by the above-mentioned electron-withdrawing power of $Pt(dien)^{2+}$, the pattern for the excited-state lifetime is more complex. This is not surprising since the excited-state lifetime is a very sensitive function of the relative energetic situation of the various electronic levels. The analysis of the trends in the values of the excited-state redox potentials is even more complex because several factors are involved, including the excited-state energy, the redox potentials of the ground state, and the change in the overall electric charge of the species.

Heavy-atom perturbers induce spin–orbit coupling (section 2.6.3), which amounts to saying that the spin-forbidden excited state *A acquires some spin-allowed character. In such a case, the excited-state energy is almost unaffected, but both the radiative and nonradiative decay processes become faster. An important consequence is a substantial decrease in the excited-state lifetime. This type of perturbation, which is well-known in molecular photochemistry [1], can also play an important role in supramolecular photochemistry. For example, room-temperature phosphorescence in fluid solution from polynuclear aromatic hydrocarbons is not generally observed, but it occurs when the molecule is included in cyclodextrins where the primary hydroxyls have been substituted by bromine [7] (section 10.5.2). Perturbation of the luminescent properties of organic molecules linked to crown ethers upon metal–ion complexation by the crown is also a quite general phenomenon [8] (section 10.2.1).

4.3   INTRODUCING NEW ENERGY LEVELS

The molecule to be perturbed is associated to another component that brings into play an energy level lower than the lowest excited energy level (*A) of the original molecule (Fig. 4.3). The new energy level may be an intrinsic level of the perturber or

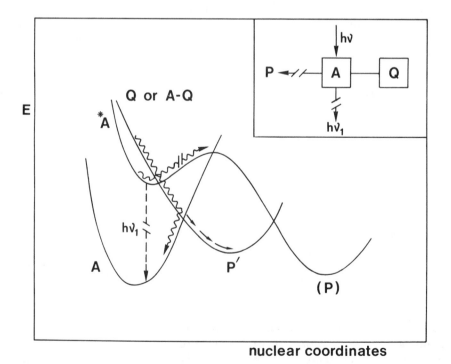

Fig. 4.3 — Perturbation by a component Q that introduces a new energy level (cf. Fig. 4.1).

it may result from the interaction of the perturber with the original molecule (for example, an intercomponent electron-transfer state, section 3.4).

This perturbation can introduce new bands in the absorption spectrum, such as the intervalence transfer bands discussed in section 3.4.1 and the charge-transfer bands discussed in section 9.2. More importantly, it can strongly modify the photochemical and photophysical behaviour of the original molecule since radiation-less decay of *A to the new, lower-lying energy level may be very fast, preventing the occurrence of luminescence and reaction of *A (Fig. 4.3). The excited-state lifetime of *A, of course, may be strongly reduced and, as a consequence, bimolecular and/or intercomponent reactions involving *A may be prevented.

In conclusion, such a type of perturbation leads to quenching of the excited-state *A and, therefore, the perturber may be called a *quencher* (Q). New products (P') deriving from the quencher or from the interaction of the original molecule with the quencher may also appear.

The assembly of a quencher to a molecule may have several applications. For example, it may eliminate a luminescent signal or protect a molecule from undesired photoreactions.

Many cases of this type of perturbation are described in the literature. A simple example concerns again the complex $Ru(bpy)_2(CN)_2$ **4.1**. In aqueous solution this molecule exhibits a ^1MLCT absorption band at 428 nm and a ^3MLCT luminescence band at 620 nm with $\tau = 250$ ns. When a $Ru(NH_3)_5^{3+}$ moiety is linked to $Ru(bpy)_2(CN)_2$ to yield the $(NC)(bpy)_2RuCNRu(NH_3)_5^{3+}$ adduct (Fig. 4.4), the

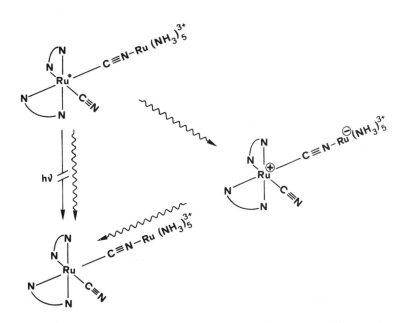

Fig. 4.4 — Quenching of the emission of $Ru(bpy)_2(CN)_2$ by adduct formation with $Ru(NH_3)_5^{3+}$.

Ru→bpy absorption band moves to 403 nm, a new broad band appears with a maximum at 644 nm, and the luminescence disappears [9]. The new band corresponds to the IT transition (section 3.4.1) from the Ru(II) ion of **4.1** to the Ru(III) ion of the Ru(NH$_3$)$_5^{3+}$ moiety. The IT level offers a fast, radiationless decay channel (k>5×10^8 s^{-1}) to the triplet Ru→bpy CT level of the component **4.1**, preventing its luminescence.

4.4   INTRODUCING NUCLEAR CONSTRAINTS

Control and tuning of the photochemical and photophysical properties can also be achieved by assembling the molecule with one or more components that modify the potential energy curves in the *nonspectroscopic* region [2] (Fig. 4.5). This can be

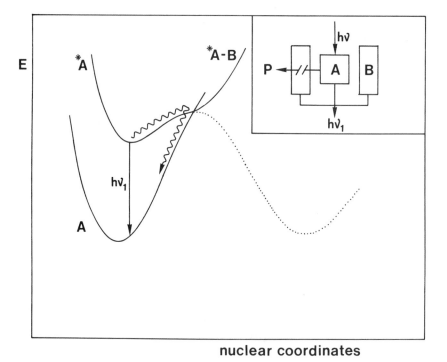

nuclear coordinates

Fig. 4.5 — Perturbation in the nonspectroscopic region of the potential energy curves. The dotted line represents the excited-state potential energy curve of the original molecule for nonspectroscopic nuclear configurations (cf. Fig. 4.1). The inset schematizes one of the ways (inclusion in a host B) in which limitations can be imposed on nuclear motions.

done by means of components that, once assembled in an appropriate way to the molecule, impose constraints to large-amplitude nuclear motions. In this way, radiationless decay processes which take place via strongly distorted molecular structures and (photo)chemical reactions can be prevented. It should be noted that this type of supramolecular assembling often causes no perturbation when the molecule lies at or near its equilibrium geometry. In such cases the absorption spectrum, the emission spectrum (if, as usual, emission occurs from an excited state which is not severely distorted), and the low-temperature lifetime are unaffected. This type of perturbation can be obtained essentially in two ways: (i) by covalent links; (ii) by host–guest interaction.

4.4.1 Covalent links

One possibility of imposing limitations to large-amplitude nuclear motions is linking together, by rigid covalent bonds, those parts of a molecule which would tend to undergo strong relative displacements in the excited state.

An illustrative example is that of a coordination compound whose ligands, which would undergo photodissociation, are linked together by rigid bridges that do not alter the coordinating ability and do not introduce low-energy excited states or redox forms. As illustrated in Fig. 4.6, in the simplest case n-unidentate ligands are replaced by a n-dentate ligand (chelate effect) [10]. A further step is the replacement of n-unidentate ligands with a n-dentate macrocyclic ligand (macrocyclic effect) [10–13]. The ultimate, and more effective, step is the encapsulation of the metal ion into a cage-type ligand (cage effect) [14–21]. Such a perturbation may not substan-

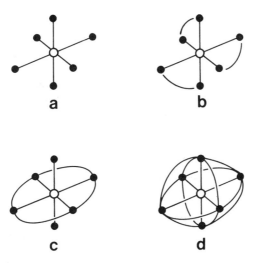

Fig. 4.6 — Complexes containing monodentate (a), chelate (b), macrocyclic (c), and cage (d) ligands.

tially modify the composition and symmetry of the first coordination sphere, which is constituted by the ligand atoms surrounding the metal ion. Thus, it may leave the absorption and luminescence spectra almost unchanged. Several examples of this perturbation will be discussed in section 11.1. We only mention here two paradigm cases to illustrate better the general concepts discussed above.

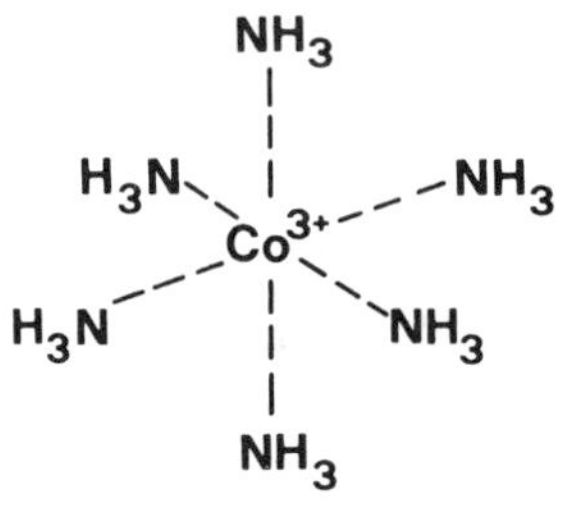

4.4 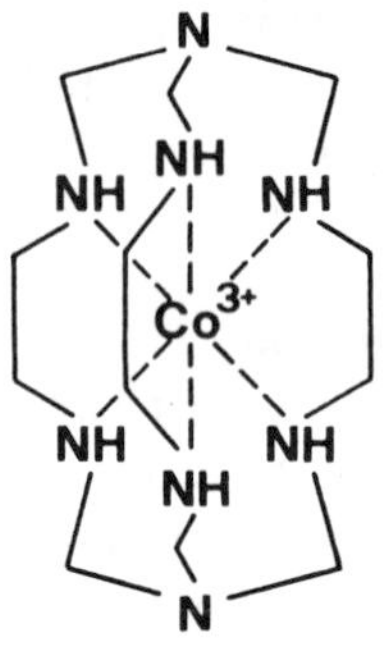   4.5

A caged version of $Co(NH_3)_6^{3+}$ (**4.4**) is the $Co(sep)^{3+}$ ion (**4.5**). The spectroscopic properties of the two complexes are quite similar because the composition and symmetry of the first coordination sphere are the same [15]. The photochemical properties, however, are completely different. Upon photoexcitation in the ligand-to-metal charge-transfer bands, the hexamine complex in acid solution undergoes fast release of the amine ligands (4.1),

$$Co(NH_3)_6^{3+} + H_3O^+ \xrightarrow{h\nu} Co^{2+} + 5NH_4^+ + \text{products} \qquad (4.1)$$

as expected because of the presence of an electron in the σ_M^* antibonding orbitals, leading to complete decomposition of the molecular structure [22]. By contrast, photoexcitation of $Co(sep)^{3+}$ causes no disruption of the structure of the complex (4.2) because the coordinating amino-groups cannot be ejected individually and the

$$Co(sep)^{3+} + H_3O^+ \xrightarrow{h\nu} \text{no reaction} \qquad (4.2)$$

metal cannot escape from the cage [23]. Thus, radiationless transitions remain the only processes available for excited state deactivation to occur (for more details, see section 11.1).

The other example concerns the photoisomerization reaction around a double bond. Stilbene (**4.6** and **4 7**) undergoes an efficient *cis⇌trans* photoisomerization

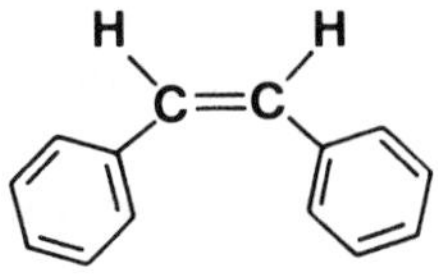

4.6 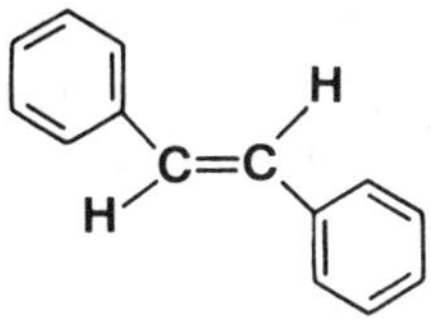   **4.7**

reaction, with the *trans* isomer also giving rise to a weak fluorescence (Φ_f=0.05) at room temperature [1]. When rotation around the double bond is prevented as in the model compounds **4.8** and **4.9**, photoisomerization cannot occur and both compounds exhibit fluorescence with unit quantum yield at room temperature [24,25].

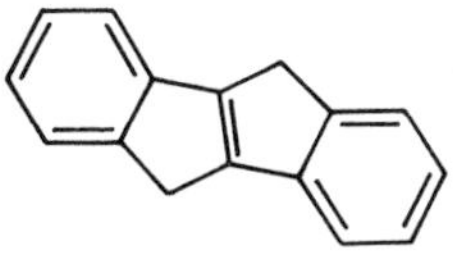

4.8 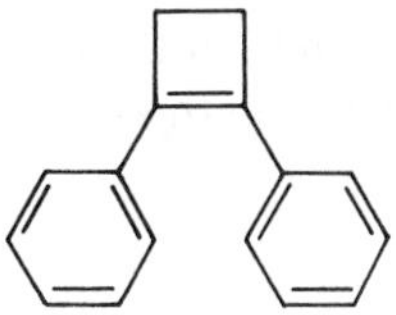   **4.9**

4.4.2 Host–guest interactions

The other way to introduce nuclear constraints is to enclose the molecule in a host, as schematized in the inset of Fig. 4.5. A wide variety of molecular hosts have been prepared in the last 20 years for selective complexation of anionic, cationic, and neutral guests [13,19–21,26–36] (Chapter 10). In systems of these types photodissociation, photoassociation, and photoisomerization processes of the guest, as well as any other extensive nuclear rearrangement, can be prevented, whereas the spectroscopic properties may or may not be affected depending on the type of host–guest interaction. A host having such an effect can be called a *blocker* (B). To illustrate the effect of blockers, we can cite two cases. The first concerns the 1:1 adducts between $Co(CN)_6^{3-}$ and protonated polyazamacrocyclic receptors [37], discussed in more detail in section 10.3.2. The photoaquation reaction (4.3) of the hexacyanocobaltate

$$Co(CN)_6^{3-} + H_3O^+ \xrightarrow{h\nu} Co(CN)_5(H_2O)^{2-} + HCN \qquad (4.3)$$

anion occurs *via* a ligand dissociation mechanism with a quantum yield of 0.30 [22]. For the adducts, the absorption spectrum is essentially the same as that of the "free" $Co(CN)_6^{3-}$ complex [38], but the quantum yield of photoreaction (4.3) is strongly reduced. The most likely explanation is that in the adducts some CN⁻ ligands are involved in hydrogen bonds with the ammonium functions (as, for example, in **4.10**) and, therefore, are prevented from dissociating [38].

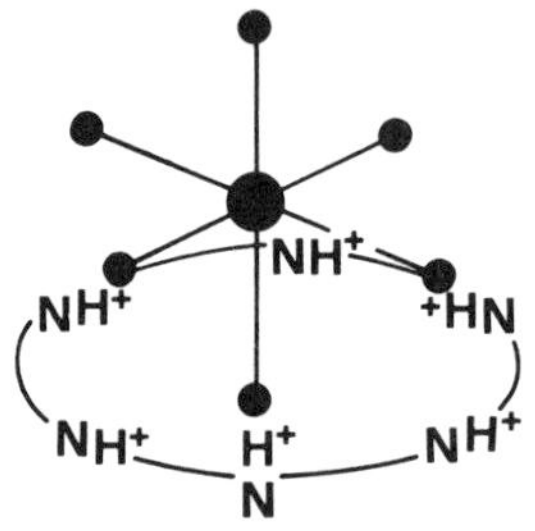

4.10

The second example concerns the photoisomerization of azobenzene (**4.11** and **4.12**). The quantum yield of the *trans→cis* photoisomerization shows a dependence on the excitation wavelength that is suppressed when rotation around the −N=N−

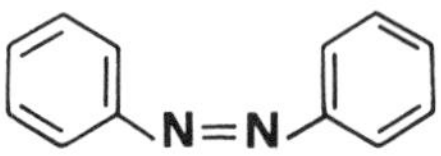

4.11 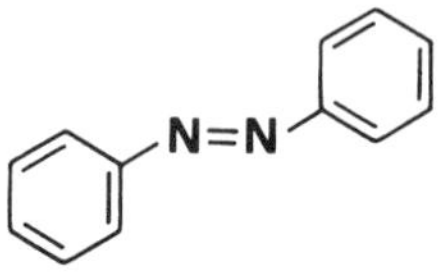   **4.12**

double bond is blocked (for example, by covalent bonds in a cyclophane-type structure [39]). When **4.12** is encapsulated in cyclodextrin cavities, the wavelength dependence disappears, indicating an effective block of the twisting motion [40]. Photochemistry and photophysics within cyclodextrin cavities is an active research field, as will be discussed in detail in section 10.5.

4.5 CONCLUSIONS

The excited state behaviour of a molecule can be controlled and tuned by perturbations induced on assembling the molecule in a supramolecular structure with appropriate components. The perturbing components may act by different mechanisms that can affect the behaviour of the electrons or nuclei of the original molecule. At least three types of perturbing components (quencher, Q; tuner, T; blocker, B) can be identified, as schematized in the insets of Figs 4.2, 4.3, and 4.5.

Supramolecular perturbations of excited-state properties can (i) provide a means to optimize the properties of the active components of photochemical molecular devices (Chapter 12) and (ii) offer the possibility of obtaining important pieces of information on the composition and structure of supramolecular species (Chapters 8–11).

REFERENCES

[1] Turro, N. J. (1978) *Modern molecular photochemistry*. Benjamin.
[2] Balzani, V., Sabbatini, N., and Scandola, F. (1986) *Chem. Rev.* **86** 319.
[3] Balzani, V., Moggi, L., and Scandola, F. (1987). In Balzani, V. (ed.) *Supramolecular Photochemistry*. Reidel, p. 1.
[4] Peterson, S. H. and Demas, J. N. (1976) *J. Am. Chem. Soc.* **98** 7880.
[5] Peterson, S. H. and Demas, J. N. (1979) *J. Am. Chem. Soc.* **101** 6571.
[6] Bignozzi, C. A. and Scandola, F. (1984) *Inorg. Chem.* **23** 1540.
[7] Femia, R. A. and Cline Love, L. J. (1985) *J. Phys. Chem.* **89** 1897.
[8] Löhr, A. G. and Vögtle, F. (1985) *Acc. Chem. Res.* **18** 65.
[9] Bignozzi, C. A., Roffia, S., and Scandola, F. (1985) *J. Am. Chem. Soc.* **107** 1644.
[10] Cotton, F. A. and Wilkinson, G. (1987) *Advanced inorganic chemistry*. Wiley.
[11] Pedersen, C. J. (1967) *J. Am. Chem. Soc.* **89** 7017.
[12] Hayard, R. C. (1983) *Chem. Soc. Rev.* 285.
[13] Pedersen, C. J. (1988) *Angew. Chem. Int. Ed. Engl.* **27** 1021.
[14] Dietrich, B., Lehn, J. M., and Sauvage J. P. (1968) *Tetrahed. Lett.* 2885.
[15] Sargeson, A. M. (1979) *Chem. Brit.* **15** 23.
[16] Grammenudi, S. and Vögtle, F. (1986) *Angew. Chem. Int. Ed. Engl.* **25** 1122.
[17] McMurry, T. J., Hosseini, M. W., Garret, T. M., Hahn, F. E., Reyes, Z. E., and Raymond, K. N. (1987) *J. Am. Chem. Soc.* **109** 7196.
[18] Belser, P., De Cola, L., and von Zelewsky, A. (1988) *J. Chem. Soc., Chem. Commun.* 1057.
[19] Lehn, J. M. (1988) *Angew. Chem. Int. Ed. Engl.* **27** 89.
[20] Cram, D. J. (1988) *Angew. Chem. Int. Ed. Engl.* **27** 1009.
[21] Ebmeyer, F. and Vögtle, F. (1989) *Angew. Chem. Int. Ed. Engl.* **28** 75.
[22] Balzani, V. and Carassiti, V. (1970) *Photochemistry of coordination compounds*. Academic.
[23] Pina, F., Ciano, M., Moggi, L., and Balzani, V. (1985) *Inorg. Chem.* **24** 844.
[24] Saltiel, J., Zafiriou, O. C., Megarity, E. D., and Lamola, A. (1968) *J. Am. Chem. Soc.* **90** 4759.
[25] Deboer, C. D. and Schlessinger, R. H. (1968) *J. Am. Chem. Soc.* **90** 803.
[26] Szejtli, J. (1982) *Cyclodextrins and their inclusion complexes*. Akademiai Kiado, Budapest.
[27] Tabushi, I. (1982) *Acc. Chem. Res.* **15** 66.
[28] Vögtle, F. (ed.) (1981) *Host–guest complex chemistry I*. Topics Curr. Chem. **98**.
[29] Vögtle, F. (ed.) (1982) *Host–guest complex chemistry II*. Topics Curr. Chem. **101**.
[30] Vögtle, F. and Weber, E. (eds) (1984) *Host–guest complex chemistry III*. Topics Curr. Chem. **121**.

[31] Raymond, K. N., Müller, G., and Matzanke, B. F. (1984) *Topics Curr. Chem.* **123** 49.

[32] Shinkai, S. (1986) *Pure Appl. Chem.* **58** 1523.

[33] Colquhoun, H. M., Stoddart, J. F., and Williams, D. J. (1986) *Angew. Chem. Int. Ed. Engl.* **25** 487.

[34] Rebek, J., Jr. (1987) *Science* **235** 1478.

[35] Diederich, F. (1988) *Angew. Chem. Int. Ed. Engl.* **27** 362.

[36] Rebek, J., Jr. (1990) *Angew. Chem. Int. Ed. Engl.* **29** 245.

[37] Dietrich, B., Hosseini, M. W., Lehn, J. M., and Session, R. B. (1981) *J. Am. Chem. Soc.* **103** 1282.

[38] Manfrin, M. F., Moggi, L., Castelvetro, V., Balzani, V., Hosseini, M. W., and Lehn, J. M. (1985) *J. Am. Chem. Soc.* **107** 6888.

[39] Rau, H. and Lüddecke, E. (1982) *J. Am. Chem. Soc.* **104** 1616.

[40] Bortolus, P. and Monti, S. (1987) *J. Phys. Chem.* **91** 5046.

5

Covalently-linked systems: photoinduced electron transfer

5.1 INTRODUCTION

The covalently-linked donor–acceptor (CLDA) systems dealt with in this chapter are supermolecules of type A–L–B, in which two active molecular components (A and B) capable of acting (in their ground or excited states) as electron donor or acceptor are covalently linked *via* a suitable connector (L). As outlined on general grounds in section 3.4, light-induced electron transfer processes in CLDA systems are: optical electron transfer (5.1), photoinduced electron transfer (5.2), charge-transfer emission (5.3), and thermal back-electron transfer (5.4). Suitably produced one-electron oxidized or reduced forms can also undergo spontaneous intercomponent electron-transfer processes, (5.5) and (5.6). With a terminology appropriate to neutral

$$A–L–B + h\nu' \rightarrow A^+–L–B^- \tag{5.1}$$

$$A–L–B + h\nu'' \rightarrow {}^*A–L–B \tag{5.2a}$$

$${}^*A–L–B \rightarrow A^+–L–B^- \tag{5.2b}$$

$$A^+–L–B^- \rightarrow A–L–B + h\nu''' \tag{5.3}$$

$$A^+–L–B^- \rightarrow A–L–B \tag{5.4}$$

$$A^+–L–B \rightarrow A–L–B^+ \tag{5.5}$$

$$A^-–L–B \rightarrow A–L–B^- \tag{5.6}$$

A–L–B species, the various radiationless electron-transfer steps can also be termed *charge separation* (5.2b), *charge recombination* (5.4), and *charge shift* (5.5 and 5.6) reactions. The process shown in (5.2b) represents oxidative quenching of the photoexcited component (electron transfer from a photoexcited donor to an acceptor). Of course, reductive quenching (that is, electron transfer from a donor to a photoexcited acceptor) is equally possible (although less frequent, see section 5.5). The earliest reports on optical [1,2] and photoinduced [3] electron transfer in CLDA systems appeared in the late sixties and early seventies. Since then, the field has been growing almost exponentially. By now, hundreds of organic and inorganic CLDA systems have been synthesized and investigated. Research on such systems has important implications with regard to diverse fields such as electron-transfer kinetics and theory, biological electron transfer, and photochemical molecular devices.

From the point of view of electron-transfer kinetics and theory, the investigation of electron transfer in CLDA systems offers mechanistic advantages over that of analogous bimolecular reactions, as it avoids some of the ambiguities inherent in the latter processes. In fact, diffusion of the reactants, which is a key step in the kinetic scheme for a bimolecular reaction (section 2.7.3, Fig. 2.10), may often level off the bimolecular rate constants, hide the unimolecular electron transfer step within the "encounter complex", and lead to a complete loss of mechanistic information. Moreover, the geometry (distance and mutual orientation) at which the actual electron transfer step occurs is always an ill-defined parameter in a bimolecular process so that, even when the unimolecular rate constant within the encounter complex is known, it represents a space-averaged parameter difficult to analyse. These complications are, in principle, avoided in a covalently-linked system in which the distance and orientation between donor and acceptor are fixed by the presence of a connector (often designated, to emphasize this role, as a "spacer"). Thus, CLDA systems are particularly suited to investigate the effects of specific factors (such as the thermodynamic driving force, distance, orientation, nature of the connector, medium, and temperature) on the electron-transfer kinetics. As a matter of fact, the study of unimolecular electron transfer in CLDA systems has proved extremely successful from a mechanistic point of view, providing the possibility of testing experimentally some of the most important predictions of current theoretical electron-transfer models (sections 5.2 and 5.3).

Electron transfer plays a crucial role in biological systems for energy transformation, such as the respiratory chain [4] and the photosynthetic process [5,6]. The key functions performed by such systems, electron transport and photoinduced charge separation, are based on specific sequences of thermal and/or photoinduced electron-transfer steps. These processes take place between donor and acceptor units that are encased in a protein matrix at distances (10–30 Å) that significantly exceed the sum of the van der Waals radii of the donor and acceptor, and are thus often designated as "long-range" electron-trasfer processes. Of particular interest in the context of supramolecular photochemistry are the processes involved in photosynthesis. The recent determination of the X-ray structures of the reaction centres of the photosynthetic bacteria *Rps. Viridis* [7] and *Rb. Sphaeroides* [8,9] has been a major breakthrough in the field, providing a sound structural basis for the interpretation of the primary events in photosynthesis [10–12]. The structure of the chromophores of the reaction centre of the *Rps. Viridis* is depicted schematically in Fig. 5.1. The key chromophores are the bacteriochlorophyll "special pair" (P), a bacteriochlorophyll monomer (BC) and a bacteriopheophytin (BP) (that are present in two structurally equivalent branches), a quinone (Q), and a four-heme *c*-type cytochrome (Cy). These chromophores are held in a fixed geometry by surrounding proteins that span

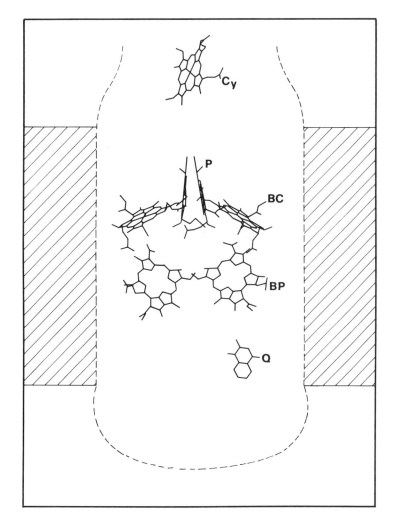

Fig. 5.1 — Arrangement of the chromophores in the reaction centre of *Rps. Viridis*, including the bacteriochlorophyll special pair (P), the bacteriochlorophyll monomer (BC), the bacteriopheophytin (BP), the quinone (Q), and the nearest heme group of the cytochrome (Cy) [7]. The shaded area and the dashed line schematically represent the membrane and the envelope of the protein matrix, respectively.

the photosynthetic membrane, so that the twofold axis of P is perpendicular to the membrane, the periplasmic face lies approximately between P and Cy, and the cytoplasmic face at the level of Q. In the reaction centre, excitation of P is followed by very fast (*ca.* 3 ps) electron transfer to the BP "primary" acceptor (whether the interposed BC plays the role of mediator in a superexchange mechanism [13] or directly intervenes as an intermediate electron acceptor [14] is still a subject of experimental debate [15–17]). The next step is fast (*ca.* 200 ps) electron transfer from BP to Q [18], followed by slower (*ca.* 270 ns) reduction of the oxidized P by the nearest heme group of Cy [19]. At that stage, transmembrane charge separation has been achieved with an efficiency approaching unity. The rate constants of the various electron transfer steps involved in charge separation are summarized in the (approximate) energy level diagram of Fig. 5.2, together with those of the non-occurring

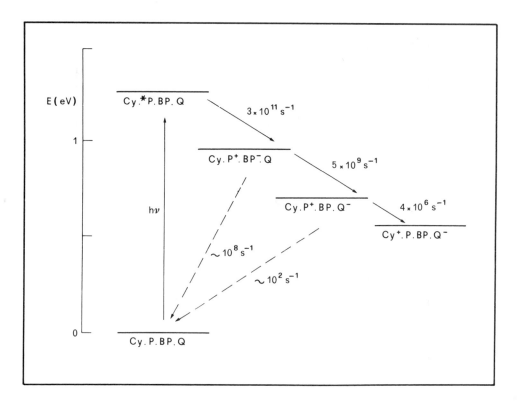

Fig. 5.2 — Energy-level diagram and kinetic parameters for the primary processes of bacterial photosynthesis in *Rps. Viridis*.

$BP^- \to P^+$ and $Q^- \to P^+$ charge recombination steps (as determined from experiments with modified reaction centres lacking the possibility of the competing forward processes) [20]. Figures 5.1 and 5.2 emphasize the fact that the achievement of efficient photoinduced charge separation in photosynthesis is based on (i) proper organization of the molecular components in the dimensions of space and energy and (ii) successful competition of forward over back electron transfer.

In order to understand a biological function based on electron transfer, the first task is, as seen above for the photosynthetic reaction centre, the dissection of the complex reaction sequence into single steps. The next important problem is to rationalize the kinetics of the various forward and back electron transfer steps by disentangling the energetic, nuclear and electronic factors that contribute in determining the rate constants. This is clearly difficult to do in the natural system, which has an extremely complex chemical structure and cannot be easily subjected to chemical and environmental modification. From such a point of view, CLDA systems can be viewed as convenient *models* for the various donor–acceptor pairs of a photosynthetic reaction centre. In principle, synthetic design may enable the chemist to reproduce or simulate some of the features of a natural donor–acceptor pair (chemical nature of the components, energetics of electron transfer, distance, orientation, degree of electronic interaction) in the artificial donor–acceptor system. On the other hand, some of these features can be changed in the model in logical and predictable ways and the effects of such changes on the electron-transfer rate constants can be checked. Thus, in addition to purely mechanistic motivations discussed above, current research on electron transfer between CLDA systems is largely driven by the concept of modelling biological electron transfer.

The study of biological systems shows that *molecular organization* and *kinetic control* are the two key features in obtaining an efficient *function*. Such features, that in biological systems come about "naturally" as a result of evolution, could be, in principle, achieved by rational synthetic design in an artificial supramolecular system. This consideration is at the root of the concept of *molecular device*, that is, an artificial supramolecular system capable of performing a useful function at the molecular level. The basic idea is not, of course, to attempt a hopeless competition with the chemical complexity of real, multifunctional biological systems, but rather to arrive at relatively simple supramolecular structures that perform a single, simple function. Chemical synthesis has made enormous progress nowadays, and the possibility of developing artificial molecular devices is not as remote as it might have seemed only a few years ago. The basic concepts relevant to the design of molecular devices featuring light-induced functions (*photochemical molecular devices*) are discussed in some detail in Chapter 12. Although several interesting light-induced functions can be considered (for example, photoswitching of electrical signals, spectral sensitization, remote photosensitization, antenna effects, light energy up-conversion), the most important one by far is that performed by photosynthesis, that is, the chemical conversion of light energy *via* photoinduced charge separation. Remarkable success has been indeed obtained in recent years by a number of laboratories in the design and synthesis of artificial three- and four-component supramolecular model systems ("triads" and "tetrads") for photoinduced charge separation (section 5.4). The possibility of developing photochemical molecular devices for practical applications is certainly an important, though futuristic, aspect that has contributed to the rapid growth of studies on photoinduced electron transfer in CLDA systems.

Before coming to the specific subject of this chapter, it should be noted that a number of other research areas, besides that of photosynthesis, are mechanistically related to CLDA systems. They involve systems in which donors and acceptors are held at a fixed distance by some kind of "matrix", rather than by a well-defined covalently-linked molecular connector. Thus, electron-transfer processes in protein –protein complexes [21–24], proteins and simple molecular species [21,22], metal-modified redox proteins [22,25,26], molecular reactants dispersed in rigid glasses or in polymers [27–30], and monolayer assemblies [31,32] have been the object of active research. For reasons of space, however, these systems are not dealt with in any detail in this monograph.

As stated at the beginning of this introduction, a large number of studies on electron transfer in CLDA systems are available. This chapter is not intended to provide an exhaustive coverage of the field, but rather to present a number of instructive examples of photoinduced electron-transfer processes in CLDA systems (5.2). Optical electron transfer (5.1), charge-transfer emission (5.3) and thermal (5.4, 5.5, and 5.6) electron-transfer processes will also be occasionally discussed in relation to the photoinduced ones. The material is discussed according to a predominantly mechanistic view, emphasizing the roles of electronic and nuclear factors in determining the kinetics of intercomponent electron-transfer processes. Section 5.2 contains some additional considerations on the kinetics of electron-transfer processes (with respect to the general concepts given in Chapter 3), that are relevant to the discussion of covalently linked donor–acceptor systems. In sections 5.3–5.6, specific examples of organic CLDA systems containing two active components ("diads") are discussed. Inorganic CLDA systems are discussed in a separate section (section 5.7) because of the peculiar problems that arise from the "complex" structure of the molecular components. Section 5.8 deals in particular with CLDA systems made up of three or more components ("triads", "tetrads", etc.), which have been specifically designed to mimic the multi-step charge separation mechanism of natural photosynthetic systems. Recent reviews on photoinduced electron-transfer in organic [33–35] and inorganic [36,37] CLDA systems and on biomimetic charge separation [35,38] are available. Finally, although synthetic aspects are not specifically stressed in the following sections, a simple look at the structural formulae should suffice to understand that a large part of the effort in the study of the CLDA systems lies in the skilful synthesis of tailor-made supramolecular structures.

5.2   KINETIC CONSIDERATIONS

According to the classical model discussed in section 3.4.4, the rate constant of an electron transfer process is given by (5.7).

$$k_{el} = \nu_N \, \kappa \, \exp(-\Delta G^{+}/RT) \tag{5.7}$$

The ν_N, κ, and ΔG^{+} terms have been discussed in some detail in section 3.4.4. This model can be used to discuss electron-transfer kinetics in any type of supramolecular system, including covalently-linked supermolecules of the A–L–B type. In this section, a few additional concepts relevant to the specific situation of CLDA systems are recalled.

In the literature on photoinduced electron transfer in CLDA systems, quantum-mechanical models are often used in alternative (or in a complementary way) to the Marcus theory. Simple quantum mechanical models treat the electron-transfer process as an activated radiationless transition between different electronic states of the supermolecule, leading to a golden-rule expression for the transition probability (5.8) [39,40].

$$k_{el} = (2\pi/\hbar) H_{AB}^2 \, \text{FCWD} \tag{5.8}$$

In this expression, H_{AB} is the same matrix element for electronic coupling between donor and acceptor discussed in section 3.4.4 (Fig. 3.9). The FCWD term is the Franck–Condon weighted density of states, that is, the sum of the products of the overlap integrals of the vibrational and solvation wavefunctions of the reactants with those of the products, suitably weighted for the Boltzmann population. In a simple approximation in which the solvent modes (average frequency, v_0) are thermally excited and treated classically ($hv_0 \ll k_BT$), and the internal vibrations (average frequency, v_i) are frozen and treated quantum mechanically ($k_BT \ll hv_i$), the FCWD term is given by (5.9) [27,40].

$$\text{FCWD} = \frac{1}{(4\pi\lambda_0RT)^{1/2}} e^{-S} \Sigma_m \frac{S^m}{m!} \exp - \left[\frac{(\Delta G^0 + \lambda_0 + mhv_i)^2}{4\lambda_0RT} \right] \qquad (5.9)$$

$$S = \lambda_i/hv_i$$

In (5.9), λ_0 and λ_i are the outer and inner reorganizational energies (section 3.4.4), and the summation extends over m, the number of quanta of the inner vibrational mode in the product state. It can be shown that, in the high temperature limit, (5.8) and (5.9) reduce to (5.10), where $\lambda = \lambda_0 + \lambda_i$ [41]. By comparison with (3.20), (3.19),

$$k_{el} = (2\pi/\hbar) \, H_{AB}^2 \, (1/4\pi\lambda RT)^{1/2} \exp[-\Delta G^0 + \lambda)^2/4\lambda RT] \qquad (5.10)$$

and (3.16), it is seen that the high temperature limit of the quantum mechanical expression corresponds to the *nonadiabatic* limit of the classical Marcus theory, in which the electronic coupling is small and the rate-determining step is electron rather than nuclear motion (section 3.4.4). In this limit the FCWD term in (5.8) corresponds to the exponential term of the classical rate constant (5.7). Besides the inherent nonadiabaticity of the quantum mechanical mode, an important difference between the quantum mechanical and the classical models is that (5.8) allows for *nuclear tunnelling* between reactant and product levels at energies lower than that of the intersection point. This difference is especially relevant to the behaviour predicted for highly exergonic reactions, for which the parabolic behaviour of the Marcus inverted region (section 3.4.4) is substituted by a linear decrease of log k_{el} with increasing driving force (energy-gap law) [41].

The reason for the widespread use of (5.8) in the discussion of electron transfer on CLDA systems is that, being inherently nonadiabatic, it explicitly states the dependence of the rate constant on the degree of donor–acceptor electronic coupling. As will be apparent in the following, in many CLDA systems investigated, the donor–acceptor distance is sufficiently large that direct (*through-space*) orbital overlap between donor and acceptor is very small, so that the nonadiabatic limit is appropriate. Under such conditions, the presence of the connector may be crucial for propagating electronic coupling between donor and acceptor (*through-bond* interaction [42,43]).

The calculation of H_{AB} in actual donor–acceptor systems is a difficult problem, but progress in this direction is rapid [44–59]. One of the simplest models to deal with through-bond interactions in covalently-bonded supramolecular systems is that of *superexchange* [34,45,60–63]. Consider, e.g., the process $A - L - B \rightarrow A^+ - L - B^-$. Figure 5.3a represents the redox orbitals of A and B together with the HOMO and the LUMO of L. The orbitals of A and B do not overlap appreciably with each other at the actual distance A–B. However, they do overlap with the HOMO and

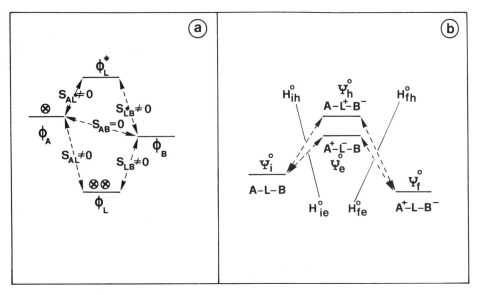

Fig. 5.3 — Orbital (a) and state (b) diagrams illustrating superexchange interaction between a donor A and an acceptor B through a simple bridging group L. In (a), ϕ_A and ϕ_B represent the donor and acceptor redox orbitals, ϕ_L and ϕ_L^* the HOMO and LUMO of the bridge, and S are overlap integrals. In (b), Ψ_h^0 and Ψ_e^0 are the hole- and electron-transfer "virtual" states. For the other symbols, see text.

LUMO of L. The overlap with the connector orbitals produces an indirect mixing of the orbitals of the two components. In terms of one-electron configurations, this amounts to considering that the initial and final states of the electron-transfer process are coupled together *via* interaction with high-energy states involving electron transfer from or to the connector (Fig. 5.3b). The two "intermediate" states can be conveniently called *electron-transfer* (A^+–L^-–B) and *hole-transfer* (A–L^+–B^-) states. It is important to realize that no electron (or hole) really hops from A to L to B in this mechanism: the electron transfer occurs in a single step from A to B. The electron- and hole-transfer configurations help to propagate the interaction, but they usually occur at very high energies and the corresponding states are never populated during the process. These states are thus often designated as "virtual states".

A more quantitative picture of the factors that affect the magnitude of the superexchange interaction can be obtained from perturbation theory. Let us label with i, f, e, and h the initial, final, electron- and hole-transfer states, respectively. Let us also use the superscript 0 (as, for example, in Ψ^0) to indicate wavefunctions, matrix elements, and energy differences relating to zero-order states, i.e., states that correspond to pure, localized electronic configurations. In the limit of small electron delocalization, the true initial and final states of the electron-transfer process are given by an appropriate admixture of the corresponding zero-order states (5.11), (5.12).

$$\Psi_i \approx \Psi_i^0 + \alpha\Psi_f^0 \qquad (5.11)$$

$$\Psi_f \approx \Psi_f^0 - \alpha \Psi_i^0 \qquad (5.12)$$

According to second-order perturbation theory, the true initial wavefunction is given by (5.13), where the H^0 and ΔE^0 terms are the matrix elements and the energy

$$\Psi_i = \Psi_i^0 + \frac{H_{if}^0}{\Delta E_{if}^0}\, \Psi_f^0 + \left[\frac{H_{ie}^0\, H_{fe}^0}{\Delta E_{ie}^0\, \Delta E_{if}^0} + \frac{H_{ih}^0\, H_{fh}^0}{\Delta E_{ih}^0\, \Delta E_{if}^0} + \frac{H_{ii}^0\, H_{if}^0}{(\Delta E_{if}^0)^2} \right] \Psi_f^0 \qquad (5.13)$$

differences between the various zero-order states. Since zero-order states. Since the first-order perturbation matrix element H_{if}^0 is equal to zero because of the negligible direct orbital overlap between A and B, the mixing coefficient is given by (5.14).

$$\alpha = \frac{H_{ie}^0\, H_{fe}^0}{\Delta E_{ie}^0\, \Delta E_{if}^0} + \frac{H_{ih}^0\, H_{fh}^0}{\Delta E_{ih}^0\, \Delta E_{if}^0} \qquad (5.14)$$

If we define the effective electronic coupling between the initial and final states as $H_{AB} = \alpha/\Delta E_{if}$ and consider $\Delta E_{if} \approx \Delta E_{if}^0$, then

$$H_{AB} \approx \frac{H_{ie}^0\, H_{fe}^0}{\Delta E_{ie}^0} + \frac{H_{ih}^0\, H_{fh}^0}{\Delta E_{ih}^0} \;. \qquad (5.15)$$

In (5.15), the energy differences ΔE_{ie}^0 and ΔE_{ih}^0 are to be taken at the geometry of the crossing point.

In practice, it is difficult to have reliable values for the quantities in (5.15) to calculate the electronic coupling in real systems (although some of these quantities could in principle be obtained from the analysis of intervalence transfer spectra, see sections 3.4.3 and 5.7.2). This expression, however, simply shows that connectors with high coupling ability ("conducting" connectors) should (i) contain relatively low-energy redox sites (relatively small values of ΔE_{ie}^0 and ΔE_{ih}^0) and (ii) have good electronic connections between these sites and the active components (relatively large H_{ie}^0 and H_{fe}^0 or H_{ih}^0 and H_{fh}^0). This is not to be taken, however, as a suggestion that connectors lacking low-energy redox sites are totally "insulating". It will be seen in the next sections, in fact, that very small values of H_{AB} are sufficient, under favourable conditions (FCWD ≈ 1) to promote reasonably fast electron transfer.

Covalently-linked donor–acceptor systems of the A–(L–)$_n$–B type in which the connector consists of a chain of identical subunits (such as, for example, polymethylene and polypeptide bridges, steroid or norbornene-type rigid spacers) have been widely used in recent studies on the distance dependence of intramolecular electron and energy transfer (sections 5.3, 5.5, 5.6.2, and 5.6.3). With such types of connectors, the above model has to be further elaborated to include interactions between the single subunits of the connector [61] (the treatment is similar to that used for electron transfer across a number of intervening solvent molecules [63]). This translates into an exponential dependence of the electronic coupling on the number of subunits of the connector, i.e., on the through-bond distance, r, between the active components (5.16).

$$H_{AB} = H_{AB}(0)\, \exp - [(\beta/2)(r - r_0)] \qquad (5.16)$$

In (5.16) r_0 represents the distance corresponding to a single subunit in the connector, and $H_{AB}(0)$ is the corresponding electronic coupling value. In these cases, the connector redox properties and the interactions betwen the subunits of the connector are described by β, while the donor–connector and connector–acceptor interactions are contained in $H_{AB}(0)$. (Analogous expressions containing the number of bonds instead of the through-bond distance can also be used [64,65].) It should be noticed that a through-space mechanism would lead to an exponential law similar to (5.16) [66,67], although with a stronger dependence (larger β) of rates on distance. The exponential dependence of H_{AB} on distance (or on number of bonds) can be translated into a corresponding relationship for the rate constant (5.17) only to the extent to which the FCWD term in (5.8) is independent of distance.

$$k_{el} = k_{el}(0) \, \exp - [\beta(r - r_0)] \tag{5.17}$$

Although there may be cases in which this happens to be approximately true, the FCWD is generally expected to depend on distance through λ_0 and, in some cases, also through ΔG^0.

5.3   ORGANIC SYSTEMS WITH STEROID-TYPE BRIDGES

An important class of organic CLDA systems is that synthesized and studied by Closs, Miller, and coworkers [64, 68–74]. In these systems, an electron donor and an electron acceptor are connected by a peculiar class of alkane bridges based on cyclohexane, decalin and androstane (Fig. 5.4). Due to their cyclic structures, such saturated bridges are rigid, thus providing CLDA systems with well-defined donor–acceptor distance and geometry (except for rotational freedom around the terminal C–C bonds). With these CLDA sytems, thermal electron-transfer processes have been studied by pulse radiolysis techniques, whereby the two radical anions $(A^--L-B$ and $A-L-B^-)$ or, depending on the experimental conditions, the two radical cations $(A^+-L-B$ and $A-L-B^+)$ can be generated in an almost statistical ratio, far away from equilibrium. In such experiments, the rate constants of the intercomponent electron-transfer reactions (5.18) and (5.19)

$$A^--L-B \rightarrow A-L-B^- \tag{5.18}$$

$$A^+-L-B \rightarrow A-L-B^+ \tag{5.19}$$

are obtained by monitoring the relaxation of the system towards thermodynamic equilibrium after the radiation pulse. In the nomenclature of Closs and Miller, process (5.18) is called *electron transfer*, while process (5.19) (that corresponds to electron transfer from B to A^+) is called *hole transfer*. The studies on this family of molecules are a remarkable example of the use of specifically designed CLDA systems to gain insight into the fundamental aspects of electron transfer processes.

The dependence of the rate of electron transfer (5.18) on the thermodynamic driving force of the process has been studied using a series of CLDA systems of type **5.1**, involving biphenyl as A, the 5-androstane skeleton as L (with β stereochemistry and the 3 and 16 positions as bridging sites), and a variety of quinones and aromatic hydrocarbons as B [64,69–71]. With these systems, it is possible to span the ΔG^0

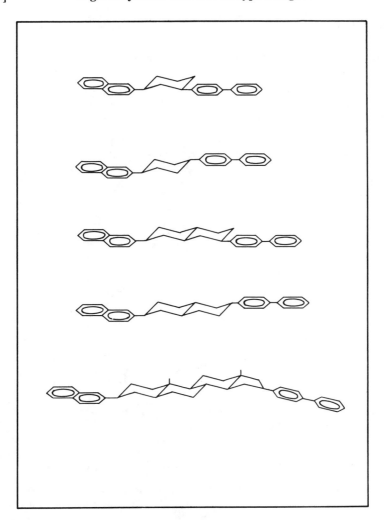

Fig. 5.4 — Biphenyl-naphthalene CLDA systems with rigid steroid-type bridges (cyclohexane, decalin, and androstane). Depending on the bridge and mode of attachment, 1–7 or 10 saturated C–C bonds separate the donor and the acceptor [64,70,72].

5.1

range from 0 to -2.4 eV for reaction (5.18). The results (Fig. 5.5) are extremely clear-cut in showing the decrease in rate constants predicted by classical and quantum mechanical models for strongly exergonic electron-transfer reactions (sections 3.4.4 and 5.2). This has actually been the first strong piece of evidence for the occurrence of the Marcus inverted region in electron transfer reactions in fluid solution. An equally striking confirmation of Marcus theory comes from the solvent dependence of the rates in the same series of compounds (Fig. 5.5). The solid curves in Fig. 5.5 represent fits to the experimental points with (5.8) and (5.9), using common values of $H_{AB} = 6.2$ cm^{-1}, $v_i = 1500$ cm^{-1}, and $\lambda_i = 0.45$ eV, and different λ_0 values for the different solvents (0.75, 0.45, and 0.15 eV, for methylhydrofuran,

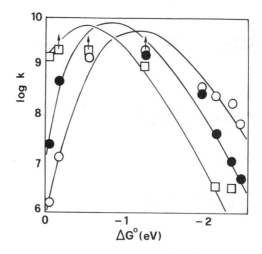

Fig. 5.5 — Free-energy dependence of the rate constant for electron transfer (5.18) in compounds of type **5.1** with biphenyl as A and various quinones and aromatic hydrocarbons as B [64,69,70]. Squares, full circles, and open circles represent data in isooctane, di-n-butyl ether, and methylhydrofuran, respectively. Arrows represent lower limiting values. The full lines represent fits to the experimental data using standard equations and appropriate parameters (see text).

di-n-butyl ether, and isooctane respectively) [64,69,70]. As predicted by the theory, the onset of the inverted behaviour (maximum in the bell-shaped curve) shifts to less negative ΔG^0 values with decreasing solvent polarity. A further check of the agreement between theory and experiment has been performed [74] by looking at the temperature dependence of the rate constant of electron transfer (5.18) in compound **5.1**, where B is a β-naphthyl group. This reaction lies in the normal Marcus region, with $\Delta G^0 = 0.05$ eV at room temperature. In 2-methyltetrahydro-furan, excellent agreement has been obtained between the experimental results and the predictions of (5.8) and (5.9), using reorganization parameters from the previous fits vs ΔG^0, and allowing for the dependence of λ_0 on temperature through the dielectric properties of the solvent [74]. Taken together, the driving force, solvent, and temperature dependences of CLDA systems of type **5.1** represent a remarkably self-consistent experimental picture, providing a clear demonstration of the validity of the basic features of the Marcus model.

In parallel studies, Closs and Miller have investigated the effect of donor–acceptor distance on the rates of the electron transfer (5.18) by using the series of A–L–B compounds shown in Fig. 5.4 [64,70,72], where A is biphenyl, B is naphthalene, and L changes from cyclohexane, through decalin to androstane. The thermodynamic driving force is small ($\Delta G^0 = 0.05$ eV) and constant. The centre-to-centre donor–acceptor distance depends on the type of connector and on the site of binding and spans the range 10.0–17.4 Å. In terms of the minimum number of C–C bonds between donor and acceptor, the series spans the range 4–10. The results show that the rate constant of electron transfer from the biphenyl anion radical to naphthalene decreases with increasing donor–acceptor distance [64,70,72]. In principle, the distance dependence of the rate comes from both the H_{AB} and FCWD terms in (5.8), the latter operating mainly through the distance dependence [75] of the λ_0 term in (5.9). Taking ν_i, λ_0, and λ_i from the previously studied androstane series **5.1**, and accounting for the distance dependence of λ_0, the rate constants can be converted into H_{AB} values. These values show an excellent correlation [70] with centre-to-centre donor–acceptor distance according to (5.16), with $\beta = 1.01$ Å^{-1} (corrected in later papers [64,72] to slightly smaller values) and $H_{AB}(0) = 1900$ cm^{-1} at $r_0 = 6$ Å (corresponding to a hypothetical one-bond connector). Satisfactory correlations are also obtained with the same equation as a function of edge-to-edge donor–acceptor distance, or with an analogous equation as a function of the number of intervening C–C bonds. In this last case, the electronic coupling decreases by a factor of *ca.* 5 for every C–C bond that is added between donor and acceptor. The fact that related experiments with non-bonded donors and acceptors in rigid glasses give smaller couplings and more pronounced attenuation with distance [27] is an indication that through-bond interaction is relevant in these CLDA systems [64,70,72].

All the compounds shown in Fig. 5.4 have an equatorial–equatorial (e–e) stereochemistry of attachment of the biphenyl and naphthyl groups to the spacer. Some rate data are also available for isomers with axial–axial (a–a), axial–equatorial (a–e) and equatorial–axial (e–a) stereochemistry [70,72]. These data are interesting because different isomers have different donor–acceptor distances but the same number of connecting bonds. The fact that isomers with widely different donor–acceptor distances have very similar H_{AB} values strongly argues for the operation of a through-bond coupling mechanism in such CLDA systems [70,72].

Closs and Miller [72] have also used the class of compounds shown in Fig. 5.4 to obtain, using pulse radiolysis, rate constants for the hole-transfer process in the corresponding radical cations (5.19). The rate constant decreases with distance in a very similar way as for the corresponding electron-transfer (5.18) processes. When the H_{AB} values for the hole transfer reactions are calculated and plotted against the donor–acceptor distance, very similar β values as for the electron-transfer case are obtained. When electron transfer (5.18) and hole transfer (5.19) are discussed in terms of a superexchange model, for obvious energetic reasons (A–L$^-$–B lower than A$^-$–L$^+$–B$^-$, and A–L$^+$–B lower than A$^+$–L$^-$–B$^+$) electron transfer occurs predominantly through electron-transfer virtual states (A–L$^-$–B) and hole transfer through hole-transfer virtual states (A–L$^+$–B). The finding of almost identical β values for the two processes points towards a substantial symmetry in the HOMO–LUMO energy diagram of these A–L–B molecules [72].

In an interesting extension of this work, Closs and coworkers have measured the rate constants for triplet–triplet intercomponent energy transfer in molecules similar to those of Fig. 5.4 [75], and have correlated the rates of energy transfer with those of electron and hole transfer [73,76]. This argument is discussed in section 6.2.1.

5.4   ORGANIC SYSTEMS WITH NORBORNYLOGOUS BRIDGES

Another important group of studies is that orginating from the collaboration of Paddon-Row, Verhoeven, and Hush. As a development of their previous studies of through-bond interactions [43,58,77,78], photoinduced and optical electron transfer in bichromophoric systems [1,79–82], and electron-transfer theory [83–86], these authors have thoroughly investigated the class of molecules depicted in Fig. 5.6 [65,87–98]. These remarkable CLDA systems involve dimethoxynaphthalene as

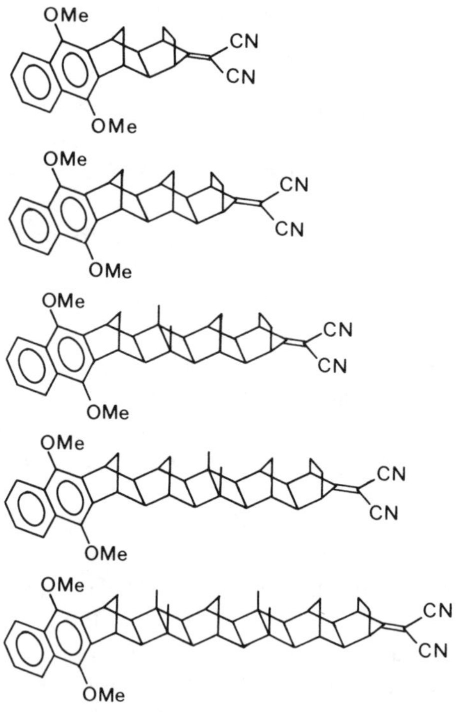

Fig. 5.6 — Dimethoxynaphthalene-dicyanovinyl CLDA systems with rigid norbornylogous bridges spanning a series of 4, 6, 8, 10, and 12 saturated C–C bonds between donor and acceptor [65,87–98].

donor (A), dicyanovinyl as acceptor (B), and norbornylogous spacers of various length as connectors (L). The spacers are saturated and rigid, and the attachments are such that no orientational freedom of the donor and acceptor fragments is possible. These CLDA systems span a range of centre-to-centre donor–acceptor distances of *ca.* 7–15 Å (edge-to-edge, 6.8–13.5 Å) and of 4 to 12 C–C bonds. The intercomponent electron-transfer processes are schematically depicted in Fig. 5.7.

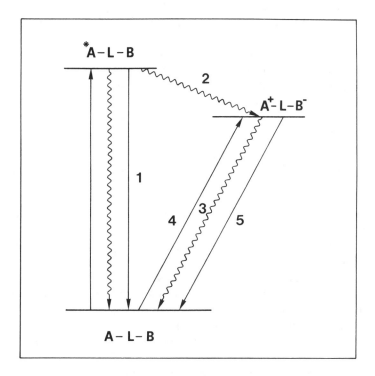

Fig. 5.7 — Radiative (straight lines) and radiationless (wavy lines) intercomponent processes taking place in the CLDA systems of Fig. 5.6. Numbering as discussed in the text.

In all the CLDA systems of Fig. 5.6, the fluorescence of the methoxynaphthalene unit (process 1 in Fig. 5.7) is substantially quenched. As shown by independent evidence (see below), this is a consequence of photoinduced electron transfer from the singlet excited state of the methoxynaphthalene unit to the dicyanovinyl unit (process 2). The kinetics of the electron transfer step has been studied in various organic solvents, by comparing the fluorescence of the CLDA systems with that of a suitable methoxynaphthalene-spacer model [65,93]. For the 4-bond bridged system, no fluorescence is observed, indicating a rate constant larger than $10^{11}\,s^{-1}$, while for the 6-bond, 8-bond, 10-bond, and 12-bond species the observation of a fluorescent emission with shortened lifetime permits the calculation of rate constants for the charge-separation step (5.2b). The rate constants decrease with increasing bridge length, being, for example, 3.3×10^{11} s^{-1}, 6.7×10^{10} s^{-1}, 1.2×10^{10} s^{-1}, and 1.3×10^{9} s^{-1}, for the 6-, 8-, 10-, and 12-bond compounds in tetrahydrofuran [65,93]. The analysis of these rate constants is complicated by the fact that, for a charge-separation reaction, two of the parameters that determine the activation free energy, namely, ΔG^0 and λ_0, depend on both solvent and distance (see (3.7), (3.8) and (3.18) in Chapter 3). In practice, calculations of the activation free energies in terms of ΔG^0, λ_0, and λ_i (taken as 0.6 eV) give, with minor differences due to solvent and distance, very small values, indicating that the photoinduced electron-transfer reactions belong to an almost activationless regime [65]. The pre-exponential factors of the rate constants (proportional to H_{AB}^2 (5.8)) exhibit good exponential dependence on the (edge-to-edge) donor–acceptor distance, as expected from (5.16), with $\beta = 0.85$ Å^{-1} (when the plot is made *vs* the number of C–C bonds in the bridge a value of $\beta = 0.98$/bond is obtained). Due to the almost activationless character of all these processes, good exponential dependencies on distance and number of bonds are also obtained using rate constants (5.17) instead of pre-exponential factors, with comparable values of β [93,98].

Interesting experiments have been performed by Paddon-Row, Warman, Verhoeven and their coworkers [88,90,92] on the same series of molecules (Fig. 5.6) using time-resolved microwave conductivity (TRMC). This technique allows the authors to detect the charge-separated A^+–L–B$^-$ product of photoinduced electron transfer (process 2 in Fig. 5.7), and to measure its dipole moment. The detection of the charge-separated states confirms the electron-transfer nature of the intra-molecular quenching processes. The measured dipole moment values are very high (26–77 D) and justify the name of "giant-dipole" states given to the charge-separated states of this class of CLDA systems [92]. The values slightly exceed those calculated on the basis of full charge separation over the edge-to-edge distance, indicating that the effective charge-separation distance is larger by *ca*. 1.5 Å than the edge-to-edge donor–acceptor distance. The decay in time of the TRMC signal can be used to monitor the disappearance of the charge-separated state and to obtain rate constants for the charge-recombination reaction (process 3 in Fig. 5.7). The rate constants decrease with donor–acceptor distance. In contrast to what happens for the charge-separation rates, however, the charge-recombination rates depend dramatically on the solvent, with those in cyclohexane being two orders of magnitude slower than those in dioxane. For each CLDA system, the charge-recombination rates (process 3) are always smaller by orders of magnitude than the charge-separation rates (process 2). These differences can be understood on the basis of the very different driving force ranges of the two processes. The moderately exergonic charge-separation rates are in the almost activationless regime ($-\Delta G^0 \approx \lambda$), whereas the highly exergonic (ΔG^0, *ca*. -3 eV) charge-recombination rates are in the Marcus-inverted region (section 3.4.4). In the latter case the FCWD terms (5.9) are small and very sensitive to solvent, since (contrary to what happens in the normal region) the solvent effects on ΔG^0 and λ_0 work in the same direction. The fact that good exponential correlations (5.17) between rate constants and donor–acceptor distance (or number of bonds) are obtained, with the same β and different $k_{el}(0)$ for various solvents) seems to indicate that the FCWD term is not substantially dependent on distance, probably because of the opposing effects of ΔG^0 and λ_0.

In some of these CLDA systems, optical electron transfer (process 4 in Fig. 5.7) and charge-transfer emission (process 5) have also been observed and studied [65,96]. Distinct electron-transfer absorption bands are observed for the 4- and 6-bond systems, whereas for the 8-, 10- and 12-bond systems the absorption spectrum is simply a superposition of those of the donor and acceptor components. This reflects the fact that the intensity of optical electron-transfer transitions is proportional to H_{AB}^2 (3.11), which decreases rapidly as the length of the bridge is increased. The charge-transfer bands exhibit strong shifts with solvent polarity, as expected on general grounds (section 3.4.1) and experimentally found for other CLDA systems [81,99]. Charge-transfer emission with low quantum yield is observed in various solvents for the 4-, 6-, 8-, and 10-bond systems, although the detection is complicated (especially for the longer systems) by overlapping with the intense residual emission of methoxynaphthalene. The lifetimes of the charge-transfer emission increase, for a given solvent, with increasing bridge length, and give a perfect agreement with the lifetimes of the charge-separated states independently measured by TRMC. The quantum yield and lifetime data can be combined to give the radiative rate constants, k_r, for charge-transfer emission, which can then be used to calculate (using expressions analogous to (3.11)) the donor–acceptor electronic coupling. The H_{AB} values so obtained (370, 112, 40, 18 cm^{-1} for the 4-, 6-, 8-, and 10-bond compounds, respectively) depend exponentially (5.16) on distance and on the number of bonds with β values of 0.88 Å^{-1} and 1.02/bond, respectively [96]. These values are in excellent agreement with those obtained in completely independent experiments (see above) from charge-recombination rates.

In a separate study, Paddon-Row, Hush, Miller and coworkers have reduced by pulse radiolysis the CLDA systems of Fig. 5.6 to the corresponding radical anions [91]. In these molecules, reduction at the dicyanovinyl group (A–L–B$^-$) is thermodynamically favoured over that at the methoxynaphthalene group (A$^-$–L–B) by *ca.* 1 eV in the solvents studied. Although a nonequilibrium mixture of both of the reduced forms is initially formed in the pulse radiolysis experiments, only dicyanovinyl-localized radical anions are detected at nanosecond time resolution. Thus, the intercomponent electron-transfer reaction from reduced methoxynaphthalene to dicyanovinyl (charge-shift reaction) is too fast to be detected in these experiments. This is consistent with calculations of the reorganizational energies in these systems, which point towards an essentially barrierless regime ($-\Delta G^0 \approx \lambda$) for these charge-shift reactions. The pulse radiolysis experiments are nevertheless interesting, since they permit the observation of the optical electron-transfer transitions A–L–B$^- \rightarrow$ A$^-$–L–B, from which the H_{AB} values can again be obtained [91]. These H_{AB} values (1300, 480, and 240 cm^{-1} for the 4-, 6-, and 8-bond radicals, respectively) are significantly larger than those measured (see above) from the charge-recombination emission of the neutral molecules (process 5 in Fig. 5.7). Actually, the values are such that the kinetic regime (section 3.4.4) of the corresponding charge-shift reactions would change from the adiabatic limit for the 4-bond anion to the nonadiabatic limit for the 8-bond anion. The difference in the H_{AB} values obtained from optical charge-transfer in the neutral molecules and in the anions (that is, for the charge-recombination and charge-shift processes) is likely due to the different orbitals of the methoxynaphthalene involved in the two processes (π and π^*, respectively) [96]. The energy gaps between these orbitals and the σ^* orbitals of the bridge are different, so that in a superexchange mechanism (section 5.2) the electron-transfer virtual state is at a lower energy for the charge-shift reaction than for charge recombination, and the electronic coupling (5.15) is larger for the former than for the latter process.

In the series of CLDA systems of Fig. 5.6 (and in the corresponding anion radicals), the electron-transfer rates (or the electronic couplings) correlate equally well with donor–acceptor distance and number of bonds in the bridge. This is due to the fact that, owing to the homologous steric configuration of the bridges, the two parameters are essentially proportional. In order to disentangle these two aspects of the problem, some elegant studies have been performed on the photoinduced electron-transfer reactions of CLDA systems differing in the stereochemistry of the bridges (Fig. 5.8) [95,97,98]. In Fig. 5.8, compounds *a* and *b* have 6- and 8-bond

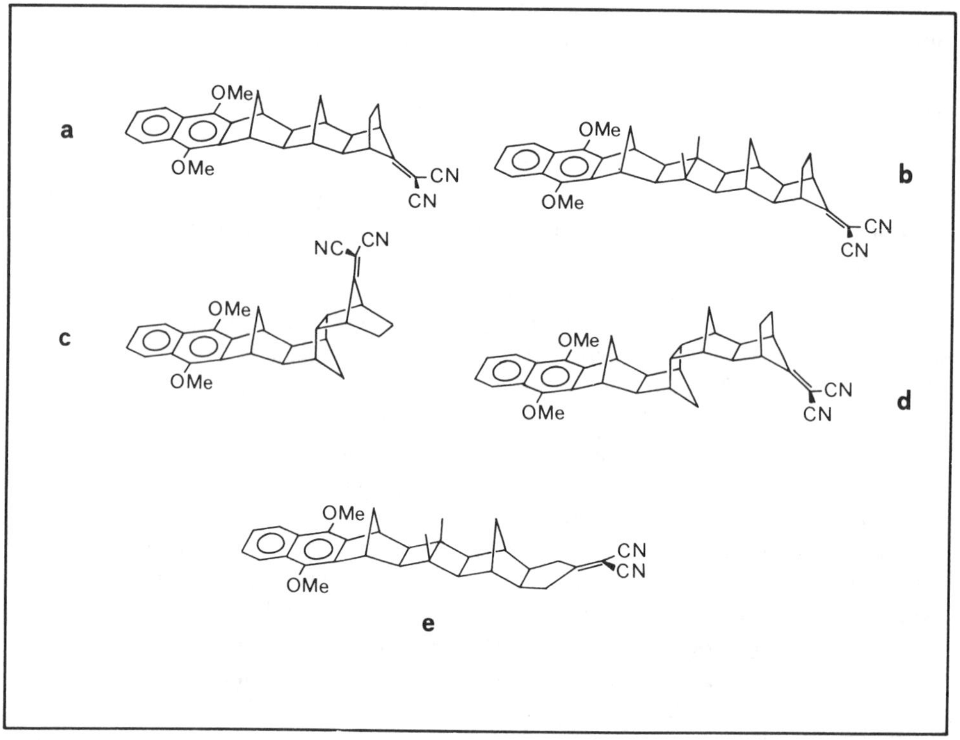

Fig. 5.8 — CLDA systems with norbornylogous bridges of different bridge stereochemistry (a–c and b–d) and geometry of the terminal cyclopentyl group (b–e) [95,97].

bridges in a stretched "all-*trans*" configuration, whereas compounds *c* and *d* have again 6- and 8-bond bridges but in a less extended configuration. Compound *e* is very similar to compound *b*, except for the fact that the cyclopentyl ring attached to the dicyanovinyl acceptor is much flatter. The rates of photoinduced electron transfer in compounds *c* and *d* are about one order of magnitude *smaller* than in compounds *a* and *b*, despite the shorter through-space donor–acceptor distance of the former systems [95]. This clearly shows that the donor–acceptor coupling is a through-bond· phenomenon. The lower efficiency of the bent bridges with respect to the stretched ones is in keeping with the original prediction by Hoffmann [100] that an all-*trans* arrangement of σ bonds is the optimum one for through-bond interactions, a prediction confirmed by calculations of various levels of sophistication, and by experiments involving photoelectron spectroscopy of symmetric bifunctional molecules with the same type of bridges [78,101]. Analogous arguments can be used to rationalize the fact that photoinduced electron transfer is five times slower in compound *e* than in compound *b*, as the flattened geometry of the cyclopentyl ring is less favourable than the bent one for interaction of the π-orbitals of the dicyanovinyl group with the bridge [97].

The studies of Paddon-Row, Verhoeven, Hush and coworkers on CLDA systems with norbornylogous bridges give a precise and self-consistent picture of long-range electron-transfer processes across saturated bridges. The mechanism for electron transfer is definitely through-bond, with an exponential dependence on the number of intervening σ bonds (or through-bond distance). With a given number of bonds, the donor–acceptor coupling is strongly influenced by the bridge configuration. The norbornylogous bridges appear to be saturated rigid connectors of remarkable "conducting" ability, allowing electron transfer to occur on the subnanosecond time scale over 15 Å distances. Some studies of electronic energy transfer in the same type of CLDA systems are dealt with in section 6.2.1.

5.5 ORGANIC SYSTEMS WITH AROMATIC BRIDGES

The series of CLDA systems **5.2**–**5.6** has been studied in considerable detail by

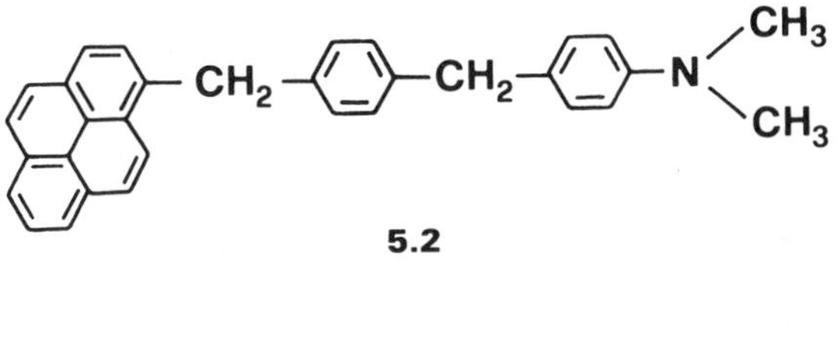

5.2

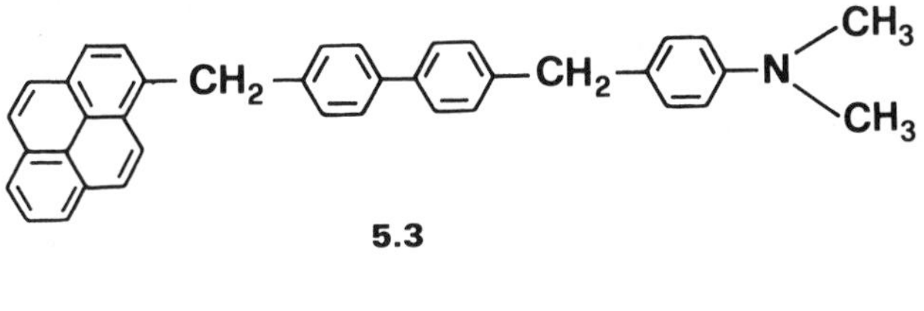

5.3

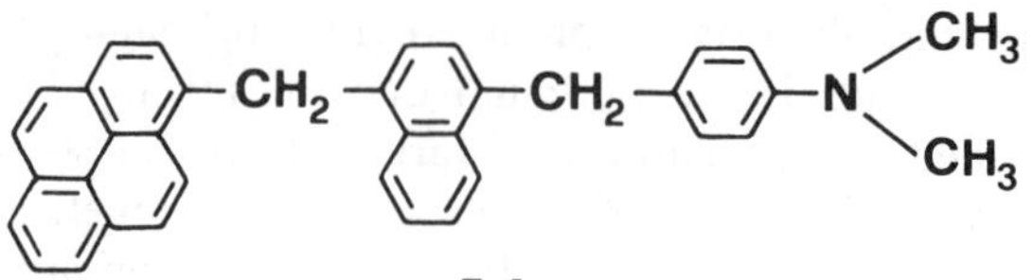

5.4

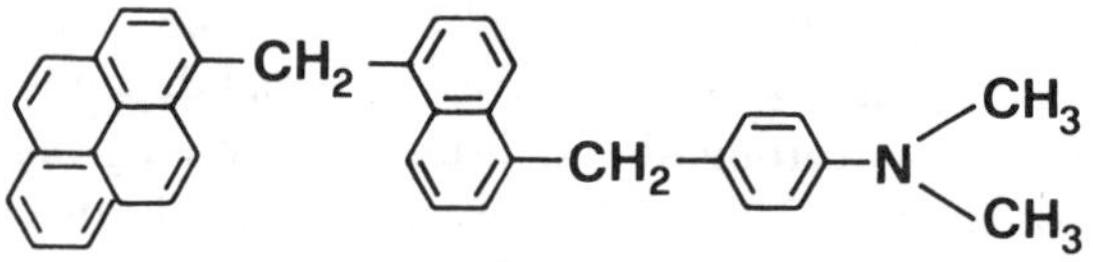

5.5

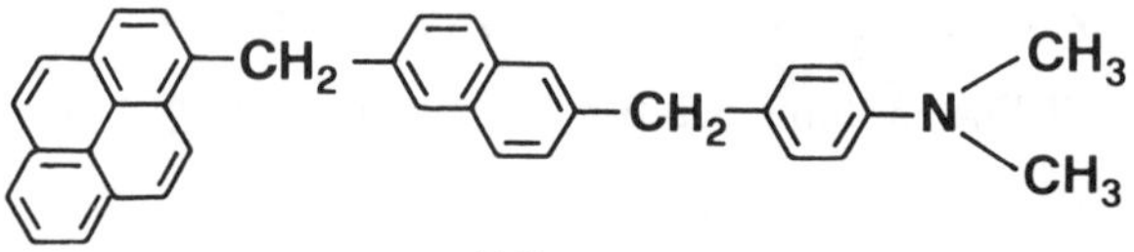

5.6

Michel-Beyerle, Heitele and their coworkers [102–105]. These systems involve pyrene as a photoexcited electron acceptor, dimethylaniline as electron donor, and bridges made up of two terminal methylene groups and various central aromatic fragments. Thus, in contrast to most of the systems discussed in sections 5.3, 5.4, and 5.5, these CLDA systems are suitable for studying reductive quenching (5.20) and subsequent charge recombination (5.21).

$$A\text{–}L\text{–}B \xrightarrow{h\nu} {}^*A\text{–}L\text{–}B \tag{5.20a}$$

$${}^*A\text{–}L\text{–}B \rightarrow A^-\text{–}L\text{–}B^+ \tag{5.20b}$$

$$A^-\text{–}L\text{–}B^+ \rightarrow A\text{–}L\text{–}B \tag{5.21}$$

The bridges in this case are not fully saturated, and sufficiently flexible to allow a considerable degree of conformational freedom.

The photophysics of systems **5.2–5.6** (and of similar systems involving anthracene instead of pyrene as donor) has been characterized in propionitrile by means of fluorescence quenching and transient absorption measurements [104]. In these systems, quenching of the pyrene fluorescence is accompanied by formation of pyrene radical anion and dimethylaniline radical cation, demonstrating the occurrence of (5.20). The rate constants for the charge-separation and charge-recombination processes have been studied as a function of temperature. This greatly helps in disentangling the various electronic, nuclear and thermodynamic factors that contribute, in the nonadiabatic kinetic limit described by (5.8) and (5.9), in determining the reaction rate. The main results of this study [104] can be summarized as follows.

(i) The donor–acceptor electronic coupling (H_{AB}) decreases slowly with distance ($\beta \approx 0.5 \text{ Å}^{-1}$ in (5.16)), as a consequence of the presence of the aromatic groups in the spacer.

(ii) According to a superexchange mechanism, the H_{AB} values are about two times larger for charge separation than for charge recombination, due to the different proximity of the hole-transfer virtual state.

(iii) The H_{AB} values are comparable or smaller than those obtained with fully saturated rigid bridges of comparable lengths (sections 5.3 and 5.4). The apparent contradiction probably arises from averaging over many unfavourable conformations in these flexible systems.

(iv) Since charge separation lies in the normal region and charge recombination in the inverted region, the former has a better nuclear factor than the latter.

(v) Differences in activation energies between compounds in the series arise also from different barriers to the bridge rearrangments required to reach the most convenient (from the electronic point of view) donor–bridge–acceptor conformation.

In polar solvents such as nitriles, a compound similar to **5.2** with anthracene as excited acceptor behaves qualitatively as described above for the pyrene-containing analogue [104]. In nonpolar solvents, on the other hand, no quenching takes place, as the charge-separated state is insufficiently stabilized to make the electron-transfer process thermodynamically allowed [102]. In solvents of intermediate polarity, a more complex situation arises, since the charge-separated state is only slightly lower in energy than the local excited singlet state of anthracene. In this case, thermal repopulation after quenching takes place, giving rise to nonexponential fluorescence decays and to a peculiar temperature dependence [105].

A study of systems similar to **5.2** and **5.3**, in which anthracene replaces pyrene, has revealed interesting kinetic behaviour related to the change in viscosity of the medium with temperature [103]. At room temperature the photoinduced electron-transfer reaction is nonadiabatic and controlled by electronic factors (as discussed above for compounds **5.2–5.6** in room-temperature propionitrile), whereas at low enough temperatures the process is adiabatic ($\kappa = 1$ in (5.7)), is insensitive to changes in electronic factors and is controlled by the solvent dielectric relaxation time (related to the inverse of v_N in (5.7)). In an intermediate temperature range, these systems exhibit a smooth transition in behaviour from the limiting nonadiabatic to the limiting adiabatic regime [103].

5.6 PORPHYRIN–QUENCHER SYSTEMS

Because of their widespread occurrence in photosynthetic reaction centres and other biological electron-transfer systems, porphyrin-type chromophores have played the leading role in the design of CLDA systems for the study of photoinduced intramolecular electron transfer. The first examples of such systems [106,107] date back to 1979. The review by Connolly and Bolton [33], that provides an exhaustive literature coverage up to 1987, lists over a hundred CLDA systems containing a porphyrin-based chromophore covalently linked to a variety of electron acceptors (quinones, bipyridinium ions, etc.), and more continue to appear at a constant rate. For reasons of space, a limited number of examples from this wide field will be selected and discussed in this section.

5.6.1 Porphyrin–quinone systems with trypticene bridges
A series of CLDA systems in which a tetraphenylporphyrin is linked to a quinone by a rigid spacer based on the trypticene structure have been studied by Wasielewski *et al.* [108,109]. A representative structure is **5.7** (the other members of the series can

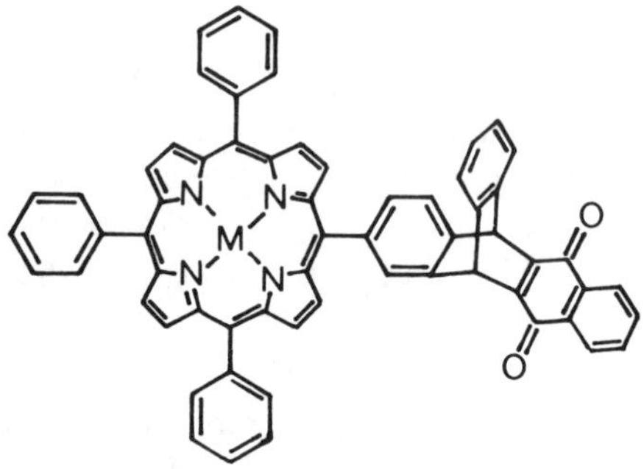

5.7

be obtained from this structure by changing the quinone (naphtho- or anthra-quinone) and by substituting the metal with hydrogens in the porphyrin). In butyronitrile, excitation of the porphyrin chromophore to the singlet state results in photoinduced charge separation (5.2b) and recombination (5.4). The charge-separation processes are moderately exergonic (0.0–0.9 eV) while charge recombination is highly exergonic (1.2–1.9 eV). The rate constants for the two processes, as obtained by picosecond absorption techniques, lie in the 1×10^9–$3 \times 10^{11}\,\mathrm{s}^{-1}$ range. When plotted against the thermodynamic driving force, the rate constants give a bell-shaped curve (of the same type as those of Fig. 5.5) with the charge separation data in the normal region and the charge recombination data in the Marcus-inverted region.

For this reason, for each CLDA system of this series, the charge-separation process is faster than charge recombination. The large overall values are consistent with the short bridge used (while the porphyrin–quinone centre-to-centre distance is *ca.* 10 Å, the phenyl-quinone edge-to-edge distance is only 2.4 Å). Similar free-energy dependences and rate constant values have been obtained by Wasielewski *et al.* [110] for related CLDA systems involving chlorophyll as the excited donor.

In most porphyrin–quinone CLDA systems, the photoinduced electron transfer is found to be strongly temperature dependent and to "shut off" abruptly at the solvent freezing point [33]. This contrasts sharply with the behaviour of photosynthetic reaction centres, in which the rate of the primary electron transfer process actually undergoes a slight increase upon cooling to 4.2 K [111]. This difference is due to the fact that in fluid solution the product of the charge-separation step is stabilized by solvent repolarization. Therefore, if the process is only slightly exergonic in fluid solution, it may become thermodynamically forbidden when the solvent repolarization is blocked by freezing the solvent into a rigid matrix. This concept has been checked by Wasielewski *et al.* [112] using a CLDA system similar to **5.7** in which, however, the quinone is substituted by a tetracyanonaphthoquinodimethane group. Due to the high electron affinity of this group, the charge-separation reaction is now highly exergonic ($\Delta G^0 = -1.49$ eV in MTHF), and the photoinduced charge separation remains efficient in MTHF rigid matrix down to 10 K.

The distance-dependence of electron-transfer rate constants has also been probed by Wasielewski *et al.* [113] using CLDA systems similar to **5.7** but with *trans*-1,2-diphenylcyclopentane and adamantane instead of trypticene as bridging groups. In this series of molecules 2, 3, and 4 saturated C–C bonds are interposed between the donor and acceptor π systems, and the edge-to-edge distances are 2.5, 3.7, and 4.9 Å. The charge-separation and -recombination rate constants decrease exponentially with distance, with β values of 2.3 and 2.1 Å^{-1}, respectively.

In a further study, Wasielewski *et al.* [114] have provided an elegant demonstration of the validity of the superexchange model (section 5.2) for the donor–acceptor interaction in this type of CLDA system. The system investigated, **5.8**, involves a

5.8

double trypticene-type bridge with an aromatic nucleus lying between the two saturated parts of the bridge. The idea is to use substitution on the aromatic nucleus (R = H, OCH$_3$) to tune its redox properties, and thus the energies of the electron- and hole-transfer virtual states, and look for effects on electronic coupling (5.15) and rates (5.8). For the charge-separation reaction (Fig. 5.9), the only viable super-

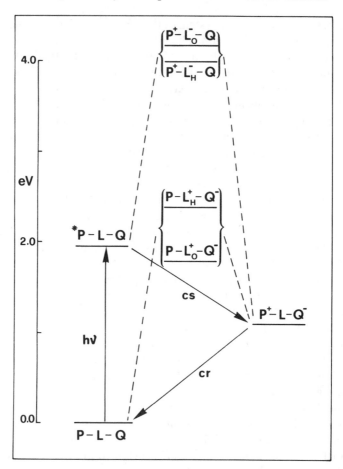

Fig. 5.9 — Energy-level diagram for photoinduced charge separation (cs) and charge recombination (cr) in the porphyrin(P)–quinone(Q) CLDA systems **5.8** [114]. Dotted lines indicate
superexchange interactions *via* charge-transfer virtual states involving the methoxy-substituted
(L_O) and unsubstituted (L_H) double trypticene bridge.

exchange pathway is that through the electron-transfer virtual state P^+–L^-–Q (the
locally excited *P–L–Q state is connected to the P–L$^+$–Q$^-$ hole-transfer state by a
two-electron exchange process that would give rise to a very small H_{ih} term in
(5.15)). In these systems the electron-transfer states are so high in energy (Fig. 5.9)
that the small difference between the methoxy-substituted and unsubstituted molecule is negligible, and the charge-separation rates are not expected to be affected
appreciably by substitution. For the charge–recombination reactions (Fig. 5.9), on
the other hand, the hole-transfer pathway is accessible and dominates over the
electron-transfer pathway for energetic reasons. In this case, the changes in the
energy of the P–L$^+$–Q$^-$ hole-transfer state induced by substitution are relevant, and
the charge-recombination rate is expected to be appreciably higher (by a factor of *ca.*
4) for the methoxy-substituted than for the unsubstituted CLDA. The experimental
results nicely confirm these expectations, with charge-separation rate constants of

1.7×10^{10} s^{-1} and 1.9×10^{10} s^{-1}, and charge-recombination rate constants of 2.5×10^{10} s^{-1} and 8.2×10^{10} s^{-1}, for the unsubstituted and methoxy-substituted molecules, respectively [114]. Experiments carried out on systems identical to **5.8** except for a different stereochemistry of attachment of the quinone (*syn* instead of *anti* with respect to the porphyrin) give differences in rate constants of both charge separation and recombination that confirm the general rule [78,100,101] that all-*trans* bridge configurations (*anti* isomer in this case) are more favourable than cisoid ones (*syn* isomer in this case) to propagate donor–acceptor interaction [114].

5.6.2 Porphyrin–quinone systems with bicyclooctane bridges

Porphyrin–quinone CLDA systems with modular rigid spacers based on bi-cyclo[2.2.2]octane structure have been mainly studied by Hopfield, Dervan, and coworkers [115–119]. In the series of systems **5.9**–**5.11** (M = Zn), the centre-to-centre porphyrin–quinone distances are 6.5, 14.8, and 18.8 Å.

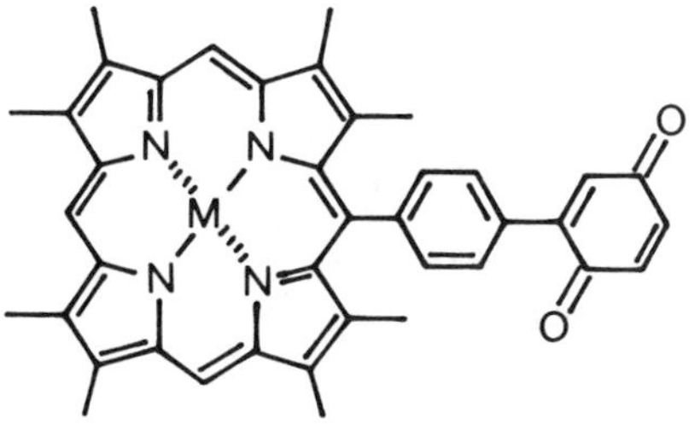

5.9

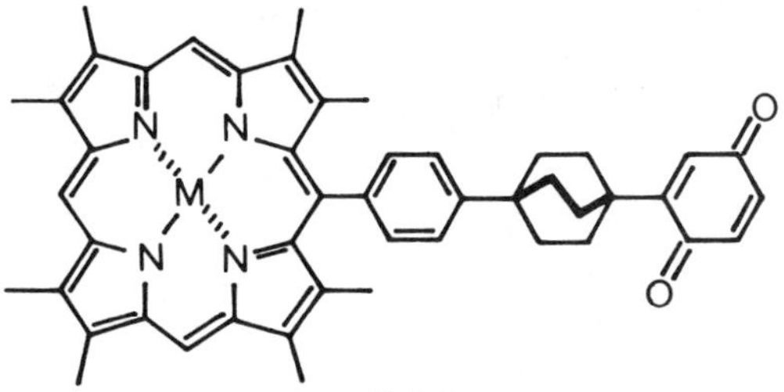

5.10

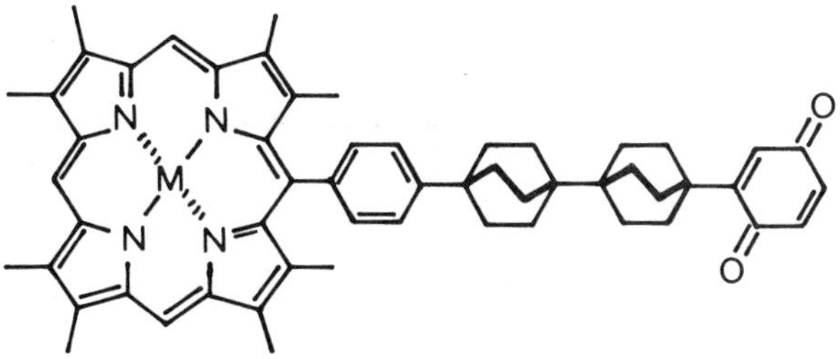

5.11

The electron-transfer rate constants from singlet excited porphyrin to the quinone (calculated from fluorescence quenching with respect to a model porphyrin lacking the quinone) decrease sharply with distance ($>10^{12}$, 1.5×10^{10}, and $\leqslant 9 \times 10^6$ s^{-1} in benzene), giving a $\beta \geqslant 1.4$ Å^{-1} in (5.17) [116]. The distance-dependence of the electronic matrix element for electron-transfer across this type of spacer has been discussed theoretically by Beratan [118]. The electron-transfer quenching processes in these systems are nearly temperature-independent down to 77 K. Low temperatures, however, seem to freeze out rotational motion around the bridge, giving rise to multiexponential decays related to an ensemble of rotational conformers [116]. The dependence of electron-transfer rate constants on the driving force was studied using a series of CLDA systems derived from the structure **5.10**, in which the redox potential of the quinone is tuned by appropriate substituents [117]. In various solvents, the observations give satisfactory agreement with the predictions of Marcus theory.

5.6.3 Porphyrin–quinone systems with polyene bridges

Wasielewski and coworkers [120] have studied the two porphyrin–quinone systems depicted in Fig. 5.10, in which the components are linked by all-*trans* polyene chains

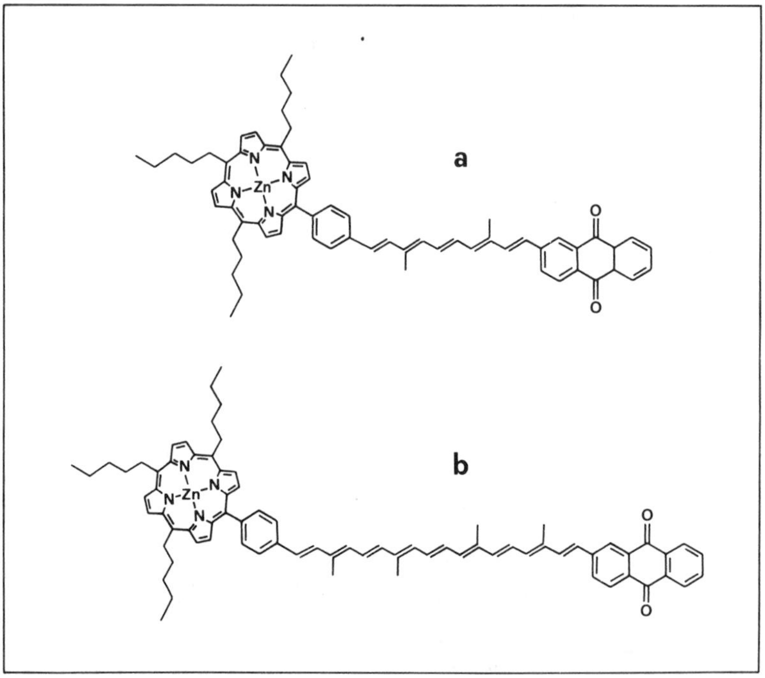

Fig. 5.10 — Porphyrin–quinone CLDA systems with polyene bridges containing 5 (a) and 9 (b) double bonds [120].

of 5 and 9 double bonds, providing rigid systems with donor–acceptor centre-to-centre distances of 25 and 35 Å. These highly conjugated bridges are designed to facilitate photoinduced electron transfer over long distances, by providing strong donor–acceptor electronic coupling. It must be pointed out, however, that this is not the only peculiar property of such conjugated bridges with respect to fully or partially saturated ones (such as those discussed in sections 5.2 and 5.3). Related to the greater electronic coupling ability is also the greater ease of oxidation and reduction and the presence of low-energy bridge-localized excited states.

The behaviour of these CLDA systems, as determined from fluorescence quenching and picosecond absorption measurements, is summarized by the mechanisms in Fig. 5.11 [120]. In the polyene-5 compound (Fig. 5.11a), charge separation

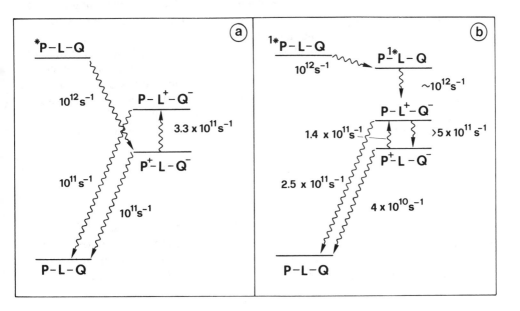

Fig. 5.11 — Mechanism and kinetics of photoinduced electron transfer in the CLDA systems of Fig. 5.10 [120].

takes place in 3 ps. Charge recombination (which is complete in about 10 ps) takes place *via* two competing pathways: direct long-range back-electron transfer, and a two-step route involving the cation radical state of the polyene. The second path probably competes with the direct one, in spite of its thermally activated character, due to better electronic factors. In the polyene-9 compound (Fig. 5.11b), the singlet excited state of the bridge lies below that of the porphyrin. Therefore the initial ultrafast ($\tau < 3$ ps) sequence of events is energy transfer from donor to the bridge and electron transfer from bridge to acceptor. Then, partitioning between recombination to the ground state and secondary electron transfer with further charge separation ($\tau = 7$ ps) takes place. As in the previous case, the final charge recombination involves a competition between two paths, dominated in this case by the activated one. It is seen that in polyene-9 compound (and to a lesser extent also in the polyene-5 case), the polyene does not act as a simple "bridge" but rather as an active component of the CLDA system. Thus, the behaviour is similar to that of the three-component CLDA systems ("triads") featuring two-step charge separation discussed in section 5.8.

Apart from the mechanistic details, the study of these CLDA systems shows that polyene bridges behave as highly conducting "molecular wires" [121,122] capable of carrying electrons in a few picoseconds over considerable distances. If the direct quinone–porphyrin charge-recombination rates for the two CLDA systems (Fig. 5.10) are compared, it is seen that the decrease in electron-transfer rates with distance is very slow (using (5.17), the two rate constants would give $\beta \approx 0.1$ Å^{-1}). A very weak distance-dependence for polyene-type bridges has been suggested by theoretical calculations [123,124]. An experimental estimate of H_{AB} across polyene bridges has been made by Launay and coworkers [125] using the intervalence transfer spectra (see section 3.4.1) of a series of mixed-valence binuclear ruthenium complexes with bis(4-pyridyl)polyenes as bridging ligands. The H_{AB} values decrease slowly in going from two to three and four double bonds, consistent (5.16) with a value of $\beta \approx 0.2$ Å^{-1}. If these values are compared with the corresponding data for fully ($\beta \approx 1$–2 Å^{-1}, sections 5.3, 5.4, and 5.6.2) and partially ($\beta \approx 0.5$ Å^{-1}, section 5.5) saturated bridges, the effect of the degree of unsaturation on the "conductivity" of the bridge is evident.

5.6.4   Porphyrin–quinone systems with flexible bridges
Porphyrin–quinone CLDA systems containing more or less flexible connectors, such as alkyl, amide, diamide, and diester bridges, have been extensively investigated [33,126,127]. The system **5.12** involving a short amide bridge has been investigated in

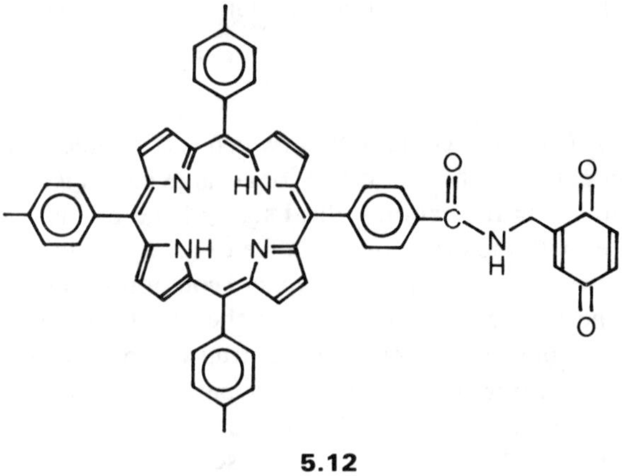

5.12

detail by Bolton, Connolly, and coworkers [128,129]. In benzonitrile, the energy of
the charge-separated state is slightly below that of the porphyrin triplet state, and
excitation of the porphyrin chromophore is followed by the sequence of events
shown in Fig. 5.12. This mechanism has been established using micro- to picosecond

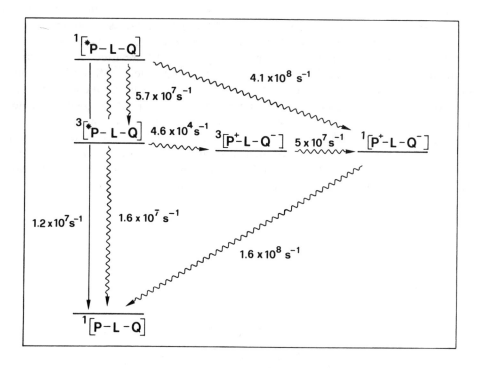

Fig. 5.12 — Mechanism and kinetics of photoinduced electron transfer in the porphyrin–
amide–quinone CLDA system **5.12** [128].

absorption measurements and fluorescence lifetimes, and is based on a comparison
between the results obtained with **5.12** and with its reduced hydroquinone analogue
[128]. The kinetics of this system has been fully characterized (Fig. 5.12). The
electron-transfer quenching (charge separation) of the singlet state is efficient
($\Phi = 0.83$), but intersystem crossing takes place to a sufficient extent ($\Phi = 0.11$) that
electron-transfer quenching of the triplet state can also be observed. Charge
recombination is thought to occur always *via* the singlet charge-separated intermediate (that in the triplet quenching pathway is reached following fast spin
rephasing of the corresponding triplet). Interesting observations are (i) the great
difference in singlet and triplet charge-separation rates, and (ii) the larger rate
constant of (singlet) charge separation over charge recombination. Point (i) is not
directly related to spin, but rather to the difference in driving force of the two

processes. Point (ii) reflects the inverted character of the highly exergonic charge-recombination step, as opposed to the normal character of the charge-separation process. Both points agree at a quantitative level with the predictions of Marcus theory [128]. A detailed study of the solvent effects on the behaviour of **5.12** has been carried out [129]. The rate constants of electron transfer from the porphyrin singlet to the quinone depend markedly on solvent, going, for example, from $4.8 \times 10^7 \text{ s}^{-1}$ in acetonitrile to $2.2 \times 10^9 \text{ s}^{-1}$ in chloroform. The solvent dependence can be satisfactorily fitted to Marcus theory, provided that solvent effects on both reorganizational energy and driving force are taken in consideration.

When bridges longer than the simple amide of **5.12** are used in porphyrin–quinone CLDA systems, the results are usually complicated by flexibility, which gives rise to a distribution of rate constants corresponding to different conformations and complicates the study of distance dependence [130,131]. At the other extreme, when conformational freedom is minimized by attaching the quinone directly to a meso-position of the porphyrin ring, as in **5.13**, the very short donor–acceptor distance may bring about other problems, such as competition between electron transfer and vibrational relaxation [132] or inadequacy of the simple Marcus two-sphere model in the calculation of reorganizational energies [133].

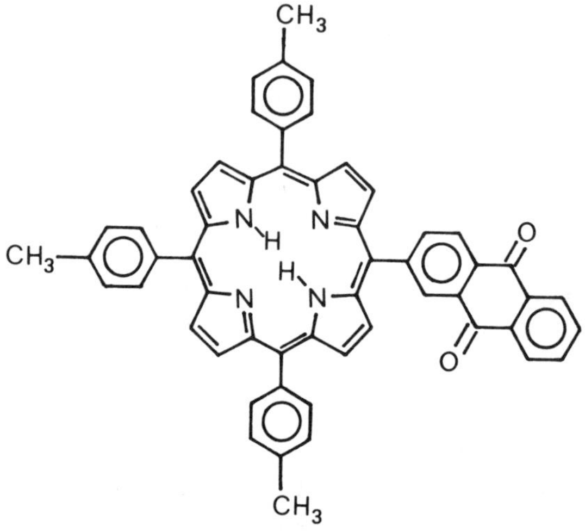

5.13

An example of a supramolecular structure that is related to (but not, strictly speaking, belonging to) porphyrin–quinone CLDA systems is a peculiar host–guest system (section 10.5) similar to **5.12**, but with a β-cyclodextrin attached to the amide bridge instead of the quinone [134]. In this system, a quinone (or other electron acceptors) can be hosted by the cyclodextrin and the photoinduced electron-transfer process can be monitored.

5.6.5 Structural effects in porphyrin–quinone systems

In many porphyrin–quinone CLDA systems, especially those with flexible bridges, much speculation has been made on the possible role of the mutual orientation of the donor and acceptor components on the photoinduced electron-transfer rates. The problem is not an easy one to deal with, since in the case of flexible bridges, conformational averaging tends to mask the phenomenon, while with rigid bridges the roles of bridge stereochemistry and donor–acceptor orientation are difficult to disentangle. A rather convincing case for orientational effects is that reported by Sakata and coworkers [135] and based on the comparison of the CLDA systems **5.14** and **5.15**. In these two systems, the bridges between the porphyrin and the quinone are rigid, the through-space donor–acceptor distance is appreciably constant, and the number of bonds separating the donor and the acceptor is the same. The only major structural difference seems to be the mutual orientation of the two π-systems

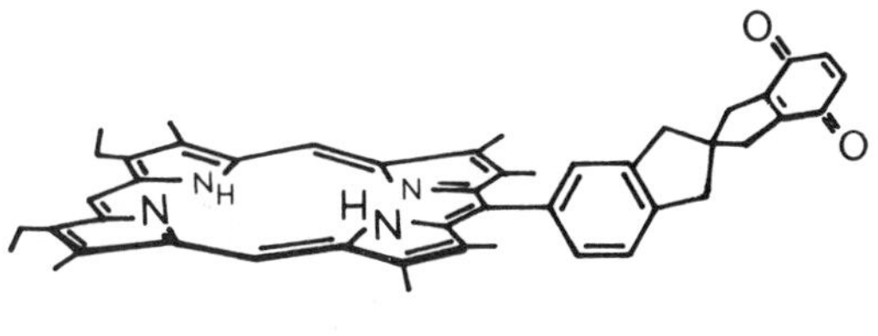

5.14

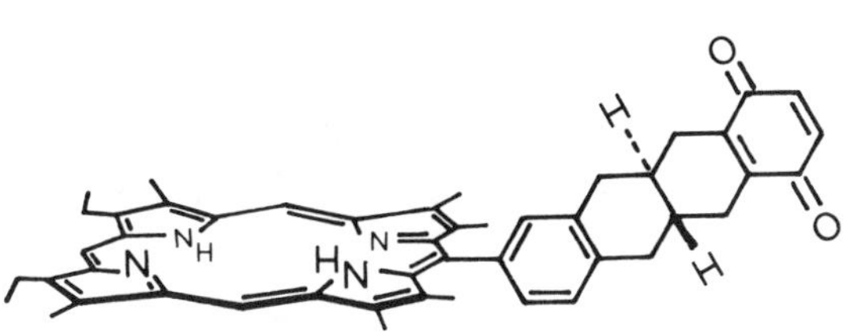

5.15

(dihedral angles of 150 and 90°). From porphyrin fluorescence quenching, the photoinduced electron-transfer process is estimated to be faster by a factor of 5 for **5.14** than for **5.15**. This can be attributed to the more favourable donor–acceptor orientation in the former case, although bridge effects cannot be completely ruled out.

Most of the studies on porphyrin–quinone systems have involved bridging through the meso positions of the porphyrin ring. An interesting synthetic alternative (reminiscent of the use of bridging ligands in the construction of CLDA systems based on metal complexes (section 5.7)) is that recently proposed by Sanders and coworkers [136]. In such a strategy (as illustrated by **5.16**) acceptors bearing a ligand

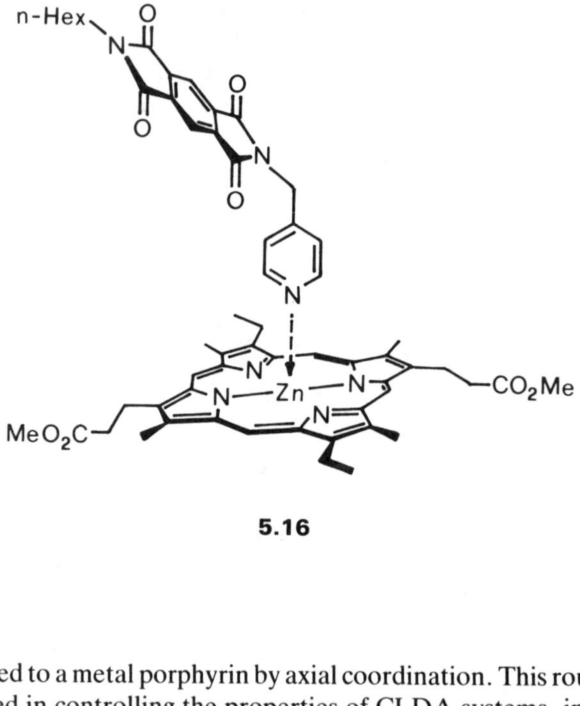

5.16

function are linked to a metal porphyrin by axial coordination. This route seems to be particularly suited in controlling the properties of CLDA systems, inasmuch as the structural parameters of the porphyrin, bridge, and acceptor can be changed independently.

5.6.6 Capped porphyrin–quinone systems

A large number of studies have been performed on porphyrin–quinone systems in which the quinone is bound to the porphyrin by two or more bridges, so that it occupies a relatively rigid position over the porphyrin plane [33]. Such systems are usually designated as "capped" porphyrin–quinone species. The system **5.17**, in which the cap quinone is bound to the porphyrin by four equivalent chains, has been

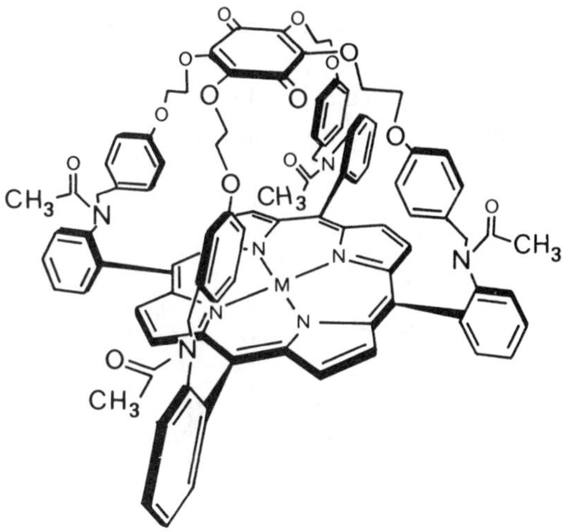

5.17

studied by Mauzerall and coworkers [137,138]. Despite its four-bridge cage structure, this CLDA system has still some conformational freedom, with two introverted and extroverted conformers connected by helical twisting of the bridges. These two conformers, in which the porphyrin–quinone distance changes from 6.5 to 8.5 Å, do not interconvert on the nanosecond time scale. Therefore, two distinct time constants for electron-transfer quenching of the porphyrin singlet in the introverted and extroverted conformers are observed ($\tau = 0.92$ ns and 6.6 ns, respectively, in dichloromethane). Analogous results are obtained for triplet electron-transfer quenching. The distance dependence is interpreted by the authors in terms of a through-space nonadiabatic electron-transfer mechanism [137]. The rate constants depend very little on temperature and on solvent dielectric properties. This last point suggests that the cage-type structure is sufficiently enclosed that the intercomponent electron-transfer reaction is largely independent of its solvent environment [138].

5.6.7 Porphyrin–viologen systems

Among the quenchers used in porphyrin–quencher CLDA systems [33], N,N'-dialkylbipyridinium dications (alkylviologens) have played, after the quinones, the second major role. A single covalently-bound methylviologen (for the structure of methylviologen, see **10.19**) quenches the excited singlet and triplet states of a porphyrin in polar solvents [139]. The quenching is modest, however, and definite proof of the quenching mechanism is difficult to obtain, although with a related porphyrin–benzylviologen system unequivocal evidence for the electron-transfer mechanism of the singlet quenching has been obtained from pulsed resonance Raman spectroscopy [140]. With 'capped" porphyrin–methylviologen CLDA systems, on the other hand, extensive singlet quenching is observed, with rapid charge-separation and charge-recombination steps [141]. With more pliable struc-tures, long-lived redox products have been detected but their origin remains unclear [142].

With two [142] and four viologens [143] bound to the same porphyrin, efficient singlet quenching is observed. The four-viologen system **5.18** has been studied in

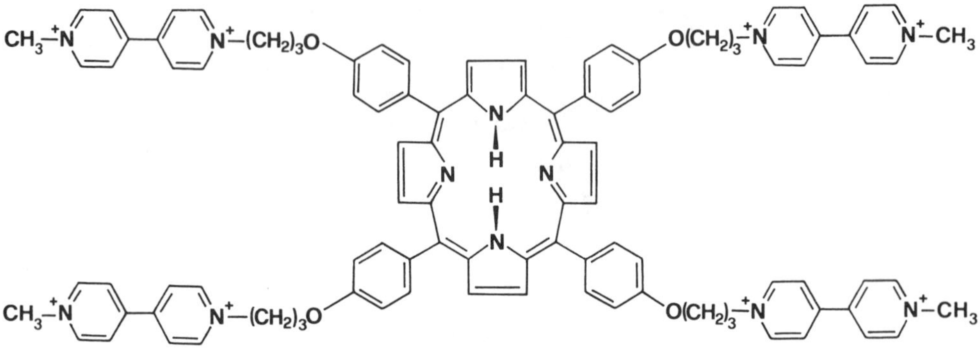

5.18

detail by Harriman, Mataga, and coworkers [143]. The porphyrin singlet decays via two routes: a fast (subnanosecond) process not accompanied by detectable redox products; a slower process (complete in a few nanoseconds and competing with intersystem crossing to the triplet) with formation of long-lived ($\tau = 6.4$ μs) redox products. The biphasic singlet quenching is presumably associated with two main conformations (or groups of conformations) of the porphyrin–viologen linkage: a first conformation corresponding to a relatively close donor–acceptor approach and giving very fast charge separation and recombination; another one corresponding to a greater donor–acceptor distance and giving rise to relatively slow charge separation and much slower charge recombination. Interestingly, the large difference in rate for charge separation and charge recombination in this second route cannot be ascribed, as in other CLDA systems, to driving force effects. Rather, it probably arises from conformational changes, since the redox products may adopt a more extended configuration (in order to minimize electrostatic repulsion) before charge recombi-nation takes place. Altogether, the results on this porphyrin tetraviologen array seem to require that electron transfer occurs, in this case, more through intervening solvent molecules rather than through the connecting bridge [143].

5.6.8 Porphyrin–host covalently-linked systems

An interesting strategy to build donor–acceptor sytems is that of covalently linking an appropriate "host" component to the donor, and to use a "guest" as acceptor. The specific features of host–guest interactions responsible for the donor–acceptor binding (Chapter 10) makes such systems interesting for several reasons. In princi-ple, acceptors can be changed more easily in such systems than in true CLDA systems. Moreover, the sensitivity of the host–guest interaction to host properties such as shape, size, and charge may bring about interesting consequences upon donor–acceptor electron transfer.

An example of such a type of system is the porphyrin–cyclodextrin–quinone system described in section 5.6.4. Another interesting example is given by the CLDA system studied by Harriman, Lehn and their coworkers [144]. This system consists of a zinc porphyrin capped by a biphenyl, with two [18]–N_2O_4 macrocyclic subunits acting as "bridges". The two macrocycles are convenient hosts for Ag(I) ions in methanol, giving rise to the stable polymetallic adduct **5.19**. In the Ag(I) complex, the fluorescence of the porphyrin chromophore is efficiently and rapidly ($\tau < 1$ ns)

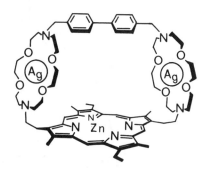

5.19

quenched by electron transfer to the silver ion. The photoproduced porphyrin cation is very long-lived ($\tau = 5$ μs). This seems to require some particular kind of rearrangement following electron transfer, such as displacement of the protoproduced silver atom from the macrocyclic ligand and related conformational changes [144]. In this respect, this system bears some analogy with the porphyrin–viologen case discussed in section 5.6.7, as a nuclear motion in response to photoinduced electron transfer is responsible for the slowing down of charge recombination with respect to charge separation.

A compound similar to **5.19** but in which the biphenyl cap has been substituted by another zinc porphyrin has also been studied [144]. In the absence of silver, extensive electronic coupling between the cofacial porphyrin rings is observed, while with coordinated Ag(I) rapid charge separation and slow charge recombination are again observed.

5.7 SYSTEMS BASED ON METAL COMPLEXES

As implied by their name, metal complexes are molecules made up of subunits (metal and ligands) that are capable, under appropriate conditions, of independent existence. The metal–ligand interactions are usually strong enough that metal complexes cannot be considered, in a strict sense, as supermolecules (section 3.2). In several cases, however, metal–ligand delocalization is small enough that reduction and oxidation processes can be considered to be localized on metal or ligands, and the excited states can be classified as metal-centred (MC), ligand-centred (LC), metal-to-ligand (MLCT) or ligand-to-metal (LMCT) charge transfer (section 2.4).

Let us consider as a prototype tris(2,2'-bipyridine)ruthenium(II), $Ru(bpy)_3^{2+}$, whose relevant properties [145] are summarized in Fig. 5.13. In the electrochemistry

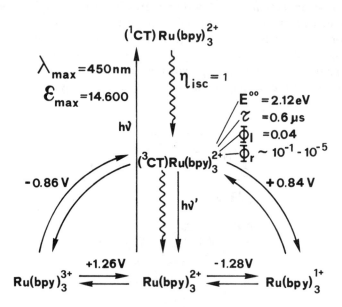

Fig. 5.13 — Schematic representation of the spectroscopic, photophysical, and redox properties of the $Ru(bpy)_3^{2+}$ chromophore [145].

of this complex, oxidation involves the metal while reduction involves the ligands. Furthermore, in the reduced form, the ligands are not collectively reduced, but the extra electron is vibronically trapped on a single ligand. Thus the one-electron oxidized form is $Ru^{III}(bpy)_3^{3+}$ and the one-electron reduced form is $Ru^{II}(bpy)_2(bpy^{\cdot -})^{2+}$. This leads to the observation of intervalence transfer (ligand–ligand optical electron-transfer) transitions in the near infrared [146], and to estimated activation energies for ligand-to-ligand electron hopping of the order of 1000 cm^{-1} in polar organic solvents [147]. The localized nature of the oxidation and reduction sites determines the properties of the long-lived excited state of these complexes, which is of the metal-to-ligand charge transfer (MLCT) type [145]. Thus, absorption and emission energies in complexes of this type can be easily predicted on the basis of the electrochemical potentials, and the radiationless decay of the MLCT state can be conveniently treated as a thermal ligand-to-metal electron-transfer process in the Marcus-inverted region [148]. Also, internal conversion between different MLCT states in a mixed-ligand complex can be considered as a simple ligand-to-ligand excited-state electron-transfer process.

It is clear from the Ru(II)-polypyridine example that metal complexes are rather peculiar molecular units which, together with many "molecular" properties, may also exhibit features (e.g., metal–ligand or ligand–ligand electron transfer) reminiscent of a "supramolecular" nature. This is due to their composite structure, in which metal and ligand subunits are relatively independent of each other with respect to electron-transfer processes and electronic excitation. Because of this composite structure, metal complexes offer many opportunities through which synthetic control and tuning of redox and excited-state properties can be achieved [149–151].

When metal complexes are used as components in a supramolecular structure, the composite structure of the "components" should be taken into consideration. In particular, the availability of various redox sites within each component is expected to enhance the variety of possible intercomponent electron-transfer processes. The situation may be further complicated when, as often happens, a chromophoric ligand is used to bridge the two metal centres. It is clear that, even for a simple system constituted by two metal complexes connected by a bridging ligand (a binuclear complex), a simple two-centre description (such as that used for two-component organic CLDA systems in sections 5.3–5.7) may be inadequate. Examples of such complex situations will be encountered in later paragraphs of this section.

5.7.1 Chromophore–quencher complexes

As porphyrins do in the field of organic CLDA systems, metal polypyridine chromophores play the leading role in inorganic CLDA systems. Most often they consist of a $4d^6$ or $5d^6$ metal such as Ru(II), Re(I) and Os(II), and one, two, or three bipyridine or phenanthroline ligands. As shown in Fig. 5.13 for the $Ru(bpy)_3^{2+}$, the common feature of such chromophores is the presence of a low-lying MLCT state of formally triplet character, easily detectable from emission. These excited states are usually good reductants and mild oxidants, so that these chromophores can be combined with various electron acceptors or donors as quenchers.

Meyer and coworkers have studied in some detail the behaviour of the Re(I) complexes $[(4,4'-X_2-bpy)Re^I(CO)_3(MQ^+)]^{2+}$ (X = NH$_2$, H, COOEt, schematic structure **5.20**) [152–154], together with that of some related Ru(II) [155] and

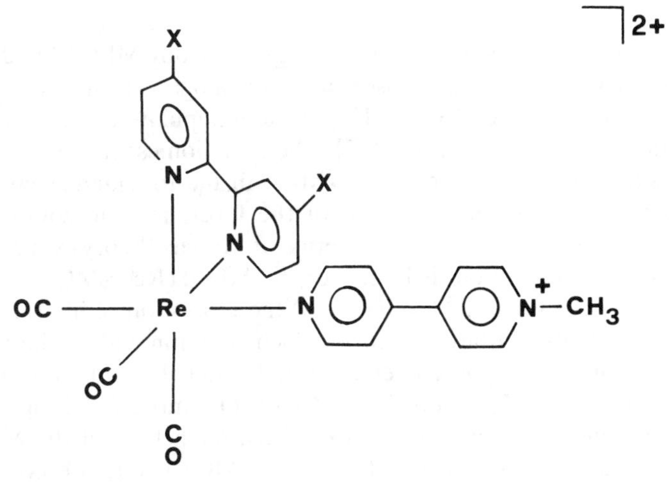

5.20

Os(II) complexes [154]. In these systems, two MLCT states are relevant, namely, MLCT(Re→bpy) and MLCT(Re→MQ$^+$). The relative energy ordering of the singlet states in the absorption spectra is MLCT(Re→bpy) < MLCT(Re→MQ$^+$), and the same ordering is expected to occur for the corresponding triplet states at Franck–Condon geometry. The energy ordering is, however, inverted after relaxation of the nuclear coordinates in fluid solution (Fig. 5.14). This is shown by the

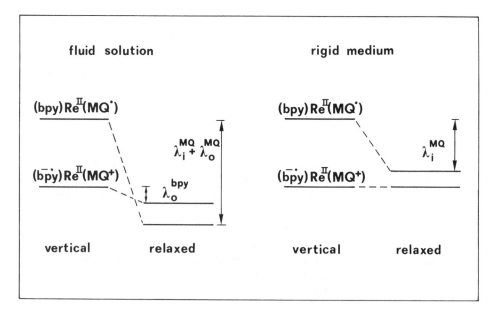

Fig. 5.14 — Schematic energy-level diagram accounting for the medium effects on the photophysics of **5.20** [152–154].

fact that, upon excitation in the low-energy Re→bpy MLCT band, the emission of the MLCT(Re→bpy) state (observable in analogous complexes lacking MQ$^+$) is quenched and the MLCT(Re→MQ$^+$) state is populated (as shown by the transient absorption of the MQ$^\cdot$ radical) [153]. The radiationless transition between the two MLCT states corresponds to a bpy→MQ$^+$ ligand-to-ligand electron transfer. Two factors contribute to the inversion of the levels in fluid solution [154]: (i) the relaxation of solvent dipoles (λ_0 in terms of Marcus theory) which, because of the greater charge separation, is larger for the MLCT(Re→MQ$^+$) state than for the MLCT(Re→bpy) one ($\lambda_0^{MQ} > \lambda_0^{bpy}$); (ii) the relaxation of internal nuclear coordinates (λ_i in terms of Marcus theory) which is dominated by the twisted-to-planar conversion of MQ$^+$ upon one-electron reduction. In contrast to the behaviour in fluid solution, when **5.20** with X = H, COOEt is embedded in rigid glasses (both at 77 K [153] and at room temperature [152]), quenching of the MLCT(Re→bpy) emission and formation of the MLCT(Re→MQ$^+$) state are no longer observed. This indicates that in a rigid medium, MLCT level inversion does not occur because, although the internal twisting motion can probably still occur [154], the solvent repolarization modes are essentially blocked (Fig. 5.14). The proof is that, when the vertical MLCT(Re→MQ$^+$)–MLCT(Re→bpy) energy difference is sufficiently small (for example when **5.20** with X = NH$_2$ is used), electron-transfer quenching of the MLCT(Re→bpy) state is efficient even in a rigid low-temperature matrix [153]. It should be noted that, from a "molecular" point of view, these systems might simply be seen as mixed-ligand Re(I) complexes and their behaviour could accordingly be analysed in conventional photophysical terms. On the other hand, the MQ$^+$ ligand is, as most alkylpyridinium ions, a good electron acceptor, so that these complexes can also be considered as CLDA systems made up of the (CO)$_3$Re(X$_2$–bpy)–$^+$ chromophore and the –MQ$^+$ quencher [152]. Strictly speaking, the supramolecular view is justified to the extent to which the redox orbital of MQ$^+$ is localized at the pyridinium moiety and does not extend over the ligand. The answer to this question is a complex one, as the MQ· radical is localized at the twisted (vertical) geometry, but becomes delocalized at the planar (relaxed) geometry. Altogether, the view of these complexes as chromophore-quencher CLDA systems seem to be more instructive. As an extreme case of chromophore-quencher complexes of this type, a Ru(II) tris-bipyridine complex carrying six peripheral methylpyridinium substituents, has been reported [156].

Analogous complexes containing an electron donor (instead of an acceptor) ligand as a quencher have been studied by Schanze, Netzel, and coworkers [157,158]. Upon excitation of [(bpy)Re^I(CO)$_3$DMABN]$^+$ (DMABN = p-(dimethylamino)benzonitrile) (**5.21**) in acetonitrile, the MLCT(Re→bpy) emission is completely quenched. As proven by transient absorption measurements, the process responsible for this result (Scheme 5.1) is a DMABN→Re electron-transfer

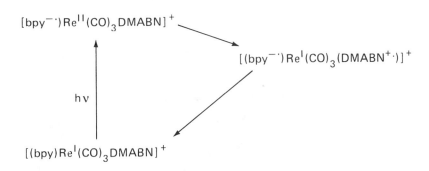

Scheme 5.1

process, and the state produced is a ligand-to-ligand charge-transfer (LLCT) excited state (direct optical detection of LLCT states has been reported in a number of cases [159–161]). When **5.21** is studied in CH$_2$Cl$_2$, no quenching of the MLCT(Re→bpy) emission is observed. This is attributed to the energetics of the system, that is

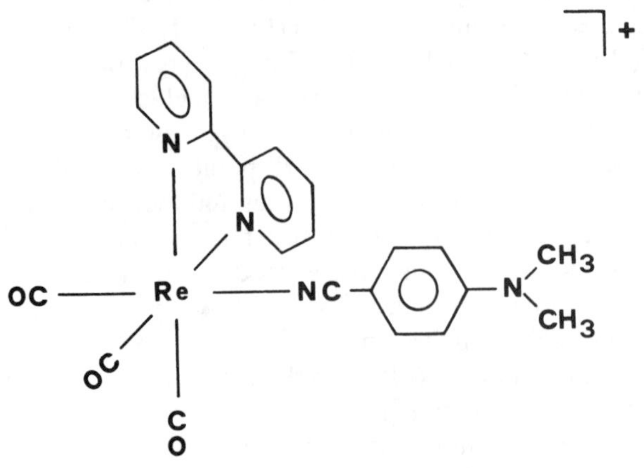

5.21

favourable (charge-separated LLCT state below the MLCT(Re → bpy) state) only in the polar solvent acetonitrile [158]. Similar results have been obtained for analogous Os(II) complexes containing various aminobenzonitriles as electron donors [157]. In this case, the decay of the LLCT state can be monitored experimentally. The change in decay rate with electron donor follows the behaviour predicted by the energy gap law for ligand–ligand charge recombination in the Marcus inverted region.

Related complexes containing a phenothiazine (PTZ) derivative as electron donor (**5.22**) have been studied by Meyer and coworkers [162,163]. Unlike the previously discussed **5.20** and **5.21** systems, these complexes can be classified

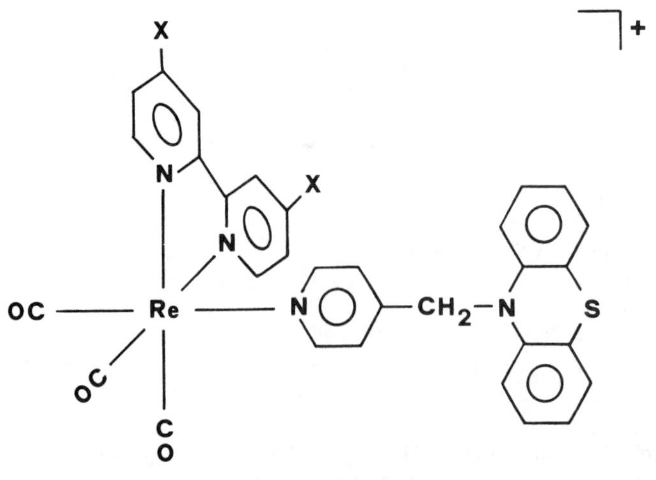

5.22

as two-component supermolecules without any ambiguity, since the quencher is attached by a methylene link to a pyridine ligand and does not belong to the coordination sphere of the metal. The behaviour of these systems in acetonitrile at room temperature [162,163] is similar to that of **5.21** and would be described by a scheme identical to Scheme 5.1 (except for the change from DMABN to py–CH$_2$–PTZ): quenching of the MLCT(Re$\rightarrow$bpy) emission by a rapid (τ<10 ns) intramolecular PTZ$\rightarrow$Re electron-transfer process. The quenching is inhibited in rigid environments, indicating that some nuclear motion (presumably folding of the pyridine–phenothiazine linkage) is required to reach a convenient conformation for electron transfer [162]. The rate constant for the bpy$\rightarrow$PTZ charge recombination, as obtained from transient absorption measurements, changes over a series of **5.22** complexes with different X substituents as predicted by the energy gap law for radiationless transitions (and for electron transfer in the Marcus inverted region with nuclear tunnelling, section 5.2). Interestingly, the slope of the log k vs ΔG^0 plots is larger for the bpy$\rightarrow$PTZ charge recombination than for the nonradiative decay of the MLCT(Re$\rightarrow$bpy) state in analogous complexes lacking the PTZ quencher fragment [163]. This is taken as an indication of the fact that (contrary to what happens for the charge-separated state) in the MLCT(Re$\rightarrow$bpy) state charge separation is incomplete and increases with excited-state energy (due to decreasing metal–ligand orbital mixing).

Systems containing a Ru(II) tris-polypyridine fragment as chromophore and a covalently-linked 2,2'-bipyridinium group as (electron-acceptor) quencher have been studied by Elliott, Kelley, and coworkers [164]. These CLDA systems have the general structure **5.23**. The series spans values of $n = 2,3,4$, and also includes

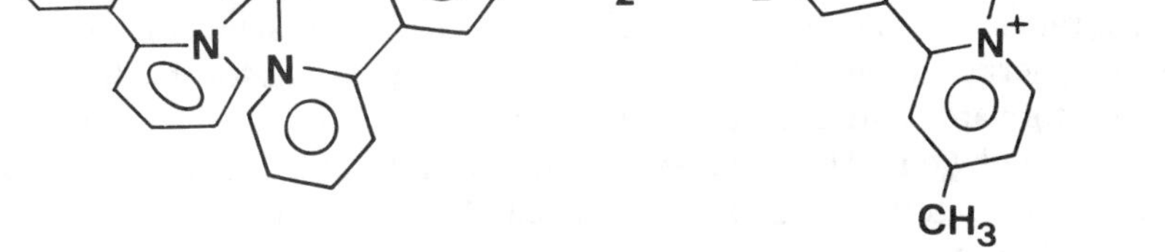

5.23

other complexes in which 4,4'-dimethyl-2,2'-bipyridine or the 4,4',5,5'-tetramethyl-2,2'-bipyridine are used as remote ligands instead of 2,2'-bipyridine. In the mixed-ligand systems, the lowest MLCT excited state is localized on the less methylated ligand. Picosecond emission and absorption measurements in acetonitrile show that electron transfer from the photoexcited Ru(II) polypyridine chromophore to the 2,2'-bipyridinium quencher occurs in these systems. The forward electron-transfer process occurs in the 80–1700 ps time scale, whereas the charge recombination appears to be always very fast ($\tau \leqslant 30$ ps). The mechanism is consistent with Scheme 5.2, in which it is assumed [164] that (i) electron transfer to the quencher only occurs

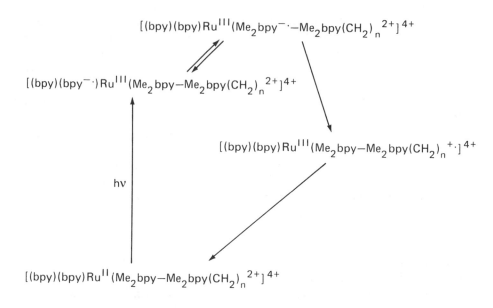

Scheme 5.2

from the adjacent (directly linked) polypyridine ligand, and (ii) interconversion between MLCT states localized on different ligands (ligand–ligand electron hopping) is fast and electron transfer to the quencher is rate-determining. The forward electron-transfer rates correlate with the driving force (tunable in the 0.2–0.6 eV range by changing n in **5.23**) as expected for reactions in the normal Marcus free-energy region with $\lambda = 0.8$ eV [164]. The reasons for the difference in rates of the forward- and back-electron transfer reactions (not discussed by the authors) are not obvious. The greater rate of the back reactions could in principle be attributed to the larger driving force (in the 2.0–1.6 eV range) of these processes, although a value of λ consistently larger than that assumed for the forward step would be required in order to keep the back reactions out of the inverted region.

An interesting possibility in the construction of chromophore–quencher complexes is to use peptide oligomers as spacers. Such a strategy has been extensively used in studies of thermal intramolecular electron transfer between aminoacids [165,166] and between metal centres [167–169]. Among oligopeptides, oligo(L-proline) is the best choice because the cyclic structure of its side chain restricts rotation and slows down ($\tau \approx 1$ s) interconversion between conformational isomers [168,169]. A series of CLDA systems of type **5.24**, in which a ruthenium(II)

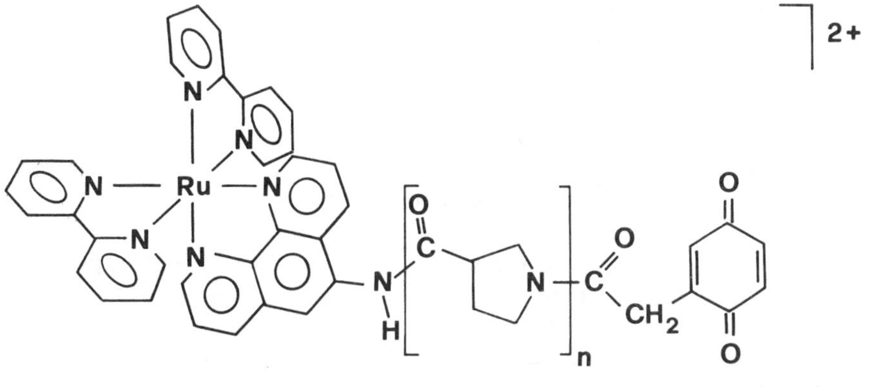

5.24

polypyridine chromophore is linked to a quinone *via* poly(L-proline) chains of various length ($n = 0,1,2,3,4$) has been investigated by Schanze and Sauer [170]. In all the complexes, substantial quenching of the Ru(II)polypyridine emission is observed in CH_2Cl_2, indicating the occurrence of electron transfer to the quinone. The emission lifetimes are, however, invariably multiexponential, suggesting the participation of several conformational isomers to the electron-transfer process. Interestingly, the electron-transfer rate constant obtained from the lifetime component with the largest amplitude (corresponding to the predominating conformer in solution) decreases markedly with an increasing number of proline residues. Based on estimates of the chromophore–quencher distance along the series [169], the rate constant data fit to the exponential dependence (5.17) with a lower limit value for β of about 0.75 Å^{-1} [170], in good agreement with typical values for rigid saturated bridges (sections 5.3, 5.4, and 5.6.2).

All the chromophore-quencher complexes discussed in the previous paragraphs contain d^6 metal polypyridine fragments as chromophores. A less conventional series of chromophore–quencher complexes is that reported by Gray and coworkers [171]. In such systems, the chromophore is a pyrazolyl-bridged Ir$_2$ dimer. This chromophore has long-lived singlet and triplet excited states of d–p^* character and behaves as a strong excited-state reductant [172]. In the CLDA system, this chromophore is covalently linked through a phosphinite ligand to various *N*-alkylpyridinium groups as electron-acceptor quenchers. Time-resolved emission and absorption measurements have been used to obtain rate constants for singlet and triplet quenching (charge separation) and for charge recombination. These three types of electron-transfer reactions, throughout the series of quenchers investigated, span a range of driving force of almost 2 eV. When plotted against the driving force, the rate constants give a clear demonstration of the parabolic dependence predicted by Marcus theory [171].

Besides CLDA systems containing an inorganic chromophore and an organic quencher, there are also cases of the reverse situation, where an organic chromophore is covalently linked to a metal complex as an electron-transfer quencher. Examples are (i) *meso*-tritolyl[*N*-(pentammineruthenium)pyridyl]-porphyrin [173],

(ii) CLDA systems containing an aromatic hydrocarbon and a Co(III) complex such as anthracene-cobalt(III)sepulchrate [174,175] (section 11.1.2) and naphthalene-hexamminecobalt(III) [176], and (iii) rhodamine complexes with Rh(I) and Ir(I) metal-containing moieties [177].

5.7.2 Binuclear metal complexes
CLDA systems in which two metal complex fragments are connected by a bridging ligand are usually designed as *binuclear* metal complexes. Extensive work on optical electron transfer (intervalence transfer) spectra of binuclear complexes has been performed [178]. This spectroscopic work is not explicitly dealt with in this section, although the analysis of intervalence transfer spectra can be useful in giving information on parameters relevant to intercomponent electron-transfer kinetics (section 3.4.1), and can be used to probe the electronic coupling ability and distance-dependent behaviour of various types of bridges [125,179–181]. The photophysical and photochemical behaviour of binuclear complexes is characterized by the wide-spread occurrence of energy- and electron-transfer intercomponent processes. Energy-transfer processes are dealt with in section 6.3. In this section are presented selected examples of photophysical mechanisms involving electron transfer in binuclear metal complexes. A more complete coverage of this field can be found in a recent review article [37].

The work on the $(NH_3)_5Ru(pz)Ru(EDTA)^+$ (**5.25**) complex (pz = pyrazine, EDTA = ethylenediaminetetraacetate) by Creutz *et al.* [182] represents one of the pioneering studies in the area. The complex **5.25** is a mixed-valence species with electronic structure $(NH_3)_5Ru^{II}(pz)Ru^{III}(EDTA)^+$. The main spectral features of this complex are a Ru(II)→pz metal-to-(bridging ligand) charge-transfer, M(BL)CT, band in the visible and a Ru(II)→Ru(III) IT band in the near infrared. Picosecond laser photolysis of the corresponding fully reduced Ru(II)–Ru(II) species or of the mononuclear $Ru(NH_3)_5(pz)^{2+}$ model compound gives rise to a transient bleaching of the ground-state absorption with lifetimes of the order of 0.1 ns, characteristic of the M(BL)CT state. With the Ru(II)–Ru(III) mixed valence species, on the other hand, a much less pronounced bleaching is observed (about 1/10 of that of the previously mentioned complexes). This small bleaching decays with a lifetime of 0.08 ns. This result is interpreted in terms of the pathway shown in Scheme 5.3. In this scheme, a ligand-to-metal electron-transfer process causes prompt

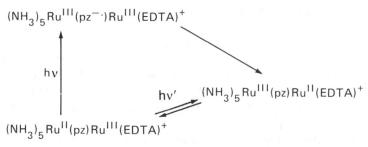

Scheme 5.3

quenching of the M(BL)CT state and population of the lowest IT excited state.

The IT state (that is considered to be responsible for the small bleaching observed) then decays back to the ground state by a metal-to-metal electron-transfer process. The observed decay rate is in satisfactory agreement with the values calculated from redox potentials and spectroscopic IT band parameters on the basis of the Hush model (section 3.4.1), provided that a frequency factor appropriate for solvent reorganizational modes is used [182]. A puzzling result, that does not easily fit into the mechanism of Scheme 5.3, is the lack of any transient bleaching following direct excitation into the IT absorption band using near infrared laser pulses. A possible rationale is offered by the authors by assuming that, upon direct population by light absorption, the IT state may decay to the ground state prior to solvent relaxation [182]. A system bearing some similarity in mechanism to the above contains Cu(II) ions covalently bound to $Ru(NH_3)_5pz^{2+}$ through the pz bridging ligand [183].

The binuclear system $(dpte)_2ClRu(bpe)Ru(bpy)_2Cl^{2+/3+}$ **5.26** has been studied, together with analogous systems involving 4,4'-bipyridine and dipyridylethane as

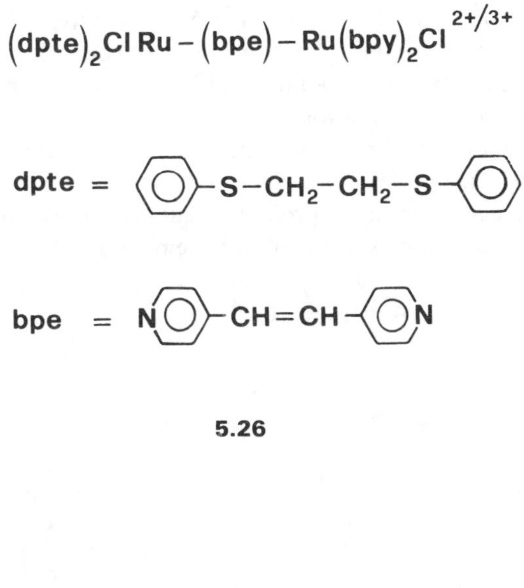

$$\left(dpte\right)_2 Cl\, Ru - \left(bpe\right) - Ru\left(bpy\right)_2 Cl^{\,2+/3+}$$

dpte = S−CH₂−CH₂−S

bpe = N CH=CH N

5.26

bridging ligands, by Curtis *et al.* [184]. The 2 + binuclear complex can be adequately represented as $(dpte)_2ClRu^{II}(bpe)Ru^{II}(bpy)_2Cl^{2+}$. The visible spectrum is dominated by Ru→bpy MLCT absorption bands. The complex exhibits an emission closely matching that of the mononuclear $Ru(bpy)_2Cl(bpe)^+$ analogue, but an appreciable reduction in lifetime is observed. The explanation proposed [184] is that excited states based on the bridging ligand (e.g., a Ru→bpe M(BL)CT state or a bpe-centred π-π* state) can be slightly lower in energy than the Ru→bpy MLCT state and can provide a reasonably efficient intramolecular quenching pathway. This last case is schematically depicted in Scheme 5.4. The key process converting the

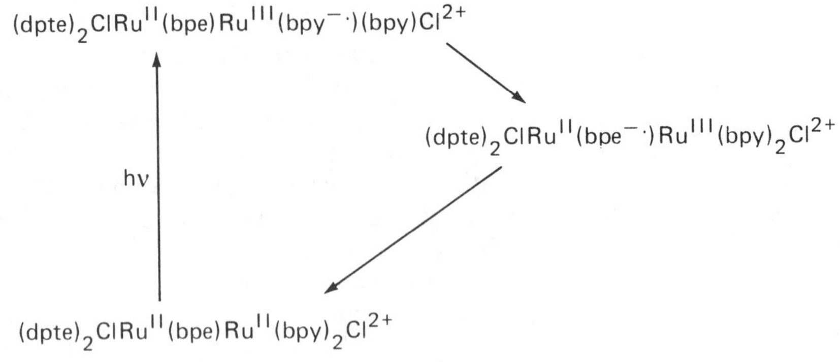

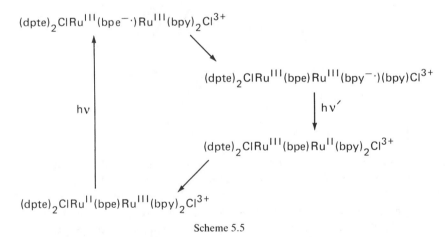

Scheme 5.4

MLCT state to the M(BL)CT one is an intramolecular ligand-to-ligand electron-transfer process similar to those described in section 5.7.1. The occurrence of a process of this kind implies that some electronic coupling is present between bpy and bpe. That ligand–ligand coupling can be sizeable in metal polypyridine-type complexes has been demonstrated by the observation of optical ligand-to-ligand charge-transfer transitions [158–160].

Given the redox properties of the two metal centres, the **5.26** 3+ ion is a mixed-valence complex of the type $(dpte)_2ClRu^{II}(bpe)Ru^{III}(bpy)_2Cl^{3+}$. Visible excitation of the complex corresponds mainly to $Ru(II) \rightarrow bpe$ M(BL)CT transitions. The interesting observation is that an emission characteristic of the $-Ru^{II}(bpy)_2-$ chromophore is present, although such a chromophore does not exist in the Ru(II)–Ru(III) complex. The suggested conclusion [184] is that a $bpe \rightarrow bpy$ ligand-to-ligand electron-transfer process occurs after excitation leading to the emitting state. This pathway is depicted in Scheme 5.5.

Scheme 5.5

There are a number of interesting points concerning this scheme. First, with respect to the ground state, the emitting state can be viewed as a "remote" MLCT state involving metal and ligands that are not directly bound to each other. Furthermore, the state reached following the emission process is *not* the ground state but rather an intervalence transfer (IT) state that must decay to the ground state by a metal-to-metal electron-transfer process. This process cannot be detected in laser photolysis, presumably because of its occurrence in the subnanosecond time scale. On the other hand, the IT absorption is clearly seen in the near-infrared spectra of the complex [184].

The series of complexes **5.27** has been studied by Meyer and coworkers [185].

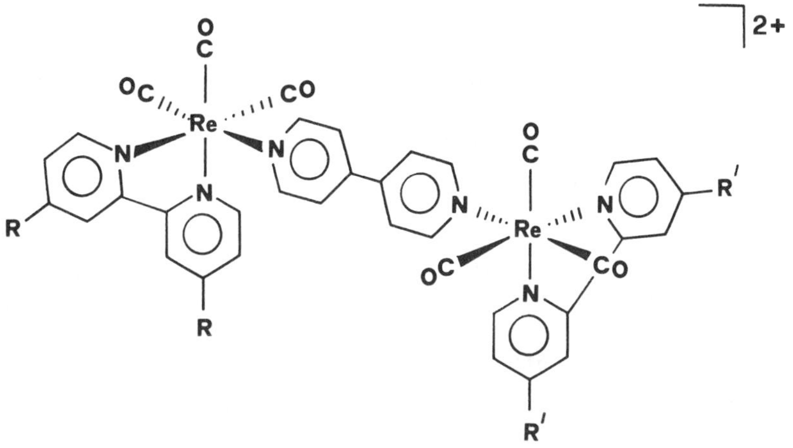

5.27

These complexes, in which both metal centres are in the +1 oxidation state, can be symmetric or asymmetric depending on the nature of the substituents on the 4,4' positions at the bpy ligands. Relevant excited states in these systems are Re→ (R_2–bpy) MLCT, Re→ (R'_2–bpy) MLCT, and Re→ (4,4'–bpy) M(BL)CT excited states. In the symmetric (R=R') complexes the relative energy ordering of the MLCT and M(BL)CT states depends on the substituents, being MLCT>M(BL)CT for R=R'=NH$_2$, MLCT≈M(BL)CT for R=R'=H, and M(BL)CT>MLCT for R=R'=COOC$_2$H$_5$, as determined by laser photolysis transient absorption measurements [185]. In the asymmetric complex in which R=H and R'=COOC$_2$H$_5$ (R'_2–bpy = dec-bpy), the Re→ (bpy) MLCT state is higher in energy than the Re→ (dec-bpy) MLCT state. Excitation of the complex in the absorption region of either chromophore always gives rise to an emission and a transient absorption signal that can be assigned to the Re→ (dec-bpy) MLCT state. Two pathways could be responsible for the quenching of the upper MLCT state and the sensitization of the lowest one, namely, (i) a direct energy-transfer process or (ii) the more complex "cascade" mechanism shown in Scheme 5.6. In the cascade mechanism, the first

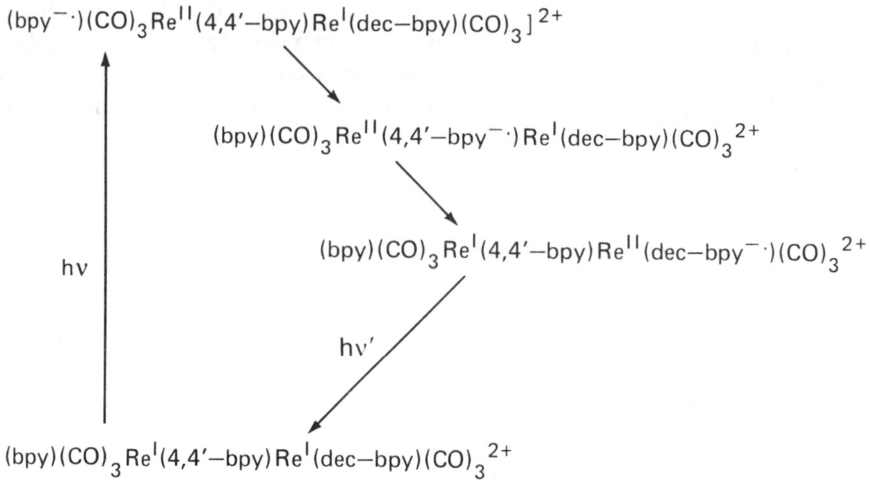

$(bpy^{-\cdot})(CO)_3Re^{II}(4,4'-bpy)Re^{I}(dec-bpy)(CO)_3]^{2+}$

$(bpy)(CO)_3Re^{II}(4,4'-bpy^{-\cdot})Re^{I}(dec-bpy)(CO)_3^{2+}$

$(bpy)(CO)_3Re^{I}(4,4'-bpy)Re^{II}(dec-bpy^{-\cdot})(CO)_3^{2+}$

$h\nu$

$h\nu'$

$(bpy)(CO)_3Re^{I}(4,4'-bpy)Re^{I}(dec-bpy)(CO)_3^{2+}$

Scheme 5.6

is a ligand-to-ligand electron transfer generating a M(BL)CT state, which then converts to the final MLCT state *via* energy transfer. The "cascade" mechanism is preferred [185] on the basis of the observation that intercomponent quenching and sensitization are completely blocked when $3,3'-(CH_3)_2-4,4'$-bpy is substituted for the 4,4'-bpy bridging ligand. The new bridge is forced to be noncoplanar by the methyl substituents, being thus a poorer electron acceptor than 4,4'-bpy. Of course, the possibility that the mechanism is direct energy transfer and that the difference between the bridging ligands lies in their ability to provide electronic coupling cannot be definitely ruled out [185].

The binuclear complexes $(CN)(bpy)_2Ru-CN-Ru(bpy)_2(CN)^{+/2+}$ **(5.28)** have been studied, together with analogous trinuclear species [186]. They contain two

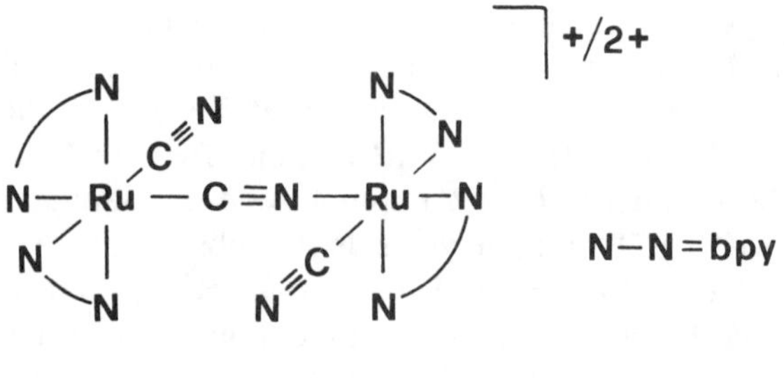

$$+/2+$$

N—N = bpy

5.28

–Ru(bpy)$_2$– chromophoric units, with cyanides as both bridging and terminal ligands. The fully reduced **5.28** species is uninteresting from the point of view of electron transfer, although the energetic asymmetry induced by the asymmetric nature of the bridging cyanide makes such a system suited to the study of intercomponent energy transfer (section 6.3.4). The 2+ **5.28** species has the $(NC)Ru^{II}(bpy)_2-CN-Ru^{III}(bpy)_2(CN)^{2+}$ mixed-valence configuration, in which the metal centre containing *N*-bonded bridging cyanide is oxidized [186]. Besides the MLCT bands characteristic of the $-Ru^{II}(bpy)_2-$ units, these complexes exhibit $Ru(II) \rightarrow Ru(III)$ IT bands in the near infrared. The intensity of these bands points toward a relatively large degree of electronic coupling between the metals (borderline between Class II and III mixed-valence behaviour, section 3.2). The binuclear and trinuclear mixed-valence complexes do not emit upon MLCT excitation. This indicates that intercomponent electron-transfer quenching of the MLCT states takes place, as shown in Scheme 5.7. The quenching step is a ligand-to-(remote metal) electron-transfer process.

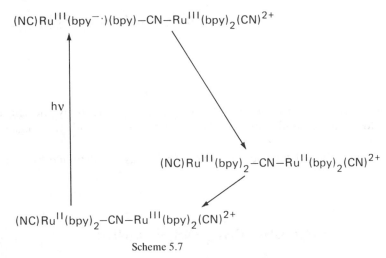

$(NC)Ru^{III}(bpy^{-\cdot})(bpy)-CN-Ru^{III}(bpy)_2(CN)^{2+}$

hν

$(NC)Ru^{III}(bpy)_2-CN-Ru^{II}(bpy)_2(CN)^{2+}$

$(NC)Ru^{II}(bpy)_2-CN-Ru^{III}(bpy)_2(CN)^{2+}$

Scheme 5.7

The binuclear complex $(bpy)_2(CO)Os(4,4'-bpy)Os(phen)(dppe)Cl^{4+}$ (**5.29**) has been studied by Schanze *et al.* [187], together with an analogous complex with dipyridylethane as bridging ligand and their reduced forms. The binuclear complex **5.29** has the mixed-valence electronic structure $(bpy)_2(CO)Os^{II}(4,4'-bpy)Os^{III}(phen)(dppe)Cl^{4+}$. The visible spectrum is dominated by $Os \rightarrow bpy$ MLCT bands. In the near infrared, an $Os(II) \rightarrow Os(III)$ IT band is observed for this complex, but not for an analogous one with the bpa (1,2-bis(4-pyridyl)ethane) bridging ligand, presumably because of the poorer metal–metal electronic coupling provided by the latter bridge. The interesting photophysical result [187] is that, upon $Os \rightarrow bpy$ MLCT excitation, emission is observed with the same energy and lifetime as in the model $Os(bpy)_2(CO)(4,4'-bpy)^{2+}$ mononuclear complex. This indicates that the $Os \rightarrow bpy$ MLCT state is not quenched in the binuclear complex, despite the presence of the adjacent Os(III) site. (A similar lack of quenching has been observed in a related Os(II)–Ru(III) complex with 4,4'–bpy as bridging ligand [188].) Thus, a quenching pathway involving a ligand-to-(remote metal) electron transfer, such as that followed by **5.28** (Scheme 5.7), is not operative in this complex. Since the differences in energetics between the two complexes are relatively minor, it appears that the difference in behaviour between **5.28** and **5.29** is

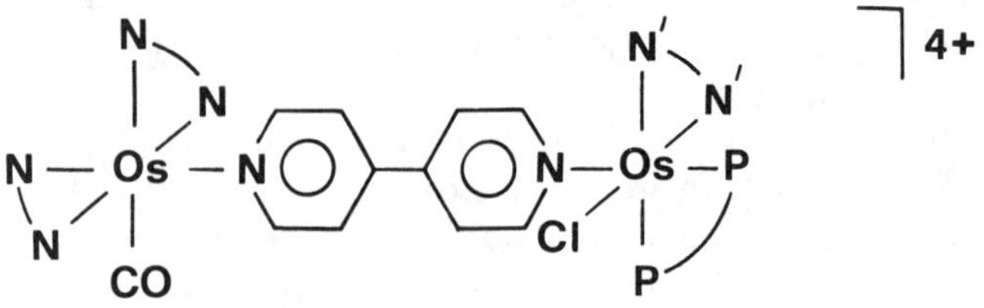

$$\left[\begin{array}{c} \text{N—Os—N} \bigcirc \text{—} \bigcirc \text{N—Os—P} \end{array} \right]^{4+}$$

$$N-N = bpy \qquad N'-N' = phen$$

$$P - P \ = \ Ph_2 P - CH = CH - PPh_2$$

5.29

mainly due to different electronic coupling provided by the shorter and more delocalizing cyanide bridge relative to the longer and more insulating 4,4'-bpy bridge.

The series of binuclear complexes **5.30** has been studied by Norton and Hurst

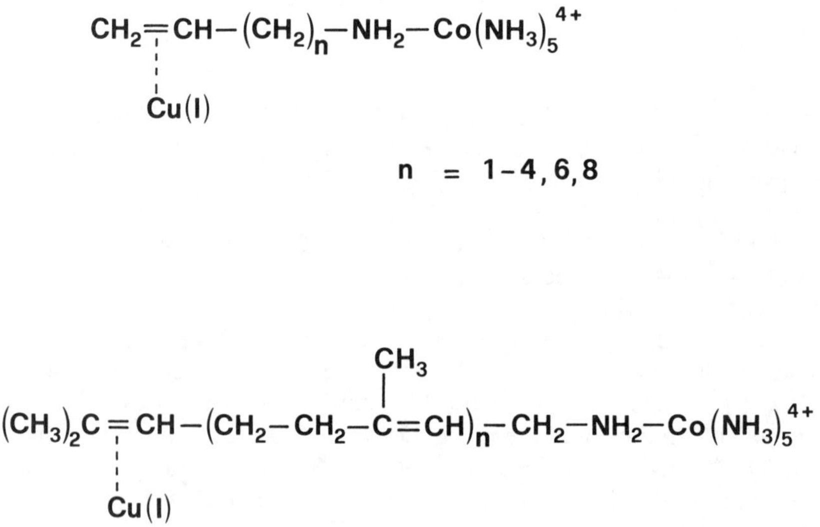

$$CH_2\text{==}CH-(CH_2)_n-NH_2-Co(NH_3)_5^{4+}$$
$$Cu(I)$$

$$n \ = \ 1-4,6,8$$

$$(CH_3)_2C = CH-(CH_2-CH_2-\overset{\overset{\displaystyle CH_3}{|}}{C}=CH)_n-CH_2-NH_2-Co(NH_3)_5^{4+}$$
$$Cu(I)$$

$$n \ = \ 1,2$$

5.30

[189], as an extension of previous work on similar systems. In these complexes, Cu(I) is bound in a π-fashion to the olefinic group on one end of the bridge, while the $Co(NH_3)_5^{3+}$ fragment is bound to an amino or a pyridine group at the other end. Saturated polymethylene or partially unsaturated polyisoprene chains of various lengths provide the chemical link between the two ends of the bridge. These complexes are formed *in situ* by reaction of cuprous ions with the pentammine-cobalt(III) complex of the bridging ligand. In these systems, thermal electron transfer from Cu(I) to Co(III), though thermodynamically allowed, is completely negligible for kinetic reasons. The binuclear complexes display an intense band in the near ultraviolet region, assigned to Cu$\rightarrow$olefin M(BL)CT transitions (5.22).

$$Cu^I[\pi\text{-}(CH_2{=}CH{-}(CH_2)_n{-}NH_2)]Co^{III}(NH_3)_5^{4+} \xrightarrow{h\nu}$$
$$Cu^{II}[\pi\text{-}(^-{\cdot}CH_2{=}CH{-}(CH_2)_n{-}NH_2)]Co^{III}(NH_3)_5^{4+} \qquad (5.22)$$

Irradiation of the complexes into this absorption band gives rise to a *photochemical* reaction, namely, redox decomposition reaction producing aqueous Co^{2+} [189]. The primary and rate-determining step is thought to be electron transfer from the reduced ethylenic group of the excited state to Co(III) (5.23), which then undergoes, as usual for Co(II) ammine complexes, prompt ligand loss. The interesting feature

$$Cu^{II}[\pi\text{-}(^-{\cdot}CH_2{=}CH{-}(CH_2)_n{-}NH_2)]Co^{III}(NH_3)_5^{4+} \rightarrow$$
$$Cu^{II}[\pi\text{-}(CH_2{=}CH{-}(CH_2)_n{-}NH_2)]Co^{II}(NH_3)_5^{4+} \qquad (5.23)$$

of these systems is that the two groups involved in the electron-transfer step are separated by more or less extended hydrocarbon chains, so that the effect of the nature and length of the electron-transfer rates can be investigated. The main experimental observations are [189]: (i) for the complexes **5.30** with polymethylene chains, the quantum yields decrease regularly with increasing number of methylene units in the chain, being relatively high (0.65) for $n = 1$ and becoming immeasurably small for $n \geq 6$; (ii) for the complexes **5.30** with polyisoprene chains, the quantum yield is close to unity and independent on the number of isoprenic groups in the chain; (iii) when the quantum yields are sufficiently high to permit laser photolysis measurements, the rate constants of the electron-transfer step are greater than 10^8 s^{-1}. In the interpretation of result (i), the authors assume that these flexible molecules adopt a fully extended equilibrium conformation under the experimental conditions used (this point is supported by molecular mechanics calculations, and by comparison of quantum yields with those of a related rigid systems of known centre-to-centre distance) [189]. Thus, the increase in the number of methylene groups translates into an increase in centre-to-centre distance. Despite the lack of direct kinetic data, the photochemical results of Norton and Hurst represent an early experimental example of a distance-dependence of electron-transfer in binuclear complexes. The strong accelerating effect of the insertion of double bonds in the methylene chain (point (ii)) is interpreted in terms of delocalized interactions between the remote redox centres proceeding via π–π interactions along the chain [189]. The fact that excited-state deactivation competes with the fast (point (iii)) intramolecular electron-transfer processes indicates that the original Cu$\rightarrow$olefin M(BL)CT state is a rather short-lived one. This led the authors to suggest a singlet state as the reactive state in these systems. Photoredox decomposition following intercomponent photoinduced electron transfer has also been observed in other binuclear complexes [190].

Photoredox reactions related to intercomponent electron tranfer can also occur in a binuclear complex by a different mechanism, namely, following direct excitation of an *optical* electron transfer transition. Examples of such a mechanism are the binuclear complexes $(CN)_5M_a-CN-M_b(NH_3)_5^-$ (**5.31**), where M_a = Fe, Ru, Os and M_b = Cr, Co, Os, studied by Vogler *et al.* [191,192,193]. In these complexes, M_a is in the 2+ and M_b in the 3+ oxidation state. All the complexes **5.31** exhibit, in addition to absorption features characteristic of the component subunits, an intense $M_a \rightarrow M_b$ IT transition (5.24) in the visible. The results obtained by irradiating the complexes

$$(CN)_5M_a^{II}-CN-M_b^{III}(NH_3)_5^- \xrightarrow{h\nu} (CN)_5M_a^{III}-CN-M_b^{II}(NH_3)_5^- \qquad (5.24)$$

in the wavelength range of this band differ sharply depending on the nature of M_b. When M_b = Os [193], the $-CN-Os^{II}(NH_3)_5$ unit present in the IT excited state is relatively inert, so that excitation is followed by efficient back electron-transfer reaction. When M_b = Cr [193] or Co [192], the $-CN-M_b^{II}(NH_3)_5$ unit present in the IT excited state is very labile, so that an efficient redox photodecomposition reaction is observed (5.25, 5.26).

$$(CN)_5Ru^{II}-CN-Co^{III}(NH_3)_5^- \xrightarrow{h\nu} (CN)_5Ru^{III}-CN-Co^{II}(NH_3)_5^- \qquad (5.25)$$

$$(CN)_5Ru^{III}-CN-Co^{II}(NH_3)_5^- \rightarrow Ru(CN)_6^{3-} + Co^{2+} + 5NH_3 \qquad (5.26)$$

In this respect, these complexes resemble the systems **5.30**. The main difference is, of course, that here the metal-to-metal electron-transfer state is reached directly by an optical process (IT excitation), whereas in the cases above described this was accomplished by a sequence of M(BL)CT excitation and intramolecular thermal electron transfer. Other examples of intervalence transfer photochemistry in cyano-bridged binuclear complexes have been reported [193,194].

5.7.3 Polynuclear metal complexes

Complexes in which three or more metal centres are connected by bridging ligands present some peculiar mechanistic aspects. One such aspect is the possibility of having sequences of electron-transfer steps leading to charge separation over non-adjacent (remote) components. This aspect is dealt with in section 5.7.4. A further aspect concerns the extent of electronic coupling between remote components and the possibility of one-step electron-transfer processes between them. In this section, some examples bearing on this point are discussed.

The series of bi- and trinuclear complexes $X(NH_3)_4Ru-NC-Ru(bpy)_2(CN)^{n+}$ and $X(NH_3)_4Ru-NC-Ru(bpy)_2-CN-Ru(NH_3)_4Y^{m+}$ (X = NH₃, py; Y = NH₃, py; m = 4–6; n = 2,3) has been studied [195,196]. The possible combinations of X and Y ligands and of oxidation states give rise to a large number of complexes within this series. The various overall oxidation states reflect the individual oxidation states of the $Ru(NH_3)_4X$ and/or $Ru(NH_3)_4Y$ subunits in the complex. For example, the three oxidation states of the trinuclear complex $py(NH_3)_4Ru-NC-Ru(bpy)_2-CN-Ru(NH_3)_5^{4+,5+,6+}$ (**5.32**) correspond to the following electronic structures:

$$py(NH_3)_4Ru^{II}-NC-Ru^{II}(bpy)_2-CN-Ru^{II}(NH_3)_5^{4+} \qquad (5.27)$$

$$py(NH_3)_4Ru^{II}-NC-Ru^{II}(bpy)_2-CN-Ru^{III}(NH_3)_5^{5+} \qquad (5.28)$$

$$py(NH_3)_4Ru^{III}-NC-Ru^{II}(bpy)_2-CN-Ru^{III}(NH_3)_5^{6+} \qquad (5.29)$$

These complexes have a remarkable variety of charge-transfer states of photochemical interest. Taking the **5.32** 5 + species as an example, the following excited states are relevant:

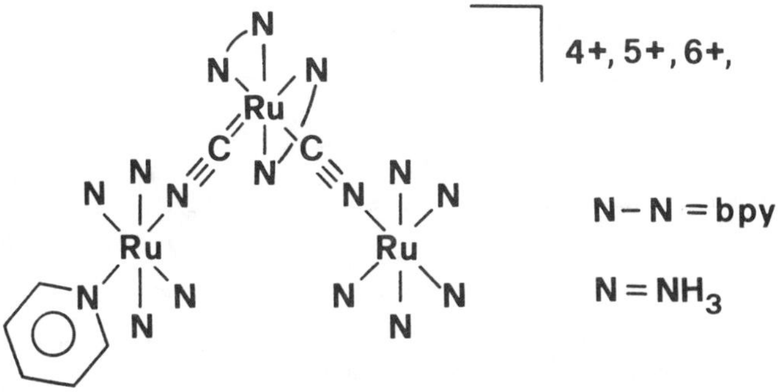

5.32

(1) Ru→ bpy MLCT:
$$py(NH_3)_4Ru^{II}-NC-Ru^{III}(bpy^-)(bpy)-CN-Ru^{III}(NH_3)_5^{5+} \qquad (5.30)$$

(2) Ru→ py MLCT:
$$(py^-)(NH_3)_4Ru^{III}-NC-Ru^{II}(bpy)_2-CN-Ru^{III}(NH_3)_5^{5+} \qquad (5.31)$$

(3) Ru→ bpy remote MLCT:
$$py(NH_3)_4Ru^{III}-NC-Ru^{II}(bpy^-)(bpy)-CN-Ru^{III}(NH_3)_5^{5+} \qquad (5.32)$$

(4) Ru→ Ru IT:
$$py(NH_3)_4Ru^{II}-NC-Ru^{III}(bpy)_2-CN-Ru^{II}(NH_3)_5^{5+} \qquad (5.33)$$

(5) Ru→ Ru remote IT:
$$py(NH_3)_4Ru^{III}-NC-Ru^{II}(bpy)_2-CN-Ru^{II}(NH_3)_5^{5+} \qquad (5.34)$$

Transitions corresponding to the various types of excited states can be easily identified in the spectra of this class of complexes. As an example, the resolution of the absorption spectrum of the **5.32** 5 + complex into various types of transitions is shown in Fig. 5.15. By selective oxidation or reduction of the various sites in the

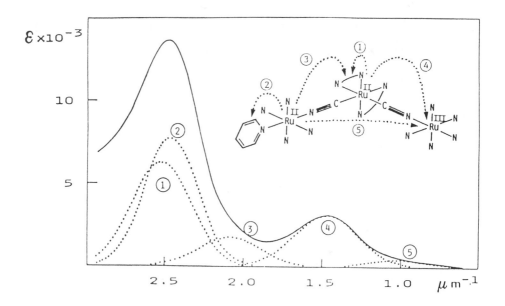

Fig. 5.15 — Resolution of the visible absorption spectrum of py$(NH_3)_4$–NC–Ru$(bpy)_2$–CN–Ru$(NH_3)_5^{5+}$ into charge-transfer transitions connecting the various redox sites of the trinuclear complex [196].

molecule, the attribution of the various types of transitions is straightforward [195,196]. Of particular interest from the spectroscopic point of view is the direct observation of *remote* MLCT and of *remote* IT, indicating that, within the limits of an essentially localized description (sections 2.2 and 2.4), sizeable electronic coupling between the various sites is present in these systems. For the remote IT transition, the intensity appears to fit a superexchange model (section 5.2) for through-bond interaction between the terminal metal centres. The $-NC-Ru(bpy)_2-CN-$ fragment is considered as the connector and the parameters required for the calculation (5.15) are taken from the band parameters of the transitions leading to the electron- (5.32) and hole- (5.33) transfer states. In this system the electron- and hole-transfer states are not "virtual" but observable states [197].

In all of the polynuclear complexes of this series, no emission can be detected following excitation in the Ru→bpy MLCT absorption band, indicating that efficient pathways are available for quenching of the MLCT state of the –Ru(bpy)$_2$– chromophore. These pathways can be easily identified on the basis of the states detected spectroscopically. For example, in the **5.32** 5 + species several electron-transfer quenching pathways are available, as shown in Scheme 5.8.

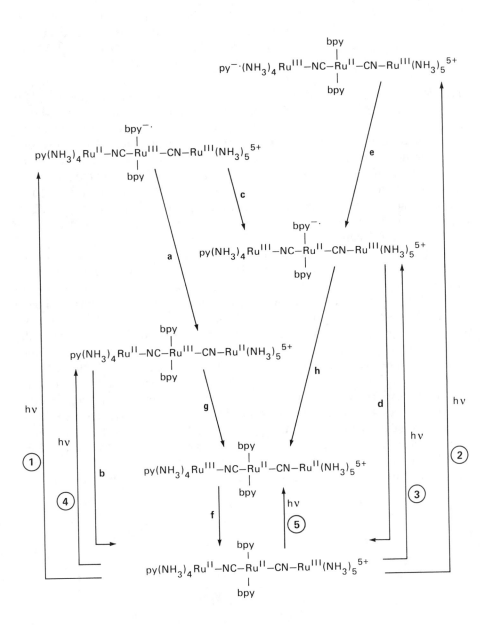

Scheme 5.8

Analogous, though simpler, schemes can be elaborated for the other trinuclear complexes in the series. Thus, a sequence of processes of the type marked *a* and *b* in Scheme 5.8, i.e., proceeding via an IT state, holds for the fully oxidized **5.32** 6 + species, while the *c* and *d* sequence, proceeding via a remote MLCT state, constitutes the quenching pathway for the fully reduced **5.32** 4 + species. In Scheme 5.8, an intramolecular electron-transfer pathway (initiated by process *e*) is also indicated for deactivation of the Ru→py MLCT excited state, although this type of state is expected to be intrinsically very short-lived [198] and could not actually use this pathway. In the complexes of this series, no transient state is detected in nanosecond laser experiments, indicating that charge recombination steps (*b,d*, and *f* in Scheme 5.8) are very fast processes. The reasons for the lack of long-lived charge separation in these systems have been discussed in terms of thermodynamic and kinetic factors [195,196]. Similar complexes containing 4-cyanopyridine instead of cyanide as bridging ligands have been studied by Sutin and coworkers [199] and found to give very similar spectroscopic and photophysical results.

Of some interest is the comparison between the **5.32** 5 + system and some related trinuclear complexes. The **5.33** and **5.34** complexes have been studied by Taube and coworkers [200] and Meyer and coworkers [201], respectively. All these complexes contain a Ru(II)- and a Ru(III)-ammine terminal unit, and a Ru(II) central bridging unit. An end-to-end remote IT absorption band is observed at 9500 cm^{-1} in **5.33**. When this energy is compared with that of the remote IT band of **5.32** (Fig. 5.15), the energy ordering is seen to be reversed with respect to the predictions of the distance dependence of reorganizational energy in the Hush–Marcus model (eqns (3.6) and (3.18)). This underlines the inadequacy of the two-spheres-in-a-dielectric-continuum model, especially when bulky bridges (in these systems, as bulky as the donor and acceptor components) are involved, and different geometries (in this case, linear and bent) of the bridging units are compared [202]. Another interesting observation is that whereas end-to-end remote IT absorption is observed in both **5.32** and **5.33**, no such band can be detected in **5.34**. Since the distance between the terminal ruthenium atoms increases in the order **5.32, 5.34, 5.33**, it is evident that through-

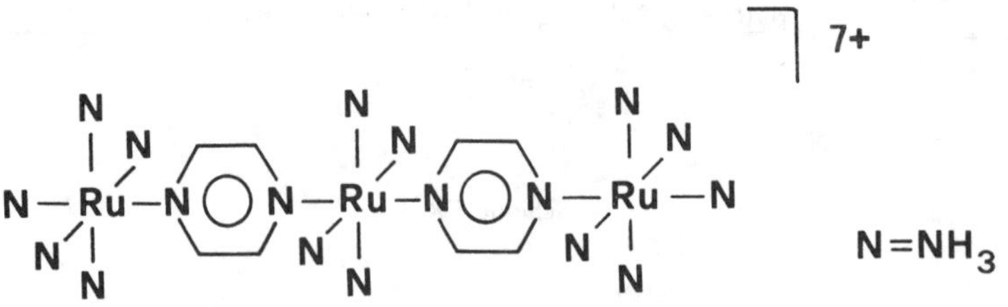

5.33

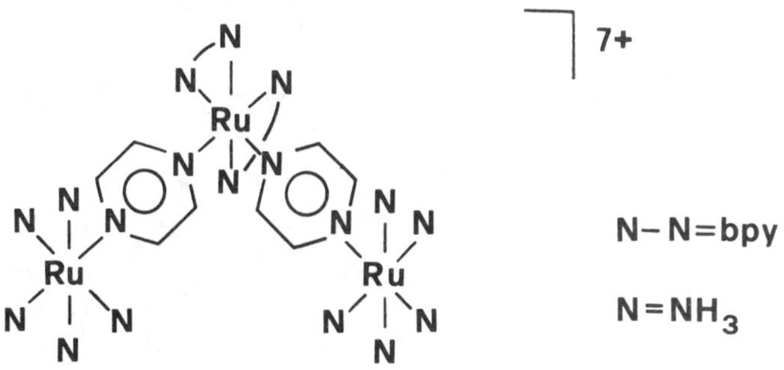

N–N=bpy

N=NH$_3$

5.34

bond rather than through-space electronic coupling is important in this comparison. In terms of superexchange (section 5.2), the difference between the structurally and energetically very similar **5.32** and **5.34** could be traced back to a weaker coupling between central and terminal ruthenium atoms through pyrazine than through cyanide. As a matter of fact, the intensity of the adjacent IT transition (which, considering the central metal complex fragment as a connector, corresponds to a transition to the hole-transfer state [197]) is much weaker in **5.34** [201] than in **5.32** (Fig. 5.15).

5.8 TRIAD AND TETRAD SYSTEMS FEATURING PHOTOINDUCED CHARGE SEPARATION

The study of two-component CLDA systems (sections 5.3–5.7) shows that photoinduced electron transfer can transiently convert light energy into chemical (redox) energy with high efficiency. In such systems, however, the back electron-transfer reaction between the reduced acceptor and the oxidized donor tends to be fast, with timescales (typically, $\tau \leqslant 10^{-8}$ s) that are much too short for any practical application. The way towards an increase in the lifetime of charge-separated species is pointed out by photosynthesis (section 5.1), where a sequence of electron-transfer steps follows light absorption, leading to localization of the oxidized and reduced equivalents on *remote* molecular components. This has led to the development, in 1983, of the first molecular "triads" for photoinduced charge separation [203,204]. The triad systems developed and extensively studied by Gust, Moore, and coworkers [38,126,204–207] have the schematic structure **5.35**. In these three-component systems, a carotenoid polyene and a quinone with a variable number of methylene spacers ($n = 1$–4) are linked through amide bridges to a porphyrin. As determined by

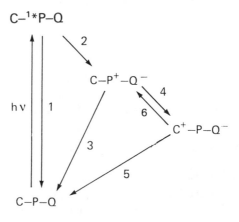

5.35

NMR measurements, the triads adopt in solution quite extended configurations [126,206]. The function of these triads is represented in Scheme 5.9, where P, Q, and C represent the porphyrin, the quinone, and the carotenoid, respectively. Light

$$C-{}^{1}*P-Q$$

Scheme 5.9

excitation of the porphyrin chromophore to its singlet state is followed by fast $(k_2 = 9.7 \times 10^9 - 1.5 \times 10^8 \text{s}^{-1}$ for $n = 1–4$, in dichloromethane) electron transfer to the quinone [126]. In competition with primary charge recombination (process 3), a secondary electron transfer from the carotenoid to the oxidized porphyrin (process 4) takes place, leading to a new charge-separated state. Charge recombination (process 5) is quite slow in such a state owing to the large spatial separation between the positive and negative changes ($\tau = 300$ and 2000 ns in dichloromethane and butyronitrile, respectively) [205]. The quantum yield of formation of the charge-separated state is a complex function of the various electron transfer rate constants ($\Phi = [k_2/(k_2 + k_1)][k_4/(k_3 + k_4)]$) and depends on the length of the methylene chain in **5.35** in a nonmonotonic fashion. Since lengthening the chain slows down both k_2 and k_3, the quantum yield of charge separation has a maximum ($\Phi \approx 0.11$ in dichloromethane at 295 K) for $n = 2$. The decay of the charge-separated state is practically independent of n, as it takes place *via* a two-step activated pathway (processes 6 and 3) rather than by direct charge recombination [207].

Another type of well-characterized triad system is that developed in 1985 by Wasielewski and coworkers [208]. This system is discussed in some detail in Chapter 12 (section 12.4.1, Fig. 12.4) in the context of "photochemical molecular devices". Some general design principles of three-component systems for photoinduced charge separation have been discussed [209] and several other triad systems have been described [36,210–219]. Among these, the triad **5.36** [219] makes use of a

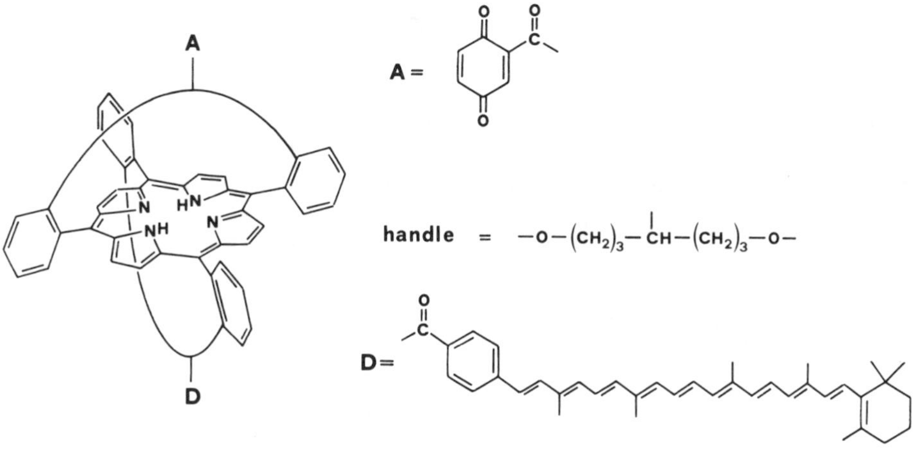

5.36

peculiar design based on a "basket handle" porphyrin (reminiscent of the capped structures discussed in section 5.6.6) in which a carotenoid donor and a quinone acceptor are attached to the handles. Thanks to the caged structure, the porphyrin fragment is highly hydrophobic (as is the carotenoid fragment), the triad is amphiphilic and can be easily incorporated in bilipid membranes. Illumination of membranes incorporating this triad system gives rise to measurable photocurrents, as a result of transmembrane photoinduced charge separation (see section 12.4.3 and Fig. 12.6 for similar experiments [220] with triads of the type **5.35**). The system **5.37** [36,215] is the fusion into a triad of two types of inorganic-chromophore–quencher

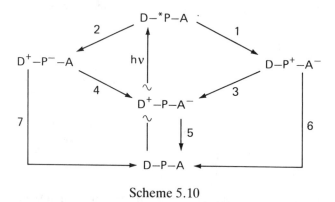

5.37

complexes (**5.22** and **5.23**) discussed in section 5.7.1. As shown in Scheme 5.10 (where P, A, and D represent the tris(bipyridine)ruthenium(II) chromophore, phenothiazine, and the bipyridinium dication, respectively), two pathways

Scheme 5.10

(processes 2 and 4, and processes 1 and 3) are thermodynamically available for attaining the charge-separated state in this system. The charge-separated state is reached within 50 ns of excitation and decays by charge recombination in about 160 ns. The quantum yield of formation of the charge-separated state, 0.26, is determined by the competition of processes 6 and 3 and/or 7 and 4 [215].

The concept of multi-step photoinduced charge separation in artificial systems has been further developed by Gust, Moore, and coworkers [38,221–223] through the synthesis of four-component systems ("tetrads"). The tetrad **5.38** is essentially an extension of the **5.35** triads, with two covalently-linked porphyrins replacing the

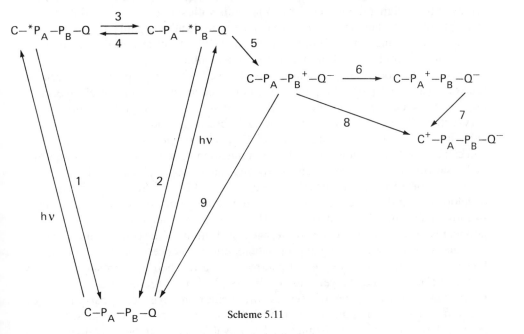

$R_1 =$ $R_2 =$

5.38

single light-absorbing chromophore [38,221]. The two porphyrins have nearly
isoenergetic excited states but differ somewhat in redox properties. The photophysical behaviour observed in anisole solution [38,221] is summarized in Scheme 5.11,
where C, Q, P_A, and P_B represent the carotenoid, the quinone, and the two

$$C-{}^*P_A-P_B-Q \overset{3}{\underset{4}{\rightleftharpoons}} C-P_A-{}^*P_B-Q$$

$$C-P_A-P_B{}^+-Q^- \overset{6}{\longrightarrow} C-P_A{}^+-P_B-Q^-$$

$$C^+-P_A-P_B-Q^-$$

$$C-P_A-P_B-Q$$

Scheme 5.11

porphyrins, respectively. After excitation, singlet energy is transferred back and forth between the porphyrins (processes 3 and 4) at a rate faster than or comparable to the rates of other processes that depopulate the singlet states. Electron transfer to the quinone from the adjacent excited porphyrin (process 5) occurs in 4 ns. The resulting state, $C-P_A-P_B^+-Q^-$, is partitioned between charge recombination (process 9) and further electron transfer to give the fully charge-separated state $C^+-P_A-P_B-Q^-$. This state is formed with $\Phi = 0.25$ and has a lifetime of 2.9 μs. The role of the $C-P_A^+-P_B-Q^-$ state depends on its energy location, which is not exactly known. Depending on whether it lies below or above $C-P_A-P_B^+-Q^-$, it can act as an actual intermediate in a two-step electron transfer (processes 6 and 7) or as a superexchange mediator in a single-step transfer (process 8) [38,221]. This problem is reminiscent of the much-debated question concerning the role of bacteriochlorophyll monomer in bacterial reaction centres (section 5.1).

A tetrad consisting of a carotenoid, a porphyrin, and two quinones has also been developed and shown to give rise to very efficient three-step charge separation [38,222,223]. This system is discussed in some detail in Chapter 12 (section 12.4.1, Fig. 12.5).

The triad and tetrad work shows that, despite the severe synthetic problems involved, continuous progress is being made in the design and synthesis of multicomponent CLDA systems to mimic various aspects of photosynthesis. Developments towards more elaborate systems are easily predictable. A biomimetic "pentad" has just been announced [224].

5.9 MACROMOLECULAR SYSTEMS

In the CLDA systems discussed in this chapter, supramolecular organization is achieved by bridging the molecular components with suitable covalent linkages. A conceptually different strategy to arrive at supramolecular organization of components is to use some kind of preformed extended structures as a *molecular scaffold* onto which the molecular components can be assembled.

In principle, polymeric structures can be used for this purpose. Research on redox polymers prepared by polymerization or copolymerization of electroactive monomers (such as, for example vinyl derivatives of $M^{III}(bpy)_3^{3+}/M^{II}(bpy)_3^{2+}$) is very active. These polymers give rise to interesting solid-state electrochemical behaviour [225], interfacial phenomena [226], solution photobehaviour [227,228], and can be used to provide electrical communication between chemical systems at a molecular level (for example, enzymes) and macroscopic electrodes [229] (for a discussion of this general problem, see section 12.7.1). Strategies have been developed to bind different types of molecular components to preformed polymeric strands of low molecular mass [230,231]. Of course, chemical attachment on a polymer is a statistical process, and average distributions of the components can only be reached by this way. Using appropriate loading sequences and characterization, however, an appreciable degree of control can be reached even in such complex systems. For example, Meyer and coworkers [232] have prepared a soluble, low molecular mass copolymer of styrene: *m, p*-chloromethylstyrene carrying three types of molecular component, with the following average characteristics: (i) each polymer strand contains *ca.* 30 functionalizable subunits; (ii) of these, 12 do not carry any substituent, 12 have covalently-bound anthryl groups, 3 have covalently-bound $Ru(bpy)_3^{2+}$ groups, and 3 have covalently-bound $Os(bpy)_3^{2+}$ groups. This particular system has been used to study energy transfer processes [232], and is discussed in some detail in section 12.5.3. Clearly, polymers of this type are interesting objects also from the point of view of photoinduced electron-transfer processes. Obviously, given the statistical characterization of the substrate, the statistical nature of the attachment processes, and the unavoidable lack of geometric definition, such systems cannot compete with simple CLDA systems for the purpose of obtaining detailed mechanistic information on photoinduced electron-transfer processes. On the other hand, the complexity of polymer-attached redox systems can be of interest for other purposes, such as photoinduced multi-electron transport, storage, and catalysis [233].

Because of its high degree of structural definition, DNA has recently attracted attention as a scaffold for the construction of well-defined multicomponent systems. Non-covalent binding modes, such as intercalation and electrostatic adsorption (section 10.6) are widely used to bind molecules or ions to DNA. Most of the studies on intercalation and adsorption are aimed at structural probing or selective recognition and cleavage (section 10.6), but organization of molecular components in such a way that intercomponent electron (and energy) transfer processes can be studied [234–236] can also be achieved in this way. An interesting and more elaborate approach, recently developed in particular by Netzel and coworkers [237,238], is that of using synthetic oligonucleotides and duplexes that contain specific chemically-modified bases. The chemical modification permits covalent attachment of selected molecular components (for example $Ru(bpy)_3^{2+}$ [237], pyrene [238], and anthraquinone [238]) to specific bases along the sequence. The use of DNA duplexes as molecular scaffolds seems to be very promising, and may constitute an initial step in the construction of macromolecules with specifically-located photoactive and redox-active subunits.

REFERENCES

[1] Verhoeven, J. W., Dirkx, I. P., and de Boer, Th. J. (1986) *Tetrahed.* **25** 4037.

[2] Creutz, C. and Taube, H. (1969) *J. Am. Chem. Soc.* **91** 3988.

[3] Okada, T., Fujita, T., Kubota, M., Masaki, S., Mataga, N., Ide, R., Sakata, Y., and Misumi, S. (1972) *Chem. Phys. Lett.* **14** 563.

[4] Salemme, F. R. (1979) In Chance, B., De Vault, D. C., Frauenfelder, H., Marcus, R. A., Schrieffer, J. R., and Sutin, N. (eds) *Tunneling in biological systems.* Academic, p. 523.

[5] Michel-Beyerle, M. E. (ed.) (1985) *Antennas and reaction centers in photo-synthetic bacteria.* Springer.

[6] Breton, J. and Vermeglio, H. (eds) (1988) *The photosynthetic bacterial reaction center — structure and dynamics,* Plenum.

[7] Deisenhofer, J., Epp, O., Miki, K., Huber, R., and Michel, H. (1984) *J. Mol. Biol.* **180** 385.

[8] Chang, C.-H., Tiede, D. M., Tang, J., Smith, U., Norris, J., and Schiffer, M. (1986) *FEBS Lett.* **205** 82.

[9] Allen, J. P., Feher, G., Yeates, T. O., Komiya, H., and Rees, D. C. (1987) *Proc. Natl. Acad. Sci. USA* **84** 5730.

[10] Deisenhofer, J. and Michel, H. (1989) Angew. Chem. Int. Ed. Engl. **28** 829.

[11] Huber, R. (1989) *Angew. Chem. Int. Ed. Engl.* **28** 848.

[12] Boxer, S. G., Goldstein, R. A., Lockhart, D. J., Middendorf, T. R., and Takiff, L. (1989) *J. Phys. Chem.* **93** 8280.

[13] Bixon, M., Jortner, J., Michel-Beyerle, M. E., Orgodnik, A., and Lersch, W. (1987) *Chem. Phys. Lett.* **140** 626.

[14] Marcus, R. A. (1987) *Chem. Phys. Lett.* **133** 471.

[15] Fleming, G. R., Martin, J. L., and Breton, J. (1988) *Nature* **333** 190.

[16] Moser, C. C., Alegria, G., Gunner, M. R., and Dutton, P. L. (1989) In Norris, J. R. Jr. and Meisel, D. (eds) *Photochemical energy conversion.* Elsevier, p. 221.

[17] Holzapfel, W., Finkele, U., Kaiser, W., Oesterhelt, D., Scheer, H., Stilz, H. U., and Zinth, W. (1989) *Chem. Phys. Lett.* **160** 1.

[18] Kirmaier, C., Holten, D., and Parson, W. W. (1985) *Biochim. Biophys. Acta* **810** 33.

[19] Holten, D., Windsor, M. W., Parson, W. W., and Thornber, J. P. (1978) *Biochim. Biophys. Acta* **501** 112.

[20] Dutton, P. L., Leigh, J. S., Prince, R. C., and Tiede, D. M. (1979) In Chance, B., DeVault, D. C., Frauenfelder, H., Marcus, R. A., Schrieffer, J. R., and Sutin, N. (eds) *Tunneling in biological systems.* Academic, p. 319.

[21] Isied, S. S. (1984) *Prog. Inorg. Chem.* **32** 43.

[22] Marcus, R. A. and Sutin, N. (1985) *Biochim. Biophys. Acta* **811,** 265.

[23] McLendon, G. (1988) *Acc. Chem. Res.* **21** 160.

[24] Dixon, D. W., Hong, X., Woehler, S. E., Mauk, A. G., and Sishta, B. (1990) *J. Am. Chem. Soc.* **112** 1082.

[25] Axup, A. W., Albin, M., Mayo, S. L., Crutchley, R. J., and Gray, H. B. (1988) *J. Am. Chem. Soc.* **110** 435, and references therein.

[26] Meade, T. J., Gray, H. B., and Winkler, J. R. (1989) *J. Am. Chem. Soc.* **111** 4353.

[27] Miller, J. R., Beitz, J. V., and Huddlestone, R. K. (1984) *J. Am. Chem. Soc.* **106** 5057.

[28] Kira, A. and Imamura, M. (1984) *J. Phys. Chem.* **88** 1865.

[29] Guarr, T., Maguire, M., and McLendon, G. (1985) *J. Am. Chem. Soc.* **107** 5104.

[30] Miloslavljevic, B. H. and Thomas, J. K. (1986) *J. Am. Chem. Soc.* **108** 2513.

[31] Kuhn, H. (1987) In Carter, F. L. (ed.) *Molecular electronic devices II.* Dekker, p. 411.

[32] Sugi, M. (1989) In Carter, F. L., Siatkowski, R. E., and Wohltjen, H. (eds) *Molecular electronic devices.* North-Holland, p. 441.

[33] Connolly, J. S. and Bolton, J. R. (1988) In Fox, M. A. and Chanon, M. (eds) *Photoinduced electron transfer.* Part D. Elsevier, p. 303.

[34] Wasielewski, M. R. (1988) In Fox, M. A. and Chanon, M. (eds) *Photoinduced electron transfer.* Part A. Elsevier, p. 161.

[35] Gust, D. and Moore, T. A. (eds) (1989) *Tetrahedron* **45** 4669–4903.

[36] Meyer, T. J. (1989) *Acc. Chem. Res.* **22** 163.

[37] Scandola, F., Indelli, M. T., Chiorboli, C., and Bignozzi, C. A. (1990) *Top. Curr. Chem.* **158**, 73.

[38] Gust, D. and Moore, T. A. (1989) *Science* **244** 35.

[39] Ulstrup, J. (1979) *Charge transfer processes in condensed media.* Springer.

[40] Jortner, J. (1976) *J. Chem. Phys.* **64,** 4860.

[41] Sutin, N. (1983) *Prog. Inorg. Chem.* **30** 441.

[42] Hoffman, R. (1971) *Acc. Chem. Res.* **4** 1.

[43] Paddon-Row, M. N. (1982) *Acc. Chem. Res.* **15** 245.

[44] Larsson, S. (1981) *J. Am. Chem. Soc.* **103** 4034.

[45] Richardson, D. E. and Taube, H. (1983) *J. Am. Chem. Soc.* **105** 40.

[46] Beratan, D. N. and Hopfield, J. J. (1984) *J. Am. Chem. Soc.* **106** 1584.

[47] Beratan, D. N., Onuchic, J. N., and Hopfield, J. J. (1984) *J. Chem. Phys.* **81** 5735.

[48] Beratan, D. N., Onuchic, J. N., and Hopfield, J. J. (1985) *J. Chem. Phys.* **83** 5325.

[49] Ohta, K., Closs, G. L., Morokuma, K., and Green, N. J. (1986) *J. Am. Chem. Soc.* **108** 1319.

[50] Beratan, D. N. (1986) *J. Am. Chem. Soc.* **108** 4321.

[51] Larsson, S. and Volosov, A. (1986) *J. Chem. Phys.* **85** 2548.

[52] Larsson, S. and Volosov, A. (1987) *J. Chem. Phys.* **87** 6623.

[53] Heitele, H., Michel-Beyerle, M. E., and Finckh, P. (1987) *Chem. Phys. Lett.* **134** 273.

[54] Hush, N. S. (1987) In Balzani, V. (ed.) *Supramolecular photochemistry.* Reidel, p. 53.

[55] Plato, M., Mobius, K., Michel-Beyerle, M. E., Bixon, M., and Jortner, J. (1988) *J. Am. Chem. Soc.* **110** 7279.

[56] Rendell, A. P. L., Backsay, G. B., and Hush, N. S. (1988) *J. Am. Chem. Soc.* **110** 8343.

[57] Scherer, P. O. J. and Fischer, S. F. (1989) *J. Phys. Chem.* **93** 1633.

[58] Falcetta, M. F., Jordan, K. D., McMurry, J. E., and Paddon-Row, M. N. (1990) *J. Am. Chem. Soc.* **112** 579.

[59] Reimers, J. R. and Hush, N. S. (1990) *Inorg. Chem.,* **29** 3686.

[60] Halpern, J. and Orgel, L. E. (1960) *Disc. Faraday Soc.* **29** 32.

[61] McConnell, H. M. (1961) *J. Chem. Phys.* **35** 508.

[62] Day, P. (1981) *Comments Inorg. Chem.* **1** 155.

[63] Miller, J. R. and Beitz, J. V. (1981) *J. Chem. Phys.* **74** 6746.

[64] Closs, G. L. and Miller, J. R. (1988) *Science* **240** 440.

[65] Oevering, H., Paddon-Row, M. N., Heppener, M., Oliver, A. M., Cotsaris, E., Verhoeven, J. W., and Hush, N. S. (1987) *J. Am. Chem. Soc.* **109** 3258.

[66] Onuchic, J. N. and Beratan, D. N. (1987) *J. Am. Chem. Soc.* **109** 6771.

[67] Cave, R. J., Baxter, D. V., Goddard, W. A., and Baldeshwieler, J. D. (1987) *J. Chem. Phys.* **87** 926.

[68] Calcaterra, L. T., Closs, G. L., and Miller, J. R. (1983) *J. Am. Chem. Soc.* **105** 670.

[69] Miller, J. R., Calcaterra, L. T., and Closs, G. L. (1984) *J. Am. Chem. Soc.* **106** 3047.

[70] Closs, G. L., Calcaterra, L. T., Green, N. J., Penfield, K. W., and Miller, J. R. (1986) *J. Phys. Chem.* **90** 3673.

[71] Miller, J. R. (1987) *Nouv. J. Chim.* **11** 83.

[72] Johnson, M. D., Miller, J. R., Green, N. J., and Closs, G. L. (1989) *J. Phys. Chem.* **93** 1173.

[73] Closs, G. L., Johnson, M. D., Miller, J. R., and Piotrowiak, P. (1989) *J. Am. Chem. Soc.* **111** 3751.

[74] Liang, N., Miller, J. R., and Closs, G. L. (1989) *J. Am. Chem. Soc.* **111** 8740.

[75] Closs, G. L., Piotrowiak, P., MacInnis, J. M., and Fleming, G. R. (1988) *J. Am. Chem. Soc.* **110** 2652.

[76] Closs, G. L., Piotrowiak, P., and Miller, J. R. (1989) In Norris, J. R. Jr. and Meisel, D. (eds) *Photochemical energy conversion.* Elsevier, p. 23.

[77] Balaji, V., Ng, L., Jordan, K. D., Paddon-Row, M. N., and Patney, H. K. (1987) *J. Am. Chem. Soc.* **109** 6957.

[78] Paddon-Row, M. N. (1989) In Liebman, J. F. and Greenberg, A. (eds) *Molecular structure and energetics.* Vol. 6. VCH Publishers, New York, p. 115.

[79] Pasman, P., Rob, F., and Verhoeven, J. W. (1982) *J. Am. Chem. Soc.* **104** 5127.

[80] Mes, G. F., van Ramesdonk, H. J., and Verhoeven, J. W. (1984) *J. Am. Chem. Soc.* **106** 1335.

[81] Mes, G. F., de Jong, B., van Ramesdonk, H. J., Warman, J. M., De Haas, M. P., and Horsman-van den Dool, L. E. (1984) *J. Am. Chem. Soc.* **106** 6524.

[82] Pasman, P., Mes, G. F., Koper, N. W., and Verhoeven, J. W. (1985) *J. Am. Chem. Soc.* **107** 5839.

[83] Hush, N. S. (1961) *Trans. Faraday Soc.* **57** 155.

[84] Hush, N. S. (1967) *Prog. Inorg. Chem.* **8** 391.

[85] Hush, N. S. (1968) *Electrochim. Acta* **13** 1004.

[86] Hush, N. S. (1985) *Coord. Chem. Rev.* **64** 135.

[87] Hush, N. S., Paddon-Row, M. N., Cotsaris, E., Oevering, H., Verhoeven, J. W., and Heppener, M. (1985) *Chem. Phys. Lett.* **117** 8.

[88] Warman, J. M., De Haas, M. P., Paddon-Row, M. N., Cotsaris, E., Hush, N. S., Oevering, H., and Verhoeven, J. W. (1986) *Nature* **320** 615.

[89] Verhoeven, J. W., Paddon-Row, M. N., Hush, N. S., Oevering, H., and Heppener, M. (1986) *Pure Appl. Chem.* **58** 1285.

[90] Warman, J. M., De Haas, M. P., Oevering, H., Verhoeven, J. W., Paddon-Row, M. N., Oliver, A. M., and Hush, N. S. (1986) *Chem. Phys. Lett.* **128** 95.

[91] Penfield, K. W., Miller, J. R., Paddon-Row, M. N., Cotsaris, E., Oliver, A. M., and Hush, N. S. (1987) *J. Am. Chem. Soc.* **109** 5061.

[92] Paddon-Row, M. N., Oliver, A. M., Warman, J. M., Smit, K. J., De Haas, M. P., Oevering, H., and Verhoeven, J. W. (1988) *J. Chem. Phys.* **92** 6958.

[93] Verhoeven, J. W., Oevering, H., Paddon-Row, M. N., Kroon, J., and Kunst, A. G. M. (1988) In Grassi, G. and Hall, D. O. (eds) *Photocatalytic production of energy-rich compounds,* Elsevier, p. 51.

[94] Kroon, J., Oliver, A. M., Paddon-Row, M. N., and Verhoeven, J. W. (1988) *Rec. Trav. Chim. Pays-Bas* **107** 509.

[95] Oliver, A. M., Craig, D. C., Paddon-Row, M. N., Kroon, J., and Verhoeven, J. W. (1988) *Chem. Phys. Lett.* **150** 366.

[96] Oevering, H., Verhoeven, J. W., Paddon-Row, M. N., and Warman, J. M. (1989) *Tetrahed.* **45** 4751.

[97] Lawson, J. M., Craig, D. C., Paddon-Row, M. N., Kroon, J., and Verhoeven, J. W. (1989) *Chem. Phys. Lett.* **164** 120.

[98] Verhoeven, J. W., Kroon, J., Paddon-Row, M. N., and Oliver, A. M. (1989) In Hall, D. O. and Grassi, G. (eds) *Photoconversion processes for energy and chemicals*, Elsevier, p. 100.

[99] Hermant, R. M., Bakker, N. A. C., Scherer, T., Krijnen, B., and Verhoeven, J. W. (1990) *J. Am. Chem. Soc.* **112** 1214.

[100] Hoffmann, R., Imamura, A., and Here, W. J. (1968) *J. Am. Chem. Soc.* **90** 1499.

[101] Paddon-Row, M. N., Oliver, A. M., Symons, M. C. R., Cotsaris, E., Wong, S. S., and Verhoeven, J. W. (1989). In Hall, D. O. and Grassi, G. (eds) *Photoconversion processes for energy and chemicals*, Elsevier, p. 79.

[102] Heitele, H. and Michel-Beyerle, M. E. (1985) *J. Am. Chem. Soc.* **107** 8068.

[103] Heitele, H., Michel-Beyerle, M. E., and Finckh, P. (1987) *Chem. Phys. Lett.* **138** 237.

[104] Finckh, P., Heitele, H., Volk, M., and Michel-Beyerle, M. E. (1988) *J. Phys. Chem.* **92** 6584.

[105] Heitele, H., Finckh, P., Weeren, S., Pollinger, F., and Michel-Beyerle, M. E. (1989) *J. Phys. Chem.* **93** 5173.

[106] Kong, J. L. Y. and Loach, P. A. (1978) In Dutton, P. L., Leigh, J. S., and Scarpa, A. (eds) *Frontiers of biological energetics*. Academic, p. 73.

[107] Tabushi, I., Koga, N., and Yanagita, M. (1979) *Tetrahed. Lett.* 257.

[108] Wasielewski, M. R. and Niemczyk, M. P. (1984) *J. Am. Chem. Soc.* **106** 5043.

[109] Wasielewski, M. R., Niemczyk, M. P., Svec, W. A., and Pewitt, E. B. (1985) *J. Am. Chem. Soc.* **107** 1080.

[110] Wasielewski, M. R., Johnson, D. G., and Svec, W. A. (1987). In Balzani, V. (ed.) *Supramolecular photochemistry*. Reidel, p. 255.

[111] Kirmaier, C. and Holten, D. (1987) *Photosyn. Res.* **13** 225.

[112] Wasielewski, M. R., Johnson, D. G., Svec, W. A., Kersey, K. M., and Minsek, D. W. (1988) *J. Am. Chem. Soc.* **110** 7219.

[113] Wasielewski, M. R. and Niemczyk, M. P. (1986). In Gouterman, M., Rentzepis, P. M., and Straub, K. D. (eds) *Porphyrins — Excited states and dynamics*. ACS Symposium Series No. 321. American Chemical Society, p. 154.

[114] Wasielewski, M. R., Niemczyk, M. P., Johnson, D. G., Svec, W. A., and Minsek, D. W. (1989) *Tetrahed.* **45** 4785.

[115] Joran, A. D., Leland, B. A., Geller, G. G., Hopfield, J. J., and Dervan, P. B. (1984) *J. Am. Chem. Soc.* **106** 6090.

[116] Leland, B. A., Joran, A. D., Felker, P. M., Hopfield, J. J., Zewail, A. H., and Dervan, P. B. (1985) *J. Phys. Chem.* **89** 5571.

[117] Joran, A. D., Leland, B. A., Felker, P. M., Zewail, A. H., Hopfield, J. J., and Dervan, P. B. (1987) *Nature* **327** 508.

[118] Beratan, D. N. (1986) *J. Am. Chem. Soc.* **108** 4321.

[119] Bolton, J. R., Ho, T. F., Liauw, S., Siemiarczuk, A., Wan, C. S. K., and Weedon, A. (1984) *J. Chem. Soc., Chem. Commun.* 559.

[120] Wasielewski, M. R., Johnson, D. G., Svec, W. A., Kersey, K. M., Cragg, D. E., and Minsek, D. W. (1989) In Norris, J. R. Jr. and Meisel, D. (eds) *Photochemical energy conversion.* Elsevier, p. 135.

[121] Arrhenius, T. S., Blanchard-Desce, M., Dvolaitzky, M., Lehn, J. M., and Malthete, J. (1986) *Proc. Natl. Acad. Sci. USA* **83** 5355.

[122] Lehn, J. M. (1988) *Angew. Chem. Int. Ed. Engl.* **27** 89.

[123] Larsson, S. (1982) *Discuss. Faraday Soc.* **74** 390.

[124] Joachim, C. (1987) *Chem. Phys.* **116** 339.

[125] Woitellier, S., Launay, J. P., and Spangler, C. W. (1989) *Inorg. Chem.* **28** 758.

[126] Gust, D., Moore, T. A., Liddell, P. A., Nemeth, G. A., Makings, L. R., Moore, A. L., Barrett, D., Pessiki, P. J., Bensasson, R. V., Rougée, M., Chachaty, C., De Schryver, F. C., Van der Auweraer, M., Holzwarth, A. R., and Connolly, J. S. (1987) *J. Am. Chem. Soc.* **109** 846, and references therein.

[127] Mataga, N. (1989) In Norris, J. R. Jr. and Meisel, D. (eds) *Photochemical energy conversion,* Elsevier, p. 32, and references therein.

[128] Schmidt, J. A., McIntosh, A. R., Weedon, A. C., Bolton, J. R., Connolly, J. S., Hurley, J. K., and Wasielewski, M. R. (1988) *J. Am. Chem. Soc.* **110** 1733.

[129] Schmidt, J. A., Liu, J. Y., Bolton, J. R., Archer, M. D., and Gadzepko, V. P. Y. (1989) *J. Chem. Soc., Faraday Trans. 1* **85** 1027.

[130] Siemiarczuk, A., McIntosh, A. R., Ho, T. F., Stillman, M. J., Roach, K. J., Weedon, A. C., Bolton, J. R., and Connolly, J. S. (1983) *J. Am. Chem. Soc.* **105** 7224.

[131] Mataga, N., Karen, A., Okada, T., Nishitani, S., Kurata, N., Sakata, Y., and Misumi, S. (1984) *J. Phys. Chem.* **88** 5138.

[132] Bergkamp, M. A., Dalton, J., and Netzel, T. L. (1982) *J. Am. Chem. Soc.* **104** 253.

[133] Cormier, R. A., Posey, M. R., Bell, W. L., Fonda, H. N., and Connolly, J. S. (1989) *Tetrahed.* **45** 4831.

[134] Gonzales, M. C., McIntosh, A. R., Bolton, J. R., and Weedon, A. C. (1984) *J. Chem. Soc., Chem. Commun.* 1138.

[135] Sakata, Y., Nakashima, S., Goto, Y., Tatemitsu, H., and Misumi, S. (1989) *J. Am. Chem. Soc.* **111** 8979.

[136] Hunter, C. A., Sanders, J. K. M., Beddard, G. S., and Evans, S. (1989) *J. Chem. Soc., Chem., Commun.* 1765.

[137] Lindsey, J. S., Delaney, J. K., Mauzerall, D. C., and Linschitz, H. (1988) *J. Am. Chem. Soc.* **110** 3610.

[138] Delaney, J. K., Mauzerall, D. C., and Lindsey, J. S. (1990) *J. Am. Chem. Soc.* **112** 957.

[139] Harriman, A., Porter, G., and Wilowska, A. (1984) *J. Chem. Soc., Faraday Trans. 2* **80** 193.

[140] McMahon, R. J., Force, R. K., Patterson, H. H., and Wrighton, M. S. (1988) *J. Am. Chem. Soc.* **110** 2670.

[141] Irvine, M. P., Harrison, R. J., Beddard, G. S., Leighton, P., and Sanders, J. K. M. (1986) *Chem. Phys.* **104** 315.

[142] Saito, T., Hirata, Y., Sato, H., Yoshida, T., and Mataga, N. (1988) *Bull. Chem. Soc. Japan* **61** 1925.

[143] Batteas, J. D., Harriman, A., Kanda, Y., Mataga, N., and Nowak, A. K. (1990) *J. Am. Chem. Soc.* **112** 126.

[144] Gubelmann, M., Harriman, A., Lehn, J. M., and Sessler, J. L. (1990) *J. Phys. Chem.* **94** 308.

[145] Juris, A., Balzani, V., Barigelletti, F., Belser, P., and Von Zelewski, A. (1988) *Coord. Chem. Rev.* **84** 85.

[146] Heath, G. A., Yellowlees, L. J., and Braterman, P. S. (1982) *Chem. Phys. Lett.* **92** 646.

[147] DeArmond, L. K., Hanck, K. W., and Wertz, D. W. (1985) *Coord. Chem. Rev.* **64** 65.

[148] Meyer, T. J. (1986) *Pure Appl. Chem.* **58** 1193.

[149] Kober, E. M., Marshall, J. L., Dressick, W. J., Sullivan, P., Caspar, J. V., and Meyer, T. J. (1985) *Inorg. Chem.* **24** 2755.

[150] Barqawi, K. R., Llobet, A., and Meyer, T. J. (1988) *J. Am. Chem. Soc.* **110** 7751.

[151] Indelli, M. T., Bignozzi, C. A., Marconi, A., and Scandola, F. (1988) *J. Am. Chem. Soc.* **110** 7381.

[152] Westmoreland, T. D., Le Bozec, H., Murray, R. W., and Meyer, T. J. (1983) *J. Am. Chem. Soc.* **105** 5952.

[153] Chen, P., Danielson, E., and Meyer, T. J. (1988) *J. Phys. Chem.* **92** 3708.

[154] Chen, P., Curry, M., and Meyer, T. J. (1989) *Inorg. Chem.* **28** 2271.

[155] Sullivan, B. P., Abruna, H., Finklea, H. O., Salmon, D. J., Nagle, J. K., Meyer, T. J., and Sprintschnik, H. (1978) *Chem. Phys. Lett.* **58** 389.

[156] Dürr, H., Thiery, U., Infelta, P. P., and Braun, A. M. (1989) *New J. Chem.* **133** 575.

[157] Perkins, T. A., Pourreau, D. B., Netzel, T. L., and Schanze, K. S. (1989) *J. Phys. Chem.* **93** 511.

[158] Perkins, T. A., Humer, W., Netzel, T. L., and Schanze, K. S. (1990) *J. Phys. Chem.* **94** 2229.

[159] Koester, V. J. (1975) *Chem. Phys. Lett.* **32** 575.

[160] Crosby, G. A., Highland, R. G., and Truesdell, K. A. (1985) *Coord. Chem. Rev.* **64** 41.

[161] Vogler, A. and Kunkely, H. (1990) *Comm. Inorg. Chem.* **9** 201.

[162] Chen, P., Westmoreland, T. D., Danielson, E., Schanze, K. S., Anthon, D., Neveux, P. E., and Meyer, T. J. (1987) *Inorg. Chem.* **26** 1116.

[163] Chen, P., Duesing, R., Tapolsky, G., and Meyer, T. J. (1989) *J. Am. Chem. Soc.* **111** 8305.

[164] Cooley, L. F., Headford, C. E. L., Elliott, C. M., and Kelley, D. F. (1988) *J. Am. Chem. Soc.* **110** 6673.

[165] Prutz, W. A., Land, E. J., and Sloper, R. W. (1981) *J. Chem. Soc., Faraday Trans. 1* **77** 281.

[166] Faraggi, M., DeFelippis, M. R., and Klapper, M. H. (1989) *J. Am. Chem. Soc.* **111** 5141.

[167] Isied, S. S. and Vassilian, A. (1984) *J. Am. Chem. Soc.* **106** 1726.

[168] Isied, S. S. and Vassilian, A. (1984) *J. Am. Chem. Soc.* **106** 1732.
[169] Isied, S. S., Vassilian, A., Magnuson, R. H., and Schwartz, H. A. (1985) *J. Am. Chem. Soc.* **107** 7432.
[170] Schanze, K. S. and Sauer, K. (1988) *J. Am. Chem. Soc.* **110** 1180.
[171] Fox, L. S., Kozik, M., Winkler, J. R., and Gray, H. B. (1990) *Science* **247** 1069.
[172] Marshall, J. L., Stobart, S. R., and Gray, H. B. (1984) *J. Am. Chem. Soc.* **106** 3027.
[173] Franco, C. and McLendon, G. (1984) *Inorg. Chem.* **23** 2370.
[174] Mau, A. W. H., Sasse, W. H. F., Creaser, I. I., and Sargeson, A. M. (1986) *Nouv. J. Chim.* **10** 589.
[175] Creaser, I. I., Hammershoi, A., Launikonis, A., Mau, A. W. H., Sargeson, A. M., and Sasse, W. H. F. (1989) *Photochem. Photobiol.* **49** 19.
[176] Osman, A. H. and Vogler, A. (1987). In Yersin, H. and Vogler, A. (eds) *Photochemistry and photophysics of coordination compounds.* Springer-Verlag, p. 197.
[177] Thorn, D. L. and Fultz, W. C. (1989) *J. Phys. Chem.* **93** 1234.
[178] Creutz, C. (1983) *Prog. Inorg. Chem.* **30** 1.
[179] Stein, C. A., Lewis, N. A., and Seitz, G. (1982) *J. Am. Chem. Soc.* **104** 2596.
[180] Lewis, N. A. and Obeng, Y. S. (1989) *J. Am. Chem. Soc.* **111** 7624.
[181] Kim, Y. and Lieber, C. M. (1989) *Inorg. Chem.* **28** 3990.
[182] Creutz, C., Kroger, P., Matsubara, T., Netzel, T. L., and Sutin, N. (1979) *J. Am. Chem. Soc.* **101** 5442.
[183] Durante, V. A. and Ford, P. C. (1975) *J. Am. Chem. Soc.* **97** 6898.
[184] Curtis, J. C., Bernstein, J. S., and Meyer, T. J. (1985) *Inorg. Chem.* **24** 385.
[185] Tapolsky, G., Deusing, R., and Meyer, T. J. (1989) *J. Phys. Chem.* **93** 3885.
[186] Bignozzi, C. A., Roffia, S., Chiorboli, C., Davila, J., Indelli, M. T., and Scandola, F. (1989) *Inorg. Chem.* **28** 4350.
[187] Schanze, K. S., Neyhart, G. A., and Meyer, T. J. (1986) *J. Phys. Chem.* **90** 2182.
[188] Loeb, B. L., Neyhart, G. A., Worl, L. A., Danielson, E., Sullivan, B. P., and Meyer, T. J. (1989) *J. Phys. Chem.* **93** 717.
[189] Norton, K. A. and Hurst, J. K. (1982) *J. Am. Chem. Soc.* **104** 5960.
[190] Malin, J. M., Ryan, D. A., and O'Halloran, T. V. (1978) *J. Am. Chem. Soc.* **100** 2097.
[191] Vogler, A., Osman, A. H., and Kunkely, H., (1985) *Coord. Chem. Rev.* **64** 159.
[192] Vogler, A. and Kunkely, H. (1975) *Ber. Bunsenges. Phys. Chem.* **79** 83.
[193] Vogler, A., Osman, A. H., and Kunkely, H. (1987) *Inorg. Chem.* **26** 2337.
[194] Vogler, A. and Kunkely, H. (1975) *Ber. Bunsenges. Phys. Chem.* **79** 301.
[195] Bignozzi, C. A., Roffia, S., and Scandola, F. (1985) *J. Am. Chem. Soc.* **107** 1644.
[196] Bignozzi, C. A., Paradisi, C., Roffia, S., and Scandola, F. (1988) *Inorg. Chem.* **27** 408.
[197] Scandola, F. (1989) In Norris, J. R. Jr. and Meisel, D. (eds) *Photochemical energy conversion.* Elsevier, p. 60.
[198] Winkler, J. R., Netzel, T. L., Creutz, C., and Sutin, N. (1987) *J. Am. Chem. Soc.* **109** 2381.

[199] Katz, N. E., Creutz, C., and Sutin, N. (1988) *Inorg. Chem.* **27** 1687.

[200] von Kameke, A., Tom, G. M., and Taube, H. (1978) *Inorg. Chem.* **17** 1790.

[201] Powers, M. J., Callahan, R. W., Salmon, D. J., and Meyer, T. J. (1976) *Inorg. Chem.* **15** 894.

[202] Balzani, V. and Scandola, F. (1988) In Fox, M. A. and Chanon, M. (eds) *Photoinduced electron transfer.* Part D. Elsevier, p. 148.

[203] Nishitani, S., Kurata, N., Sakata, Y., Misumi, S., Karen, A., Okada, T., and Mataga, N. (1983) *J. Am. Chem. Soc.* **105** 7771.

[204] Gust, D., Mathis, P., Moore, A. L., Liddell, P. A., Nemeth, G. A., Lehman, W. R., Moore, T. A., Bensasson, R. V., Land, E. J., and Chachaty, C. (1983) *Photochem. Photobiol.* **37S** S46.

[205] Moore, T. A., Gust, D., Mathis, P., Mialocq, J. C., Chachaty, C., Bensasson, R. V., Land, E. J., Doizi, D., Liddell, P. A., Nemeth, G. A., and Moore, A. L. (1984) *Nature* **307** 630.

[206] Chachaty, C., Gust, D., Moore, T. A., Nemeth, G. A., Liddell, P. A., and Moore, A. L. (1984) *Org. Magn. Reson.* **22** 39.

[207] Gust, D., Moore, T. A., Makings, L. R., Liddell, P. A., Nemeth, G. A., and Moore, A. L. (1986) *J. Am. Chem. Soc.* **108** 8028.

[208] Wasielewski, M. R., Niemczyk, M. P., Svec, W. A., and Pewitt, E. B. (1985) *J. Am. Chem. Soc.* **107** 5562.

[209] Beddard, G. S. (1986) *J. Chem. Soc., Faraday Trans. 2* **82** 2361.

[210] Sanders, G. M., van Dijk, M., van Veldhuizen, A., and van der Plas, H. C. (1986) *J. Chem. Soc., Chem. Commun.* 1311.

[211] Sessler, J. L., Hugdahl, J., and Johnson, M. R. (1986) *J. Org. Chem.* **51** 2828.

[212] Liddell, P. A., Barrett, D., Makings, L. R., Pessiki, P. J., Gust, D., and Moore, T. A. (1986) *J. Am. Chem. Soc.* **108** 5350.

[213] Sessler, J. L. and Johnson, M. R. (1987) *Rec. Trav. Chim. Pays-Bas* **106** 222.

[214] Cowan, J. A., Sanders, J. K. M., Beddard, G. S., and Harrison, R. J. (1987) *J. Chem. Soc., Chem. Commun.* 55.

[215] Danielson, E., Elliott, C. M., Merkert, J. W., and Meyer, T. J. (1987) *J. Am. Chem. Soc.* **109** 2519.

[216] Sessler, J. L., Johnson, M. R., Lin, T. Y., and Creager, E. E. (1988) *J. Am. Chem. Soc.* **110** 3659.

[217] Hofstra, U., Schaafsma, T. J., Sanders, G. M., van Dijk, M., van der Plas, H. C., Johnson, D. G., and Wasielewski, M. R. (1988) *Chem. Phys. Lett.* **151** 169.

[218] Sessler, J. L., Johnson, M. R., and Lin, T. Y. (1989) *Tetrahed.* **45** 4767.

[219] Momenteau, M., Loock, B., Seta, P., Bienvenue, E., and d'Epenoux, B. (1989) *Tetrahed.* **45** 4893.

[220] Seta, P., Bienvenue, E., Moore, A. L., Mathis, P., Bensasson, R. V., Liddell, P., Pessiki, P. J., Joy, A., Moore, T. A., and Gust, D. (1985) *Nature* **316** 653.

[221] Gust, D., Moore, T. A., Moore, A. L., Makings, L. R., Seely, G. R., Ma, X. C., Trier, T. T., and Gao, F. (1988) *J. Am. Chem. Soc.* **110** 7567.

[222] Gust, D., Moore, T. A., Moore, A. L., Barrett, D., Harding, L. O., Makings, L. R., Liddell, P. A., De Schryver, F. C., Van der Auweraer, M., Bensasson, R. V., and Rougée, M. (1988) *J. Am. Chem. Soc.* **110** 321.

[223] Gust, D., Moore, T. A., Moore, A. L., Seely, G. R., Liddell, P. A., Barrett, D., Harding, L. O., Ma, X. C., Lee, S. J., and Gao, F. (1989) *Tetrahed.* **45** 4867.

[224] Gust, D., Moore, T. A., Moore, A. L., Lee, S. J., Bittersman, E., Rehms, A. A., Belford, R. E., Luttrull, D. K., DeGraziano, J., Ma, X. C., Gao, F., and Trier, T. T. (1990) *Abstracts of the XIIIth IUPAC Conference on Photochemistry.*

[225] Jernigan, J. C., Surridge, N. A., Zvanut, M. E., Silver, M., and Murray, R. W. (1989) *J. Phys. Chem.* **93** 4620.

[226] Jernigan, J. C. and Murray, R. W. (1990) *J. Am. Chem. Soc.* **112** 1034.

[227] Olmsted III, J., McClanahan, S. F., Danielson, E., Younathan, J. N., and Meyer, T. J. (1987) *J. Am. Chem. Soc.* **109** 3297.

[228] Ennis, P. M. and Kelly, J. M. (1989) *J. Phys. Chem.* **93** 5735.

[229] Degani, Y. and Heller, A. (1989) *J. Am. Chem. Soc.* **111** 2357.

[230] Kaneko, M. and Nakamura, H. (1987) *Macromolecules* **20** 2265.

[231] Younathan, J. N., McClanahan, S. F., and Meyer, T. J. (1989) *Macromolecules* **22** 1048.

[232] Strouse, G. F., Worl, L. A., Younathan, J. N., and Meyer, T. J. (1989) *J. Am. Chem. Soc.* **111** 9101.

[233] Meyer, T. J. (1989). In Norris, J. R. Jr. and Meisel, D. (eds) *Photochemical energy conversion.* Elsevier, p. 75.

[234] Fromherz, P. and Rieger, B. (1986) *J. Am. Chem. Soc.* **108** 5361.

[235] Barton, J. K., Kumar, C. V., and Turro, N. J. (1986) *J. Am. Chem. Soc.* **108** 6391.

[236] Purugganan, M. D., Kumar, C. V., Turro, N. J., and Barton, J. K. (1988) *Science* **241** 1645.

[237] Telser, J., Cruickshank, K. A., Schanze, K. S., and Netzel, T. L. (1989) *J. Am. Chem. Soc.* **111** 7221.

[238] Telser, J., Cruickshank, K. A., Morrison, L. E., Netzel, T. L., and Chan, C. (1989) *J. Am. Chem. Soc.* **111** 7226.

6

Covalently-linked systems: electronic energy transfer

6.1 INTRODUCTION

Electronic energy transfer in supramolecular systems lies at the heart of important natural phenomena (for example, photosynthesis) as well as of practical applications (section 12.5). This chapter deals with electronic energy transfer in covalently-linked donor–acceptor (CLDA) systems (6.1),

$$*A–L–B \rightarrow A–L–*B \tag{6.1}$$

that is, supramolecular species in which an energy donor ($*A$) and an energy acceptor (B) are bound together by some kind of connector (L). Examples of intercomponent energy-transfer processes in supramolecular structures other than CLDA systems can be found in Chapters 7–11. The mechanistic issues encountered in dealing with energy transfer in supramolecular systems are in some respects similar to those found in other types of more complex systems not dealt with in this monograph, such as crystalline solids [1], polymers [2], solid-state inclusion compounds [3], organized molecular assemblies [4] and ordered phases [5].

Given the inhomogeneous nature of the studies reviewed, it has been impossible to find a completely satisfactory criterion to organize the material. As far as possible, however, CLDA systems with similar structural characteristics have been grouped together. Early work on energy transfer in organic CLDA systems has already been reviewed [6]. A survey of energy transfer in CLDA systems based on coordination compounds is available [7].

6.2 SUPRAMOLECULAR SYSTEMS BASED ON ORGANIC MOLECULAR COMPONENTS

6.2.1 Systems with rigid or conformationally-restricted bridges

Some of the rigid bridges originally developed for systematic studies of photoinduced and thermal intercomponent electron-transfer reactions (sections 5.3 and 5.4) have subsequently been used to build up CLDA systems suitable for the study of intercomponent energy-transfer processes. For this purpose, Closs, Miller, and coworkers [8–10] have synthesized a series of CLDA systems based on steroid-type bridges. The CLDA systems are identical to those of Fig. 5.4, except for the presence of a benzophenone group replacing the biphenyl fragment. In these systems, the triplet energy levels are such that benzophenone and naphthalene behave as triplet-energy donor and acceptor, respectively. Following pulsed laser excitation of the benzophenone chromophore, the decay of the benzophenone triplet absorption and the simultaneous build-up of the naphthalene triplet absorption are observed. As usual for triplet–triplet processes, this energy transfer occurs by an *exchange* mechanism (section 3.4.5). The rate of the energy-transfer process falls off with increasing number of C–C bonds separating the donor and the acceptor. As for electron-transfer processes (5.17) the fall-off is exponential, with a β coefficient of 2.6/bond [8]. This value is about twice those obtained for electron transfer and hole transfer in the radical anions and cations, respectively, of analogous CLDA systems [11,12]. The stronger distance-dependence of triplet energy transfer can be explained by the following arguments [9,10]. For a unimolecular exchange energy-transfer process, the rate constant is given by a Golden Rule expression identical to that for electron (and hole) transfer (5.8). The only difference lies in the nature of the electronic matrix element, which is a two-electron exchange integral in the case of energy transfer, and a one-electron resonance integral in the case of electron (or hole) transfer. As shown in the orbital diagram of Fig. 6.1, electron transfer is the transfer of an electron between the LUMOs of the donor and acceptor, hole transfer is the transfer of an electron between the HOMOs, and exchange energy transfer is a double electron transfer involving LUMOs and HOMOs. In such a simple view, the matrix element for an exchange energy-transfer process is expected to be pro-portional to the product of the matrix elements for electron and hole transfer. Since the electron- and hole-transfer processes depend exponentially on distance (with almost identical β values) [11,12], an exponential dependence is also expected for the energy-transfer process, with a β value roughly twice those of the other two processes. A further complete check of this model can be made [9] by plotting the rate constants of energy transfer for the various CLDA of this series against the product of the corresponding electron- and hole-transfer rate constants (after correction of the latter rate constants for the distance dependence of the solvent reorganizational energy). The plot obtained is indeed linear. Thus, the intuitively simple view of exchange energy transfer as a simultaneous electron and hole transfer is substantially correct. Information from energy transfer can be used to learn about electronic coupling in electron (or hole) transfer, and *vice versa*.

The norbornylogous bridges shown in Fig. 5.6 have been used by Verhoeven, Paddon-Row, and coworkers [13,14] in the synthesis of CLDA systems suitable for singlet–singlet energy transfer. They are identical to those of Fig. 5.6, except for a carbonyl group replacing the dicyanovinyl fragment. In these systems, the fluorescence of the dimethoxynaphthalene component is quenched to a different extent depending on the number of C–C bonds in the bridge. The quenching constant decreases exponentially with the number of bonds in the bridge, becoming immeasurably small for the 12-bond system. The β coefficient is 1.59/bond, again

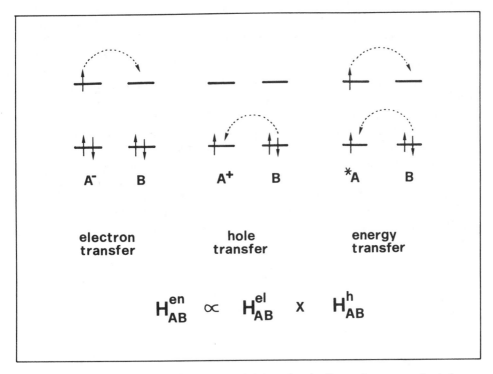

Fig. 6.1 — Relationship between the electronic interactions leading to electron transfer, hole transfer, and energy transfer [9].

about twice the value obtained for electron transfer across the same bridges in similar CLDA systems [15–19] (section 5.4). As far as the mechanism is concerned, singlet–singlet energy transfer can in principle occur by either coulombic or exchange mechanisms (section 3.4.5). The exponential dependence of energy-transfer rate constants on distance (rather than the $1/(r_{AB})^6$ dependence required by a coulombic mechanism [20]) points towards an *exchange* mechanism [13]. The objection [21] that exchange is an inherently short-range mechanism is based on a conventional through-space view and is not appropriate for CLDA systems of this type, where the bridge acts to propagate exchange interaction over relatively long distances [13].

Singlet–singlet and triplet–triplet energy transfer have been studied by Rubin, Speiser, and coworkers, using CLDA systems of type **6.1** [22] and **6.2** [23,24], in

6.1

which the aromatic component is the donor and the diketone component is the acceptor. In **6.2**, bridging chains of various length ($n = 2$–6) are used. Despite the presence of flexible methylene chains, the double-bridged structure severely restricts the overall flexibility of these systems, leading to a relatively narrow distribution of intercomponent distances for each species [23]. Upon excitation of the aromatic chromophore, these systems exhibit dual fluorescence, arising from both the aromatic and the diketone components. This indicates the occurrence of a partially efficient singlet–singlet energy-transfer process. The dependence of such a process on distance and temperature has been used to identify the mechanism as electron exchange. Triplet–triplet energy transfer has also been demonstrated to occur in **6.2**, by comparing the intensity of the diketone phosphorescence upon light absorption by the diketone or the benzenoid chromophores [24]. Fast (3×10^{10} s^{-1}) triplet–triplet energy transfer has also been observed by Maki *et al.* [25] in a CLDA system in which an anthrone donor and a naphthalene acceptor are coupled by a spiran linkage.

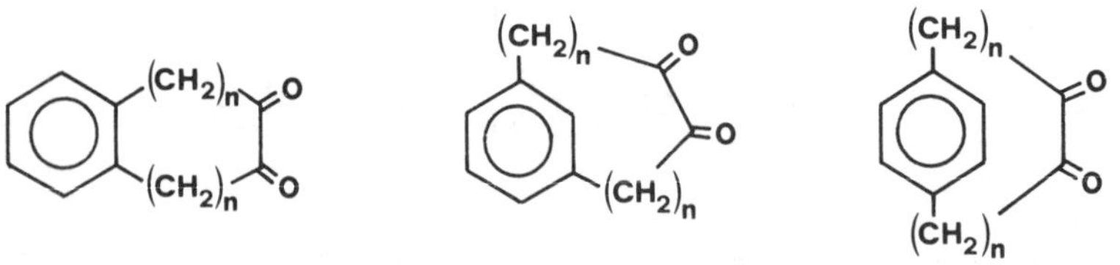

6.2

In the CLDA system **6.3** studied by Morrison and coworkers [26], a singlet–singlet energy-transfer process is used to sensitize the photochemistry of the ketone

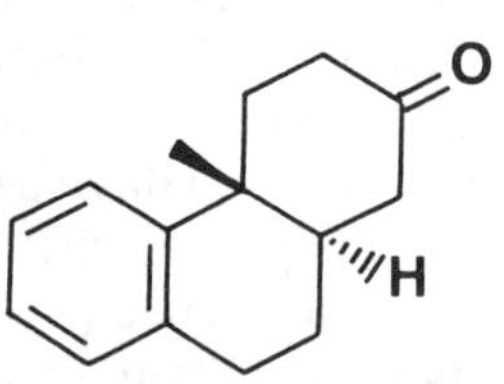

6.3

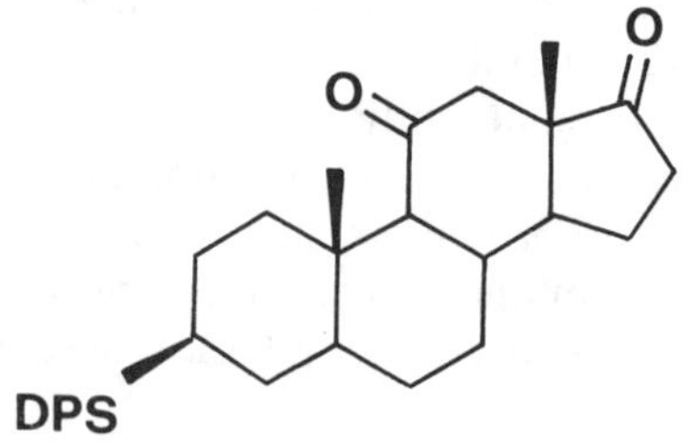

6.4

group with light absorbed by the benzenoid chromophore. In an interesting development of this work, the three-component system **6.4**, which contains the DPS (dimethylphenylsiloxy) group as chromophore and two (11-keto) and (17-keto) ketone functions, has been studied [27,28]. When the 17-keto group is irradiated directly, a typical singlet reaction (epimerization) and a typical triplet reaction (photoreduction) are observed in a *ca.* 2:1 ratio. This is consistent with the relatively inefficient intersystem crossing yield of cyclopentanones. When, on the other hand, the DPS chromophore is irradiated, photoreduction of the 17-keto group is observed as the only photoreaction. This requires that when excitation occurs through the DPS group, the 17-keto triplet is formed without the intermediacy of the 17-keto singlet. The explanation is given in terms of the three-step mechanism shown in Scheme 6.1: (i) the initial energy-transfer step to the 11-keto group is of the singlet–singlet type (as in two-component systems like **6.3**); (ii) the 11-keto group (as most cyclohexanones) undergoes facile intersystem crossing; (iii) the final energy-transfer step from 11-keto to 17-keto groups is of the triplet–triplet type. In the energy-transfer pathway of this three-component system, the intermediate component acts as a *singlet–triplet switch* [28].

Energy transfer across polyene bridges has been studied by Effenberger, Wolf,

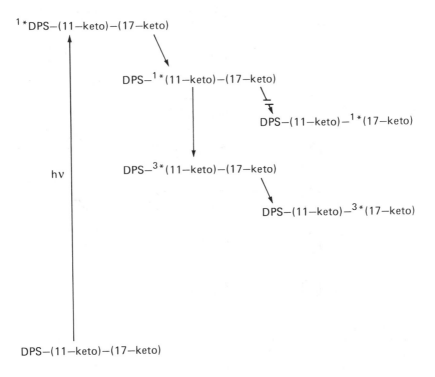

Scheme 6.1

and coworkers [29] in a CLDA system similar to those shown in Fig. 5.10, but having tetraphenylporphyrin and anthracene (or naphthalene) on the two sides of the polyene chain. In such systems, light excitation of the anthryl group is followed by partial energy transfer to the porphyrin, as shown by sensitization of the porphyrin fluorescence. Since no change is observed in the decay of the emissions compared with the uncoupled components, it is suggested that energy transfer takes place *via* population of a delocalized supramolecular excited state (in competition with the luminescent S_1 level of the anthryl group) which then relaxes into the luminescent S_1 state of the porphyrin [29].

In the extensive series of CLDA systems shown in Fig. 6.2 a zinc-substituted and

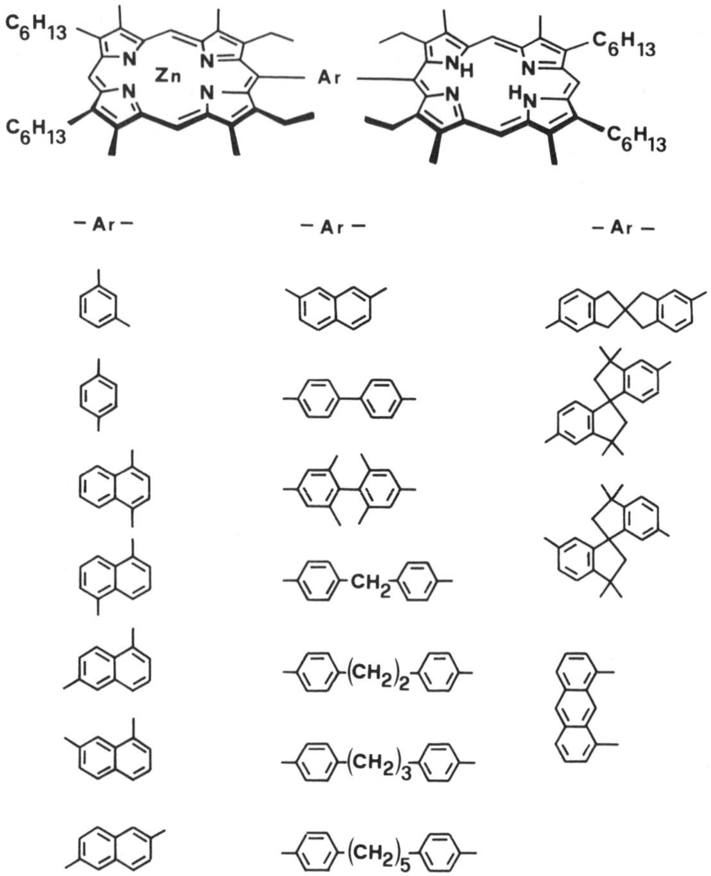

Fig. 6.2 — Zn-porphyrin and porphyrin chromophores bridged by various connectors [30].

an unsubstituted porphyrin are linked together by various rigid bridges [30]. These CLDA systems are designed to investigate in some detail the role of the mutual orientation of the components in intercomponent energy transfer. All of the systems undergo singlet–singlet energy transfer from the metalated to the free-base porphyrin, as shown by time-resolved fluorescence spectroscopy. The effect of mutual orientation of the porphyrin rings on the rate constants is in good quantitative agreement with the predictions of Förster theory [20], indicating that (except perhaps for some systems with very short donor–acceptor separation) the singlet–singlet energy transfer occurs by a coulombic mechanism. Resonance Raman studies of analogous diporphyrin compounds with *p*- and *m*-phenylene bridges have given a clear indication that the two components are weakly interacting in their excited states, so that excitation can be considered to be component-localized on a timescale of *ca*. 1 ps [31].

An "oblique" diporphyrin system (**6.5**) has been synthesized and studied by Sauvage and coworkers [32]. In this CLDA system, the intercomponent distance and angle are not far from those found in the bacterial reaction centres between the special pair and bacteriochlorophyll monomer, and between bacteriochlorophyll monomer and bacteriopheophytin. Light absorption in the Soret region (where the spectrum is virtually identical to that of the corresponding monomers, indicating weak interchromophoric interaction) excites in comparable amounts the zinc and free-base porphyrin. Efficient intercomponent energy transfer occurs in this system, as shown by the fact that the fluorescent emission of the zinc porphyrin is completely quenched, while the intensity of the free base fluorescence is almost doubled. Intercomponent electron transfer, which is thermodynamically less favoured than energy transfer (estimated ΔG^0, -0.07 eV and -0.17 eV, respectively), does not occur to any appreciable extent in this system.

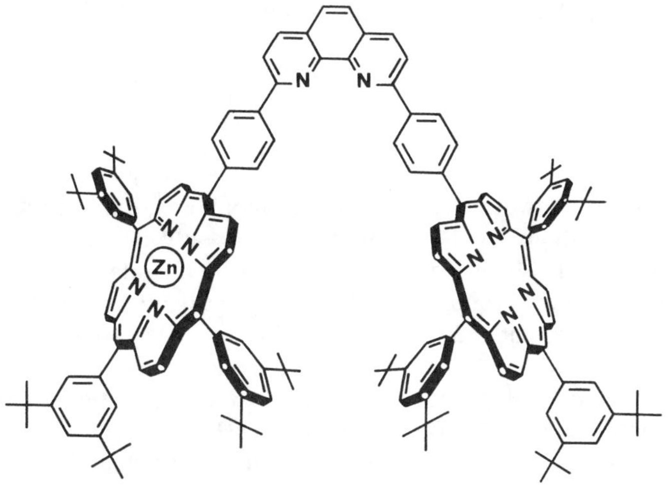

6.5

Sessler *et al.* [33], have studied three-component systems made of a zinc porphyrin (ZnP), a free-base porphyrin (P), and a quinone (Q) covalently linked. When the quinone is linked to the free-base porphyrin, a very fast energy-electron transfer sequence is observed (*ZnP–P–Q → ZnP–*P–Q → ZnP–P$^+$–Q$^-$). When the opposite situation obtains, electron transfer between the terminal free-base porphyrin and the quinone (*P–ZnP–Q → P$^+$–ZnP–Q$^-$) occurs, but at a smaller rate.

6.2.2 Systems with flexible bridges

Energy-transfer processes between coumarin-type components linked by flexible bridges have been investigated by Valeur and coworkers [34–37]. For the supramolecular species **6.6** (where n may be 3, 4, 8, or 12), emission and excitation spectra showed that light absorption by the 7-hydroxycoumarin chromophoric group is followed by fluorescence from the 4-trifluoromethyl-7-aminocoumarin moiety [34,35]. The energy-transfer process, which occurs with high efficiency, is probably favoured by the flexibility of the chain and, in some solvents, by the formation of a hydrogen bond between the oxygen and the NH group of the chain. The rate of the energy-transfer process in propylene glycol at 25°C, measured by phase-modulation fluorimetry, was found to be $\sim 2 \times 10^{10}\,\mathrm{s}^{-1}$, and slightly sensitive to the number of methylene groups [36]. The distance distribution of the two components has been calculated for the various supramolecular species and the results obtained, when used in the frame of the Förster (coulombic) energy-transfer model, give values for energy-transfer efficiency in good agreement with those obtained from experiments in DMF solution at 25°C.

6.6

Energy transfer between two identical chromophores, that cannot be observed by conventional methods, can be followed *via* time-resolved fluorescence anisotropy measurements [38]. This method was used to measure the rate of excitation hopping between two 2-naphthyl groups linked by $-(CH_2)_n-$ alkane chains ($n = 3, 5, 7$, and 12) [39]. The rate constant in 2-methyltetrahydrofuran decreases from $1.8 \times 10^8\,\mathrm{s}^{-1}$ for $n = 3$ to $6.8 \times 10^6\,\mathrm{s}^{-1}$ for $n = 12$. The experimental values were found to be in fair agreement with those calculated on the basis of a conformational analysis of the system and the Förster mechanism for energy transfer.

Green plants and photosynthetic bacteria employ antenna pigments to harvest sunlight and channel energy to a reaction centre where chemical reaction originates. These antenna units contain 200–300 closely associated chlorophyll molecules encased in protein. Energy transfer takes place by the coulombic mechanism and trapping by the reaction centre is essentially quantitative. In an attempt to mimic this process, synthetic porphyrin arrays have been investigated [40–44]. Symmetrical dimers show some degree of interaction between the porphyrin rings, as indicated by changes in intensity and position of the Soret absorption bands. For asymmetrical dimers (that is, for zinc and free-base porphyrins covalently linked) singlet–singlet transfer can occur between the two porphyrin rings. The efficiency of this process depends upon the mutual orientation of the two rings and the nature of the central cations [41,42]. In a pentameric array of porphyrin molecules **6.7** efficient singlet–singlet energy transfer occurs from peripheral zinc porphyrins to a central free-base porphyrin [42,43] (antenna effect, section 12.5.2). Investigations on triplet–triplet energy transfer in asymmetric dimers have also been performed [44].

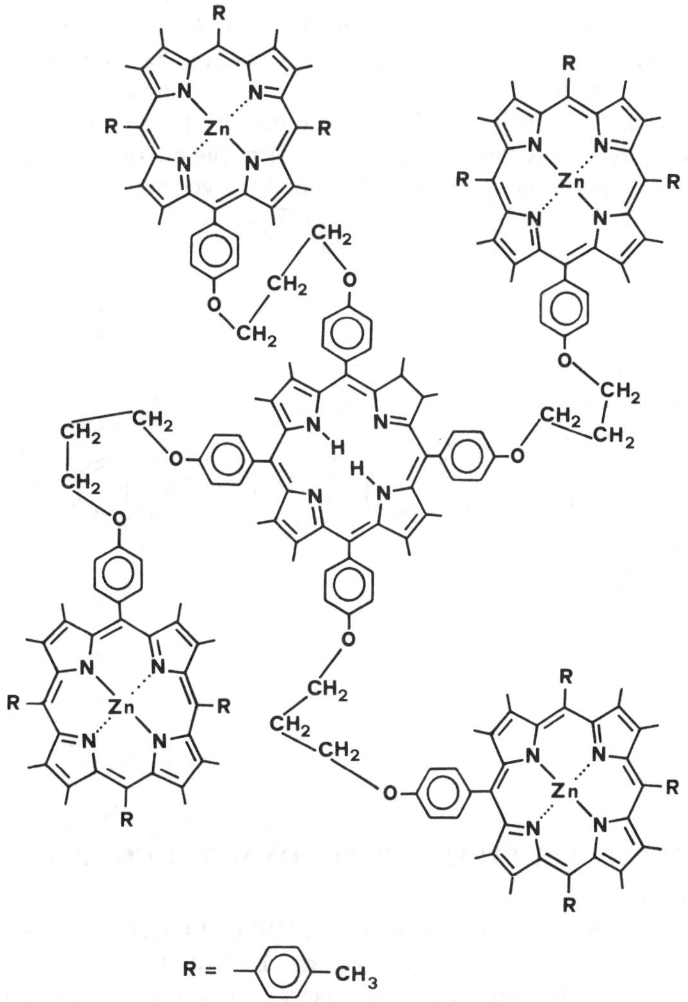

$$R = \text{—}\bigcirc\text{—} CH_3$$

6.7

In natural photosynthetic membranes, carotenoid molecules play the roles of singlet energy donors to, and triplet energy acceptors from, chlorophyll molecules. Such antenna and protection functions can be mimicked in artificial diads [45,46], triads [47], and tetrads [48]. Rather interesting is the tetrad system described by Gust *et al.* [48] (see also section 5.8), which consists of two covalently-linked porphyrins, one bearing a carotenoid polyene and the other a naphthoquinone electron-acceptor. In such a system the following processes take place: (i) singlet energy transfer between the two porphyrins; (ii) singlet energy transfer from carotenoid to porphyrins; (iii) carotenoid quenching of porphyrin triplet states; (iv) multistep photoinduced electron transfer.

Energy transfer in the supramolecular species **6.8** (where $n = 0$ or 2), which consists of covalently-linked porphyrin and cyanine dye, has been studied in detail

[49]. When $n = 0$ the cyanine dye absorbs in the 450–470 nm region and energy transfer occurs with 80% efficiency from the dye to the free-base porphyrin. When $n = 2$, the cyanine dye absorbs at lower energies (650–780 nm region) and energy transfer occurs with 80% efficiency from the free-base porphyrin to the dye. When the free-base porphyrin is replaced by zinc porphyrin, the energy-transfer efficiencies are much lower, presumably because of competing electron-transfer quenching. The energy-transfer efficiencies fall within the range predicted by Förster theory, but an exchange mechanism cannot be ruled out [49].

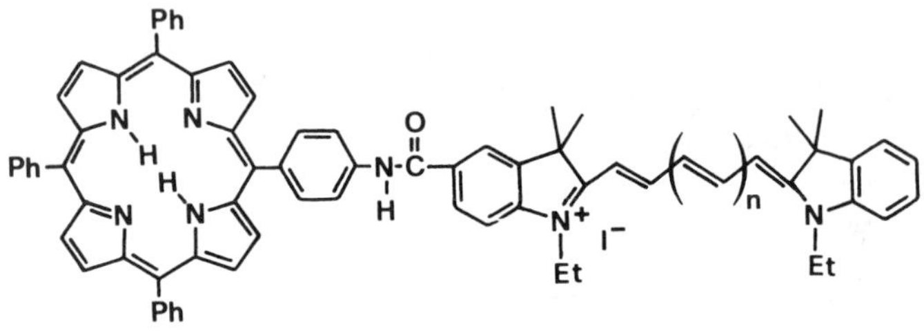

6.8

6.3 SUPRAMOLECULAR SYSTEMS BASED ON METAL COMPLEXES

6.3.1 Introduction

In the past 15 years a great number of Ru(II) [50] and Os(II) [51] polypyridine complexes have been synthesized and characterized. Most of them exhibit spectroscopic, photophysical, photochemical, and electrochemical properties appropriate to playing the role of building blocks for the construction of supramolecular systems for energy migration purposes. In the studies performed so far such building blocks have been connected in two fundamentally different ways, that is (1) by chemical bridges R that link together ligands belonging to distinct complexes, or (2) by bridging ligands BL that directly coordinate distinct metal-containing chromophoric fragments. A recent review on photoinduced electron- and energy-transfer processes in polynuclear complexes is available [7].

6.3.2 Complexes linked through bridges

6.3.2.1 Identical complexes

Identical complexes can be linked together to give binuclear species of the type

$(L)_2Ru(L'-R-L')Ru(L)_2^{4+}$, where $L'-R-L'$ (**6.9**) is made of two Mebpy chelating units covalently linked by a connector R, and L is either bpy or Me_2bpy [52–56]. Except for the case of the R_c linker, the absorption spectra, emission spectra, luminescence lifetimes, luminescence quantum yields, and electrochemical properties of each binuclear species are practically identical to those of the respective parent component. These results indicate that the interaction between the two components is very small. The symmetrical tetranuclear complex $[(Me_2bpy)_2Ru(L'-R_b-L')]_3Ru^{8+}$ again shows spectroscopic and redox properties very similar to those of the parent component, except for the luminescence quantum yield which is 40% lower [55]. This different behaviour has been tentatively attributed to a smaller radiative rate constant. The behaviour of the species with the $L'-R_c-L'$ bridge is quite different because the connector R_c is conjugated to the L' ligands. The bridge is therefore directly involved in the first reduction process and in the low-energy excited states which are a luminescent ^3MLCT level and a ^3LC level, in thermal equilibrium [56]. In this case, the properties of the dinuclear complex are different from those of the mononuclear parent.

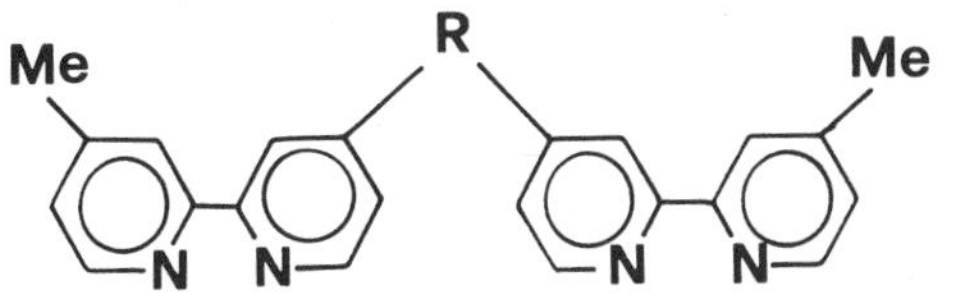

$$R_2 \quad (CH_2)_2$$
$$R_3 \quad (CH_2)_3$$
$$R_5 \quad (CH_2)_5$$
$$R_{12} \quad (CH_2)_{12}$$
$$R_a \quad CH_2-CHOH-CH_2$$
$$R_b \quad CH_2-C_6H_4-CH_2$$
$$R_c \quad CH=CH-C_6H_4-CH=CH$$

6.9

The tripod tribpy ligand **6.10** can coordinate one, two, or three $Ru(bpy)_2^{2+}$ units to give mono-, bi-, and trinuclear complexes of general formula $[(bpy)_2Ru]_n(tribpy)^{2n+}$ ($n = 1-3$) [57]. The spectroscopic, photophysical, and electrochemical properties of the three complexes are independent of the number of Ru(II) units, indicating that there is no appreciable interaction between the metal centres. The luminescence exhibited by the free tripod ligand at 77 K is not present in the trinuclear complex, as expected because of the fast conversion of upper excited states to the lowest (luminescent) ^3MLCT state within each metal-containing unit. In the mononuclear $Ru(bpy)_2(tribpy)^{2+}$ complex, where two bpy-type arms of **6.10** are free, ligand-centred luminescence could be expected to occur. In fact, only the ^3MLCT luminescence is observed, and quantitative measurements show that energy transfer from the ligand-localized levels of the noncoordinating arms to the metal-containing unit is very efficient [57]. The tribpy-type ligands can also be used to obtain hemicaged complexes, as discussed in section 11.1.5.

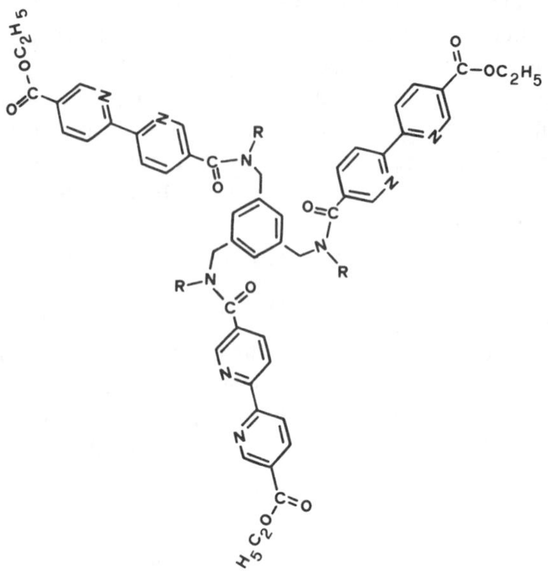

6.10

6.3.2.2 Complexes of different metals

The visible absorption spectrum of the binuclear complex $(bpy)_2Ru(L'-R_2-L')PtCl_2^{2+}$ is coincident with the sum of the spectra of the $Ru(bpy)_2(L'-R_2-L')^{2+}$ and $Pt(bpy)Cl_2$ components [52]. The binuclear complex shows two reversible one-electron reduction processes at potentials close to those found for the components. The luminescence lifetime and quantum yield are practically equal to those of the Ru-based unit. This behaviour indicates that there is no appreciable interaction between the two components.

In the $(bpy)_2Ru(L'-R_a-L')Os(bpy)_2^{4+}$ complex selective excitation of the two fragments is not possible due to spectral overlap, so that the emission spectrum consists of Ru-based ($\lambda_{max} = 615$ nm) and Os-based ($\lambda_{max} = 720$ nm) emissions [58]. With respect to the emissions of the two free components, a decrease of the intensity of the Ru-based emission and an increase in that of the Os-based emission were observed, indicating that intercomponent energy transfer takes place. The time-dependence of the luminescence signals at 610 and 800 nm showed that the decay of the Ru-based emission and the rise of the Os-based emission coincide, as expected in the case of energy transfer. The ratio of the pre-exponential factors of the two time-dependent processes indicates that the intercomponent energy-transfer process has almost unit efficiency. The absorption of the Os-moiety overlaps substantially the emission of the Ru-moiety, and calculations of the energy-transfer rate constant according to the coulombic model yielded a value consistent with that obtained from the experiments [58].

An interesting series of heterotetranuclear complexes of general formula $[(bpy)_2Ru(L'-R-L')]_3Fe^{8+}$, where R (**6.9**) is R_2, R_5, R_{12}, or R_b, has been studied by

Schmehl *et al.* [59]. These tetranuclear complexes, constituted by a central Fe^{2+} ion coordinated to three $(bpy)_2Ru(L'-R-L')^{2+}$ moieties, were obtained in solution by *in situ* complexation of $(bpy)_2Ru(L'-R-L')^{2+}$ with Fe(II). The solutions exhibit emissions with double exponential decays. The fast component of the decay corresponds to the emission from Ru-based units complexed to iron, and the long component comes from excess monomeric Ru(II) complexes present in solution. The reason for the faster decay of the emission of the polynuclear complexes with respect to mononuclear ones is, according to the authors, energy transfer from the ruthenium-containing moieties to the iron centre [59]. This conclusion is based on the following observations: (i) bimolecular quenching does not occur in solutions containing $Ru(bpy)_3^{2+}$ and $Fe(bpy)_3^{2+}$ at comparable concentrations; (ii) the quenching behaviour mirrors the formation of the tetranuclear species; (iii) the relative importance of the fast decay increases with increasing Fe:Ru molar ratio; (iv) the luminescence decay of the corresponding homometallic tetranuclear species $[(bpy)_2Ru(L'-R-L')]_3Ru^{8+}$ does not exhibit a short-lived component. The experimental results do not enable a definite mechanism to be established (coulombic or exchange) of the energy-transfer process. No systematic dependence of the energy-transfer rate on the length of the hydrocarbon chain of the bridge was observed.

Preliminary observations on a trinuclear complex obtained by coordination of two $Ru(bpy)_2^{2+}$ and one $Os(bpy)_2^{2+}$ units to a tripod ligand of the type of **6.10** indicate that energy transfer takes place from the Ru-based to the Os-based components [60].

6.3.2.3 *Complexes of different ligands*

The binuclear $(Me_2bpy)_2Ru(L'-R_b-L')Ru(dec-bpy)_2^{4+}$ and tetranuclear $[(dec-bpy)_2Ru(L'-R_b-L')]_3Ru^{8+}$ systems, where dec-bpy is 4,4'-dicarboxyethyl bipyridine, have been studied by Schmehl and coworkers [55]. For the binuclear complex the absorption spectrum and the redox properties are consistent with those expected for noninteracting components. The emission spectrum is very similar to that of the $Ru(dec-bpy)_2(L'-R_b-L')^{2+}$ component, which contains the lowest excited state of the system, while no emission is observed from the $Ru(Me_2bpy)_2(L'-R_b-L')^{2+}$ fragment. Excitation spectra at 77 K show that quenching of the luminescence of the higher energy $Ru(Me_2bpy)_2(L'-R_b-L')^{2+}$ component is accompanied by sensitization of the emission of the lower energy $Ru(dec-bpy)_2(L'-R_b-L')^{2+}$ unit. This indicates that energy transfer, which is exergonic by ~ 0.12 eV, successfully competes with the electron-transfer quenching, which is also slightly exergonic (~ 0.07 eV). The behaviour of the tetranuclear systems $[(dec-bpy)_2Ru(L'-R_b-L')]_3Ru^{8+}$ is quite similar to that of the binuclear one.

For the binuclear $(biq)_2Ru(L'-R_b-L')Ru(bpy)_2^{4+}$ species, spectroscopic and redox properties clearly indicate that the fragment containing the $Ru(biq)_2^{2+}$ unit is that with the lowest MLCT energy [61]. Overlap with the other MLCT bands ($Ru \rightarrow bpy$ and $Ru \rightarrow L'-R_b-L'$) prevents selective excitation of this fragment. The emission spectrum exhibits both $Ru \rightarrow biq$ and $Ru \rightarrow bpy$ (or $Ru \rightarrow L'-R_b-L'$) MLCT emissions. However, the $Ru \rightarrow bpy$ (or $Ru \rightarrow L'-R_b-L'$) emission is strongly quenched with respect to that of the mononuclear $Ru(bpy)_2(L'-R_b-L')^{2+}$ complex (at 77 K, $\tau = 230$ ns and 4150 ns for the bi- and mononuclear species, respectively). The excitation spectrum of the Ru→biq CT emission matches very closely the absorption spectrum. An analysis of the time-dependence of the emission at 13 K shows that the risetime of the Ru→biq emission coincides with the decay of the Ru→bpy (or Ru→L'–R_b–L') emission. These results demonstrate that quenching occurs by intercomponent energy transfer. An important question concerns the mechanism of the energy-transfer process: the small spectral overlap between the donor emission and acceptor absorption and the spin-forbidden character of the transition involved might suggest an exchange mechanism. However, in view of the partial singlet character of the formally triplet states, a coulombic energy-transfer mechanism cannot be ruled out. Actually, the energy-transfer rate constant calculated with the coulombic model is in good agreement with the experimental value ($k = 8.9 \times 10^5$ s^{-1} at 13 K) [61].

The investigation of the binuclear complex $(Me_2bpy)_2Ru(L'-R_b-L')$–$Ru(Me_2bpy)(CN)_2^{2+}$ has given considerable information on the energy-transfer mechanisms [62]. In this complex the higher energy centre is $(Me_2bpy)_2$–$Ru(L'-R_b-L')^{2+}$ and the lower energy one is $(L'-R_b-L')Ru(Me_2bpy)(CN)_2$. Luminescence is observed from both components, but excitation spectra show that energy transfer from the $(Me_2bpy)_2Ru(L'-R_b-L')^{2+}$ to the $(L'-R_b-L')Ru(Me_2bpy)(CN)_2$ moiety contributes to the emission of the latter. The energy-transfer process exhibits both a solvent dependence and a temperature dependence. An accurate study of the temperature dependence of the energy-transfer rate between 200 and 300 K showed that more than one decay path is introduced in the excited $(Me_2bpy)_2Ru(L'-R_b-L')^{2+}$ species when it is linked to the other fragment. Since electron-transfer quenching is thermodynamically disfavoured, the results have been interpreted as indicating that both exchange and coulombic energy-transfer paths are operative at high temperature. Cooling of the solution results in a decrease in the energy-transfer rate until a constant value is obtained. Such a low-temperature plateau is thought to correspond to coulombic energy transfer [62].

6.3.3 Metal-containing fragments bridged by polyimine ligands

6.3.3.1 Introduction
A quite efficient way to build up polynuclear metal complexes is that based on the use of polyimine ligands to bridge metal-containing building blocks. The structural formulae and abbreviations of the most commonly-used bridging ligands (BL) are shown in Fig. 6.3.

6.3.3.2 Symmetrical homometallic binuclear complexes
Symmetrical homometallic binuclear complexes are not of direct interest from the point of view of energy transfer. We shall discuss briefly their behaviour to understand the perturbation caused by bi- and polynucleation on a chromophoric unit. Table 6.1 summarizes the properties of some mononuclear $Ru(bpy)_2(BL)^{n+}$ and binuclear $(bpy)_2Ru(BL)Ru(bpy)_2^{n+}$ complexes, where the bridging ligand BL is bpm [63–68], bpt$^-$ [69], 2,5-dpp [70,71], BiBzIm^{2-} [72,73], 2,3-dpp [66–68,70,71,74], ppz [75], HAT [76], dpq [65–67,77], tppq [65], bidpq [65,78]. (The nonequivalence of the two chelating sites of bpt$^-$, discussed in the next section, is

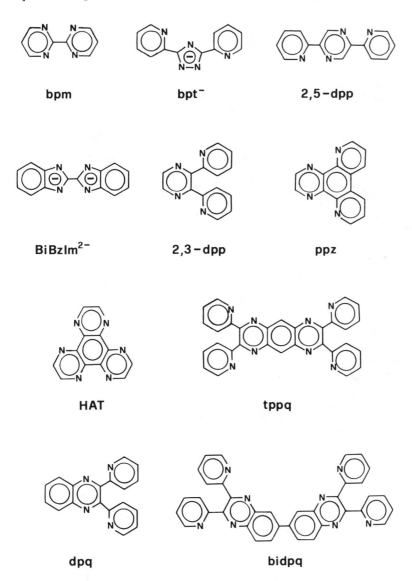

bpm   bpt⁻   2,5−dpp

BiBzlm^{2-} 2,3−dpp   ppz

HAT   tppq

dpq   bidpq

Fig. 6.3 — Structural formulae of the commonest polyimine bridging ligands.

irrelevant to the present discussion.) The structure of the 2,3-dpp complex is schematized in Fig. 6.4 and the absorption spectra of the mono- and binuclear 2,3-dpp complexes are shown in Fig. 6.5. For the mononuclear complexes in all cases, with the exception of BL = bpt⁻ and BiBzIm^{2-}, the first reduction takes place at the bridging ligand, and the lowest (luminescent) excited state is a metal-to-(bridging ligand) charge-transfer (M(BL)CT) excited state. The behaviour of the binuclear complexes can be easily rationalized if they are seen as mononuclear

Table 6.1 — Spectroscopic, photophysical, and redox properties of mono- and binuclear Ru(II)–polyimine complexes[a]

	λ^{abs}_{max}[b] (nm)	λ^{em}_{max} (nm)	τ (ns)	$E^{ox\,c}$ (V)	$E^{red\,c}$ (V)	Ref.[d]
Ru(bpy)$_2$(bpm)$^{2+}$	480	710[e]	76[e]	+1.40	−1.02	67
(bpy)$_2$Ru(bpm)Ru(bpy)$_2^{4+}$	606[ef]	795[fg]	—	+1.53	−0.41	67
Ru(bpy)$_2$(bpt)$^+$	480	678	160	+0.85	−1.47	69
(bpy)$_2$Ru(bpt)Ru(bpy)$_2^{3+}$	453	648	100	+1.04	−1.40	69
Ru(bpy)$_2$(2,5-dpp)$^{2+}$	486	695	620	+1.33	−1.03	70
(bpy)$_2$Ru(2,5-dpp)Ru(bpy)$_2^{4+}$	585	824	155[h]	+1.37	−0.53	70
Ru(bpy)$_2$(BiBzImH$_2$)$^{2+}$ [i]	463	683	161	+1.04	−1.66	73
(bpy)$_2$Ru(BiBzIm)Ru(bpy)$_2^{2+}$	512	733	60	+0.76	−1.49	73
Ru(bpy)$_2$(2,3-dpp)$^{2+}$	475sh	691	380	+1.31	−1.06	70
(bpy)$_2$Ru(2,3-dpp)Ru(bpy)$_2^{4+}$	527	802	125	+1.38	−0.67	70
Ru(bpy)$_2$(ppz)$^{2+}$	474	700	200	+1.37	−1.11	75
(bpy)$_2$Ru(ppz)Ru(bpy)$_2^{4+}$	573	820	<50	+1.35	−0.67	75
Ru(bpy)$_2$(HAT)$^{2+}$	484sh[f]	745[f]	105[f]	+1.56	−0.84	76
(bpy)$_2$Ru(HAT)Ru(bpy)$_2^{4+}$	572[f]	825[f]	148[f]	+1.53	−0.49	76
Ru(bpy)$_2$(dpq)$^{2+}$	517	766	71	+1.42	−0.77	67
(bpy)$_2$Ru(dpq)Ru(bpy)$_2^{4+}$	603	822	<20	+1.47	−0.37	67
Ru(bpy)$_2$(tppq)$^{2+}$	573	—	—	+1.42	−0.42	65
(bpy)$_2$Ru(tppq)Ru(bpy)$_2^{4+}$	642	—	—	+1.45	−0.16	65
Ru(bpy)$_2$(bidpq)$^{2+}$	525	770	<60	+1.41	−0.72	78
(bpy)$_2$Ru(bidpq)Ru(bpy)$_2^{4+}$	528	780	<60	+1.42	−0.67	78

[a]Deaerated CH$_3$CN solution at room temperature, unless otherwise noted; [b]lowest energy maximum; [c]$E_{1/2}$ values vs SCE; [d]for other references, see text; [e]propylene carbonate; [f]water; [g][68]; [h]aerated solution; [i]the deprotonated species can only be obtained in very basic solutions [72].

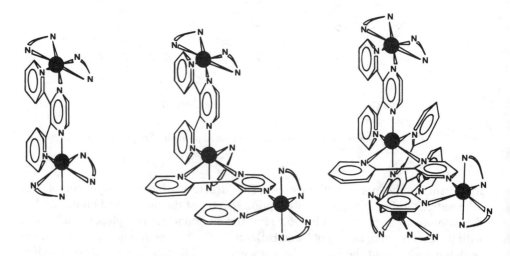

Fig. 6.4 — Schematic representation of bi-, tri-, and tetranuclear complexes of the 2,3-dpp bridging ligand; N–N stands for bpy or biq [70].

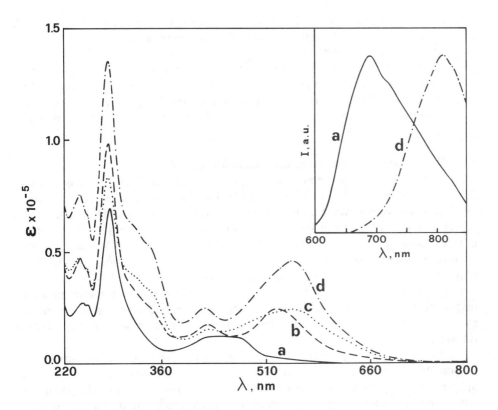

Fig. 6.5 — Absorption spectra of Ru(2,3-dpp)(bpy)$_2^{2+}$ (a), (bpy)$_2$Ru(2,3-dpp)Ru(bpy)$_2^{4+}$(b), (bpy)Ru[(2,3-dpp)Ru(bpy)$_2$]$_2^{6+}$ (c), and Ru[(2,3-dpp)Ru(bpy)$_2$]$_3^{8+}$ (d); the inset shows the corrected luminescence spectra of the mono- and tetra-metallic complexes [70].

(bpy)$_2$Ru(BL)$^{(n-2)+}$ species which carry the electron-acceptor Ru(bpy)$_2^{2+}$ substituent on the other side of the bridge. This substitution withdraws electronic charge firstly from the bridging ligand, to a smaller extent from the metal, and to an even smaller extent from the terminal bpy ligands. Thus in all cases, except for BL = bpt$^-$ and BiBzIm^{2-}, it is expected (and generally found) that on binucleation (i) the first (BL-centred) reduction potential becomes much less negative, (ii) the first (metal-centred) oxidation potential becomes slightly more positive, and the lowest M(BL)CT singlet (seen in absorption) and triplet (seen in emission) excited states move to lower energies. Similar results have been obtained for analogous systems where bpy is replaced by other polypyridine-type ligands [66,70,71,77] and/or Ru is replaced by Os [71,79]. For BL = bpt$^-$, the first reduction potential does not concern the bridge but the terminal bpy ligands and therefore, on binucleation, it is much less affected than the oxidation potential; as a consequence, the lowest energy absorption and emission bands, which correspond to Ru → bpy transitions, move to higher energies [69]. Similar behaviour is observed for complexes containing imidazole-type bridging ligands [72,73]. Several symmetrical homometallic binuclear complexes containing peripheral CO ligands have also been investigated [68,80-87], but their behaviour is of limited interest in the present context.

While the general behaviour of the symmetrical homometallic binuclear complexes can be rationalized qualitatively, the quantitative data (including the separation between the oxidation potentials of the two metal centres) depend on subtle electronic and nuclear factors. This is particularly true for properties like excited-state lifetimes and luminescence quantum yields which depend on dynamic parameters.

6.3.3.3 Unsymmetrical homometallic binuclear complexes

The properties of the homometallic complexes $(L)_2M(BL)M(L')_2^{4+}$, where $BL =$ 2,3- and 2,5-dpp, M = Ru, L = bpy, and $L' = biq$, have been investigated in detail [70,71]. These complexes exhibit a broad absorption in the visible region, which receives contributions from $Ru \rightarrow L$, $Ru \rightarrow L'$, and two different (because of the different coordination environment of the two metal ions) $Ru \rightarrow BL$ charge-transfer transitions. On the basis of the properties of the separated components and of the observed electrochemical behaviour, the lowest-energy excited state at room temperature can be assigned to a triplet $(bpy)_2Ru \rightarrow BL$ CT excited state. Luminescence originates only from this level, with comparable efficiencies regardless of the excitation wavelength, which implies the occurrence of intercomponent energy transfer.

In the binuclear complex $(bpy)_2Ru(bpt)Ru(bpy)_2^{3+}$ encountered in the previous section, the 1 (or 2) and 4 nitrogens of the triazole ring (shown in Fig. 6.3) are inequivalent [88]. As a consequence, the levels of the two $Ru(bpy)_2^{2+}$ fragments are not isoenergetic and energy transfer can be expected to occur. In CH_2Cl_2 solution this complex exhibits both luminescence (Table 6.1) and ligand photodissociation [69]. The excited state responsible for luminescence is the lowest triplet $Ru \rightarrow bpy$ CT level centred on the Ru-containing component which is attached to the N^1 (or N^2) position. No evidence has been found for luminescence from the lowest CT excited state of the other component. This shows that either the energy difference between the two levels is too small to be noticed, or energy transfer from the higher to the lower level takes place. Interestingly the observed photoreaction takes place from the lowest ^3MC level, which is localized on the component attached to the N^4 position [69].

6.3.3.4 Heterometallic binuclear complexes

In the $(bpy)_2Ru(2,3-dpp)Os(bpy)_2^{4+}$ complex the energy levels of the two components are definitely different and energy transfer from the Ru-based fragment to the Os-based fragment can be expected. However, no luminescence could be observed for this complex up to 850 nm, presumably because the emission from the Ru component is quenched by intercomponent energy transfer and that from the Os component lies outside the spectral region examined [79].

Preliminary investigations show that the complexes $(phen)_2Ru(2,3-dpp)Fe(CN)_4$ and $(phen)_2Ru(2,3-dpp)Fe(CN)_4^+$ emit in room temperature fluid solution [67]. This result is rather surprising, as the Fe-based fragments are expected to quench the emission of the ruthenium chromophore on the basis of the well-known behaviour of analogous Fe(II) and Fe(III) complexes in bimolecular processes [89,90]. A tentative explanation of this puzzling result based on the non-planar structure of the 2,3-dpp bridge has been advanced [67]. The absorption spectrum of $(bpy)_2Ru(bpm)PtCl_2^{2+}$ is quite similar in the visible region to that of the parent homobinuclear ruthenium complex, $(bpy)_2Ru(bpm)Ru(bpy)_2^{4+}$, except for a slight blue shift of the low-energy M(BL)CT band [91]. On the other hand, in contrast to the homobinuclear Ru complex, the electrochemical behaviour indicates little communication between the metal centres; $(bpy)_2Ru(bpm)PtCl_2^{2+}$ could thus be expected to emit, but no information is available in this regard.

The complexes $(CN)_5Fe(BL)Co(CN)_5^{5-}$, $(CN)_5Fe(BL)Rh(NH_3)_5$, and $(NH_3)_5Ru(BL)Rh(NH_3)_5^{5+}$, where BL = pyrazine (pz), 4-cyanopyridine (4-CNpy), or 4,4′-bipyridine (4,4′-bpy), have been studied by Petersen and coworkers [92,93]. The aim of this study was to connect a highly absorbing and photochemically stable chromophore ($(CN)_5Fe(BL)-$ or $(NH_3)_5Ru(BL)-$) and a nonabsorbing and photo-labile fragment ($-(BL)Co(CN)_5$ or $-(BL)Rh(NH_3)_5$), and to look for energy transfer from the former to the latter. These systems, however, display much more complex photochemical behaviour than expected on these simple grounds. All the complexes containing the $(CN)_5Fe(BL)-$ chromophore undergo a bleaching of the M(BL)CT band with high quantum yield ($\Phi = 0.6-0.01$), indicating the occurrence of a photocleavage reaction at the Fe–BL bond (6.2).

$$(CN)_5Fe(BL)Co(CN)_5^{5-} + H_2O \xrightarrow{h\nu} Fe(CN)_5(H_2O)^{3-} + Co(CN)_5(BL)^{2-} \qquad (6.2)$$

This result is unexpected, since mononuclear $(CN)_5Fe(BL)^{3-}$ species of comparable M(BL)CT energy are known to be virtually photostable [94]. Apparently, the Fe→BL CT state has a different reactivity in the binuclear complexes and is not quenched by intramolecular energy transfer.

For the $(NH_3)_5Ru(BL)Rh(NH_3)_5^{5+}$ complexes, the photochemical behaviour was found to depend on the nature of the bridging ligand BL [92,93]. With BL = pz, practically no photochemistry was observed ($\Phi < 10^{-5}$). With BL = 4-CNpy, moderate yields ($\Phi = 10^{-2}-10^{-4}$) of Rh–BL bond cleavage (6.3)

$$(NH_3)_5Ru(BL)Rh(NH_3)_5^{5+} + H_2O \xrightarrow{h\nu} Ru(NH_3)_5(BL)^{2+} + Rh(NH_3)_5(H_2O)^{3+}$$
$$(6.3)$$

were observed. With BL = 4,4′-bpy, competitive Rh–BL (6.3) and Ru–BL (6.4)

$$(NH_3)_5Ru(BL)Rh(NH_3)_5^{5+} + H_2O \xrightarrow{h\nu} Ru(NH_3)_5(H_2O)^{2+} + Rh(NH_3)_5(BL)^{3+}$$
$$(6.4)$$

bond cleavage of moderate quantum yields ($\Phi = 10^{-2} - 10^{-3}$) were observed. Since the Rh–BL bond cleavage can be considered a reaction characteristic of MC states of Rh(III) [95], its occurrence upon Ru→BL CT excitation is taken as an indication of energy transfer from the Ru-based to the Rh-based fragment [93]. As discussed by the authors, however, there are problems in accounting for the dependence of the behaviour on BL, as the energy of the M(BL)CT donor state follows the order 4-CNpy > pz > 4,4′-bpy, whereas that of the energy-transfer efficiency would be 4-CNpy = 4,4′-bpy ≫ pz. An alternative mechanism based on BL→Rh electron transfer followed by reaction at the labile Rh(II) centre was considered less plausible.

The bimetallic complex $(bpy)_2Ru(bpt)RhH_2(PPh_3)_2^{2+}$ undergoes photochemistry from a Rh-based MC excited state and emission from a Ru→bpy CT excited state. The emission quantum yield is invariable at longer wavelength where no photochemistry occurs. With the onset of photochemistry, the emission quantum yield decreases. These results have been interpreted in terms of intercomponent energy transfer [96].

The complex $(bpy)_2Ru(bpm)Re(CO)_3Cl^{2+}$ emits at 77 K from the lowest-energy Ru→bpm M(BL)CT state regardless of the Re→bpm or Ru→bpm nature of the M(BL)CT bands irradiated. This behaviour has been interpreted in terms of energy transfer from the Re-containing fragment to the emissive M(BL)CT state of the Ru-containing fragment [85]. In the $(bpy)_2Ru(2,3-dpp)Re(CO)_3Cl^{2+}$ compound luminescence occurs again from the Ru moiety [68].

In the binuclear complex $(tpy)(bpy)Os(BL)Ru(bpy)_2(H_2O)^{4+}$, where BL = 4,4′-bipyridine and tpy = 2,2′:6,2″-terpyridine, Ru→bpy CT excitation is followed by rapid and efficient energy transfer to the lower energy Os→tpy CT state [97]. When the pH of the solution is raised, the dominant form of the complex becomes $(tpy)(bpy)Os(BL)Ru(bpy)_2(OH)^{3+}$ and the emission of the Os(tpy) moiety is quenched.

6.3.3.5   Complexes of higher nuclearity

Polyimine complexes of higher nuclearity have been known for several years, [63,78,98–102], but only recently have luminescent species been obtained and intercomponent energy-transfer processes have been investigated.

The trinuclear $[Ru(bpy)_2]_3HAT^{6+}$ species, obtained by attaching three $Ru(bpy)_2^{2+}$ units around the central symmetrical HAT ligand shown in Fig. 6.3, is luminescent in fluid solution at room temperature [76,103]. Emission originates from the lowest level which is a triplet Ru→HAT CT excited state. The three chromophoric units, however, are identical and energy transfer cannot be investigated.

The trinuclear complexes $(bpy)Ru[(bpm)Re(CO)_3Cl]_2^{2+}$ and $(bpy)Ru(HAT)$ $[Re(CO)_3Cl]_2^{2+}$ and the tetranuclear complex $Ru[(bpm)Re(CO)_3Cl]_3^{2+}$ have been investigated [104]. An intriguing result of this study is that while the trinuclear HAT-bridged complex exhibits emission only at 77 K, the trinuclear and tetranuclear bpm-bridged complexes were found to emit at room temperature, though very weakly, with a relatively long lifetime. The emission is tentatively assigned as a M(BL)CT luminescence of the Ru-containing fragment.

The trinuclear complex $(bpy)Ru[(BiBzIm)Ru(bpy)_2]_2^{2+}$ is luminescent in CH_3CN solution at room temperature ($\lambda_{max} = 734$ nm, $\tau = 58$ ns, $\Phi_e = 2.2 \times 10^{-4}$)

[73]. The analogous tetranuclear complex is isolated as a Ru^{III}–Ru^{II} mixed-valence species and is not luminescent.

Trinuclear complexes of general formula $(bpy)Ru[(BL)Ru(L)_2]_2^{6+}$, where L = bpy or biq and BL = 2,3-dpp or 2,5-dpp, have been investigated [105]. Their properties are summarized in Table 6.2, where the data of the tetra- and hepta-nuclear complexes discussed below have also been gathered (for the parent binuclear complexes, see sections 6.3.3.2 and 6.3.3.3). The structural formulae of the tri- and tetranuclear complexes of 2,3-dpp and their absorption spectra are shown in Figs 6.4 and 6.5.

As one can understand on looking at the schematic formula **6.11**, these trinuclear complexes contain two nonequivalent Ru sites, one central (Ru_c) and two peripheral (Ru_p). For each complex four different types of (proximate) metal-to-ligand charge-transfer transitions are therefore expected: $Ru_c \rightarrow bpy$, $Ru_c \rightarrow BL$, $Ru_p \rightarrow BL$, and $Ru_p \rightarrow L$. Two broad bands are observed for each complex in the visible region (Fig. 6.5). The low-energy band receives contributions mainly from CT transitions involving the bridging ligand. The bielectronic nature of the first oxidation wave indicates that oxidation first occurs at the two peripheral Ru centres. This result suggests that the lowest excited state, which is responsible for the emission, corresponds to a $Ru_p \rightarrow BL$ transition. Excitation spectra show that the luminescent level is populated with the same efficiency regardless of the initially-populated CT level, which implies the occurrence of intercomponent energy-transfer processes having unit efficiency.

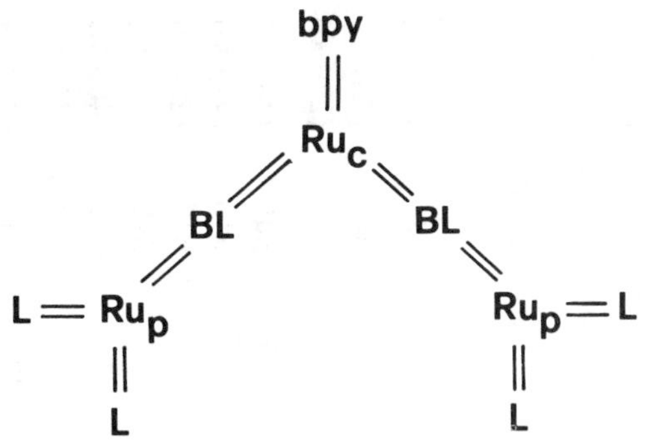

bpy
||
Ru$_c$

BL   BL

L = Ru$_p$ Ru$_p$ = L
|| ||
L   L

6.11

In the homometallic tetranuclear complex, $Ru[(2,3-dpp)Ru(bpy)_2]_3^{8+}$ (Figs 6.4 and 6.5) the energy levels of the central unit are higher than the corresponding levels of the peripheral units [106,107]. This is consistent with the fact that the central Ru site, bound to three 2,3-dpp ligands that are stronger π-acceptors than bpy, is more difficult to oxidize than the peripheral Ru sites (bound to one 2,3-dpp and two bpy ligands). In this system, energy transfer occurs from the central unit to the peripheral luminescent units. The analogous tetrametallic complexes with biq as peripheral ligand and/or 2,5-dpp as bridging ligand (Table 6.2) exhibit similar behaviour [70,71]. The homometallic heptanuclear complex $Ru[(2,3-dpp)Ru(bpy)(2,3-dpp)Ru(bpy)_2]_3^{14+}$ (Fig. 6.6) has recently been prepared by using the "complexes as ligands" strategy illustrated by (6.5) and (6.6) [108]:

Table 6.2 — Spectroscopic, photophysical, and redox properties for some tri-, tetra-, and heptanuclear polyimine metal complexes[a]

	Absorption[b]		Emission			Electrochem.[c]	
	λ_{max} (nm)	(ε_{max}) (M^{-1}cm^{-1})	λ_{max} (nm)	τ^d (ns)	$\Phi \times 10^2$	E^{ox} (V)	E^{red} (V)
(bpy)Ru[(2,3-dpp)Ru(bpy)$_2$]$_2$$^{6+}$	545	(23500)	804	75(80)	0.1	+1.48	−0.55
(bpy)Ru[(2,5-dpp)Ru(bpy)$_2$]$_2$$^{6+}$	595	(28100)	831	65		+1.45[e]	−0.48
(bpy)Ru[(2,3-dpp)Ru(biq)$_2$]$_2$$^{6+}$	546	(28700)	773	140		+1.60[e]	−0.47[e]
(bpy)Ru[(2,5-dpp)Ru(biq)$_2$]$_2$$^{6+}$	591	(22900)	805	120(190)	0.6	+1.57[e]	−0.47[e]
Ru[(2,3-dpp)Ru(bpy)$_2$]$_3$$^{8+}$ [f]	545	(46000)	811	50(60)	0.1	+1.50[g]	−0.56
Ru[(2,3-dpp)Ru(biq)$_2$]$_3$$^{8+}$	610	(41500)	795	130(190)	0.1	+1.58[g]	(−0.6)[g]
Ru[(2,5-dpp)Ru(bpy)$_2$]$_3$$^{8+}$	649	(25400)	810[h]	158		+1.40[g]	−0.58
Ru[(2,5-dpp)Ru(biq)$_2$]$_3$$^{8+}$	699	(30500)	799[h]	100		+1.67[g]	−0.52
Os[(2,3-dpp)Ru(bpy)$_2$]$_3$$^{8+}$	549	(40000)	875	(18)		+1.25	−0.55
Ru[(2,5-dpp)Os(bpy)$_2$]$_3$$^{8+}$	618	(24600)				+0.92	−0.64
Ru[(2,3-dpp)Ru(bpy)(2,3-dpp)Ru(bpy)$_2$]$_3$]$^{14+}$	547	(76200)	808	(80)	0.09	+1.38[g]	−0.58[g]

[a]Room temperature, deaerated CH$_3$CN solution [70,71,105,107,108]; [b]lowest energy maximum; [c]$E_{1/2}$ values *vs* SCE; [d]aerated solution (deaerated value in parentheses); [e]dielectronic wave; [f]see also [106]; [g]trielectronic wave; [h]uncorrected spectrum.

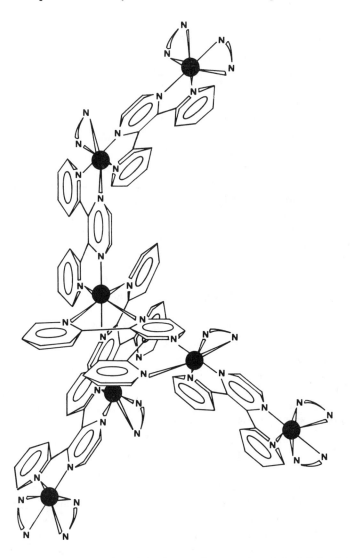

Fig. 6.6 — Schematic representation of the heptanuclear complex of the 2,3-dpp bridging ligand; N–N stands for bpy [108].

$$RuCl_3 + 3\ Ru(2,3\text{-dpp})_2(bpy)^{2+} \rightarrow Ru[(2,3\text{-dpp})Ru(bpy)(2,3\text{-dpp})]_3^{8+} \qquad (6.5)$$

$$Ru[(2,3\text{-dpp})Ru(bpy)(2,3\text{-dpp})]_3^{8+} + 3\ Ru(bpy)_2Cl_2$$
$$\rightarrow Ru[(2,3\text{-dpp})Ru(bpy)(2,3\text{-dpp})Ru(bpy)_2]_3^{14+} \qquad (6.6)$$

The absorption spectrum of this complex shows extremely intense LC bands ($\varepsilon_{max} \sim 300\,000$ M^{-1}cm^{-1}) in the near ultraviolet region and a broad and intense band ($\varepsilon_{max}\ 76\,000$ M^{-1}cm^{-1}) in the visible region. The latter absorption receives contributions from 5 different types of (proximate) metal-to-ligand CT transitions. The luminescence band (Table 6.2) is due to the peripheral Ru→2,3-dpp triplet CT level. The excitation spectrum in the visible region shows that the luminescent level is populated with the same efficiency regardless of the excitation wavelength, again implying an intercomponent energy-transfer process of unit efficiency.

The hetero-tetranuclear $Os[2,3-dpp)Ru(bpy)_2]_3^{8+}$ complex, shown in Fig. 12.10 and discussed in more detail in section 12.5.2, consists of a central $Os(2,3-dpp)_3^{2+}$ core surrounded by three $Ru(bpy)_2^{2+}$ peripheral units [107]. This complex exhibits emission in room temperature fluid solution from a triplet Os→2,3-dpp M(BL)CT state. This assignment is consistent with the electrochemical behaviour (Table 6.2): reduction first takes place at the 2,3-dpp ligand and oxidation first occurs at the central osmium core. The important observation is that the luminescence from the Os→2,3-dpp CT state is obtained with the same efficiency regardless of the Ru→bpy, Ru→2,3-dpp or Os→2,3-dpp type of excitation. This result indicates that intramolecular energy transfer from the peripheral Ru-containing units to the central Os-containing core occurs with unit efficiency (antenna effect, section 12.5.2). As far as the energy transfer mechanism is concerned, an exchange triplet–triplet mechanism is likely, but coulombic singlet–singlet energy transfer cannot be excluded.

In the hetero-tetranuclear $Ru[(2,5-dpp)Os(bpy)_2]_3^{8+}$ complex [71] the lowest excited state is expected to be a triplet Os→2,5-dpp CT level. As for the dinuclear Ru–Os complexes discussed in section 6.3.3.4, luminescence was not observed presumably because it lies outside the spectral region examined.

The general characteristics of the oligonuclear polyimine-bridged complexes can be summarized as follows: (1) very intense absorption spectra in the uv and visible regions (Fig. 6.5); (2) luminescence both at 77 K and in fluid solution at room temperature (Table 6.2); (3) very rich electrochemical behaviour (for example, $(bpy)Ru[(2,3-dpp)Ru(biq)_2]_3^{6+}$ can be reversibly reduced by 9 electrons in the potential window $-0.47/-1.79$ V [70], and the heptanuclear complex shown in Fig. 6.6 contains 7 oxidizable metal centres and 21 reducible bipyridine-type sites [108]). Because of such characteristics, these systems are of outstanding interest not only from the point of view of energy transfer and its applications (section 12.5), but also for investigations in the fields of photo-, chemi-, and electrochemiluminescence, electrochemistry, spectroelectrochemistry, intervalence transfer, photosensitization, and multielectron transfer catalysis. Particularly interesting might be the use of such complexes as luminescent probes and photochemical cleavers for biological applications (section 10.6).

6.3.4 Metal-containing fragments bridged by cyanide ligands

6.3.4.1 Oligonuclear ruthenium complexes
The complexes $(CN)(bpy)_2Ru–CN–Ru(bpy)_2(CN)^+$ and $(CN)(bpy)_2–Ru–CN–Ru(bpy)_2–NC–Ru(bpy)_2(CN)^{2+}$ contain two or three $–Ru(bpy)_2–$ units, with cyanides both as bridging and terminal ligands [109,110]. The bonding mode of the bridging cyanides is determined by the synthetic procedure used. For the binuclear complex there is obviously only one possible structure (although two

possibilities would arise for analogous binuclear complexes with different polypyridine ligands on the two metal centres [109]), but for the trinuclear complexes three linkage isomers are possible, only one of which (that shown above) has actually been studied. It should be noted that the various –Ru(bpy)$_2$– chromophoric units differ for the *C*- or *N*-bonded nature of the bridging cyanides. This affects to some extent the energies of the Ru → bpy CT states that decrease (by ~ 1500 cm^{-1} for each step) in the order NC–Ru–CN > NC–Ru–NC > CN–Ru–NC. In these systems, a single emission attributable to the lowest energy chromophore is observed. Although no selective population of the various MLCT excited states is possible due to overlapping absorption bands, the exact correspondence between excitation and absorption spectra points towards a very efficient energy-transfer process from the higher-energy chromophores to the lowest-energy emitting one, as expected for short and conjugated bridges like –CN–.

The photophysical behaviour of these polychromophoric Ru(II) species is represented in Scheme 6.2, using the trinuclear complex as an example [109]. As

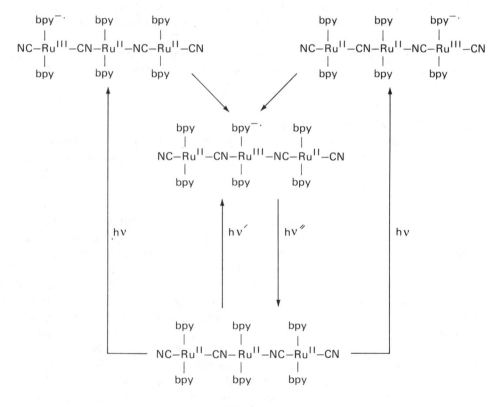

Scheme 6.2

pointed out in Chapter 5, in such schemes only the electronic configurations (not the actual energy levels) are indicated. The steps converting the higher MLCT states to the lowest one are most probably energy-transfer processes, although a two-step metal-to-metal, ligand-to-ligand electron-transfer sequence cannot be strictly ruled out. As far as the mechanism of energy transfer is concerned, a singlet–singlet process is unlikely in view of the fast and efficient intersystem crossing that characterizes Ru(II)-polypyridine complexes [111]. For the more plausible triplet–triplet pathway, both coulombic and exchange mechanisms should be considered. Realistic calculations using the Förster relationship (coulombic mechanism) are impossible because of the unavailability of relevant parameters. Overall, given the strong electronic coupling provided by the bridging cyanide, an exchange Dexter-type mechanism seems to be more likely [109].

A related trinuclear complex where the two bpy ligands of the central Ru ion bear two carboxylic groups has also been studied [112]. This complex is better suited to investigate intramolecular energy transfer since the presence of the carboxylic groups on the ligands of the central chromophoric unit has the effect of further lowering the MLCT levels of this unit. This leads to sizeable shifts in MLCT absorption and thus to the possibility to address the individual chromophores with light of different wavelength. Emission from the central chromophore is again observed with constant efficiency, independent of the nature of the excited chromophore. The application of this trinuclear complex as an antenna-sensitizer photochemical molecular device is discussed in section 12.7.2.

One-electron oxidation of the binuclear and trinuclear complexes discussed above gives mixed-valence species whose photochemical behaviour is dealt with in section 5.7.

6.3.4.2 Ruthenium–chromium complexes

There are a number of recent studies on energy transfer in polynuclear complexes in which Cr(III) species have been used as energy-accepting fragments. The main reason for this choice lies in the peculiar light-emitting properties of Cr(III) complexes [113–115].

Octahedral Cr(III) complexes have a quartet ground state $^4A_{2g}$ belonging to the electronic configuration t_{2g}^3. These complexes exhibit relatively weak ligand-field (MC) bands in the near ultraviolet and visible range. The lowest spin allowed excited state is $^4T_{2g}$ arising from the $t_{2g}^2 e_g$ configuration, while the lowest spin-forbidden state is ^2E_g arising from the t_{2g}^3 configuration. The quartet excited state is usually very reactive towards ligand dissociation, lives in the subnanosecond time scale, and undergoes relatively efficient intersystem crossing to the doublet [115]. Because of its intraconfigurational character, on the other hand, the doublet state is a MC excited state with rather peculiar properties: (i) its energy depends only slightly on the ligands, through the nephelauxetic effect; (ii) it is essentially unreactive with respect to ligand dissociation; (iii) it has a long (μs to ms) lifetime; (iv) its emission has a very narrow bandshape. These properties make the doublet state easily observable and suggest Cr(III) complexes as convenient light-emitting ("luminophoric") fragments to detect intercomponent energy transfer in polynuclear complexes.

While being good light emitters, Cr(III) complexes have weak absorption in the visible because of the symmetry-forbidden nature of their low-energy MC bands. To exploit their luminophoric properties, Cr(III) complexes can be coupled to strong light-absorbing (chromophoric) energy donors as the Ru(II) complexes. A number of such Ru(II)–Cr(III) chromophore–luminophore systems have been recently investigated.

The binuclear species NC–Ru(bpy)$_2$–NC–Cr(CN)$_5^{2-}$ (illustrated in Fig. 12.9 and discussed in detail in section 12.5.1) and the trinuclear complex (CN)$_5$Cr–CN–Ru(bpy)$_2$–NC–Cr(CN)$_5^{4-}$ have been studied in DMF solution where the Cr(CN)$_6^{3-}$ luminophore is known [116] to be a good emitter. The two complexes give rise to very similar results [117]. Visible absorption, that exclusively excites the –Ru(bpy)$_2$– chromophore, gives rise to efficient emission from the (CN)$_5$Cr–CN– luminophore, demonstrating the occurrence of exchange energy transfer from the Ru$\rightarrow$bpy MLCT triplet state to the Cr doublet state. The lack of the photosubstitutional lability (characteristic of Cr(III) quartet photochemistry [113,114]) in the polynuclear complexes shows that energy transfer does not proceed via the upper quartet state. The energy-transfer processes, that in these systems are exergonic by ~ 3000 cm^{-1}, occur on the subnanosecond time scale. The processes are 100% efficient, leading to a greater efficiency of population of the emitting state relative to that (50%) obtained upon direct excitation of the bare luminophore in its spinallowed MC bands. The behaviour of these polynuclear complexes shows some of the ways in which the performance of a luminophore can be improved by covalent coupling to a chromophore: spectral sensitization, antenna effect, enhanced luminescence yields, and photoprotection [117].

The absorption spectra of the long-lived doublet state of these polynuclear complexes exhibit an interesting new type of transition [117], that is the intervalence transfer from Ru(II) to *excited* Cr(III) illustrated by equation (6.7):

$$NC\text{–}Ru^{II}(bpy)_2\text{–}NC\text{–}{^*Cr^{III}}(CN)_5^{2-} \xrightarrow{h\nu} NC\text{–}Ru^{III}(bpy)_2\text{–}NC\text{–}Cr^{II}(CN)_5^{2-} \quad (6.7)$$

The presence of this IT state above the Cr(III) doublet is responsible for the failure to observe doubly-excited *Cr(III)–Ru(II)–*Cr(III) species upon two-photon absorption by the trinuclear complex (as could have been expected on the basis of two successive absorption/energy transfer sequences).

In a very recent investigation the heterometallic trinuclear complex (CN)Cr(cyclam)–CN–Ru(bpy)$_2$–NC–Cr(cyclam)(CN)$^{4+}$ has also been synthesized [118]. This constitutes one of the very few cases, among the polynuclear complexes of photochemical and photophysical interest, for which an X-ray structure is available (Fig. 6.7). The photophysical behaviour of this complex closely parallels that of (CN)$_5$Cr–CN–Ru(bpy)$_2$–NC–Cr(CN)$_5^{4-}$, except for emission which can also be observed in aqueous solution ($\Phi = 5.3 \times 10^{-3}$).

The (NH$_3$)$_5$Cr–NC–Ru(bpy)$_2$–CN–Cr(NH$_3$)$_5^{6+}$ complex has also been studied [119]. This system exhibits efficient energy transfer from the Ru$\rightarrow$bpy CT triplet to the Cr(III) doublet, as shown by quenching of the CT emission and sensitization (77 K) of the Cr(III) phosphorescence. Energy transfer is estimated to be exoergonic by ~ 5000 cm^{-1}. Energy transfer in analogous complexes containing various Rh(III) ammine fragments instead of the Cr(III) one has also been investigated [119].

A specific type of Ru(II)–Cr(III) chromophore–luminophore complex,

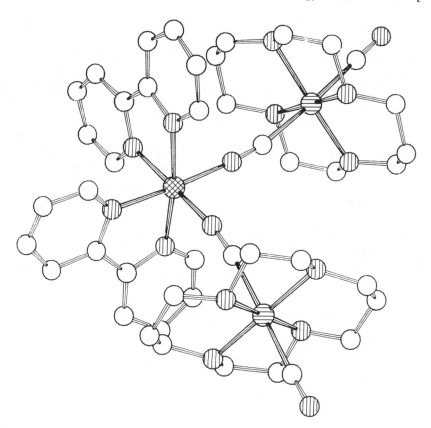

Fig. 6.7 — Skeletal representation of the X-ray structure of Ru(bpy)$_2$(CN)$_2$[Cr(cyclam)CN]$_2^{4+}$ [118].

Ru(bpy)(CN)$_3$–CN–Cr(NH$_3$)$_5^+$, has been synthesized recently [120], with the aim of using second-sphere interactions at the free cyanides of the Ru(bpy)(CN)$_4^{2-}$ moiety [121] to tune the energy gap between the Ru→bpy MLCT state and the Cr(III) doublet. Efficient quenching and sensitization have been observed even at the lowest driving forces attainable (~ 1000 cm^{-1}).

6.3.4.3   *Cobalt–chromium complexes*

The complex (CN)$_5$Co–CN–Cr(NH$_3$)$_5$ has been studied in aqueous solution [122]. The photochemical and photophysical properties of the mononuclear model compounds are as follows [113–115]. Upon excitation to the first spin-allowed $^1T_{1g}$ state, Co(CN)$_6^{3-}$ gives rise to an efficient photoaquation reaction ($\Phi = 0.30$), believed to occur in the lowest triplet $^3T_{1g}$ state reached by intersystem crossing. The Co(III) complex is practically nonemitting in aqueous solution at room temperature. On the other hand, the Cr(NH$_3$)$_6^{3+}$ model component gives, upon excitation to the lowest spin-allowed state $^4T_{2g}$, both an efficient photoaquation ($\Phi = 0.47$) and emission. The emission is a typical doublet ^2E_g phosphorescence, while the photoaquation is probably a quartet reaction. The comparison between the absorption spectrum of the $(CN)_5Co-CN-Cr(NH_3)_5$ binuclear complex and those of the components shows that excitation at 313 nm involves almost exclusively the Co(III)-based fragment, whereas light of 436 nm exclusively excites the Cr(III)-based unit. The results obtained upon irradiation at these wavelengths are clear-cut. Irradiation at 313 nm leads to simultaneous bridge cleavage (6.8), release of ammonia (6.9),

$$(CN)_5Co-CN-Cr(NH_3)_5 + H_2O \xrightarrow{h\nu} Co(CN)_5H_2O^{2-} + Cr(NH_3)_5(CN)^{2+} \qquad (6.8)$$

$$(CN)_5Co-CN-Cr(NH_3)_5 + H_2O \xrightarrow{h\nu} (CN)_5Co-CN-Cr(NH_3)_4H_2O + NH_3 \qquad (6.9)$$

and Cr(III) doublet emission. Visible excitation at 436 nm, on the other hand, gives rise to release of ammonia (6.9) and Cr(III) doublet emission. The photocleavage quantum yield at 313 nm is only 25% of that expected on the basis of the photoaquation of the mononuclear cobalt complex, while the ammonia release and phosphorescence yields are 75% of the corresponding values for 436 nm excitation. This is clear evidence for the occurrence of energy transfer from the Co(III)- to the Cr(III)-based unit. As far as the detailed nature of the energy-transfer process is concerned, several pathways are energetically allowed in this system, including $^1T_{1g}$-Co(III) or $^3T_{1g}$-Co(III) as energy donors and $^4T_{2g}$-Cr(III) or ^2E_g-Cr(III) as energy acceptors. On the basis of several lines of evidence, among which the coincidence between the extent of quenching of the Co(III) photoreaction and the sensitization of the Cr(III) photoreaction at 313 nm, $^3T_{1g}$-Co(III) $\rightarrow ^4T_{2g}$-Cr(III) transfer appears to be favoured [122].

6.3.4.4 Chromium–chromium complexes

The binuclear complex $(CN)(cyclam)Cr-CN-Cr(CN)_5^-$ has been studied recently [123] with the aim of looking for energy transfer between two Cr(III) doublet states. The $(CN)Cr(cyclam)-CN-$ fragment exists as an independent species, $Cr(cyclam)(CN)_2^+$, that in DMF emits a structured phosphorescence at 720 nm with a lifetime of 330 μs [124]. In polynuclear complexes in which this unit is bound to another metal without quenching capability, the emission is still expected to be long-lived and easily detectable. A reasonable model for the $-CN-Cr(CN)_5$ fragment is the $Cr(CN)_5(NH_3)^{2-}$ complex, that is known to emit in DMF solutions at 777 nm with a lifetime of 40 μs [125]. Thus, in the binuclear complex energy transfer from the cyclam-based fragment to the cyanide-based one is exergonic by ~ 1000 cm^{-1}.

In the binuclear complex, overlap between the absorption spectra does not permit independent excitation of the two Cr(III) centres. Irrespective of the excitation wavelength, however, the emission expected from the (CN)Cr(cyclam)–CN– fragment is completely quenched, while emission from the $-CN-Cr(CN)_5$ fragment ($\lambda_{max} = 778$ nm; $\tau = 80$ μs) is observed in DMF. This strongly suggests that exchange energy transfer from the $^2E_g(O_h)$ state of the Cr-cyclam-based unit to that of the Cr-cyanide-based one occurs with high efficiency in this system. The excitation spectrum of the 778 nm emission resembles, but is not identical to, the absorption spectrum. This distortion is expected in view of the intersystem crossing efficiency, that is presumably non-unitary and different at the two Cr(III) centres. No risetime in the emission is observed upon laser excitation, indicating that the energy transfer is fast ($k > 10^8$ s^{-1}). This is quite reasonable since analogous Cr(III)–Cr(III) bimolecular energy-transfer processes, in which the exchange interaction is expected to be much weaker, have rate constants of the order of 10^8 M^{-1} s^{-1} [126].

6.3.4.5 Cobalt–cobalt complexes

The two linkage isomers $(CN)_5Co–NC–Co(NH_3)_5$ and $(CN)_5Co–CN–Co(NH_3)_5$, which differ in the bonding mode of the bridging cyanide, exhibit quite different photochemical behaviour [127]. For both complexes the absorption spectrum is a simple superposition of the spectra of their model subunits $Co(CN)_5(CH_3CN)^{2-}$ and $Co(NH_3)_5CN^{2+}$, and $Co(CN)_6^{3-}$ and $Co(NH_3)_5(CH_3CN)^{3+}$, respectively, pointing towards relatively small mutual perturbation of the fragments. The spectra are such that, in both cases, selective excitation of the $Co(CN)_5$-based chromophore can be achieved with ultraviolet light and of the $Co(NH_3)_5$-based one with visible light.

For $(CN)_5Co–NC–Co(NH_3)_5$, excitation in the ultraviolet range gives rise to a very efficient ($\Phi = 0.28$) photocleavage reaction (6.10),

$$(CN)_5Co–NC–Co(NH_3)_5 + H_2O \quad \xrightarrow{h\nu \,(366\,nm)} \quad Co(CN)_5(H_2O)^{2-} + Co(NH_3)_5(CN)^{2+}$$

$$(6.10)$$

wheras visible excitation has very little photochemical consequence ($\Phi \sim 10^{-3}$). This is the behaviour expected on the basis of the known photoreactivity of the two fragments in the binuclear complex.

For the linkage isomer $(CN)_5Co–CN–Co(NH_3)_5$, on the contrary, both ultra-violet and visible excitation give rise to only small photoreaction quantum yields (ca. 10^{-3} and 10^{-5}, respectively). Thus, the expected photoreactivity of the $Co(CN)_5$-based chromophore appears to be efficiently quenched in this binuclear complex presumably by energy transfer from the lowest MC triplet level of the $Co(CN)_5$–CN– component to the lowest MC triplet level of the –CN–$Co(NH_3)_5$ one [127]. The difference in behaviour between the two linkage isomers could in principle be attributed to differences in either the rate of bond cleavage in the excited $Co(CN)_5$–(CN)– (–(CN)– = –CN– or –NC–) chromophore, or the rate of inter-component energy transfer. The authors [127] favour the latter explanation and tentatively attribute the difference in energy-transfer rate constants to symmetry factors arising from different tetragonal splittings of the d-orbitals in the two isomers.

6.3.5 Other systems

The luminescent ^{3}MLCT excited state of $Ru(bpy)_3^{2+}$ can be quenched by anthracene (An) in fluid solution with formation of triplet anthracene, ^{3}An [128]. This process can be exploited to cause a very efficient reduction of methylviologen, and subsequent H$_2$ evolution, using visible light. A supramolecular species where a $Ru(bpy)_3^{2+}$-type chromophore and anthracene are covalently linked has been prepared by Meyer and coworkers [129] making use of **6.12** as a chelating ligand for Ru^{2+}. When the species obtained in this way, $Ru(bpyCH_2OCH_2An)_3^{2+}$, is excited in its ^{1}MLCT band, rapid ($\tau < 5$ ns) and efficient ($\eta \sim 100\%$) energy transfer takes place with quenching of the ^{3}MLCT luminescence of the Ru moiety and formation of the absorption spectrum of ^{3}An. This excited state has a lifetime of 6 μs in CH_3CN at room temperature. The energy-transfer process continues to occur rapidly even in a rigid matrix at 100 K. A further use of $Ru(bpy)_3^{2+}$ and anthracene moieties in energy-transfer processes [130] is described in section 12.5.3.

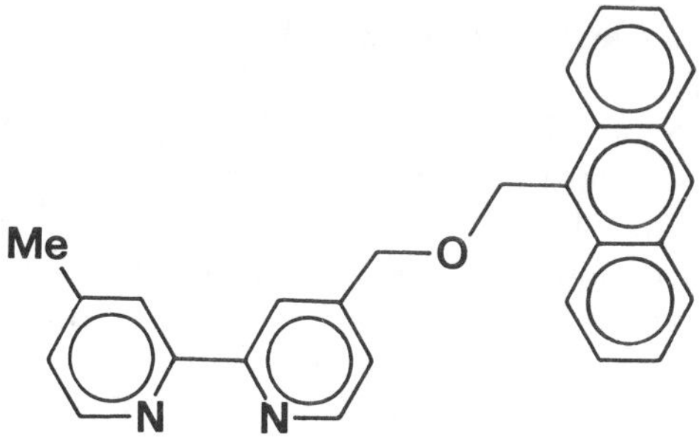

6.12

REFERENCES

[1] Blasse, G. (1986) *Rec. Trav. Chim. Pays-Bas* **105** 143.

[2] Holyle, C. E. and Torkelson, J. M. (ed.) (1987) *Photophysics of polymers.* ACS Symposium Series No. 358.

[3] Eaton, D. F., Caspar, J. V., and Tam, W. (1989) In Norris, J. R., Jr. and Meisel, D. *Photochemical energy conversion.* Elsevier, p. 122.

[4] Yamazaki, I., Tamai, N., and Yamazaki, T. (1990) *J. Phys. Chem.* **94** 516.

[5] Ringsdorf, H., Schlarb, B., and Venzmer, J. (1988) *Angew. Chem. Int. Ed. Engl.* **27** 113.

[6] De Schryver, F. C., Boens, N., and Put, J. (1977) *Adv. Photochem.* **10** 359.

[7] Scandola, F., Indelli, M. T., Chiorboli, C., and Bignozzi, C. A. (1990) *Topics Curr. Chem.* **158**, 73.

[8] Closs, G. L., Piotrowiak, P., McInnis, J. M., and Fleming, G. R. (1988) *J. Am. Chem. Soc.* **110** 2657.

[9] Closs, G. L., Johnson, M. D., Miller, J. R., and Piotrowiak, P. (1989) *J. Am. Chem. Soc.* **111** 3751.

[10] Closs, G. L., Piotrowiak, P., and Miller, J. R. (1989) In Norris, J. R., Jr. and Meisel, D. (eds) *Photochemical energy conversion.* Elsevier, p. 23.

[11] Closs, G. L. and Miller, J. R. (1988) *Science* **240** 440.

[12] Johnson, M. D., Miller, J. R., Green, N. D., and Closs, G. L. (1989) *J. Phys. Chem.* **93** 1173.

[13] Oevering, H., Verhoeven, J. W., Paddon-Row, M. N., Cotsaris, E., and Hush, N. S. (1988) *Chem. Phys. Lett.* **143** 488.

[14] Oevering. H., Verhoeven, J. W., Paddon-Row, M. N., Cotsaris, E., and Hush, N. S. (1988) *Chem. Phys. Lett.* **150** 179.
[15] Penfield, K. W., Miller, J. R., Paddon-Row, M. N., Cotsaris, E., Oliver, A. M., and Hush, N. S. (1987) *J. Am. Chem. Soc.* **109** 5061.
[16] Paddon-Row, M. N., Oliver, A. M., Warman, J. M., Smit, K. J., De Haas, M. P., Oevering, H., and Verhoeven, J. W. (1988) *J. Chem. Phys.* **92** 6958.
[17] Verhoeven, J. W., Oevering, H., Paddon-Row, M. N., Kroon, J., and Kunst, A. G. M. (1988) In Grassi, G. and Hall, D. O. (eds) *Photocatalytic production of energy-rich compounds*. Elsevier, p. 51.
[18] Oevering, H., Verhoeven, J. W., Paddon-Row, M. N., and Warman, J. M. (1989) *Tetrahed.* **45** 4751.
[19] Verhoeven, J. W., Kroon, J., Paddon-Row, M. N., and Oliver, A. M. (1989) In Hall, D. O. and Grassi, G. *Photoconversion processes for energy and chemicals*. Elsevier, p. 100.
[20] Turro, N. J. (1978) *Modern molecular photochemistry*. Benjamin.
[21] Speiser, S. and Rubin, M. B. (1988) *Chem. Phys. Lett.* **150** 177.
[22] Getz, D., Ron, A., Rubin, M. B., and Speiser, S. (1980) *J. Phys. Chem.* **84** 768.
[23] Hassoon, S., Lustig, H., Rubin, M. B., and Speiser, S. (1984) *J. Phys. Chem.* **88** 6367.
[24] Speiser, S., Hassoon, S., and Rubin, M. B. (1986) *J. Phys. Chem.* **90** 5085.
[25] Maki, A. H., Weers, J. G., Hilinski, E. F., Milton, S. V., and Rentzepis, P. M. (1984) *J. Chem. Phys.* **80** 2288.
[26] Morrison, H., Pallmer, M., Loeschen, R., Pandey, B., Muthuramu, K., and Maxwell, B. (1986) *J. Org. Chem.* **51** 4676.
[27] Wu, Z. and Morrison, H. (1989) *Photochem. Photobiol.* **50** 525.
[28] Wu, Z. and Morrison, H. (1989) *J. Am. Chem. Soc.* **111** 9267.
[29] Effenberger, F., Sclosser, H., Bauerle, P., Maier, S., Port, H., and Wolf, H. C. (1988) *Angew. Chem. Int. Ed. Engl.* **27** 281.
[30] Osuka, A., Maruyama, K., Yamazaki, I., and Tamai, N. (1990) *Chem. Phys. Lett.* **165** 392.
[31] Greiner, S. P., Winzenburg, J., von Maltzan, B., Winscom, C. J., and Mobius, K. (1989) *Chem. Phys. Lett.* **155** 93.
[32] Chardon-Noblat, S., Sauvage, J. P., and Mathis, P. (1989) *Angew. Chem. Int. Ed. Engl.* **28** 593.
[33] Sessler, J. L., Johnson, M. R., and Lin, T. Y. (1989) *Tetrahed.* **45** 4767.
[34] Bourson, J., Mugnier, J., and Valeur, B. (1982) *Chem. Phys. Lett.* **92** 430.
[35] Mugnier, J., Pouget, J., Bourson, J., and Valeur, B. (1985) *J. Luminescence* **33** 273.
[36] Mugnier, J., Valeur, B., and Gratton, E. (1985) *Chem. Phys. Lett.* **119** 217.
[37] Valeur, B., Mugnier, J., Pouget, J., Bourson, J., and Santi, F. (1989) *J. Phys. Chem.* **93** 6073.
[38] Bauman, J. and Fayer, M. D. (1986) *J. Chem. Phys.* **85** 4087, and references therein.
[39] Ikeda, T., Lee, B., Kurihara, S., Tazuke, S., Ito, S., and Yamamoto, M. (1988) *J. Am. Chem. Soc.* **110** 8299.

[40] Dubowchik, G. M. and Hamilton, A. D. (1986) *J. Chem. Soc., Chem. Commun.* 1391.

[41] Regev, A., Galili, T., Levanon, H., and Harriman, A. (1986) *Chem. Phys. Lett.* **131** 140.

[42] Harriman, A. (1987) In Balzani, V. (ed.) *Supramolecular photochemistry.* Reidel, p. 207.

[43] Davila, J., Harriman, A., and Milgron, L. R. (1987) *Chem. Phys. Lett.* **136** 427.

[44] Levanon, H., Regev, A., and Das, P. K. (1987) *J. Phys. Chem.* **91** 14.

[45] Moore, A. L., Joy, A., Tom, R., Gust, D., Moore, T. A., Bensasson, R. V., and Land, E. J. (1982) *Science* **216** 982.

[46] Wasielewski, M. R., Liddell, P. A., Barrett, D., Moore, T. A., and Gust, D. (1986) *Nature* **322** 570.

[47] Liddell, P. A., Barrett, D., Makings, L. R., Pessiki, P. J., Gust, D., and Moore, T. A. (1986) *J. Am. Chem. Soc.* **108** 5350.

[48] Gust, D., Moore, T. A., Moore, A. L., Makings, L. R., Seely, G. R., Ma, X., Trier, T. T., and Gao, F. (1988) *J. Am. Chem. Soc.* **110** 7567.

[49] Lindsey, J. S., Brown, P. A., and Siesel, D. A. (1989) *Tetrahed.* **45** 4845.

[50] Juris, A., Balzani, V., Barigelletti, F., Campagna, S., Belser, P., and von Zelewsky, A. (1988) *Coord. Chem. Rev.* **84** 85.

[51] Kober, E. M., Caspar, J. V., Sullivan, B. P., and Meyer, T. J. (1988) *Inorg. Chem.* **27** 4587.

[52] Sahai, R., Baucom, D. A., and Rillema, D. P. (1986) *Inorg. Chem.* **25** 3843.

[53] Furue, M., Kuroda, N., and Sano, S. (1988) *J. Macromol. Sci. Chem.* **A25** 1263.

[54] Furue, M., Kuroda, N., and Nozakura, S. (1986) *Chem. Lett.* 1209.

[55] Wacholtz, W. F., Auerbach, R. A., and Schmehl, R. H. (1987) *Inorg. Chem.* **26** 2989.

[56] Shaw, J. R., Webb, R. T., and Schmehl, R. H. (1990) *J. Am. Chem. Soc.* **112** 1117.

[57] De Cola, L., Belser, P., Ebmeyer, F., Barigelletti, F., Vögtle, F., von Zelewsky, A., and Balzani, V. (1990) *Inorg. Chem.* **29** 495.

[58] Furue, M., Kinoshita, S., and Kushida, T. (1987) *Chem. Lett.* 2355.

[59] Schmehl, R. H., Auerbach, R. A., Wacholtz, W. F., Elliott, C. M., Freitag, R. A., and Merkert, J. W. (1986) *Inorg. Chem.* **25** 2440.

[60] De Cola, L., Barigelletti, F., and Belser, P., unpublished observations.

[61] Schmehl, R. H., Auerbach, R. A., and Wacholtz, W. F. (1988) *J. Phys. Chem.* **92** 6202.

[62] Ryu, C. K. and Schmehl, R. H. (1989) *J. Phys. Chem.* **93** 7961.

[63] Hunziker, M. and Ludi, A. (1977) *J. Am. Chem. Soc.* **99** 7370.

[64] Dose, E. V. and Wilson, L. J. (1978) *Inorg. Chem.* **17** 2660.

[65] Rillema, D. P. and Mack, K. B. (1982) *Inorg. Chem.* **21** 3849.

[66] Petersen, J. D. (1987) In Yersin, H. and Vogler, A. (eds) *Photochemistry and photophysics of coordination compounds.* Springer-Verlag, p. 147.

[67] Petersen, J. D. (1987) In Balzani, V. (ed.) *Supramolecular photochemistry.* Reidel, p. 135, and references therein.

[68] Kalyanasundaram, K. and Nazeeruddin, M. K. (1990). *Inorg. Chem.* **29** 1888.
[69] Barigelletti, F., De Cola, L., Balzani, V., Hage, R., Haasnoot, J. G., Reedijk, J., and Vos, J. G. (1989) *Inorg. Chem.* **28** 4344.
[70] Denti, G., Campagna, S., Sabatino, L., Serroni, S., Ciano, M., and Balzani, V. (1990) *Inorg. Chem.*, in press.
[71] Denti, G., Campagna, S., Sabatino, L., Serroni, S., Ciano, M., and Balzani, V. (1990) In Pelizzetti, E. and Schiavello, M. (eds) *Photochemical conversion and storage of solar energy*, in press.
[72] Haga, M. (1980) *Inorg. Chim. Acta* **45** L183.
[73] Rillema, D. P., Sahai, R., Matthews, P., Edwards, A. K., Shaver, R. J., and Morgan, L. (1990) *Inorg. Chem.* **29** 167.
[74] Braunstein, C. H., Baker, A. D., Strekas, T. C., and Gafney, H. D. (1984) *Inorg. Chem.* **23** 857.
[75] Fuchs, Y., Lofters, S., Dieter, T., Shi, W., Morgan, R., Strekas, T. C., Gafney, H. D., and Baker, A. D. (1987) *J. Am. Chem. Soc.* **109** 2691.
[76] Masschelein, A., Kirsch-De Mesmaeker, A., Verhoeven, C., and Nasielski-Hinkens, R. (1987) *Inorg. Chim. Acta* **129** L13.
[77] Wallace, A. W., Murphy, W. R., Jr., and Petersen, J. D. (1989) *Inorg. Chim. Acta* **166** 47.
[78] Rillema, D. P., Callahan, R. W., and Mack, K. B. (1982) *Inorg. Chem.* **21** 2589.
[79] Kalyanasundaram, K. and Nazeeruddin, M. K. (1989) *Chem. Phys. Lett.* **158** 45.
[80] Zulu, M. M. and Lees, A. J. (1988) *Inorg. Chem.* **27** 1139.
[81] Zulu, M. M. and Lees, A. J. (1988) *Inorg. Chem.* **27** 3325.
[82] Zulu, M. M. and Lees, A. J. (1989) *Inorg. Chem.* **28** 85.
[83] Campagna, S., Denti, G., De Rosa, G., Sabatino, L., Ciano, M., and Balzani, V. (1989) *Inorg. Chem.* **28** 2565.
[84] Shoup, M., Hall, B., and Ruminski, R. R. (1988) *Inorg. Chem.* **27** 200.
[85] Vogler, A. and Kisslinger, J. (1986) *Inorg. Chim. Acta* **115** 193.
[86] Juris, A., Campagna, S., Bidd, I., Lehn, J. M., and Ziessel, R. (1988) *Inorg. Chem.* **27** 4007.
[87] Lin, R. and Guarr, T. F. (1990) *Inorg. Chim. Acta* **167** 149.
[88] Hage, R., Haasnoot, J. G., Nieuwenhuis, H. A., Reedijk, J., De Ridder, D. J. A., and Vos, J. G., *J. Am. Chem. Soc.*, in press.
[89] Lin, C. T., Bottcher, W., Chou, M., Creutz, C., and Sutin, N. (1976) *J. Am. Chem Soc.* **98** 6536.
[90] Creutz, C. and Sutin, N. (1976) *Inorg. Chem.* **15** 496.
[91] Sahai, R. and Rillema, D. P. (1986) *Inorg. Chim. Acta* **118** L35.
[92] Gelroth, J. A., Figard, J. E., and Petersen, J. D. (1979) *J. Am. Chem. Soc.* **101** 3649.
[93] Moore, K. J., Lee, L., Figard, J. E., Gelroth, J. A., Stinson, A. J., Wohlers, H. D., and Petersen, J. D. (1983) *J. Am. Chem. Soc.* **105** 2274.
[94] Figard, J. E. and Petersen, J. D. (1978) *Inorg. Chem.* **17** 1059.
[95] Ford, P. C., Hintze, R. E., and Petersen, J. D. (1975) In Adamson, A. W. and Fleischauer, P. D. (eds) *Concepts of inorganic photochemistry*. Wiley, p. 203.
[96] MacQueen, D. B. and Petersen, J. D. (1990) *Coord. Chem. Rev.* **97** 249.

[97] Loeb L. B., Neyhart, G. A., Worl, L. A., Danielson, E., Sullivan, B. P., and Meyer, T. J. (1989) *J. Phys. Chem.* **93** 717.

[98] Sahai, R. and Rillema, D. P. (1986) *J. Chem. Soc., Chem. Commun.* 1133.

[99] Sahai, R., Morgan, L., and Rillema, D. P. (1988) *Inorg. Chem.* **27** 3495.

[100] Toma, H. E., Auburn, P. R., Dodsworth, E., Golovin, M. N., and Lever, A. B. P. (1987) *Inorg. Chem.* **26** 4257.

[101] Toma, H. E. and Lever, A. B. P. (1986) *Inorg. Chem.* **25** 176.

[102] Toma, H. E., Santos, P. S., and Lever, A. B. P. (1988) *Inorg. Chem.* **27** 3850 .

[103] Kirsch-De Mesmaeker, A., Jacquet, L., Masschelein, A., Vanhecke, F., and Heremans, K. (1989) *Inorg. Chem.* **28** 2465.

[104] Sahai, R., Rillema, D. P., Shaver, R., Wallendael, S. V., Jackman, D. C., and Boldaji, M. (1989) *Inorg. Chem.* **28** 1022.

[105] Campagna, S., Denti, G., Sabatino, L., Serroni, S., Ciano, M., and Balzani, V. (1989) *Gazz. Chim. Ital.* **119** 415.

[106] Murphy, W. R., Jr., Brewer, K. J., Gettliffe, G., and Petersen, J. D. (1989) *Inorg. Chem.* **28** 81.

[107] Campagna, S., Denti, G., Sabatino, L., Serroni, S., Ciano, M., and Balzani, V. (1989) *J. Chem. Soc., Chem. Commun.* 1500.

[108] Denti, G., Campagna, S., Sabatino, L., Serroni, S., Ciano, M., and Balzani, V. (1990) *Inorg. Chim. Acta*, **176**, 175.

[109] Bignozzi, C. A., Roffia, S., Chiorboli, C., Davila, J., Indelli, M. T., and Scandola, F. (1989) *Inorg. Chem.* **28** 4350.

[110] Scandola, F., Bignozzi, C. A., Chiorboli, C., Indelli, M. T., and Rampi, M. A. (1990) *Coord. Chem. Rev.* **97** 299.

[111] Meyer, T. J. (1986) *Pure Appl. Chem.* **58** 1193.

[112] Amadelli, R., Argazzi, R., Bignozzi, C. A., and Scandola, F. (1990) *J. Am. Chem. Soc.* **112** 7099.

[113] Balzani, V., and Carassiti, V. (1970) *Photochemistry of coordination compounds*. Academic.

[114] Zinato, E. (1975) In Adamson, A. W. and Fleischauer, P. D. (eds) *Concepts of inorganic photochemistry*. Wiley, p. 143.

[115] Forster, L. S. (1990) *Chem. Rev.* **90** 331.

[116] Wasgestian, H. F. (1972) *J. Phys. Chem.* **76** 1947.

[117] Bignozzi, C. A., Indelli, M. T., and Scandola, F. (1989) *J. Am. Chem. Soc.* **111** 5192.

[118] Bignozzi, C. A., Chiorboli, C., Indelli, M. T., and Scandola, F., manuscript in preparation.

[119] Lei, Y. and Endicott, J. F., private communication.

[120] Rampi, M. A., Checchi, L., and Scandola, F., manuscript in preparation.

[121] Scandola, F. and Indelli, M. T. (1988) *Pure Appl. Chem.* **60** 973.

[122] Kane-Maguire, N. A. P., Allen, M. M., Vaught, J. M., Hallock, J. S., and Heatherington, A. L. (1983) *Inorg. Chem.* **22** 3851.

[123] Chiorboli, C., Indelli, M. T., Bignozzi, C. A., and Scandola, F., manuscript in preparation.

[124] Miller, D. B., Miller, P. K., and Kane-Maguire, N. A. P. (1983) *Inorg. Chem.* **22** 3831.

[125] Riccieri, P., and Zinato, E., *Inorg. Chem.*, in press.

[126] Endicott, J. F., Lessard, R. B., Lei, Y., and Ryu, C. K. (1987) In Balzani, V. (ed.) *Supramolecular Photochemistry*. Reidel, p. 167.

[127] Nishizawa, M. and Ford, P. C. (1981) *Inorg. Chem.* **20** 2016.

[128] Mau, A. W. H., Johansen, O., and Sasse, W. H. F. (1985) *Photochem. Photobiol.* **41** 503, and references therein.

[129] Boyde, S., Strouse, G. F., Jones, W. E., Jr., and Meyer, T. J. (1989) *J. Am. Chem. Soc.* **111** 7448.

[130] Strouse, G. F., Worl, L. A., Younathan, J. N., and Meyer, T. J. (1989) *J. Am. Chem. Soc.* **111** 9101.

7

Structural changes in photoflexible systems

7.1 INTRODUCTION

The molecular mechanism of several photoregulated biological processes, such as vision and phototropism, is based on photochromic molecules embedded in macromolecular matrices. The photoisomerization of a photochromic component induces variations in the conformation of a macromolecule and triggers subsequent reactions [1–3]. Photochromic compounds are distributed widely in living organisms, the most prominent among them being the flavins and carotenoids [3].

Photochromic molecules and their photoisomerization reactions have been the object of extensive investigation in the field of molecular photochemistry [4]. Incorporation of a photochromic component into a supramolecular structure can lead to artificial photoresponsive species that may be quite valuable as model systems for theoretical studies and as photochemical molecular devices (Chapter 12). The basic requirements for the design of an artificial photoresponsive supramolecular system are as follows: (a) a component of the supramolecular system must be able to absorb light; (b) as a consequence of light excitation, the chromophoric component (or another component) must undergo a structural change; (c) such a molecular structural change must cause a "functional" change (that is a change in some properties relevant to a function) in another component or in the whole supramolecular structure.

7.2 PHOTOACTIVE COMPONENTS

As we will see later on, the most widely used chromophores in artificial supramolecular systems are aromatic azo [5] and spiropyran [6] compounds. Therefore, we shall recall the principal characteristics of the photoinduced processes that cause conformational changes in these two families of compounds. We shall also recall the principles of photoinduced conformational changes in olefin-type compounds, which are widely diffuse in nature [7], and in the so-called twisted-intramolecular-charge-transfer (TICT) compounds [8], which are currently the object of extensive investigations. These four classes of molecules satisfy quite well the requirements needed for photoactive components of supramolecular species since their photo-isomerization reactions cause large structural changes and occur with high efficiency and high reversibility.

7.2.1 Olefin-type compounds

The photoinduced *cis-trans* isomerization of olefin-type compounds is an important natural process (for example, for vision [7,9,10]) and an extensively investigated reaction in molecular photochemistry [11]. In general, the *trans* isomer is the thermodynamically more stable form for free molecules, and the thermal *cis → trans* isomerization can be slow or fast, depending on the compound and the experimental conditions. Both *trans → cis* and *cis → trans* isomerizations can usually be obtained by irradiation with light of appropriate wavelength. The reaction mechanism is excited state twisting about the $-C=C-$ double bond.

The best studied example of a double-bond twist in the excited state is stilbene, whose *trans* and *cis* isomers are shown as **7.1** and **7.2**. The intimate mechanism of stilbene photoisomerization has been the object of much controversy. Following Förster's suggestion [12] that the lowest singlet state S_1 and the ground state S_0 possess an energy maximum for the perpendicular conformation and, consistent with the results of triplet sensitization experiments [13], it was claimed that isomerization upon direct excitation takes place in the excited triplet [14,15]. Later Saltiel and coworkers [13] showed that a singlet state of energy lower than that reached by direct excitation can be populated ("phantom" singlet state, $^1P^*$). In agreement with this finding, Orlandi and Siebrand [16] demonstrated that the ground state $S_0(A_g)$ of *trans*-stilbene correlates with a doubly-excited state $S_2(A)$ of *cis*-stilbene, and *vice versa*, so that an energy minimum at the double-bond twisted geometry can be predicted because the states are of the same symmetry and interact, giving rise to a strongly avoided crossing. The optical transition between the ground state and the doubly-excited singlet state is forbidden by one-photon spectroscopy, but its presence has been confirmed by two-photon spectroscopy [17]. From the large body of nano- and picosecond time-resolved studies [18–22], the dynamic behaviour of stilbene isomerization in non-polar solvents can be summarized by the scheme shown in Fig. 7.1. Deactivation of excited *trans*-stilbene occurs across an energy barrier of ~ 12 kJ mol^{-1} in about 70 ps towards the double-bond twisted conformation ($^1P^*$ state), a local minimum from which deactivation to the ground state is very rapid (3 ps). Coming from the *cis* side, $^1P^*$ formation is faster than its decay, so that this state can be studied by laser flash spectroscopy.

7.1   **7.2**

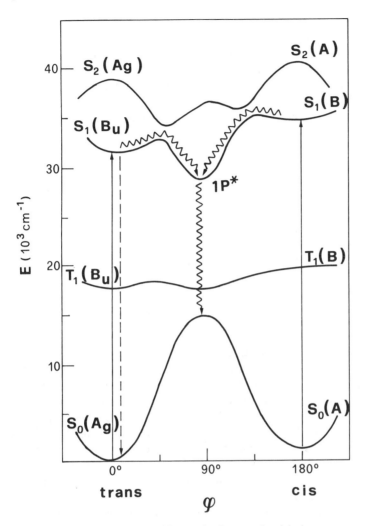

Fig. 7.1 — Schematic potential energy diagram for the ground and the low-energy excited states
of stilbene as a function of the torsional coordinate.

The photochemical and photophysical behaviour of α,ω-diphenylpolyenes has
recently been reviewed [11] and the similarity between the $^1P^*$ state and the TICT
states (section 7.2.4) has been emphasized [8].

7.2.2 Azobenzene and related compounds
Azobenzene is isostructural with stilbene. As for stilbene, the *trans* isomer (**7.3**) of
azobenzene is more stable than the *cis* isomer (**7.4**). Azobenzene and nearly all its
monosubstituted derivatives are coloured since the principal absorption band ($\pi\pi^*$)
in the ultraviolet region (~ 320 nm) is accompanied by a weak band ($n\pi^*$) near
450 nm. On conversion to the *cis* isomer, the $\pi\pi^*$ band shifts to shorter wavelengths

and there is an increase in the intensity of the $n\pi^*$ absorption. Consequently, these compounds appear to deepen in colour upon *trans → cis* isomerization [5]. Most of the simple azobenzenes, other than those bearing amino or hydroxy substituents, are sufficiently stable in the *cis* form to be isolated. The main feature of the photoisomerization reaction in homogeneous solution is a marked wavelength effect on the quantum yield (see Table 10.4). For the *trans → cis* process excitation in the visible $S_1(n\pi^*)$ band is about twice as effective as excitation in the ultraviolet $S_2(\pi\pi^*)$ band [23].

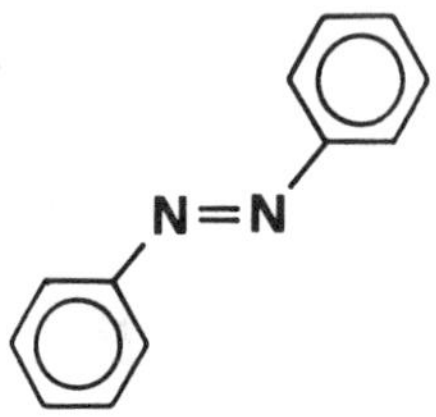

7.3 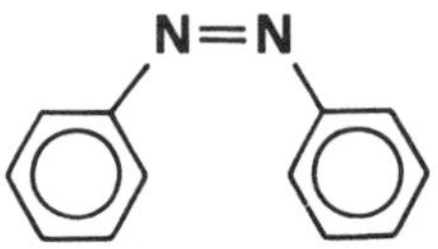   **7.4**

It has since long been known [5] that for azobenzene there are two possible isomerization coordinates: (i) twisting about the $-N=N-$ double bond, as in the case of stilbene, and (ii) in-plane inversion at one of the two nitrogen atoms. The in-plane mechanism is responsible for the dark isomerization, whereas it is not yet clear whether photoisomerization takes place exclusively *via* inversion [23–25], or *via* twisting and inversion depending on the excited state populated by light excitation [26,27]. Photoinduced isomerization of azobenzene involves a large structural rearrangement and a big change in the dipole moment. The isomerization causes a decrease in the distance between the *para* carbon atoms from about 9.0 Å in the *trans* form to 5.5 Å in the *cis* form. The *trans* form is planar and has no dipole moment, whereas the *cis* form is nonplanar and exhibits a dipole moment of 3.0 D [28]. As we will see later, such changes may strongly affect the properties of supermolecules containing the azobenzo group as a component.

7.2.3 Spiropyrans and related compounds

The term spiropyran is used to denote very generally a molecule containing a 2H-pyran ring in which the number-2 carbon atom of the ring is involved in a spiro linkage [6]. Spiropyrans can exist in two basic forms illustrated by **7.5** and **7.6** for 1′,3′,3′-trimethylspiro-[2H-1-benzopyran-2,2′-indoline], commonly called BIPS: the colourless parent spiropyran structure **7.5** and the intensely colored merocyanine structure **7.6**. These two forms can be very efficiently and reversibly interconverted by light. For many spiropyrans the equilibrium between the two forms can also be strongly displaced by changing the temperature. Because of their photochromic and thermochromic properties spiropyrans have since long been the object of extensive academic and industrial research. An excellent review up to 1969 is available [6], and a continuous flow of papers and patents on this subject is coming out in the scientific and technical literature.

7.5

7.6

Nanosecond and picosecond studies have been carried out by several groups in order to elucidate the primary events of the photoisomerization processes in some important families of spiropyrans. In 1962 Fischer and coworkers [29] proposed that the first product of the spiropyran→merocyanine photoisomerization reaction for BIPS was a species, labelled X, in which the C_{spiro}-O bond is broken but the orthogonal parent geometry is maintained. Several years later, for 6-nitro-BIPS, the transient absorption spectrum 10 ps after ultraviolet excitation was found to exhibit a dominant band centered at 440 nm and a weaker band at 585 nm, both attributed to X [30,31]. The precursor of X was thought to be a vibrationally-excited triplet of the spiropyran and the decay of X was found to occur in the nanosecond time scale. Different results and interpretations, however, have also been reported [32–40]. In particular, the participation of triplet states in the photochemical reaction has been a matter of debate. A recent investigation [41] has concerned a family of 6-nitro-BIPS in which a triplet quencher (that is, a naphthyl group) or a triplet sensitizer (a benzophenone group) were covalently linked to the nitrogen atom of the spiropyran structure. Identical transient absorption spectra were obtained for all the compounds examined, indicating that the triplet state of spiropyran makes a minor contribution to the generation of merocyanine, in agreement with the results obtained from photostationary state experiments. Such a triplet state, in fact, is probably responsible for the photodegradation ("fatigue") of this photochromic system [41]. A transient absorption spectroscopic investigation on BIPS with a time resolution of 0.42 ps has also been performed [42]. During the pump pulse, an induced transient absorption is observed only below 500 nm, with a maximum at 440 nm. Such a transient absorption decays within ~5 ps, while a broad band at ~550 nm, which is the region of merocyanine absorption, appears with a rise time of ~1.3 ps. The latest recorded spectrum (104 ps after excitation) closely resembles that of the merocyanine products and is assigned to the merocyanine isomer distribution at that time. From these results, and consistent with theoretical calculations [43], the reaction is described as a barrierless rotation around the C_4-C_{4a} bond on the excited singlet potential energy surface towards a nonplanar cisoid form, followed by internal conversion and rapid attainment of planarity. Regardless of the intimate photoreaction mechanism, photoisomerization of spiropyran-type compounds causes not only a colour change, but also a large change in the conformation of the molecule and in its charge distribution. As a consequence, the interaction with other species, linked to or surrounding the molecule, will also change upon light excitation, opening the way to the design of photoresponsive systems.

7.2.4   TICT compounds

The concept of twisted-intramolecular-charge-transfer (TICT) states [8,44–46] was first introduced by Grabowski and coworkers [47,48] to account for the anomalous dual fluorescence of N,N-dimethylaminobenzonitrile (DMABN, **7.7**) observed by Lippert *et al.* [49] in polar solvents. Contrary to earlier proposals involving ground- or excited-state complex formation, it was shown that the long-wavelength "anomalous" fluorescence band originates from a twisted amine configuration. Such a band, in fact, is missing for planar model compounds like **7.8** where twisting of the amine group is blocked, whereas model compounds that are twisted in the ground state, like **7.9**, lack the "normal" short-wavelength fluorescence.

7.7   **7.8**   **7.9**

Experimentally, the large charge separation of TICT states manifests itself by a strong red shift of the TICT fluorescence in polar solvents [50], or by its sizeable response to applied electric fields [51].

According to the accepted model, TICT states are accessible in systems possessing a sufficiently strong electron donor, D, and a sufficiently strong electron acceptor, A, weakly coupled. The classical arrangement for TICT compounds is schematized in Fig. 7.2 [8]. The formation of the TICT state involves (i) light absorption to an excited state, which has a planar conformation and only partial CT character, and (ii) intramolecular twisting in the excited state, which can be viewed as an adiabatic photoreaction proceeding on the S_1 hypersurface towards the excited-state energy minimum at $\varphi = 90°$. At such a perpendicular geometry there is zero π overlap between the sub-systems (for **7.7**, between the lone pair of the amino group and the benzene π^* orbital), so that a full electric charge is transferred from D to A. Fluorescence emission from the TICT states is a radiative electron-transfer process which can be discussed [52] in terms of the Marcus inverted region (section 3.4.4). The simple picture of the two moieties rotating against each other under the influence of a driving force is the essential feature of TICT formation, but many more internal degrees of freedom and also solvent interactions intervene [8,44–46,53–59].

The number of compounds able to form TICT states has rapidly grown in the last ten years [8,44,60]. The TICT family now encompasses many aromatic amines, pyrroles, and carbazoles, some biaryls, rhodamine, coumarin, and arylidene dyes, organosilicon and organoboron compounds, and metal complexes.

From a theoretical viewpoint, the prerequisite for TICT state formation is the

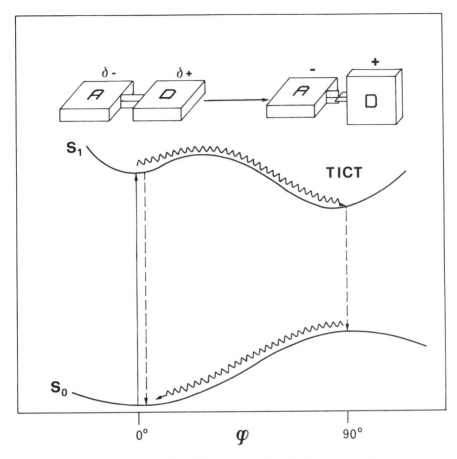

Fig. 7.2 — Schematic representation of the ground- and excited-state potential energy curves
for a TICT compound [46].

decoupling of the π system [8]. In a valence bond description it can be shown that
systems with small π coupling possess biradicaloid states with a minimum interaction
for the perpendicular (small overlap) geometry ("minimum overlap rule") [61]. This
means that at $\varphi = 90°$ the ground state exhibits a maximum and the excited state a
minimum. Even compounds that are nonpolar and highly symmetric like 9,9'-
bianthryl (**7.10**), which has a perpendicular configuration in the ground state, form
very polar TICT states upon excitation [62,63]. The symmetry-breaking process
(that is, the charge separation between the identical moieties) is caused in these cases
by solvent perturbations. The similarity between the light-induced charge separation
with symmetry breaking in these species and in the bacterial photosynthetic reaction
centre has been discussed by Rettig [44] who has also pointed out the relationship
between TICT states and $^1P^*$ states involved in the photochemical twisting of
olefin-type compounds [8,64].

7.10

7.3 SUPRAMOLECULAR SYSTEMS

As mentioned in section 7.2, when a photoisomerizable molecule is a component of a suitable supramolecular system, the photoinduced molecular structural change may trigger a chemical function of another component or of the whole supermolecule. In artificial supramolecular systems, the easiest way to exploit the effect of a photo-induced molecular structural change is to connect the photoisomerizable molecule to other components of the supermolecule by covalent bonds. This is the case, for example, of photoisomerizable components covalently linked to crown ethers or cyclodextrins. However, exploitation of intermolecular forces is also possible, as in the case of photoisomerizable components enclosed in micelles.

In the following sections, we will examine the behaviour of some important families of photoresponsive supramolecular systems whose responsiveness is photoinduced by a molecular conformational change.

7.3.1 Systems involving crown ethers
Crown ethers [65,66] are macrocyclic polyethers that can play the role of multiden-tate ligands towards a variety of metal ions and organic cations. Since the formation constant of the resulting complex depends on the fitting of the cavity to the size of the metal ion, crown ethers can be designed to be selective [66–68]. Crown ethers are flexible, so that their conformation can be easily changed if they are covalently bound to components that undergo structural changes. Therefore, the coordinating ability of supramolecular systems that contain both crown ethers and photoisomerizable components may be controlled by light. Most of the work in this field has been carried out by Shinkai and coworkers [69,70].

The ion-binding ability of crown ether components in supramolecular systems can also be controlled by other means such as acid-base processes [70], redox processes [71–73], coordination of heavy-metal ions in remote sites [71,74,75], and temperature [76]. An excellent review on redox-responsive macrocyclic receptor molecules containing transition-metal redox centers has been recently published [71].

7.3.1.1 *Crown ethers with an intramolecular bridge*
A direct way to transmit the photoinduced structural change of a chromophoric component to a crown ether is to bridge the crown ether ring by the chromophoric unit. The *trans*-azobenzene capped crown ether **7.11** binds preferably small metal

cations such as Li$^+$ and Na$^+$. On excitation with ultraviolet light, the *trans*→*cis* photoisomerization reaction of the azobenzene component takes place, yielding the structure **7.12** which preferably binds large metal cations like K$^+$ and Rb$^+$, indicating a photoinduced expansion of the crown ether cavity [77]. When the azobenzene bridge is replaced by the 2,2′-azopyridine bridge, in the *trans* conformation the pyridine nitrogens are directed towards the crown plane and can coordinate

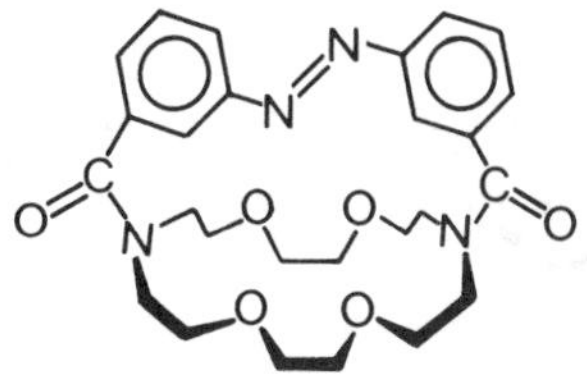

7.11 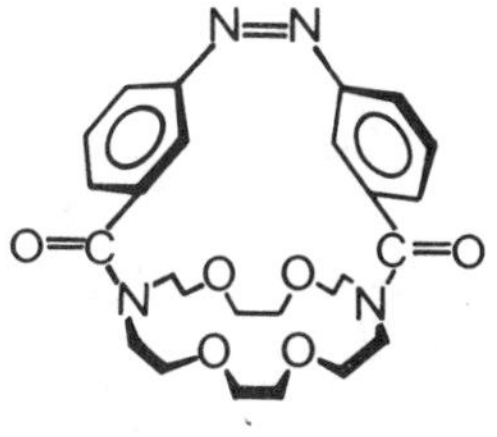   **7.12**

metal ions enclosed in the crown ether ring, whereas this is not possible in the *cis* conformation of the bridge [78]. As a consequence, the ability to extract heavy metal ions (Cu^{2+}, Ni^{2+}, Co^{2+}, and Hg^{2+}) from aqueous solution to the organic phase can be photocontrolled. The transformation of **7.13** into its isomer **7.14** and the reverse reaction can also be driven by light [79]. The two photoisomers exhibit different affinities towards alkali metal ions and can thus be used for selective extraction.

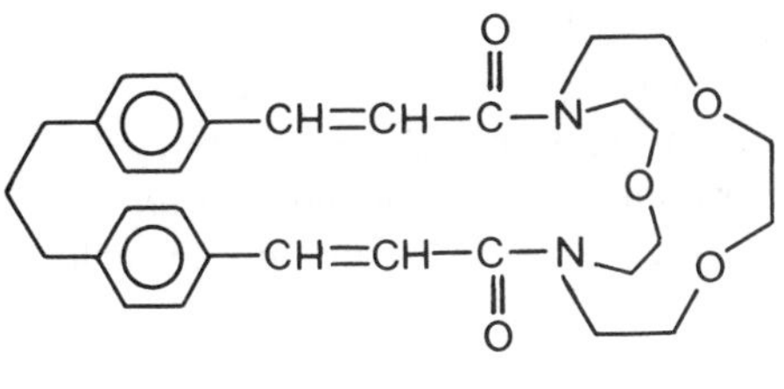

7.13 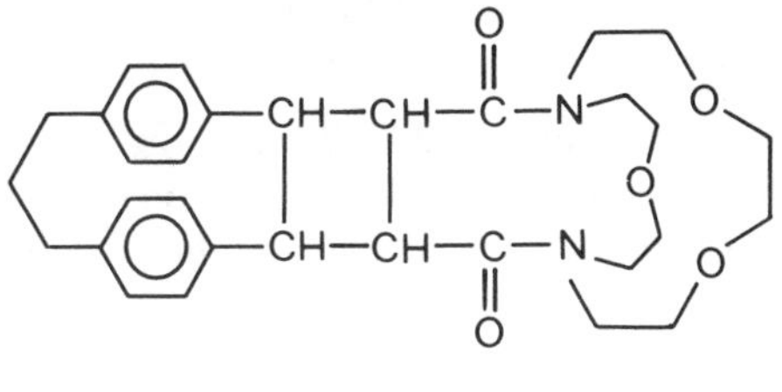   **7.14**

7.3.1.2 *Cylindrical and phane-type crown ethers*

When two macrocyclic ligands are linked together by two (or more than two) pillars, cylindrical structures are obtained which are suitable for linear recognition or as ditopic coreceptors [80,81]. The binding selectivity is governed not only by the size of the macrocyclic ligands, but also by the length of the pillars. Insertion of azobenzene-photoisomerizable components in the pillars yields "photoelastic" cylindrical structures whose binding selectivity can be photocontrolled [70]. The cylindrical crown ether **7.15**, which contains two azobenzene components in the *trans* conformation, is able to extract polymethylenediammonium ions, H$_3$N$^+$(CH$_2$)$_n$NH$_3^+$, with $n = 10$ and 12, but barely extracts the same type of ion with $n = 4$ and 6. Upon ultraviolet irradiation, a photostationary state with 35% *trans-trans* **7.15** and 65% *cis-cis* **7.16** is reached, and preference for extraction of polymethylenediammonium ions with

$n = 4$ and 6 is exhibited, as expected because of the shorter distance between the two coordinating crown ether components [82].

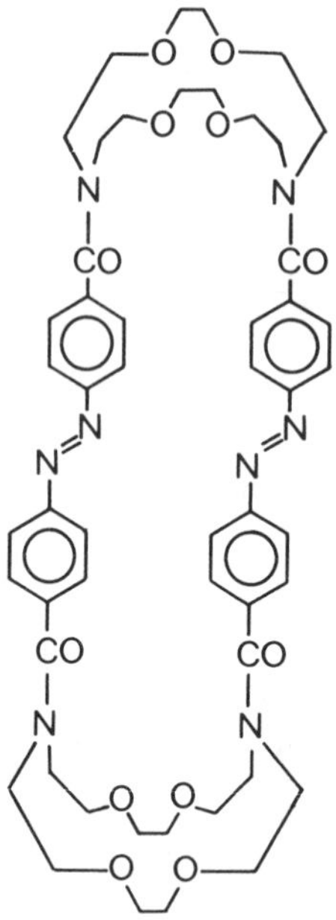

7.15 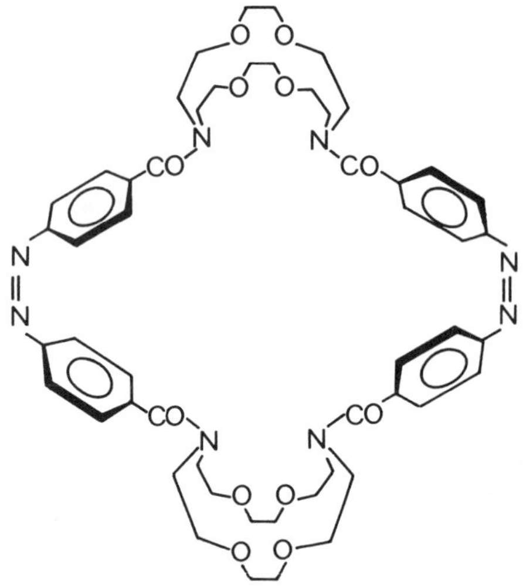   **7.16**

The compound **7.17** lacks any affinity towards alkali metal cations, but affinity towards Rb$^+$ and Cs$^+$ appears on ultraviolet irradiation, showing that the *trans* → *cis* isomerization of the chromophoric units allows the formation of an ionophoric loop

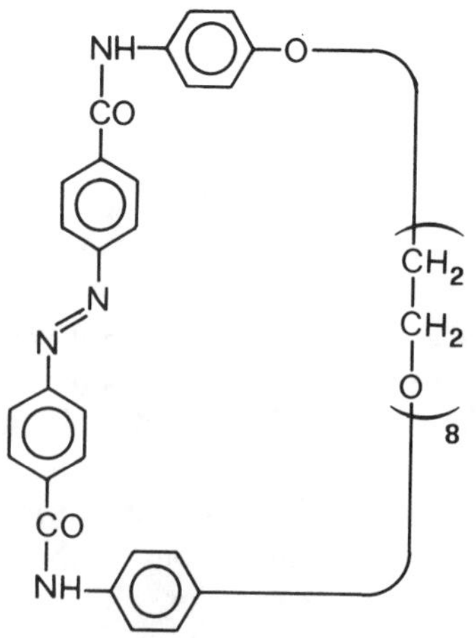

7.17

in the polyoxyethylene chain [83]. The reverse effect is obtained with the azobenzo-
phanes **7.18** and **7.19** which lose their affinity for alkali metal cations on photoisomerization [84]. These systems, however, are not at all satisfactory from the point
of view of photoreversibility because of the steric strain caused by the linkage in the

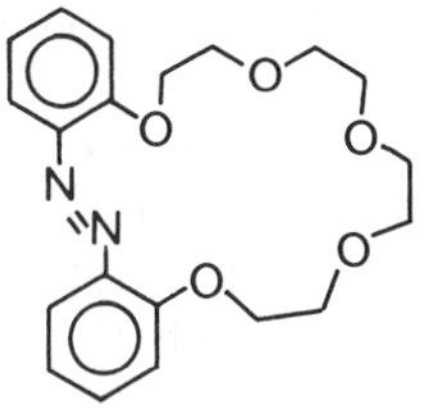

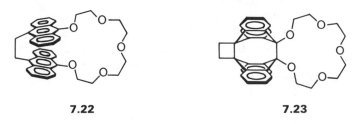

7.18 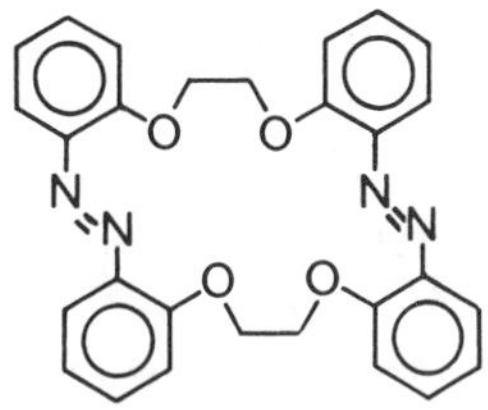   **7.19**

2,2'-positions of azobenzene. When linking takes place in the 4,4'-positions, more
flexible and reversible systems are obtained. Compounds **7.20** lack any affinity for
metal ions, whereas their *cis* photoisomers **7.21** exhibit a strong affinity for Na^+
($n = 1$), K^+ ($n = 2$), and Rb^+ ($n = 3$) [85]. The thermal **7.21** → **7.20** isomerization is
fully suppressed on metal ion coordination. These systems are rather interesting
because an "all-or-nothing" change in the ion-binding ability takes place upon
photoisomerization.

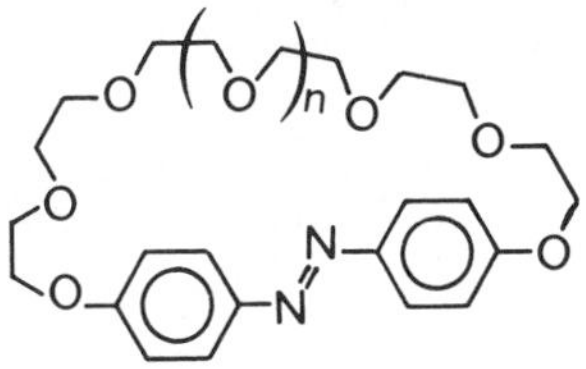

7.20

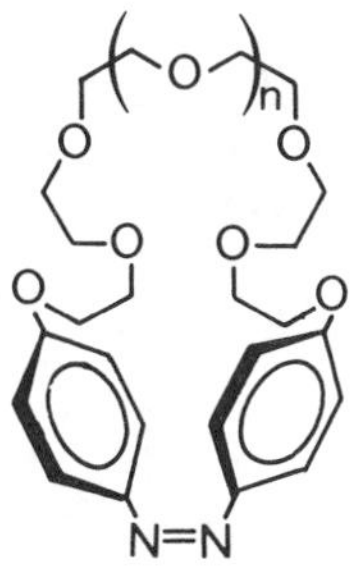

7.21

The crown ether **7.22** contains two anthracene components which can give rise to
intramolecular photoisomerization (**7.23**). In the presence of metal ions, the rate of
the photoreaction is unaffected, but the reverse thermal reaction becomes much
slower [86]. A similar "cation-lock" effect is described in section 10.2.1.

7.22   **7.23**

7.3.1.3 *"Butterfly" and capped crown ethers*

In the previous examples, photocontrol of the coordinating ability of the crown ether is achieved by conformational changes in the crown ether ring brought about by photoisomerization of a chromophoric component. An alternative way to reach the same goal is exploitation of the photoinduced change in a chromophoric component to modify the distance and the arrangements of two coordinating components, so as to create or destroy a cooperative coordination effect [70].

Shinkai and coworkers [87–89] synthesized a series of azobis(crown ethers) such as **7.24** whose *trans-cis* isomerization is like the motion of a butterfly. The photostationary state in absence of metal ions is attained when the concentration of the *cis* isomer **7.25** reaches 52%. Small metal ions that can fit the size of the crown ring do not affect the stationary-state concentrations, but large metal ions like Rb^+ and Cs^+ remarkably displace the equilibrium towards the *cis* isomer (98% for high Rb^+ concentrations). These and other results indicate that the *cis* form **7.25** can give stable sandwich-type 1:1 complexes with metal ions that are too large to fit the cavity of a single ring. Ion extraction and ion transport can thus be photocontrolled by using these supramolecular species as "phototweezers" (section 12.6.2). The effects of bulky substituents on the metal ion selectivity was also investigated [89].

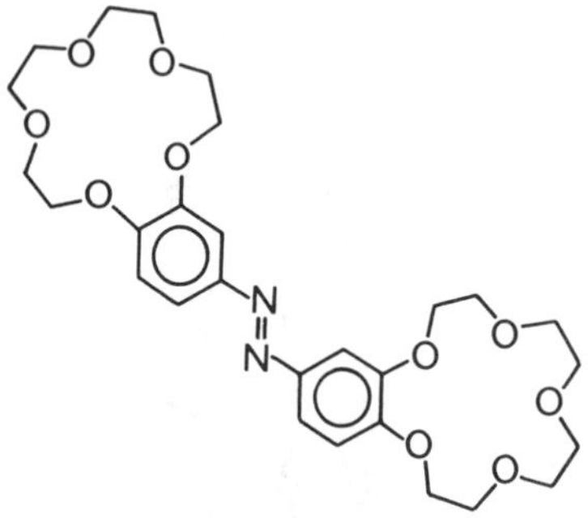

7.24 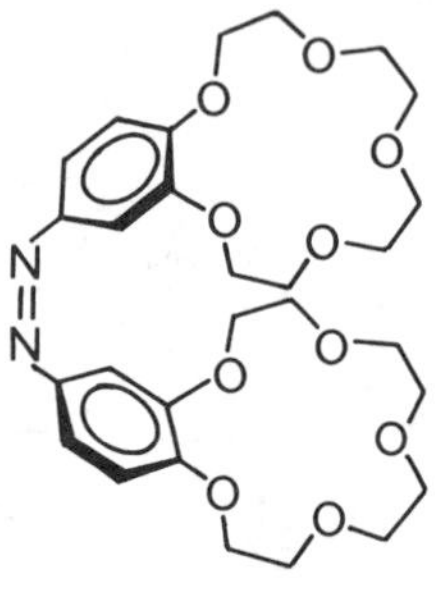   **7.25**

Crown ethers containing an anionic cap in a suitable position exhibit superior ion-binding properties (see, for example [90]). Therefore, the coordinating ability of a crown ether can be photocontrolled by a conformational change which modifies the reciprocal position of the crown and its anionic cap [70]. Several photoresponsive systems of this type have been synthesized [91,92], once more making use of the *trans → cis* photoisomerization of an azobenzene molecular component. Compound **7.26**, which contains a phenoxide anion with a nitro group in the 5 position to lower the pK_a value, in the *cis* configuration **7.27** may extract Ca^{2+} in neutral aqueous solution. Using this system, it is therefore possible to obtain a light-driven Ca^{2+} transport across a liquid membrane [92].

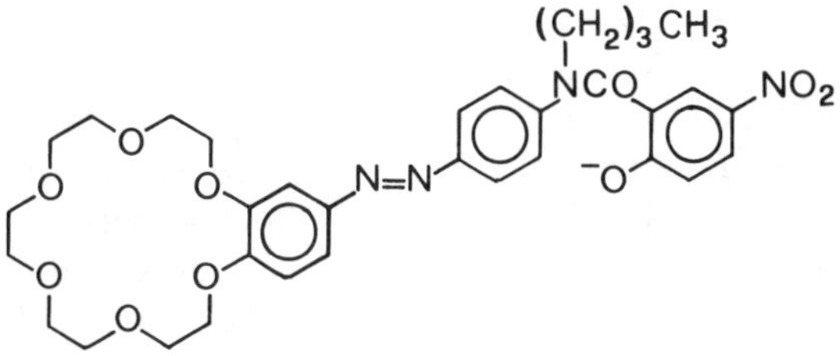

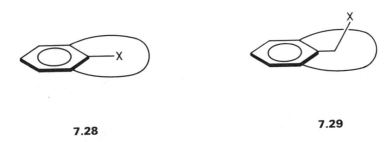

7.26

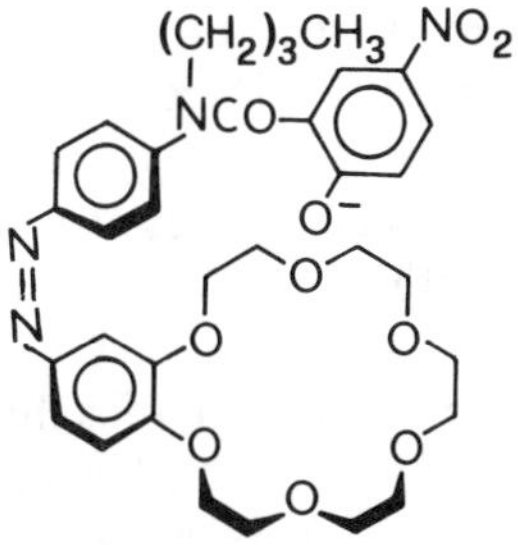

7.27

A particular case of capping is that of the crown ethers which bear an intra-annular substituent X, like **7.28** and **7.29**. Such a substituent can prevent, promote, and even switch ion coordination, depending on its specific nature. When the

7.28

7.29

substituent is a phenylazo group, the quite interesting possibility arises of changing its position by light excitation. This happens for the compound **7.30** where a methoxyphenylazo substituent occupies the crown cavity and prevents complexation of Na$^+$. Irradiation with ultraviolet light causes formation of the *cis* isomer **7.31**. The azo group is therefore removed from the ring and Na$^+$ ion can be coordinated [93].

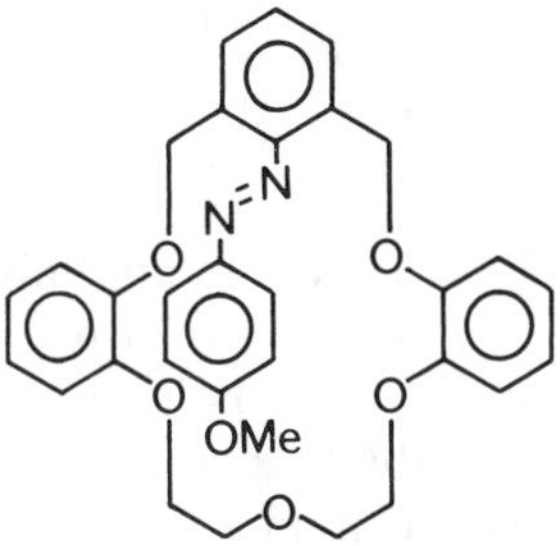

7.30

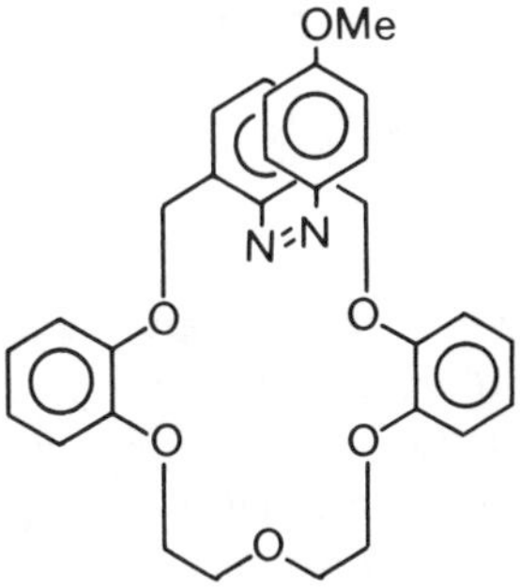

7.31

7.3.2 Systems involving cyclodextrins

As shown in section 10.5, cyclodextrins (CDs) and their derivatives are quite effective host species for a variety of guest molecules in aqueous solution [94,95]. As in the case of crown ethers, their binding ability may be controlled by photoinduced conformational changes of a component of the supramolecular structure. Ueno *et al.* [96,97] synthesized the azobenzene-capped β-CD **7.32** which can photoisomerize to **7.33**. They found that the binding ability of **7.32** towards various substrates is smaller than that of β-CD, whereas that of **7.33** is larger [96]. In particular **7.33** can bind 4,4′-bipyridine, whereas **7.32** cannot. These results suggest that the *trans→ cis* photoisomerization of the azobenzene cap expands the CD cavity. Because of the different depth of their cavities, **7.32** and **7.33** exhibit different catalytic activity towards the rate of hydrolysis of *p*-nitrophenylacetate. This permits on–off control of the catalytic activity [98] (for another example of photocontrol of catalytic activity, see [99]).

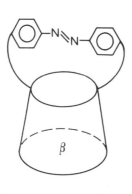

7.32   **7.33**

Owing to its large cavity (section 10.5), γ-CD has a poor binding ability for many organic molecules. Azobenzene-appended γ-CD, with the azobenzene group in the stable *trans* form, exhibits a higher binding ability. This may be due to a capping effect (**7.34**) or to a spacer effect (**7.35**). The binding ability is further enhanced upon *trans→ cis* photoisomerization of the azobenzene component [100]. In a γ-CD derivative containing two anthracene moieties as substituents, light excitation causes intramolecular photodimerization with disappearance of any binding ability, presumably because the anthracene photodimer penetrates the cavity and makes it too shallow for inclusion of other guests [101].

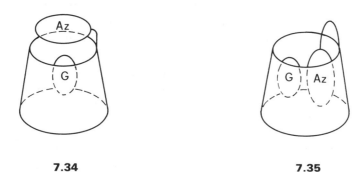

7.34   **7.35**

7.3.3 Systems involving metal complexes

In principle, photoinduced conformational changes can be used to turn on or off the electronic coupling between two components of a supramolecular structure. As shown in section 3.2, in mixed-valence metal complexes the electronic coupling can be readily evaluated from the characteristics of the intervalence transfer bands. This is the reason why much current effort in this field is focused on the possibility of bridging two metal complexes by a photoisomerizable component.

Launay and coworkers [102–105] have discussed in some detail the theoretical aspects of switching on/off (photo)induced electron transfer in systems of type A–L–A$^+$, where A is a metal-complex moiety like $Ru(NH_3)_5^{2+}$, A$^+$ is its one-electron oxidized form, and L is a photoisomerizable bridging ligand like bipyridyl-butadiene, py$-CH=CH-CH=CH-$py. Calculations showed that for the triad system composed by the three above-mentioned components, the effect of double-bond photoisomeritazion on the electronic coupling parameter is weak in spite of the marked difference in the geometries of the *trans-trans* **7.36** and *trans-cis* **7.37**

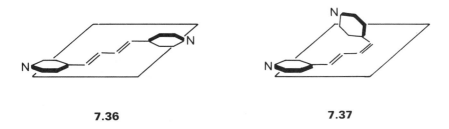

7.36   **7.37**

isomers. As an attempt to explore the behaviour of TICT-type bridges, they investigated the photophysical properties of $Ru(NH_3)_5(DMABN)^{2+}$, $Ru(NH_3)_5(DMABN)^{3+}$, and $Ru(bpy)_2(Cl)(DMABN)^+$ [106], where DMABN is the dual-luminescence emitter dimethylaminobenzonitrile **7.7** discussed in section 7.2.4. For $Ru(NH_3)_5(DMABN)^{2+}$ in alcohol solution at room temperature, the characteristic dual DMABN luminescence was observed upon excitation in the DMABN band at 290 nm, with the intensity ratio between the "anomalous" 498 and the "normal" 343 nm emission maxima slightly different from that obtained for free DMABN. On excitation in the Ru→DMABN CT band at 342 nm, the ligand-centred luminescence was replaced by a very weak band with a maximum at 393 nm, attributed to a MLCT excited state. For $Ru(NH_3)_5(DMABN)^{3+}$, the dual ligand-centred emission was again observed (with a reverse intensity ratio compared to the free ligand), and a band at 435 nm was also present when excitation was performed at 380 nm. Since this complex shows absorption bands up to 700 nm, the reported luminescence results are quite surprising. The result obtained for $Ru(bpy)_2Cl(DMABN)^+$ are also quite peculiar since as many as four different types of emissions (DMABN fluorescence and phosphorescence, and MLCT bands involving DMABN and bpy) were apparently observed at 77 K. While these experiments seem to have been performed carefully, we should also recall that the dual emission previously reported for other mixed-ligand Ru(II)-polypyridine complexes [107,108] was later shown to be due to decomposition of the sample or very small amounts of impurities [109].

As a further step towards the design of molecular switching devices, Launay and coworkers [110] investigated the photophysical properties of the potentially bridging ligand N,N'-bis(4-cyanophenyl)piperazine **7.38**. The crystal structure shows that the molecule is almost planar in the ground state. The luminescence behaviour is very sensitive to solvent polarity and in butanol the dual luminescence characteristic of TICT compounds is observed.

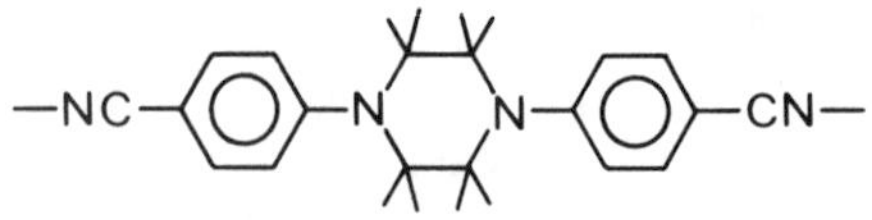

7.38

7.3.4 Other systems
Several other photoresponsive systems of considerable interest have been reported.

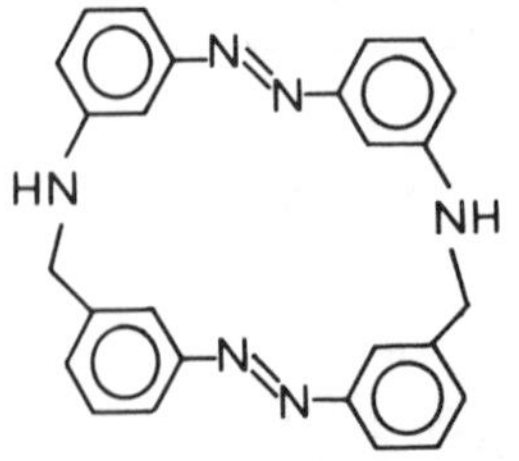

7.39

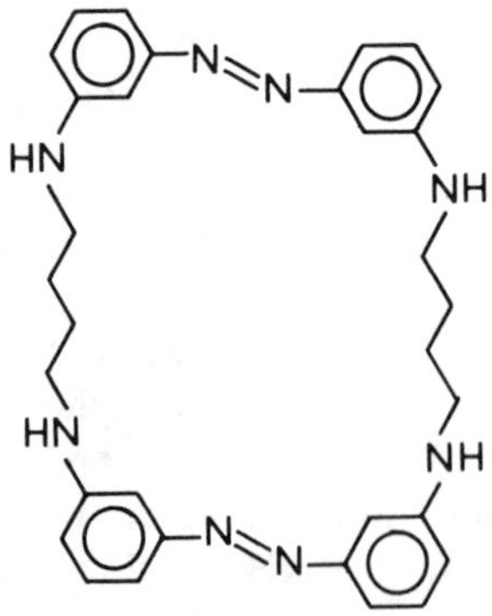

7.40

Vögtle and coworkers [111] have synthesized a number of azamacrocycles (for example, **7.39** and **7.40**) and the very interesting macrobicyclic species **7.41**. All these compounds exhibit the photoisomerization reaction characteristic of azobenzene (section 7.2.2) and therefore their cavity size can be photocontrolled. For the cage-type structure **7.41** all possible configurational isomers (*E,E,E*; *E,E,Z*; *E,Z,Z*; *Z,Z,Z*) have been isolated.

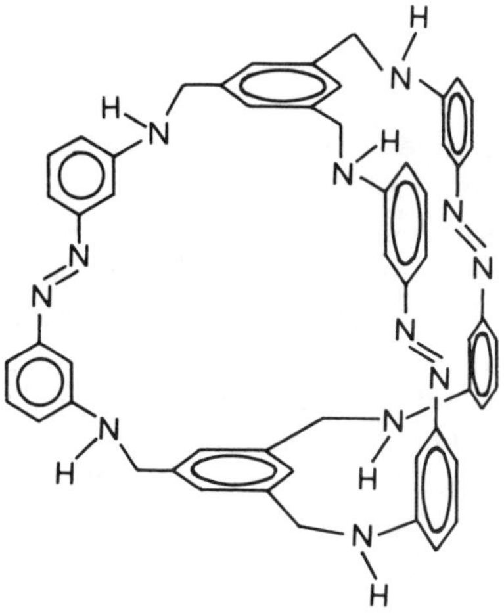

7.41

The planar *trans* isomer of **7.42** does not bind Zn^{2+} ions, but upon photoisomerization to the *cis* form, cooperation between the two iminodiacetic chelating groups becomes possible and Zn^{2+} ions are complexed [112]. The interconversion of the *cis* and *trans* isomers is reversible and the possible application of this class of compounds as a photoresponsive ion pump has been suggested.

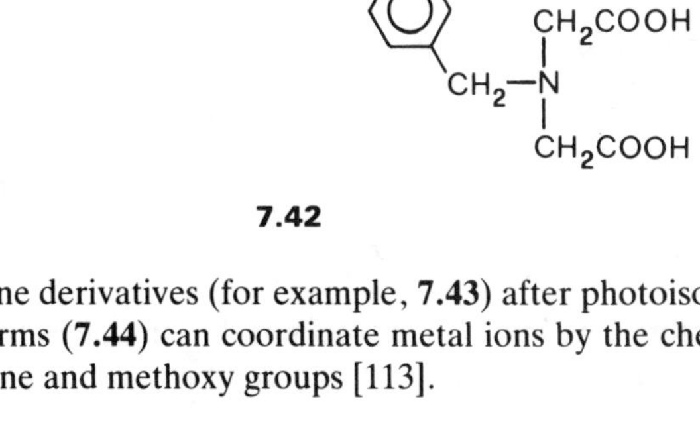

7.42

Spironaphthoxazine derivatives (for example, **7.43**) after photoisomerization to their merocyanine forms (**7.44**) can coordinate metal ions by the chelating moiety consisting of the ketone and methoxy groups [113].

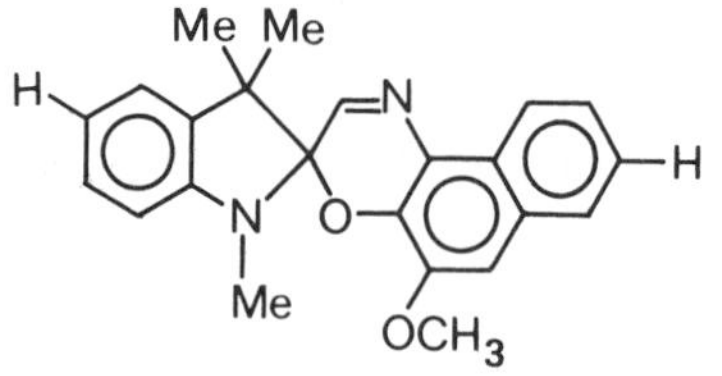

7.43 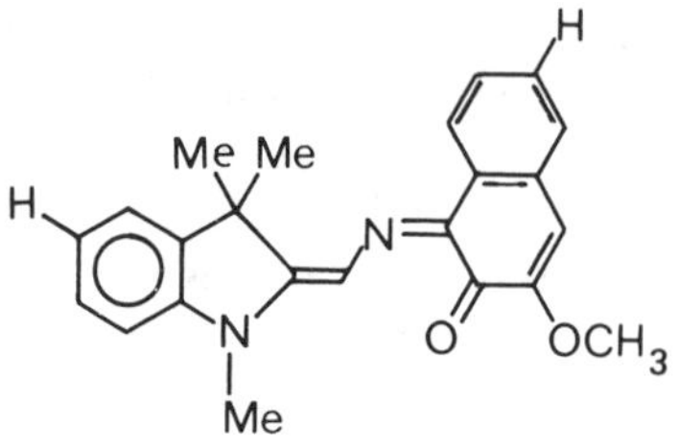   **7.44**

The spiropyran derivative **7.45**, bearing a long alkyl group, is insoluble in aqueous solution. On ultraviolet irradiation it is converted into the coloured merocyanine form **7.46** which is also insoluble in water and can be reconverted into **7.45** by visible light or by a thermal reaction. The merocyanine form **7.46**, because of its zwitterionic nature, exhibits a much larger affinity towards salts like KCl and NaCl compared to **7.45**. This system can thus be used as a photoactivated carrier for transportation of alkali metal salts through a 1-octanol liquid membrane [114]. The photoisomerization of a spiropyranindoline moiety incorporated into carbocyclic rings of different size has also been investigated [115].

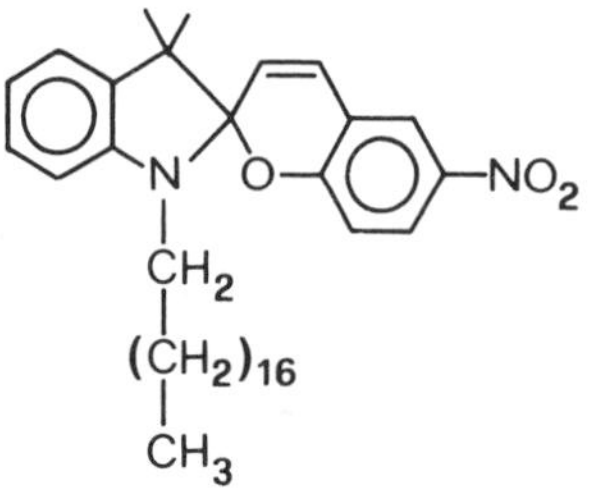

7.45 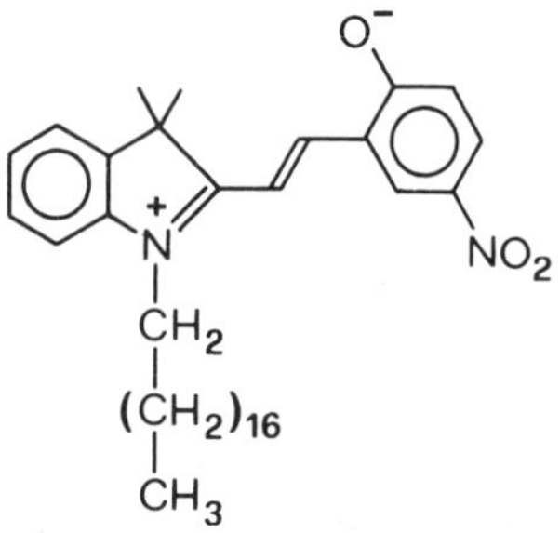   **7.46**

A particularly promising switchable chromophoric system is the dihydroazulene (**7.47**) ⇄ vinylheptafulvene (**7.48**) studied by Daub and coworkers [116]. The two compounds exhibit quite different redox properties so that in photoelectrochemical experiments it is possible to switch the current intensity by light excitation.

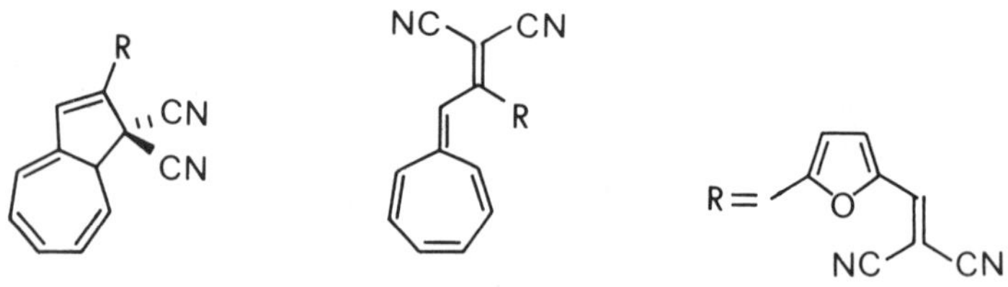

7.47   **7.48**

A quite interesting multi-mode chemical transducer has been reported by Shimidzu *et al.* [117]. It consists of the combination of a quinone and an azo compound. This system exhibits both electrochromism and photochromism and may therefore be used for dual-mode memory.

7.4 MICELLES, MEMBRANES, LIQUID CRYSTALS, AND POLYMERS

Photoinduced conformational changes of photochromic units can be used to modify the physical properties of micelles, membranes, polymers, and polypeptides. A detailed discussion of this topic is outside the scope of this book. We shall only recall some processes which are related to those discussed in the previous sections. Several reviews on light-induced effects on micelles [118,119], membranes [120,121], polypeptides [121,122], and polymers [28,122–124] have been published.

7.4.1 Photoeffects on micelles

Photorheological effects in micellar solutions have recently met considerable interest in view of potential actinometric applications. The photochemical reaction of an additive can cause a change in its solubility in the micellar solution and, as a consequence, a change in the solution viscosity [125,126]. Alternatively, a change in viscosity may be caused by the transformation of rodlike micelles into small spherical aggregates following the photochemical transformation of a polar additive, solubilized into the rodlike micelles, into a nonpolar compound [127]. In the first approach, the photochemical reactions used were the photodimerization of 9-anthracene derivatives [125], the *cis → trans* photoisomerization of 4-carboxystilbene [126], and the photocyclization of *N*-methyl-*N,N*-diphenylamine [126]. In the second approach, the photodecomposition of 3-methyl-phenyl-azo-phenylsulphone to alkylbenzenes was used [127].

The change in polarity caused by the phototransformation of polar merocyanine to nonpolar spiropyran leads to abrupt micelle formation in merocyanines derivatized with long alkyl chains [128]. This is due to the different critical micelle concentration values of the polar merocyanine and nonpolar spiropyran species. When disk-like micelles composed of surfactant containing *trans*-azobenzene are irradiated, formation of the *cis*-azobenzene isomer causes the transformation of the disk-like micelles into rodlike micelles [128]. Photocontrol of micellar catalysis for a surfactant bearing an azobenzene head group has also been reported [129].

7.4.2 Photocontrol of membrane functions

Model membranes containing photoresponsive components have been investigated by several research groups in an attempt to mimic photobiological processes. The photoinduced transportation of metal ions through liquid membranes by photochromic carriers has been discussed in sections 7.3.1 and 7.3.4. Crown ethers linked to azobenzene groups can also be immobilized in a polymer matrix [130]. In such a case, the contraction of the azobenzene unit produced by the *trans → cis* photoisomerization must be compensated by the elongation of the flexible crown ether moiety, with a consequent change in its coordination ability towards metal ions. Actually, Cs^+ and K^+ ions, which are bound to the polymer in the dark, are rapidly released in solution upon ultraviolet irradiation.

Bis-(crown ether)s consisting of two crown ether rings linked by an azobenzene moiety (section 7.3.1.3) undergo *trans → cis* photoisomerization even when they are entrapped in plasticized poly(vinyl chloride) membranes. In such systems, ultraviolet irradiation causes a reversible change in the membrane potential related to the change in the binding ability of the membrane towards alkali cations [131].

Langmuir–Blodgett (LB) films based on TCNQ (7,7,8,8-tetracyanoquinodimethane) can be switched by introducing an azobenzene unit in the amphiphilic columns [132]. The tentative explanation for this result is that the geometrical change caused by the *trans → cis* photoisomerization of azobenzene enhances the ordering of the TCNQ columns. Incorporation of the photoreactive benzylammonium salts in LB films permits photodestabilization of the multilayer structure [133]. The spiropyran-merocyanine photoisomerization can also take place in LB films [134]. The use of LB films for molecular electronics, including optical memories based on photochromic spiropyrans, photoinduced aggregation, and photovoltaic effects, has been reviewed by Sugi [135].

The fluidity of phospholipid membranes may be controlled by the photoisomerization of azobenzene containing amphiphiles incorporated in the membrane [136]. Embedding azobenzene containing surfactants in phospholipid vesicles perturbs the membrane structure and yields small channels for water permeation [137]. When light excitation causes the *trans → cis* isomerization of the azobenzene moiety, the channels are further enlarged, as shown by measurements of release of bromothymol blue from the membrane.

The transliposomal release of aminoacid can be photoinduced by light taking advantage of the conversion of spiropyran derivatives in the zwitterionic merocyanine forms which extract zwitterionic amino acids by mutual charge neutralization [138]. Reversible photoinduced changes of the membrane functions have also been observed in polypeptide membranes containing azobenzene groups in a side chain [139].

Destabilization of bilayer vesicles, which offer a spatially and temporally selective method of release of chemical and biological reagents, can be obtained by the photoinduced photopolymerization of the lipid components [140]. Nylon capsules coated with a photoresponsive synthetic membrane containing an azochromophore release NaCl on irradiation with ultraviolet light [141].

Polymer membranes, prepared from poly(hydroxyethyl methacrylate) containing small amounts of azobenzene groups in the side chains are permeable to insulin and lysozyme; upon ultraviolet irradiation (and the consequent *trans → cis* isomerization of the azobenzene units) the permeability is strongly decreased [142]. Photoinduced changes in membrane potentials can be obtained *via* the isomerization of azobenzene modified crown ethers [143] or spiropyran derivatives [144,145].

Rather interesting are the bilayer lipid membranes (BLM), comprising a bilayer of lipid molecules with their polar, hydrophilic groups oriented towards the aqueous solution and their hydrophobic hydrocarbon chains forming the liquid interior of the membrane. At the BLM/solution interface, the polar groups are organized resembling an ordered array in a crystal. A BLM may be considered as a limiting structure of a lamellar micelle and is similar to the so-called smectic soap mesophase of liquid crystals [146]. Unmodified BLMs are uninteresting from a photochemical viewpoint. However, a number of compounds of photochemical relevance may be incorporated in BLMs, including supramolecular species (for example, covalently-linked systems) and polycrystalline semiconductors (for example, CdS) [147]. The mechanism of light action is photoinduced charge separation, discussed in Chapter 5.

7.4.3 Photoresponsive liquid crystals

Photochemical and photophysical investigations involving liquid crystals follow three main directions: (i) the use of fluorescent probes to investigate the structure of liquid crystals (see, for example [148,149]); (ii) the effect of the regular structure of liquid crystals on the photochemical and photophysical behaviour of molecules (see, for example [150–152]); (iii) the influence of a photoreaction on the structure of the liquid crystal. We shall mention a few details on this last topic because it is pertinent to the content of this chapter.

Ringsdorf and coworkers [153–156] have prepared photoreactive mesogens and pre-mesogens by incorporation of photolabile moieties into polymers. The photo-destruction of mesophases has been achieved by using 1-benzoyliminopyridiliumy-lide derivatives which undergo a drastic conformational change and/or cleavage [153]. Photogeneration of mesophases was obtained by photoelimination of bulky *tert*-butyloxycarbonyl groups from azobenzene derivatives [153]. Photochromic effects have also been obtained with liquid crystal copolymers containing spiro-pyrans [153,155] and liquid crystal homopolymers containing salicylidene aniline [153]. Stabilization of columnar structures of substituted discotic azacrowns can be obtained photochemically by means of the $(2 + 2)$ photocycloaddition of pendant cinnamoyl side groups (Fig. 7.3) [133]. Photoisomerization of a small amount of

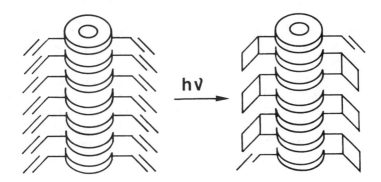

Fig. 7.3 — Schematic representation of light-induced intracolumnar cross-linking [133].

4-butyl-4'-methoxyazobenezene mixed with 4-cyano-4'-*n*-pentylbiphenyl can trigger a phase transition which leads to amplified image recording [157,158].

An excellent review on polymeric liquid crystals and on their use as models for functional biomembranes has appeared [159].

7.4.4 Photoresponsive polymers

The need for polymer materials is currently changing from structural materials to functional materials. Molecules which exhibit photoinduced structural changes, suitably linked to polymer chains, can provide light control of various physical properties and chemical functions. Azobenzene and spiropyran units, because of their large structural changes and dipole-moment variations upon photoisomerization, are extensively employed to obtain photoresponsive synthetic polymers [28,124].

The *viscosity* of a polymer system is directly related to the polymer conformation. Solutions of poly(methyl methacrylate) having spiro-benzopyran side groups undergo a reversible viscosity decrease upon light excitation, attributed to intramolecular solvation by the methacrylate ester side groups of the polar merocyanines produced by irradiation [32,160].

Changes of dipole moments of pendant groups can strongly affect the *solubility* of polymers. It is therefore possible to control the solubility of polystyrene in cyclohexane by appending a small amount of azobenzene [161] or spiropyran [162] pendant groups to the polymer chains. Irradiation with ultraviolet light causes precipitation of polymer, while resolubilization can be achieved by irradiation with visible light. A flash photolysis study of azobenzene-containing polystyrene systems [163] showed that the *trans → cis* isomerization of the azobenzene groups essentially takes place during the 20 ns flash, whereas conformational changes of the macromolecules take place over the microsecond time scale and polymer aggregation and precipitation occurs several hundred milliseconds after the flash. For the same polymer, the sol-gel *phase transition* can also be controlled by photoisomerization of the photochromic side chains [164]. Another property that can be photocontrolled is the *wettability* of polymer surfaces by water [165]. Detailed laser photolysis studies on the photoinduced aggregation in polymers containing spiropyran units have also been reported [36].

Segmental motions of polymer side chains [166,167] and the free volume distribution around a polymer main chain [168] of polymers can be investigated by using TICT molecules as fluorescence probes.

Photoresponsive *metal ion chelation* by polymers which contain azobenzene units linked to crown ethers has been discussed in the previous section. Reversible *pH changes* can also be photoinduced in some polymers since the conformational change induced by light can influence the dissociation of acid groups [161].

Polymeric composite films consisting of poly(vinyl chloride)/azobenzene liquid crystals/LiClO$_4$/(12-crown-4) show photoresponsive ionic conductivity that can be used for image storage [169].

Light-induced conformational changes of photochromic polymers may produce changes of dimensions of bulk polymers and can thus generate reversible *photomechanical effects*. A discussion covering the literature up to 1982 was reported by Smets [123]. More recent but less detailed reviews have also appeared [28,121,124]. The degree of *swelling* of a polymer membrane composed of poly(2-hydroxyethyl methacrylate) with azobenzene groups in the side chains can be reversibly photoregulated by irradiating with ultraviolet or visible light [170]. Photoinduced reversible variations of the *surface area* have been obtained for monolayers consisting of polymers which contain spiropyran units [171] and for films containing azobenzene [172] and β-carotene or cyanostilbene [173] units.

From the point of view of conformational effects, photoresponsive polypeptides

[121] represent especially interesting systems because of their existence in definite ordered structures such as α-helix and β-structures. In addition, their structures are more relevant to the proteinic nature of biological photoreceptors. In polypeptides, conformations and conformational changes can be directly detected by circular dicroism spectroscopy. The most common photochromic component used to induce light effects in polypeptides is the azobenzene group (see, for example [174–176]), but stilbene [177] and spiropyran [178] units have also been employed. Light excitation of the chromophoric group can lead to drastic conformational changes, for example from left-handed helix to right-handed helix [179] or from random coil to α-helix [178]. Detailed studies on polypeptides containing azobenzene groups suggest that the simple geometry variation during the photoisomerization of the chromophore is insufficient to induce appreciable conformational changes in the main chain of the macromolecule. The driving force for photoregulation seems to result from a polarity variation of the environment around the polypeptide backbone, as a consequence of the different polarity and hydrophobicity between the *trans* and the *cis* isomers [174,175,180].

REFERENCES

[1] Erlanger, B. F. (1976) *Annu. Rev. Biochem.* **45** 267.

[2] Martinek, K. and Berezin, I. V. (1979) *Photochem. Photobiol.* **29** 637.

[3] Erlanger, B. F. (1983) In Montagnoli, G. and Erlanger, B.F. (eds) *Molecular models of photoresponsiveness.* Plenum, p. 1.

[4] Brown, G. H. (ed.) (1971) *Photochromism.* Wiley.

[5] Ross, D. L. and Blanc, J. (1971) In Brown, G. H. (ed.) *Photochromism.* Wiley, p. 471.

[6] Bertelson, R. C. (1971) In Brown, G. H. (ed.) *Photochromism.* Wiley, p. 45.

[7] Häder, D. P. and Tevini, M. (1987) *General photobiology.* Pergamon.

[8] Rettig, W. (1988) In Liebman, J.F. and Greenberg, A. (eds) *Modern models of bonding and delocalization.* VCH Publishers, p. 229.

[9] Warshel, A. (1976) *Nature* **260** 679.

[10] Honig, B. (1982) In Alfano, R. (ed.) *Biological events probed by ultrafast laser spectroscopy.* Academic, p. 281.

[11] Allen, M. T. and Whitten, D. G. (1989) *Chem. Rev.* **89** 1691, and references therein.

[12] Th. Förster (1952) *Z. Elektrochem.* **56** 716.

[13] Saltiel, J., D'Agostino, J., Megarity, E. D., Metts, L., Neuberger, K. R., Wrighton, M. S., and Zafiriou, O. C. (1973) *Org. Photochem.* **3** 1.

[14] Gegiou, D., Muszkat, K. A., and Fischer, E. (1968) *J. Am. Chem. Soc.* **90** 12.

[15] Gegiou, D., Muszkat, K. A., and Fischer, E. (1968) *J. Am. Chem. Soc.* **90** 3907.

[16] Orlandi, G. and Siebrand, W. (1975) *Chem. Phys. Lett.* **30** 352.

[17] Hohlneicher, G. and Dick, B. (1984) *J. Photochem.* **27** 215.

[18] Hochstrasser, R. M. (1980) *Pure Appl. Chem.* **52** 2683.

[19] Courtney, S. H. and Fleming, G. R. (1985) *J. Chem. Phys.* **83** 215.

[20] Syage, J. A., Felker, P. M., and Zewail, A. H. (1984) *J. Chem. Phys.* **81** 4706.
[21] Doany, F. E., Hochstrasser, R. M., Greene, B. I., and Millard, R. R. (1985) *Chem. Phys. Lett.* **118** 1.
[22] Langkilde, F. W., Wilbrandt, R., Negri, F., and Orlandi, G. (1990) *Chem. Phys. Lett.* **165** 66.
[23] Bortolus, P. and Monti, S. (1979) *J. Phys. Chem.* **83** 648.
[24] Monti, S., Orlandi, G., and Palmieri, P. (1982) *Chem. Phys.* **71** 87.
[25] Bortolus, P. and Monti, S. (1987) *J. Phys. Chem.* **91** 5046, and references therein.
[26] Rau, H. and Luddecke, E. (1982) *J. Am. Chem. Soc.* **104** 1616.
[27] Rau, H. (1984) *J. Photochem.* **26** 221.
[28] Kumar, G. S. and Neckers, D. C. (1989) *Chem. Rev.* **89** 1915.
[29] Heiligman-Rim, R., Hirshberg, Y., and Fischer, E. (1962) *J. Phys. Chem.* **66** 2470.
[30] Krysanov, S. A. and Alfimov, M. V. (1982) *Chem. Phys. Lett.* **91** 77.
[31] Krysanov, S. A. and Alfimov, M. V. (1984) *Laser Chem.* **4** 129.
[32] Irie, M., Menju, A., and Hayashi, K. (1979) *Macromolecules* **12** 1176.
[33] Gehrtz, M., Bräuchle, Chr., and Voitländer, J. (1982) *J. Am. Chem. Soc.* **104** 2094.
[34] Kellmann, A., Lindqvist, L., Monti, S., Tfibel, F., and Guglielmetti, R. (1983) *J. Photochem.* **21** 223.
[35] Kalisky, Y., Orlowski, T. E., and Williams, D. J. (1983) *J. Phys. Chem.* **87** 5333.
[36] Kalisky, Y. and Williams, D. J. (1984) *Macromolecules* **17** 292.
[37] Kellmann, A., Lindqvist, L., Monti, S., Tfibel, F., and Guglielmetti, R. (1985) *J. Photochem.* **28** 547.
[38] Monti, S., Orlandi, G., Kellmann, A., and Tfibel, F. (1986) *J. Photochem.* **33** 81.
[39] Kellmann, A., Lindqvist, L., Tfibel, F., and Guglielmetti, R. (1986) *J. Photochem.* **35** 155.
[40] Lenoble, C. and Becker, R. S. (1986) *J. Phys. Chem.* **90** 62.
[41] Tamaki, T., Sakuragi, M., Ichimura, K., and Aoki, K. (1989) *Chem. Phys. Lett.* **161** 23.
[42] Ernsting, N. P. (1989) *Chem. Phys. Lett.* **159** 526.
[43] Zerbetto, F., Monti, S., and Orlandi, G. (1984) *J. Chem. Soc., Faraday Trans. 2* **80** 1513.
[44] Rettig, W. (1986) *Angew. Chem. Int. Ed. Engl.* **25** 971.
[45] Lippert, E., Rettig, W., Bonačič-Koutecký, V., Heisel, F., and Miehé, J. A. (1987) *Adv. Chem. Phys.* **68** 1.
[46] Rettig, W. (1988) *Appl. Phys. B* **45** 145.
[47] Rotkiewicz, K., Grellmann, K. H., and Grabowski, Z. R. (1973) *Chem. Phys. Lett.* **19** 315; **21** 212.
[48] Grabowski, Z. R., Rotkiewicz, K., Siemiarczuk, A., Cowley, D. J., and Baumann, W. (1979) *Nouv. J. Chim.* **3** 443.
[49] Lippert, E., Luder, W., and Boos, H. (1962). In Mangini, A. (ed.) *Advances in molecular spectroscopy*. Pergamon, p. 443.
[50] Tseng, J. C. C. and Singer, L. A. (1989) *J. Phys. Chem.* **93** 7092.

[51] Baumann, W., Schwager, B., Detzer, N., Okada, T., and Mataga, N. (1988) *J. Phys. Chem.* **92** 3742.

[52] Grabowski, Z. R. (1987) In Balzani, V. (ed.) *Supramolecular photochemistry*. Reidel, p. 319.

[53] Okada, T., Mataga, N., and Baumann, W. (1987) *J. Phys. Chem.* **91** 760.

[54] Siemiarczuk, A. and Ware, W. R. (1987) *J. Phys. Chem.* **91** 3677.

[55] Law, K. Y. (1987) *J. Phys. Chem.* **91** 5184.

[56] Herbich, J., Dobkowski, J., Rullière, C., and Nowacki, J. (1989) *J. Luminescence* **44** 87.

[57] Nag, A., Kundu, T., and Bhattacharyya, K. (1989) *Chem. Phys. Lett.* **160** 257.

[58] Hrnjez, B., Yazdi, P. T., Fox, M. A., and Johnston, K. P. (1989) *J. Am. Chem. Soc.* **111** 1915.

[59] Nag, A. and Bhattacharyya, K. (1990) *J. Chem. Soc., Faraday Trans.* **86** 53.

[60] Meyer, M., Mialocq, J. C., and Perly, B. (1990) *J. Phys. Chem.* **94** 98.

[61] Bonačič-Koutecký, V., Koutecký, J., and Michl, J. (1987) *Angew. Chem. Int. Ed. Engl.* **26** 170.

[62] Rettig, W. (1987). In Balzani, V. (ed.) *Supramolecular photochemistry*. Reidel, p. 329.

[63] Mataga, N., Yao, H., Okada, T., and Rettig, W. (1989) *J. Phys. Chem.* **93** 3383.

[64] Rettig, W. and Majenz, W. (1989) *Chem. Phys. Lett.* **154** 335.

[65] Pedersen, C. J. (1967) *J. Am. Chem. Soc.* **89** 7017.

[66] Pedersen, C. J. (1988) *Angew. Chem. Int. Ed. Engl.* **27** 1021.

[67] Cram, D. J. (1988) *Angew. Chem. Int. Ed. Engl.* **27** 1009.

[68] Vögtle, F. (ed.) (1981) *Host–guest complex chemistry I. Topics Curr. Chem.* **38**.

[69] Shinkai, S. (1983). In Montagnoli, G. and Erlanger, B. F. (eds) *Molecular models of photoresponsiveness*. Plenum, p. 325.

[70] Shinkai, S. and Manabe, O. (1984) *Topics Curr. Chem.* **121** 67.

[71] Beer, P. D. (1989) *Chem. Soc. Rev.* **18** 409.

[72] Shinkai, S., Inuzuka, K., Miyazaki, O., and Manabe, O. (1985) *J. Am. Chem. Soc.* **107** 3950.

[73] Beer, P. D., Blackburn, C., McAleer, J. F., and Sikanyika, H. (1990) *Inorg. Chem.* **29** 378.

[74] Rebek, J. Jr., Trend, J. E., Wattley, R. V., and Chakravorti, S. (1979) *J. Am. Chem. Soc.* **101** 4333.

[75] Beer, P. D. and Rotlin, A. S. (1988) *J. Chem. Soc., Chem. Commun.* 52.

[76] Warshawsky, A. and Kahana, N. (1982) *J. Am. Chem. Soc.* **104** 2663.

[77] Shinkai, S., Nakaji, T., Nishida, Y., Ogawa, T., and Manabe, O. (1980) *J. Am. Chem. Soc.* **102** 5860.

[78] Shinkai, S., Kounot, T., Kusano, Y., and Manabe, O. (1982) *J. Chem. Soc., Perkin Trans.* 2 2741.

[79] Akabori, S., Kumagai, T., Habata, Y., and Sato, S. (1988) *J. Chem. Soc., Chem. Commun.* 661.

[80] Lehn, J. M. (1988) *Angew. Chem. Int. Ed. Engl.* **27** 89.

[81] Vögtle, F., Wallon, A., Müller, W. M., Werner, U., and Nieger, M. (1990) *J. Chem. Soc., Chem. Commun.* 158.

[82] Shinkai, S., Honda, Y., Kusano, Y., and Manabe, O. (1982) *J. Chem. Soc., Chem. Commun.* 848.

[83] Shinkai, S., Honda, Y., Minami, T., Ueda, K., Manabe, O., and Tashiro, M. (1983) *Bull. Chem. Soc. Japan* **56** 1700.

[84] Shiga, M., Takagi, M., and Ueno, K. (1980) *Chem. Lett.* 1021.

[85] Shinkai, S., Minami, T., Kusano, Y., and Manabe, O. (1983) *J. Am. Chem. Soc.* **105** 1851.

[86] Yamashita, I., Fujii, M., Kaneda, T., Misumi, S., and Otsubo, T. (1980) *Tetrahed. Lett.* **21** 541.

[87] Shinkai, S., Ogawa, T., Nakaji, T., and Manabe, O. (1980) *J. Chem. Soc., Chem. Commun.* 375.

[88] Shinkai, S., Nakaji, T., Ogawa, T., Shigematsu, K., and Manabe, O. (1981) *J. Am. Chem. Soc.* **103** 111.

[89] Shinkai, S., Ogawa, T., Kusano, Y., Manabe, O., Kikukawa, K., Goto, T., and Matsuba, T. (1982) *J. Am. Chem. Soc.* **104** 1960.

[90] Wierenga, W., Evans, B. R., and Woltersom, J. A. (1979) *J. Am. Chem. Soc.* **101** 1334 .

[91] Shinkai, S., Shigematsu, K., Ogawa, T., Minami, T., and Manabe, O. (1980) *Tetrahed. Lett.* **21** 4463.

[92] Shinkai, S., Minami, T., Kusano, Y., and Manabe, O. (1982) *J. Am. Chem. Soc.* **104** 1967.

[93] Shinkai, S., Miyazaki, K., and Manabe, O. (1985) *Angew. Chem. Int. Ed. Engl.* **24** 866.

[94] Saenger, W. (1980) *Angew. Chem. Int. Ed. Engl.* **19** 344.

[95] Szejtli, J. (1982) *Cyclodextrins and their inclusion complexes.* Akademiai Kiado, Budapest.

[96] Ueno, A., Yoshimura, H., Saka, R., and Osa, T. (1979) *J. Am. Chem. Soc.* **101** 2779.

[97] Ueno, A., Saka, R., and Osa, T. (1979) *Chem. Lett.* 841.

[98] Ueno, A., Takahashi, K., and Osa, T. (1981) *J. Chem. Soc. Chem., Commun.* 94.

[99] Ueno, A., Takahashi, K., and Osa, T. (1980) *J. Chem. Soc. Chem., Commun.* 831.

[100] Ueno, A., Tomita, Y., and Osa, T. (1983) *Tetrahed. Lett.* **24** 5245.

[101] Moriwaki, F., Ueno, A., Osa, T., Hamada, F., and Murai, K. (1986) *Chem. Lett.* 1865.

[102] Joachim, C. and Launay, J. P. (1986) *Chem. Phys.* **109** 93.

[103] Launay, J. P. (1987) In Carter, F. L. (ed.) *Molecular electronic devices II.* Dekker, p. 39.

[104] Joachim, C. and Launay, J. P. (1988) In Carter, F. L., Siatkowski, R. E., and Wohltjen, H. (eds) *Molecular electronic devices.* North-Holland, p. 149.

[105] Launay, J. P., Woitellier, S., Sowinska, M., Tourrel, M., and Joachim, C. (1988) In Carter, F. L., Siatkowski, R. E., and Wohltjen, H. (eds) *Molecular electronic devices.* North-Holland, p. 171.

[106] Sowinska, M., Launay, J. P., Mugnier, J., Pouget, J., and Valeur, B. (1987) *J. Photochem.* **37** 69.

[107] Juris, A., Barigelletti, F., Balzani, V., Belser, P., and von Zelewski, A. (1982) *Isr. J. Chem.* **22** 87.

[108] Cocks, A. T., Wright, R., and Seddon, K. R. (1982) *Chem. Phys. Lett.* **85** 369.

[109] Belser, P., von Zelewsky, A., Juris, A., Barigelletti, F., and Balzani, V. (1984) *Chem. Phys. Lett.* **104** 100.

[110] Launay, J. P., Sowinska, M., Leydier, L., Gourdon, A., Amouyal, E., Boillot, M. L., Heisel, F., and Miehé, J. A. (1989) *Chem. Phys. Lett.* **160** 89.

[111] Losensky, H. W., Spelthann, H., Ehlen, A., Vögtle, F., and Bargon, J. (1988) *Angew. Chem. Int. Ed. Engl.* **27** 1189.

[112] Blank, M., Soo, L. M., Wassermann, N. H., and Erlanger, B. F. (1981) *Science* **214** 70.

[113] Tamaki, T. and Ichimura, K. (1989) *J. Chem. Soc., Chem. Commun.* 1477.

[114] Shimidzu, T. and Yoshikawa, M. (1983) *J. Membrane Science* **13** 1.

[115] Winkler, J. D. and Deshayes, K. (1987) *J. Am. Chem. Soc.* **109** 2190.

[116] Daub, J., Saltbeck, J., Knöchel, T., Fischer, C., Kunkely, H., and Rapp, K. M. (1989) *Angew. Chem. Int. Ed. Engl.* **28** 1494.

[117] Shimidzu, T., Iyoda, T., and Honda, K. (1988) *Pure Appl. Chem.* **60** 1025.

[118] Wolff, T., Suck, T. A., Emming, C. S., and von Bünau, G. (1987) *Progr. Colloid Polym. Sci.* **73** 18.

[119] von Bünau, G. and Wolff, T. (1988) *Adv. Photochem.* **14** 273.

[120] Pepe, I. M. and Gliozzi, A. (1983) In Montagnoli, G. and Erlanger, B. F. (eds) *Molecular models of photoresponsiveness.* Plenum, p. 377.

[121] Pieroni, O., Fissi, A., and Ciardelli, F. (1986) *Photochem. Photobiol.* **44** 785.

[122] Houben, J. L. and Rosato, N. (1983) In Montagnoli, G. and Erlanger, B. F. (eds) *Molecular models of photoresponsiveness.* Plenum, p. 313.

[123] Smets, G. , (1983) *Adv. Polymer Sci.* **50** 17.

[124] Irie, M. (1990) *Adv. Polymer Sci.* **94** 27.

[125] Wolff, T. and von Bünau, G. (1984) *J. Photochem.* **24** 37.

[126] Wolff, T., Schmidt, F., and von Bünau, G. (1989) *J. Photochem. Photobiol. A: Chemistry* **48** 435.

[127] Wittenbeck, P., Franzke, D., and Wokaun, A. (1990) *J. Photochem. Photobiol. A: Chemistry* **53** 343.

[128] Tazuke, S., Kurihara, S., Yamaguchi, H., and Ikeda, T. (1987) *J. Phys. Chem.* **91** 249.

[129] Shinkai, S., Matsuo, K., Sato, M., Sone, T., and Manabe, O. (1981) *Tetrahed. Lett.* **22** 1409.

[130] Shinkai, S., Kinda, H., and Manabe, O. (1982) *J. Am. Chem. Soc.* **104** 2933.

[131] Shinkai, S., Ishihara, M., and Manabe, O. (1985) *Polymer J.* **17** 1141.

[132] Tachibana, H., Nakamura, T., Matsumoto, M., Komizu, H., Manda, E., Niino, H., Yabe, A., and Kawabata, Y. (1989) *J. Am. Chem. Soc.* **111** 308.

[133] Häubling, C., Mertesdorf, C., and Ringsdorf, H. (1989) In Hall, D. O. and Grassi, G. (eds) *Photoconversion processes for energy and chemicals.* Elsevier, p. 51..

[134] Polymeropoulos, E. E. and Möbious, D. (1979) *Ber. Bunsenges. Phys. Chem.* **83** 1215.

[135] Sugi, M. (1988) In Carter, F. L., Siatkowski, R. E., and Wohltjen, H. (eds) *Molecular electronic devices*. North-Holland, p. 441.

[136] Yamaguchi, H., Ikeda, T., and Tazuke, S. (1988) *Chem. Lett.* 539.

[137] Kano, K., Tanaka, Y., Ogana, T., Shimomura, M., Okahata, Y., and Kunitake, T. (1981) *Photochem. Photobiol.* **34** 322.

[138] Sunamoto, J., Iwamoto, K., Mohri, Y., and Kominato, T. (1982) *J. Am. Chem. Soc.* **104** 5502.

[139] Kinoshita, J., Sato, M., Takizawa, A., and Tsujita, Y. (1986) *Macromolecules* **19** 51.

[140] Frankel, D. A., Lamparski, H., Liman, U., and O'Brien, D. F. (1989) *J. Am. Chem. Soc.* **111** 9262.

[141] Okahata, Y. (1986) *Acc. Chem. Res.* **19** 57.

[142] Ishihara, K. and Shinohara, I. (1984) *J. Polymer Sci.: Polymer Lett. Ed.* **22** 515.

[143] Anzai, J., Sasaki, H., Ueno, A., and Osa, T. (1983) *J. Chem. Soc., Chem. Commun.* 1045.

[144] Bellobono, I. R., Marcandalli, B., Selli, E., and Calgari, S. (1984) *Photogr. Sci. Eng.* **28** 162.

[145] Anzai, J., Hasebe, Y., Ueno, A., and Osa, T. (1987) *Bull. Chem. Soc. Japan* **60** 1515.

[146] Tien, H. T. (1974) *Bilayer liquid membranes (BLM): theory and practice.* Dekker.

[147] Tien, H. T. (1988) In Carter, F. L., Siatkowski, R. E., and Wohltjen, H. (eds) *Molecular electronic devices*. North-Holland, p. 209.

[148] Johansson, L. B. A., Molotkowsky, J. G., and Bergelson, L. D. (1987) *J. Am. Chem. Soc.* **109** 7374.

[149] Arcioni, A., Tarroni, R., and Zannoni, C. (1988). In Samorí, B. and Thulstrup, E. W. (eds) *Polarized spectroscopy of ordered systems*. Kluwer, p. 421.

[150] Weiss, R. G., Treanor, R. L., and Nuñez, A. (1988) *Pure Appl. Chem.* **60** 999.

[151] Das, S., Lenoble, C., and Becker, R. S. (1987) *J. Am. Chem. Soc.* **109** 4349.

[152] Gregg, B. A., Fox, M. A., and Bard, A. J. (1989) *J. Phys. Chem.* **93** 4227.

[153] Cabrera, I., Engel, M., and Ringsdorf, H. (1989) In Hall, D. O. and Grassi, G. (eds) *Photoconversion processes for energy and chemicals*. Elsevier, p. 68.

[154] Cabrera, I., Krongauz, V., and Ringsdorf, H. (1987) *Angew. Chem. Int. Ed. Engl.* **26** 1178.

[155] Cabrera, I., Krongauz, V., and Ringsdorf, H. (1988) *Mol. Cryst. Liq. Cryst.* **155** 221.

[156] Eich, M., Wendorff, J. H., Reck, B., and Ringsdorf, H. (1987) *Makromol. Chem. Rapid. Commun.* **8** 59.

[157] Tazuke, S., Kurihara, S., and Ikeda, T. (1987) *Chem. Lett.* 911.

[158] Kurihara, S., Ikeda, T., and Tazuke, S. (1988) *Jap. J. Appl. Phys.* **27** L1791.

[159] Ringsdorf, H., Schlarb, B., and Venzmer, J. (1988) *Angew. Chem. Int. Ed. Engl.* **27** 188.

[160] Irie, M. and Hayashi, K. (1979) *J. Macromol. Sci. Chem.* **A13** 511.

[161] Irie, M. and Tanaka, H. (1983) *Macromolecules* **16** 210.

[162] Irie, M., Iwayanagi, T., and Taniguchi, Y. (1985) *Macromolecules* **18** 2418.

[163] Irie, M. and Schnabel, W. (1985) *Macromolecules* **18** 394.

[164] Irie, M. and Iga, R. (1985) *Macromol. Chem. Rapid Commun.* **6** 403.

[165] Ishihara, K., Hamada, N., Kato, S., and Shinohara, I. (1983) *J. Polym. Sci.: Polym. Chem. Ed.* **21** 1551.

[166] Hayashi, R., Tazuke, S., and Frank, C. W. (1987) *Macromolecules* **20** 983.

[167] Tazuke, S., Guo, R.K., and Hayashi, R. (1988) *Macromolecules* **21** 1046.

[168] Tazuke, S., Guo, R.K., and Hayashi, R. (1989) *Macromolecules* **22** 729.

[169] Kimura, K., Suzuki, T., and Yokoyama, M. (1989) *J. Chem. Soc., Chem. Commun.* 1570.

[170] Ishihara, K., Hamada, N., Kato, S., and Shinohara, I. (1984) *J. Polymer Sci.: Polymer Chem. Ed.* **22** 121.

[171] Vilanove, R., Herret, H., Gruler, H., and Rondelez, F. (1983) *Macromolecules* **16** 825.

[172] Blair, H. S., Pague, H. I., and Riordan, J. E. (1980) *Polymer* **21** 1195.

[173] Blair, H. S. and Law, T. K. (1980) *Polymer* **21** 1475..

[174] Houben, J. L., Fissi, A., Bacciola, D., Rosato, N., Pieroni, O., and Ciardelli, F. (1983) *Int. J. Biol. Macromol.* **5** 94.

[175] Ciardelli, F., Pieroni, O., Fissi, A., and Houben, J. L. (1984) *Biopolymers* **23** 1423.

[176] Kinoshita, J., Sato, M., Takizawa, A., and Tsujita, Y. (1984) *J. Chem. Soc., Chem. Commun.* 929.

[177] Fissi, A., Houben, J. L., Rosato, N., Lopes, S., Pieroni, O., and Ciardelli, F. (1982) *Makromol. Chem. Rapid. Commun.* **3** 29.

[178] Ciardelli, F., Fabbri, D., Pieroni, O., and Fissi, A. (1989) *J. Am. Chem. Soc.* **111** 3470.

[179] Ueno, A., Anzai, J., Osa, T., and Kadoma, Y. (1977) *Bull. Chem. Soc. Japan* **50** 2995.

[180] Pieroni, O., Fissi, A., Houben, J. L., and Ciardelli, F. (1985) *J. Am. Chem. Soc.* **107** 2990..

8

Ion pairs

8.1 INTRODUCTION

Several chemical species (for example, most metal complexes) are ionic and therefore can give rise to ion association phenomena. This may lead to supramolecular species which maintain most of the properties of the components and also exhibit new properties characteristic of the ensemble (Chapter 3). Since ion pairing mainly concerns metal complexes, our discussion will be focused on such systems, but it can be easily generalized to ion pairs involving organic molecules.

In the field of coordination chemistry, true ion pairs are often called *outer-sphere* ion pairs, to signify that the coordination spheres of the two components remain intact. In some cases, however, association occurs after dissociation of a ligand from one of the two complexes, followed by coordination of a (bridging) ligand which originally belongs to the other complex [1]. Such species are called *inner-sphere* ion pairs: being covalently-linked systems, they are discussed in Chapters 5 and 6. In this chapter, only *outer-sphere* ion pairs will be treated.

In the field of organic chemistry, a distinction is usually made between contact ion pairs and solvent-separated ion pairs [2–4]. Only contact ion pairs can be considered supramolecular species since solvent-separated ion pairs are too loosely bound to exhibit properties characteristic of the ensemble.

To a first approximation, the encounter of two molecular ions and the association that may follow can be described as the encounter (and association) of two charged spheres. It should be pointed out, however, that real molecular ions, including coordination compounds, do not have spherical shapes. For example, six-coordinated, octahedral metal complexes like $Ru(bpy)_3^{2+}$ appear spherical at first sight but, in reality, they have large "pockets" generated by the propeller-like arrangement of the three bpy ligands. In most cases, structure-specific effects can be ignored or, at most, treated as a refinement of the hard-sphere model. This chapter deals with such systems, while ion pairs where structure-specific interactions play an important role are dealt with in Chapter 10.

8.2   ELECTRONIC ENERGY LEVELS

In ion pairs, as in any supramolecular system (section 3.2), *inter*component charge-transfer (CT) transitions can occur besides electronic transitions within each parent ion. Such transitions (and the corresponding bands) can be called ion-pair charge transfer (IPCT) [3,5] and should be distinguished from *intra*component CT transitions.

When the two associated ions are coordination compounds, the IPCT transitions can also be called outer-sphere charge-transfer (OSCT) [5] or second-sphere charge-transfer (SSCT) [6], to emphasize that the transfer involves a species outside the first coordination sphere of the metal. As schematized in Fig. 8.1, there may be four types

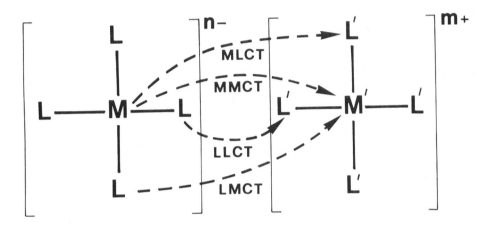

Fig. 8.1 — Types of outer-sphere charge-transfer transitions in ion pairs of coordination compounds.

of such CT transitions: (i) metal-to-metal (MMCT), (ii) ligand-to-ligand (LLCT), (iii) metal-to-ligand (MLCT), and (iv) ligand-to-metal (LMCT). Examples of all four cases are known [6–10]. MMCT transitions are often called intervalence transfer (IT) transitions [5–7,11,12] (section 3.4.1).

Depending on the specific nature of the parent ions, the CT levels of the ion pair may lie at high or low energy. Three limiting cases of interest will now be described [6,12,13].

(a) The IPCT levels lie at very high energy (Fig. 8.2a). In such a case, the absorption spectrum of the ion pair is essentially the sum of the spectra of the two isolated ions and no CT photochemical reaction can take place. However, radiation-less transitions between levels centred on the different partners (that is, intercomponent energy transfer, section 3.4.5) can be induced. This may cause a cascade nonradiative deactivation to the lowest excited state of the system, with quenching of the excited state processes (for example, photochemical reactions and luminescence) of one of the two partners and sensitization of those of the other one.

(b) The IPCT levels lie in the energy range of the low-energy spectroscopic levels of the two isolated partners (Fig. 8.2b). In some spectral regions, the ion pair may

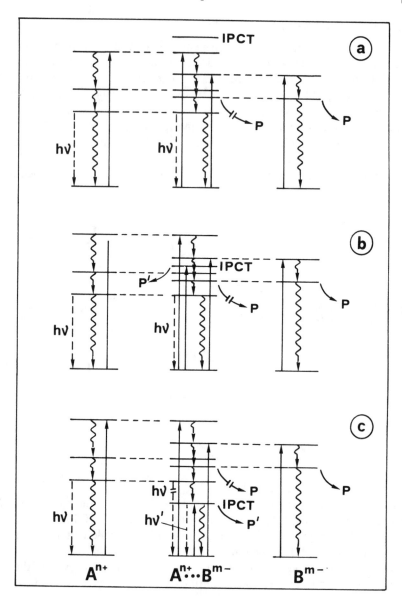

Fig. 8.2 — Schematic representation of three limiting cases of interaction in ion pairs.

show a somewhat stronger absorption than that corresponding to the sum of the absorption spectra of the two isolated ions. Excitation in the IPCT band or in higher-energy bands may cause either a photoreaction of CT origin or radiationless deactivation to the lowest excited state of the system. Excitation below the IPCT levels may be followed by radiationless deactivation to the lowest-energy excited state, which is essentially localized on one of the two ions. From this level, luminescence and/or reaction can take place as in case (a).

(c) An IPCT level is the lowest excited state of the system (Fig. 8.2c). A new, broad absorption band is likely to appear at low energy. Radiationless deactivation to the lowest excited state can take place, with complete quenching of the photochemical and luminescence properties of the two partners. New photochemical properties resulting from the lowest IPCT excited state may or may not appear, depending on the stability of the primary electron-transfer products. A CT luminescence band can also appear in some favourable cases.

In terms of potential energy curves, an IPCT transition can be represented as E_{op} in Fig. 8.3. Light excitation of the $A^{n+} \ldots B^{m-}$ ion pair in the IPCT band (or in

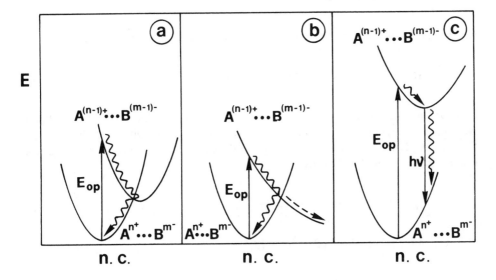

Fig. 8.3 — Photoinduced electron-transfer processes in ion pairs (n.c. means nuclear coordinates): (a) no permanent chemical change; (b) permanent chemical change because of decomposition or scavenging of $A^{(n-1)+}$ and/or $B^{(m-1)-}$; (c) charge-transfer luminescence.

higher energy levels followed by radiationless decay) leads to $A^{(n-1)+} \ldots B^{(m-1)-}$, an electronic isomer of the original species. The curves representing the two isomeric species are displaced along the nuclear coordinates because of solvent repolarization and changes in the intramolecular bond distances upon electron transfer. As happens for other supramolecular species (Chapter 3), three different situations can be imagined. If the displacement along the nuclear coordinates is noticeable and/or the energy difference between the two electronic isomers is not too large, the two curves will cross. In such a case, excitation is followed by a rapid back electron-transfer reaction leading to the original $A^{n+} \ldots B^{m-}$ ion pair (Fig. 8.3a), unless $A^{(n-1)+}$ and/or $B^{(m-1)-}$ undergo an irreversible reaction or are scavenged by another species (Fig. 8.3b). If the displacement along the nuclear coordinates is small and/or the energy difference between the two electronic isomers is large, the curves will not cross (Fig. 8.3c). In such a case $A^{(n-1)+} \ldots B^{(m-1)-}$ behaves more properly as an electronic excited state of the supramolecular species and deactivation of the IPCT level can also take place by emission if this level is the lowest excited state of the system.

8.3 CORRELATIONS BETWEEN OPTICAL AND THERMAL QUANTITIES

As we have seen in section 3.4.4, the rate constant for an intercomponent thermal electron-transfer reaction is given by (8.1),

$$k_{el} = \kappa v_N \exp(-\Delta G^{\pm}/RT) \tag{8.1}$$

where $\Delta G^{\pm}$ can be classically expressed by the Marcus quadratic equation (8.2) [14]

$$\Delta G^{\pm} = \Delta G^{\pm}(0)\{1 + [\Delta G^0/4 \, \Delta G^{\pm}(0)]\}^2 \tag{8.2}$$

as a function of the standard free energy change ΔG^0 and of the intrinsic barrier $\Delta G^{\pm}(0)$ (Fig. 2.11). (Equation (8.2) corresponds to (3.16), with $\Delta G^{\pm}(0) = \lambda/4$.) In the classical Hush model [15] the energy of the optical electron transfer transition is related to ΔG^0 and $\Delta G^{\pm}(0)$ by (8.3)

$$E_{op} = 4 \, \Delta G^{\pm}(0) + \Delta G^0 \tag{8.3}$$

that, taken together with (8.2), yields (8.4)

$$\Delta G^{\pm} = E_{op}^2/4(E_{op} - \Delta G^0) \tag{8.4}$$

which relates the free energy of activation with the free energy change and the energy of the maximum of the IPCT band. Furthermore, the integrated intensity of the optical transition is related by (8.5) to the interaction energy H_{AB}, whose value,

$$\varepsilon_{max} \, \Delta v_{1/2} = (r(\text{Å})^2 \, H_{AB}^2)/(4.20 \times 10^{-4} \, E_{op}) \tag{8.5}$$

through (3.19) and (3.20) discussed in section 3.4.4, governs the pre-exponential factor of the rate constant for the thermal electron-transfer process (8.1). These relationships between thermal and optical electron transfer parameters have been applied and discussed only in a few cases [16]. An empirical incremental system proposed for the evaluation of ΔG^0 from spectroscopic data and *vice versa* succeeds quite well in correlating the available experimental results [17–19].

The most extensive collection of homogeneous experimental data is that concerning the ion-pairs formed by $M(CN)_6^{4-}$ anions ($M = Fe, Ru, Os$) with $Ru(NH_3)_5L^{3+}$

cations (L = pyridine or substituted pyridine) [5]. For these systems, the intrinsic barrier $\Delta G^{+}(0)$ should be constant and the optical energy is thus expected to increase linearly with increasing ΔG^{0} (8.3). This expectation is fully confirmed and the intercept of the linear plot is in fair agreement with the value of $4 \Delta G^{+}(0)$ obtained from the intrinsic barriers of the corresponding self-exchange electron-transfer reactions [16]. Linear plots of E_{op} vs ΔG^{0} have also been obtained for the $[Eu \subset 2.2.1]^{3+} \ldots M(CN)_6^{4-}$ (M = Fe, Ru, Os) ion pairs in aqueous solution [20,21], and for the solid salts $[Ru(NH_3)_6][Fe(CN)_5L]$ (L is CO, $(CH_3)_2SO$, or a nitrogen heterocycle) [22].

Equations (8.1) and (8.3) allow us to calculate the rate constant for the thermal electron-transfer reaction from the free energy change, the optical transition energy, and the assumption of an adiabatic behaviour ($\kappa = 1$). When this method is applied [16] to the ion pairs between $M(CN)_6^{4-}$ (M = Fe, Ru) and $Ru(NH_3)_5py^{3+}$, $Co(NH_3)_6^{3+}$, and $[Eu \subset 2.2.1]^{3+}$, the ratios of the rate constants for the reactions involving $Fe(CN)_6^{4-}$ or $Ru(CN)_6^{4-}$ are 1.4×10^6, 2.0×10^6, 6.7×10^6, in reasonable agreement with the constancy predicted by Marcus theory (section 3.4.4), although their magnitude is not that expected from the rates of the $Fe(CN)_6^{4-/3-}$ and $Ru(CN)_6^{4-/3-}$ self-exchange reactions.

When correlations between optical and thermal quantities are used, there are problems with the absolute values of the rate constants [16]. In a case where a measured value is available, that is for the $Ru(NH_3)_5py^{3+} \ldots Fe(CN)_6^{4-}$ ion pair, the calculated rate constant (1×10^6 s^{-1}) is about 10^3 times higher than the measured rate constant (1.7×10^3 s^{-1}). As a possible reason for this disagreement, it has been suggested that the ion pairs for optical and thermal electron transfer may have different intimate structures. A similar problem arises when the results obtained in static and dynamic quenching processes are compared [23].

It should be noted that for both the $M(CN)_6^{4-} \ldots Ru(NH_3)_5L^{3+}$ [5] and $M(CN)_6^{4-} \ldots [Eu \subset 2.2.1]^{3+}$ [20] systems the value of H_{AB} calculated by (8.5) is larger than 100 cm^{-1}, indicating that the electron-transfer reactions (in the optical ion-pair configurations) are in the adiabatic regime (section 3.4.4).

8.4 PHOTOINDUCED PROCESSES

The photoinduced phenomena taking place in an outer-sphere ion-pair system cannot be treated with the kinetic scheme of Fig. 2.10. The association constants, in fact, are generally in the range 10^1–10^3 M^{-1}, which means that under the usual experimental conditions substantial amounts of both free and paired species may be present. In these cases, a more complete kinetic scheme (Fig. 8.4) must be used [12]. The behaviour of the system can be studied by monitoring the luminescence of the excited states and/or the chemical (transient or net) changes produced by light absorption. Continuous or pulsed light excitation can be used.

Table 8.1 shows a selected collection of data concerning photoinduced electron-transfer processes in ion pairs. As one can see, CT bands have been observed in a large number of cases (for a more extended compilation, see [7]). Even when a CT band is not observed, the occurrence of a photoreaction and/or electron-transfer quenching assures that a low-lying CT level must be present, although its spectro-scopic observation is prevented by the high intensity tails of the bands corresponding

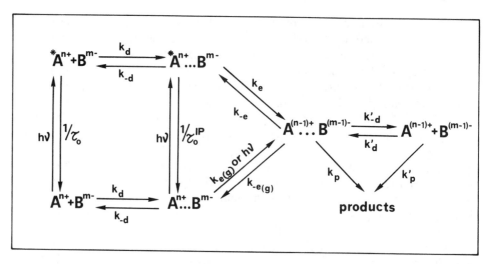

Fig. 8.4 — Kinetic scheme for the discussion of photoinduced phenomena in ion pairs.

Table 8.1 — Outer-sphere ion pairs[a]

Cation	Anion	CT band[b]	Luminescent quenching	Photo-reaction	Ref.
$Co(NH_3)_6^{3+}$ [c]	$Fe(CN)_6^{4-}$	440(300)	—	—	16
$Co(en)_3^{3+}$	$Fe(CN)_6^{4-}$	430(105)	—	yes	16,24
$Co(NH_3)_6^{3+}$ [c]	$Ru(CN)_6^{4-}$	342(243)	—	yes	25,26
$Co(sep)^{3+}$	$C_2O_4^{2-}$	275	—	yes	27
$Co(C_5H_5)_2^+$	$Co(CO)_4^-$	520[d]	—	yes	28
$Ni(tim)^{2+}$	$Ni(mnt)^{2-}$ [c]	840[e]	—	—	29
$Rh(bpy)_3^{3+}$	$Fe(CN)_6^{4-}$ [c]	480(61)	—	yes	10
$Ru(NH_3)_6^{3+}$	$Fe(CN)_6^{4-}$ [c]	744[e]	—	—	30
$Ru(NH_3)_6^{3+}$	$Fe(CN)_5py^{3-}$ [c]	725[e]	—	—	22
$Ru(NH_3)_5py^{3+}$ [c]	$Fe(CN)_6^{4-}$ [c]	910(33)	—	—	5,31
$Ru(NH_3)_5Cl^{2+}$	$Ru(CN)_6^{4-}$	510(20)	—	yes	32
$Ru(bpy)_3^{2+}$ [c]	$Fe(CN)_6^{4-}$	—	yes	—	33–36
$Ru(bpy)_3^{2+}$	$Ni(mnt)^{2-}$	—	yes	—	35
$Ru(bpy)_3^{2+}$ [c,f]	$Mn(OH)PW_{11}O_{39}^{6-}$	—	yes	—	37,38
$Ru(bpy)_3^{2+}$ [f]	$Co(H_2O)SiW_{11}O_{39}^{6-}$	—	yes [g]	—	37,38
$Os(NH_3)_5Cl^{2+}$	$Fe(CN)_6^{4-}$ [c]	438	—	yes	26
$Os(5-Clphen)_3^{2+}$	$Fe(CN)_6^{4-}$	—	yes [h]	—	33
$[Eu \subset 2.2.1]^{3+}$	$Fe(CN)_6^{4-}$ [c]	530(110)	yes	—	20
MV^{2+}	$Fe(CN)_6^{4-}$	530(50)	—	—	39
MV^{2+} [c]	$Zn(mnt)_2^{2-}$ [c]	460(404)[i]	—	—	40,41
MV^{2+}	$Ir(CO)_2(mnt)^-$ [c]	470[j]	—	yes	42
MV^{2+} [c]	$C_2O_4^{2-}$ [c]	310–400	—	yes	43

[a]aqueous solutions, and cation as the oxidizing species, unless otherwise noted; [b]wavelength (nm) and molar absorption coefficient ($M^{-1}cm^{-1}$) of the maximum; [c]other cations and/or anions are also discussed in the original papers; [d]dichloromethane; [e]solid state; [f]the oxidant species is the anion; [g]two exponential decays corresponding to the dynamic and static components; [h]emission from both free and ion-paired $Os(5-Clphen)_3^{2+}$ species; [i]dimethyl sulphoxide; [j]acetonitrile.

to localized transitions. The observed photoreaction refers in most cases to the formation of permanent products deriving from the dissociation of one of the two primary photoproducts.

8.4.1 Dynamic and static quenching

The scheme shown in Fig 8.4 [33,35,38,44] differs from the simpler scheme used in section 2.6.5 for dynamic quenching (Fig. 2.10) because it also takes into account light absorption and light emission by the ion pairs. In the scheme of Fig. 8.4, k_d and k_{-d} are the diffusion-controlled rate constants for the formation and dissociation of the precursor complex, k_e, k_{-e}, $k_{e(g)}$, and $k_{-e(g)}$ are the rate constants for electron transfer in the precursor and successor complex, and τ_0 and τ_0^{IP} are the lifetimes due to radiative and nonradiative (except electron transfer) deactivation of the free and paired excited state. In the case of weak interaction, the molar absorption coefficients of the free and paired species and the emission wavelengths and lifetimes (if the quenching rate constant is excluded) of the free and paired luminescent moieties can be taken to be same. An implicit assumption of the scheme of Fig. 8.4 is that the precursor complex of the dynamic quenching coincides with the excited preformed ion pair. It has been pointed out by several authors [38,45–48] that this last assumption may be incorrect, especially for nonspherical systems and for species that can undergo copenetration. An example of copenetration is given by the ion pairs between Ru(II)–polypyridine complexes (for example, $Ru(bpy)_3^{2+}$) and hetero-polytungstate anions (for example, $Mn(OH)PW_{11}O_{39}^{6-}$) [38,47].

For the simple case of electron-transfer quenching without ion-pair formation, the well-known relationships (8.6) and (8.7) [49]

$$I_0/I = 1 + k_q\tau_0[B^{m-}] \tag{8.6}$$

$$\tau_0/\tau = 1 + k_q\tau_0[B^{m-}] \tag{8.7}$$

between luminescence intensity and lifetime in the absence (I_0 and τ_0) and in the presence (I and τ) of the quencher are obeyed and thus the intensity and lifetime Stern–Volmer plots coincide. When ion pairs are formed, the scheme of Fig. 8.4 yields the complicated equation (8.8) for the relationship between I_0/I and $[B^{m-}]$ [33,35,37].

$$I_0/I = \frac{(1 + k_q\tau_0[B^{m-}])(1 + K[B^{m-}])}{1 + \alpha[B^{m-}]} \tag{8.8}$$

$$\alpha = \frac{k_{-d}K + (k_r^{IP}/k_r)\{(1/\tau_0 + k_d[B^{m-}])K + k_d\}}{1/\tau_0^{IP} + k_e + k_{-d}}$$

In this equation, K is the ion-pair formation constant, taken as k_d/k_{-d}, k_r^{IP} and k_r are the rate constants for emission by the ion pair and by the free species $*A^{n+}$, respectively, and k_q is given by (8.9).

$$k_q = \frac{k_d k_e}{k_e + k_{-d}} \cdot \frac{1 + 1/(\tau_0^{IP} k_e)}{1 + 1/\{\tau_0^{IP}(k_e + k_{-d})\}} \tag{8.9}$$

Equation (8.8) obviously reduces to (8.6) when ion pairs are not present, but it is clear that the kinetic behaviour of a photoinduced electron-transfer process between species giving rise to ion pairs may be quite complicated [33,35,37,38]. Some interesting limiting cases are as follows.

(a) When $k_e \ll 1/\tau_0^{IP}$, the trivial case of no intensity and lifetime quenching obtains.

(b) When $k_e \ll k_{-d}$, the presence of ion pairs is irrelevant to the quenching process because the precursor complexes $*A^{n+} \ldots B^{m-}$ obtained by direct excitation of the $A^{n+} \ldots B^{m-}$ ion pairs undergo dissociation before being able to react. Under such conditions, the quenching process obeys the laws of dynamic quenching, equations (8.6) and (8.7). An example of this behaviour is given by the ion-pairs between $[Eu \subset 2.2.1]^{3+}$ and $Ru(CN)_6^{4-}$, where dissociation of the ion-pairs is much faster than electron-transfer quenching [48].

(c) When k_e is much larger than $1/\tau_0^{IP} + k_{-d}$, the A^{n+} species that are excited while paired with B^{m-} undergo static quenching with rate constant k_e, whereas the excited free species $*A^{n+}$ are quenched at a diffusion controlled rate (that is, $k_q = k_d$) via dynamic formation of encounters with B^{m-}. The observable behaviour of the system will depend on the relative values of k_e, $1/\tau_0 + k_d[B^{m-}]$, and $1/\tau_{res}$ (where τ_{res} is the time resolution of the equipment used to monitor the luminescence lifetime). Two interesting sub-cases are as follows.

(c-1) When $1/k_e$ is shorter than τ_{res}, the decay of the excited ion pairs cannot be followed and the luminescence emission decreases with time according to a single exponential decay which is related to the dynamic quenching ($1/\tau = 1/\tau_0 + k_d[B^{m-}]$). On the other hand, under stationary excitation conditions both the dynamic and static processes are effective in quenching the emission intensity. It follows that the I_0/I and τ_0/τ vs $[B^{m-}]$ Stern–Volmer plots are no longer coincident, with the intensity plot showing an upward curvature (at constant ionic strength) caused by the increasing importance of the static component as $[B^{m-}]$ increases. Under such conditions, $\alpha[B^{m-}]$ in (8.8) may be much smaller than unity even for sufficiently high B^{m-} concentrations, so that (8.8) reduces to the following quadratic equation:

$$I_0/I = (1 + k_d \tau_0 [B^{m-}])(1 + K[B^{m-}]) = 1 + (k_d \tau_0 + K)[B^{m-}] + k_d \tau_0 K[B^{m-}]^2 \tag{8.10}$$

This case has often been observed [33–38,44,49,50]. For example, the quenching of $Ru(bpy)_3^{2+}$ by $PtCl_4^{2-}$ in DMF and by $Mo(CN)_8^{4-}$ in water clearly shows an upward curvature with increasing quencher concentration [36]. In these systems, comparison between the data obtained for dynamic (8.7) and overall (8.10) quenching allows the evaluation of the ion-pair association constant, K.

(c-2) When $1/k_e$ is longer than τ_{res} (but still much shorter than $\tau_0 + k_d[B^{m-}]$), the intensity of the luminescence emission decreases with time according to two distinct exponentials related to the decays of $*A^{n+}$ and $*A^{n+} \ldots B^{m-}$. Such conditions, however, are very difficult to obtain because (i) the time window between τ_0 and τ_{res} may be narrow, (ii) when k_e becomes relatively small it will be hardly larger than $1/\tau_0^{IP} + k_{-d}$, and (iii) sufficiently high concentrations of both free $*A^{n+}$ and paired $*A^{n+} \ldots B^{m-}$ species need to be present to obtain observable signals. In particular, for small K values a large amount of B^{m-} has to be used to have enough ion pairs, but this increases the rate of the dynamic quenching (whose rate constant is $\tau_0 + k_d[B^{m-}]$) which can thus approach the rate of the static quenching, k_e. When the two rates are sufficiently close, the luminescence decay may appear to be exponential within the experimental uncertainty and its value may decrease with increasing $[B^{m-}]$ up to a saturation value given by the rate of the intercomponent electron-transfer step, k_e. Clear examples of double exponential decay and of saturation of the luminescence decay with increasing $[B^{m-}]$ have been found for the electron-transfer quenching of Ru(II)-polypyridine complexes by heteropolytungstate anions [38] (for more details, see section 8.4.2) and for the quenching of Ir(III)-quinolinesulphonate by Cr(bpy)$_3^{3+}$ [44].

(d) When $k_e > 1/\tau_0 + k_{-d}$ (so as to ensure static quenching) and $k_d[B^{m-}]\tau_0 \ll 1$ (so as to make negligible the dynamic quenching), (8.10) reduces to (8.11)

$$I_0/I = (1 + K[B^{m-}]) \tag{8.11}$$

and the lifetime of the free $*A^{n+}$ species is not quenched. Whether or not the decay of the excited ion pairs can be seen will again depend on the relative values of $1/k_e$ and τ_{res}. The above conditions, however, are difficult to obtain because $k_e > 1/\tau_0 + k_{-d}$ means that τ_0 cannot be too small and the dynamic quenching is diffusion controlled. Thus, $k_d[B^{m-}]\tau_0$ will be much smaller than unity only when the concentration of free B^{m-} is small, but at the same time there should be a non-negligible amount of ion pairs, which implies a large K.

It should also be recalled that the ionic strength of the solution affects both the ion-pair association constants and the diffusion and dissociation rate constants [51], with profound effects on the I_0/I and τ_0/τ vs $[B^{m-}]$ plots [34,36,37,49].

A quite interesting ion-pair system, which is worthwhile discussing in some detail, is constituted by the $[Eu \subset 2.2.1]^{3+}$ cryptate and $M(CN)_6^{4-}$ complexes (M = Fe, Ru, Os) [20,21,48]. This system exhibits both IPCT bands and excited-state electron-transfer quenching and can therefore be used to illustrate better the connections among electron-transfer quenching, thermal electron transfer, and ion-pair spectroscopy (Fig. 8.5). In aqueous solution, the luminescence emission of $[Eu \subset 2.2.1]^{3+}$ ($\lambda_{max} = 615$ nm, $\tau_0 = 215$ μs) is quenched by $M(CN)_6^{4-}$ complexes [20]. At low quencher concentration the quenching process is dynamic since the intensity (8.6) and lifetime (8.7) Stern–Volmer plots coincide. On thermodynamic grounds it is possible to show [48] that the quenching takes place by an electron-transfer reaction (Fig. 8.5). The bimolecular quenching constants are 5×10^8, 2×10^8 and 6×10^8 M^{-1}s^{-1} for the Fe, Ru, and Os cyanide, respectively. When relatively concentrated solutions of $[Eu \subset 2.2.1]^{3+}$ and $M(CN)_6^{4-}$ are used, formation of ion pairs clearly occurs, as shown by the appearance of a new, low-energy broad absorption band in the visible region [20] (case (c) in section 8.2, Fig. 8.2c). The maximum of this new band moves to higher energies in the series Fe(530 nm), Os(450

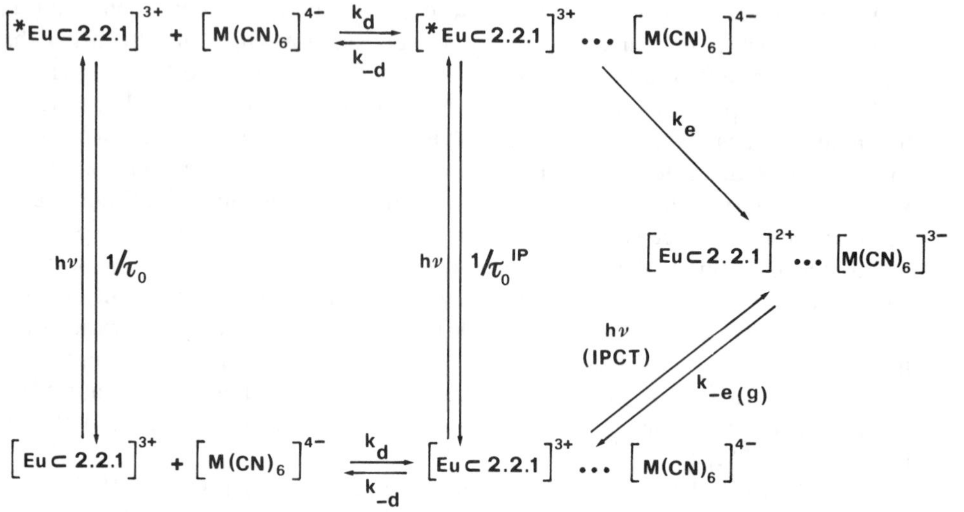

Fig. 8.5 — Processes occurring on photoexcitation of the $[Eu \subset 2.2.1]^{3+} \ldots [M(CN)_6]^{4-}$ system [20].

nm), and Ru(434 nm); this parallels the oxidation potentials of the cyanide complexes, as expected (8.3) for IPCT transitions. Since no net chemical effect is observed upon excitation in the IPCT band, the photoinduced IPCT reaction must be followed by a fast, back electron-transfer reaction which leads the system to its original situation. In concentrated solutions a substantial fraction of $[Eu \subset 2.2.1]^{3+}$ is paired with $M(CN)_6^{4-}$, but static quenching does not play an important role, as mentioned under case (b) above, because the rate constants for dynamic quenching are much smaller than the diffusion-controlled rate constant ($k_d \sim 7 \times 10^9 \, M^{-1}s^{-1}$), so that dissociation of the encounter complex (rate constant k_{-d} in Fig. 8.5) is much faster than electron transfer quenching (k_e). The thermal ($k_{-e(g)}$) and the light-induced (k_e) electron-transfer reactions (Fig. 8.5) are expected to exhibit the same intrinsic barrier and orbital overlap because the $^5D_0$ excited state of the europium complex has the same equilibrium geometry and the same orbital configuration as the $^7F_0$ ground state. These reactions, however, are not expected to behave in the same way as far as spin is concerned. The possible spin values are 3 for the ground-state ion pair, 3 and 4 for the product pair, and 2 for the excited-state ion pair. Thus, the reductive quenching process is spin-forbidden while the thermal process is spin-allowed. A very large influence of the spin factor, however, cannot be expected in these systems because of the presence of heavy atoms.

8.4.2 Direct measurements of electron-transfer rates
One of the goals in studying photoinduced electron-transfer reactions in ion pairs is the possibility of measuring the rate constant of the intrinsic act of electron transfer. This goal, however, is not easy to reach, and only a few papers are reported in the

literature where rate constants of photoinduced electron-transfer reactions in ion pairs have been directly measured or evaluated [33,37,38,44].

The best example is probably that involving the $RuL_3^{2+} \ldots MW_{11}^{6-}$ ion pairs in aqueous solutions, where RuL_3^{2+} is $Ru(bpy)_2(4,4'-Cl_2bpy)^{2+}$, or $Ru(bpy)_2(biq)^{2+}$ (bpy = 2,2'-bipyridine, biq = 2,2'-biquinoline) and MW_{11}^{6-} is $Mn(OH)PW_{11}O_{39}^{6-}$ or $Co(H_2O)SiW_{11}O_{39}^{6-}$ [37,38]. By an appropriate choice of the excited RuL_3^{2+} complex and of the MW_{11}^{6-} quencher it has been possible to tune the rate of the electron-transfer quenching process. Other favourable conditions were the relatively long lifetime of the excited states (125–400 ns) compared with the time resolution of the available equipment for lifetime measurements (2 ns), and the high charge product (-12) which implies high association constants (of the order of $10^4\, M^{-1}$) and small k_{-d} values (of the order of $10^6\, s^{-1}$). The most interesting results are summarized in Table 8.2 and can be described on the basis of the scheme of Fig. 8.4, excluding formation of permanent products. The $Ru(bpy)_3^{2+} \ldots MnW_{11}^{6-}$ ion pair corresponds to the case of the most exergonic electron-transfer reaction. The luminescence decay was strictly exponential (within the time resolution of the equipment used) and the τ_0/τ vs $[MW_{11}^{6-}]$ plot was linear, yielding a rate constant for the dynamic quenching practically equal to the calculated diffusion rate constant. The I_0/I vs $[MW_{11}^{6-}]$ plot exhibited an upward curvature and was fitted by (8.10). For the less exergonic $Ru(bpy)_3^{2+} \ldots CoW_{11}^{6-}$ system, a double exponential decay was observed. Taking the longer decay, the τ_0/τ vs $[CoW_{11}^{6-}]$ plot was linear, yielding again a dynamic quenching constant practically equal to the diffusion rate constant. The I_0/I vs $[CoW_{11}^{6-}]$ plot showed an upward curvature and was fitted by (8.8). The double exponential decay yielded experimental decay constants $8.9 \times 10^7\, s^{-1}$ and 9.6×10^6 s^{-1}, related to the static and dynamic quenching paths. For the $Ru(bpy)_2(4,4'-Cl_2bpy)^{2+} \ldots CoW_{11}^{6-}$ system, the rate of the electron-transfer step had to be smaller because of the smaller exergonicity. The luminescence emission was found to decay according to a single exponential, but the τ_0/τ vs $[CoW_{11}^{6-}]$ plot exhibited a downward curvature reaching a plateau for $\tau_0/\tau = 1.7$. This shows that the quenching limiting factor at high CoW_{11}^{6-} concentrations becomes the rate of the intramolecular step, k_e, that from the lifetime of the plateau values was estimated to be $\sim 4 \times 10^6\, s^{-1}$. Finally, for the $Ru(bpy)_2(biq)^{2+} \ldots CoW_{11}^{6-}$ system the electron-transfer step is noticeably endergonic and using the Marcus relationship its rate constant is estimated to be much smaller than τ_0^{IP}. This explains why there is no quenching effect.

8.4.3 Photochemical reactions

Light absorption by an ion pair can lead to excited levels of the isolated partners or to IPCT levels. The subsequent behaviour of the system will depend on the relative position of the various levels, as discussed in section 8.2. A most interesting and quite common case is that corresponding to Fig. 8.2c, where an IPCT level is lower in energy than the localized excited states. Light excitation of such an ion pair leads, directly or indirectly, to the formation of the $A^{(n-1)+} \ldots B^{(m-1)-}$ species, that can still be an ion pair. Unless the energy of $A^{(n-1)+} \ldots B^{(m-1)-}$ is lower than that of $A^{n+} \ldots B^{m-}$ (in such a case, the reaction would also have occurred thermally), the $A^{(n-1)+} \ldots B^{(m-1)-}$ species can only appear as a more or less short-lived transient since it must either go back to $A^{n+} \ldots B^{m-}$ or undergo some competitive process

Table 8.2 — Electron transfer quenching of $^*RuL_3^{2+}$ complexes by $Mn(OH)PW_{11}O_{39}^{6-}$ (MnW_{11}^{6-}) and $Co(H_2O)SiW_{11}O_{39}^{6-}$ (CoW_{11}^{6-})[a]

$^*RuL_3^{2+} \ldots MW_{11}^{6-}$	$\Delta G(eV)$[b]	$k_q(M^{-1}s^{-1})$[c]	$k_d(M^{-1}s^{-1})$[d]	$k_e(s^{-1})$[e]	Remarks
$^*Ru(bpy)_3^{2+} \ldots MnW_{11}^{6-}$	−0.34	2.1×10^{10}	2.2×10^{10}	$(\geq 5 \times 10^8)$	Static and dynamic quenching. One exponential decay.
$^*Ru(bpy)_3^{2+} \ldots CoW_{11}^{6-}$	−0.20	2.3×10^{10}	2.2×10^{10}	8.5×10^7	Static and dynamic quenching. Two exponential decays.
$^*Ru(bpy)_2(4,4'\text{-}Cl_2bpy)^{2+} \ldots CoW_{11}^{6-}$	−0.09	f	—	$\sim 4 \times 10^6$	τ_0/τ reaches a plateau as $[CoW_{11}^{6-}]$ increases.
$^*Ru(bpy)_2(biq)^{2+} \ldots CoW_{11}^{6-}$	+0.35	$<3 \times 10^8$	1.1×10^{10}	$(<5 \times 10^5)$	No observable quenching.

[a]from [37,38]; aqueous solution, room temperature; for the ionic strengths, see the original papers; [b]free energy change for the photoinduced electron-transfer process within the ion pair; RuL_3^{2+} is oxidized; [c]experimental rate constant for dynamic quenching obtained from the τ_0/τ Stern–Volmer plots; in the case of two exponential decays, the slower one was used; [d]calculated diffusion rate constant; [e]rate constant for electron transfer within the ion pair; [f]non-linear τ_0/τ Stern–Volmer plots.

leading to permanent photoproducts (for example, reaction with a scavenger). Continuous photolysis experiments can only reveal, of course, permanent products while flash experiments with appropriate time resolution may allow us to observe the formation and decay of $A^{(n-1)+}\ldots B^{(m-1)-}$ as well as of other transients.

8.4.3.1 Metal complexes

As shown in Table 8.1, photoinduced reactions in ion pairs of metal complexes are frequent. In most cases, only the formation of permanent products deriving from secondary reactions has been observed. The importance of secondary reactions to obtain a net photochemical effect (Fig. 8.3b) is well-illustrated by the behaviour of the ion pairs of $Co(NH_3)_6^{3+}$ and $Co(sep)^{3+}$ with $Ru(CN)_6^{4-}$ [26]. In both cases excitation in the CT band leads to the formation of the one-electron reduction product of the Co complex and the one-electron oxidation product of the Ru complex ((8.12) and (8.13)),

$$Co^{III}(NH_3)_6^{3+}\ldots Ru^{II}(CN)_6^{4-} \xrightarrow{h\nu} Co^{II}(NH_3)_6^{2+}\ldots Ru^{III}(CN)_6^{3-} \qquad (8.12)$$

$$Co^{III}(sep)^{3+}\ldots Ru^{II}(CN)_6^{4-} \xrightarrow{h\nu} Co^{II}(sep)^{2+}\ldots Ru^{III}(CN)_6^{3-} \qquad (8.13)$$

but only for the first system does a net photoreaction take place, with formation of Co^{2+}, NH_3, and $Ru^{III}(CN)_6^{3-}$ ($\Phi = 0.034$ in DMSO) [25]. This is in agreement with the known lability of $Co^{II}(NH_3)_6^{2+}$ [52], which undergoes ligand release in competition with back electron transfer, and with the inertness of the cage-type $Co^{II}(sep)^{2+}$ complex, which cannot undergo ligand dissociation and is therefore involved in the back electron-transfer reaction with unit efficiency. This different behaviour of the otherwise similar $Co(NH_3)_6^{3+}$ and $Co(sep)^{3+}$ complexes can be observed in photochemical, pulse radiolysis, and electrochemical reduction processes [6], as discussed in section 11.1.2.

By using suitable ligands, it is possible to obtain ion pairs between complexes of the same metal in two different oxidation states, which exhibit IPCT bands [26,32,53,54]. For the $Ru^{III}(NH_3)_5Cl^{2+}\ldots Ru^{II}(CN)_6^{4-}$ ion pair [32], excitation in the CT band in aqueous solution leads to the formation of the binuclear complex $(NH_3)_5Ru^{III}NCRu^{II}(CN)_5^-$ with $\Phi = 2\times10^{-3}$. The reaction is thought to proceed via the rapid aquation (in competition with back electron transfer, Fig. 8.3b) of the $Ru^{II}(NH_3)_5Cl^+$ primary photoproduct, with successive replacement of the labile H_2O ligand by $Ru^{II}(CN)_6^{4-}$ and oxidation of the $(NH_3)_5Ru^{II}NCRu^{II}(CN)_5^{2-}$ product so obtained by the other primary photoproduct, $Ru^{III}(CN)_6^{3-}$.

Brightly coloured salts are formed by $Co(CO)_4^-$, $Mn(CO)_5^-$, and $V(CO)_6^-$ with metallocenium and pyridinium cations [28]. X-ray crystallography of such salts has given interesting information on the spatial cation/anion orientations, the interionic separations, and other structural details. Such ion pairs persist in nonpolar solvents, where light excitation leads to radical pairs that can be observed directly by time-resolved spectroscopy. Phosphine ligands can scavenge the labile $Co(CO)_4$ and $Mn(CO)_5$ radicals, leading to a variety of well-defined products.

Simple ion pairs are obtained from the association of positively-charged coordination compounds (for example, $Co(NH_3)_6^{3+}$, $Co(sep)^{3+}$, $Ru(NH_3)_5py^{3+}$) with halide, $C_2O_4^{2-}$, $B(C_6H_5)_4^-$, and citrate anions [27,55–59]. When both the cation and the anion can undergo reversible redox processes (for example, $Ru(NH_3)_5py^{3+}$ and I^-), only transient products can be observed. When the complex (for example, $Co(NH_3)_6^{3+}$) and/or the anion (for example, $C_2O_4^{2-}$) undergo irreversible redox processes, a net photoreaction is obtained (see also section 11.1.2).

Systems that are borderline between the outer-sphere ion pairs discussed in this section and inner-sphere systems have also been reported [60]. $Mo(CN)_8^{4-}$, as well as other cyanide complexes, gives rise to ion pairs (or more probably to cyano-bridged compounds) with a variety of Cu^{2+} species and other cations like Fe^{3+} and UO_2^{2+}. The IPCT bands extend to the visible and, in some cases, to the near-infrared. Photochemical investigations on the $Mo(CN)_8^{4-}\ldots Cu^{2+}$ system have shown that irradiation in the IPCT bands leads to formation of free cyanide, as expected because of the known kinetic lability of $Mo(CN)_8^{3-}$. The relevance of these studies for photocatalysis and spectral sensitization has also been discussed [60,61].

The primary (transient) photoproducts obtained upon CT excitation of ion pairs have seldom been observed. For the $Ru(NH_3)_5py^{3+}\ldots Fe(CN)_6^{4-}$ ion pair in aqueous solution, excitation with a 20 ns pulse of 1060 nm light gives no permanent photoproduct but leads to the transient formation of $Ru(NH_3)_5py^{2+}$ and $Fe(CN)_6^{3-}$ with low yield [31]. The following reaction mechanism has been proposed:

$$Ru(NH_3)_5py^{3+}\ldots Fe(CN)_6^{4-} \underset{k_b'}{\overset{h\nu}{\rightleftharpoons}} {}^{(FC)}Ru(NH_3)_5py^{2+}\ldots Fe(CN)_6^{3-} \qquad (8.14)$$

$$^{(FC)}Ru(NH_3)_5py^{2+}\ldots Fe(CN)_6^{3-} \overset{k_{rel}}{\rightarrow} Ru(NH_3)_5py^{2+}\ldots Fe(CN)_6^{3-} \qquad (8.15)$$

$$\begin{array}{r} k_{-d}\nearrow Ru(NH_3)_5py^{2+}\ldots Fe(CN)_6^{3-} \qquad (8.16)\\ Ru(NH_3)_5py^{2+} + Fe(CN)_6^{3-} \underset{k_d}{\overset{}{\diagup}} \\ k_b\searrow Ru(NH_3)_5py^{3+}\ldots Fe(CN)_6^{4-} \qquad (8.17) \end{array}$$

Reaction (8.14) represents the photoexcitation step with formation of the Franck–Condon charge-transfer excited state. Relaxation of the Franck–Condon excited state proceeds by two competing pathways: return to the ground state of the original ion pair (k_b') or formation of the relaxed electronic isomer (k_{rel}). The low quantum yield requires that $k_{rel} \ll k_b'$, suggesting that the relaxation to the original ion-pair occurs *via* relatively high-frequency (10^{12}–10^{13} s^{-1}) inner-sphere modes before the change in solvent polarization (which is determined by the dielectric relaxation frequency of water, $\sim 10^{11} s^{-1}$) can occur. Since the reaction of $Ru(NH_3)_5py^{2+}$ with $Fe(CN)_6^{3-}$ is not diffusion-controlled, the rate of cage escape (k_{-d}) for the two partners of the electronic isomer is faster than the rate of back electron transfer (k_b) within the cage [16,31].

In principle, energy transfer can occur in an excited ion pair. This is apparently the case for the ion pairs formed by $[Tb \subset 2.2.1]^{3+}$ with $Cr(CN)_6^{3-}$ and other similar systems [48], where excitation of the lanthanide cryptate is followed by a quenching reaction. In such systems electron-transfer quenching can be ruled out for thermo-dynamic reasons, whereas electronic energy transfer is energetically allowed. In practice, however, most of the observed quenching takes place after several dissociation–association steps of the ion pair since the rate of energy transfer is smaller than the dissociation rate constant.

Ground-state ion pairs of $Cu(NN)_3^+$ (NN = 2,9-dimethyl-1,10-phenanthroline or 2,9-diphenyl-1,10-phenanthroline) with anions in CH_2Cl_2 solution promote the quenching of the $*Cu(NN)_3^+$ luminescence *via* an exciplex-type mechanism (Chapter 9) [62]. The luminescent MLCT level of $Cu(NN)_3^+$ is, in fact, a Cu(II) complex which exhibits a great tendency towards coordination of a fifth ligand. The quenching ability of the anions follows the order $BPh_4^- < PF_6^- < BF_4^- < ClO_4^- < NO_3^-$, which is consistent with the increasing donor strength along the series. Exciplexes are also formed with Lewis bases such as DMF and DMSO [63].

8.4.3.2 *Organic compounds*

In contrast to the situation for coordination compounds, most organic species are stable as neutral molecules so that the photochemistry of ion pairs is less developed in this field. Most of the reported studies concern pyridinium-type cations, and particularly the 1,1'-dimethyl-4,4'-dipyridinium dication (for the structure formula, see **10.19**), usually called methylviologen (MV^{2+}) or paraquat (PQ^{2+}). Some of the ion-pair systems involving MV^{2+} are shown in Table 8.1.

MV^{2+} is an interesting species, widely used as an electron relay in model systems for the photochemical conversion and storage of solar energy [46,64,65]. In aqueous solution it forms ion pairs with a broad range of electron donors [66], including the anions used as sacrificial species in the photochemical conversion schemes, such as EDTA [67,68] and $C_2O_4^{2-}$ [43]. It seems likely that ion pairs indeed play an important role [69] in the $Ru(bpy)_3^{2+}-MV^{2+}-EDTA$ system extensively used for photogeneration of hydrogen from water [65]. MV^{2+} forms a 1:1 complex ($K_{eq} = 21$ M^{-1}) with $C_2O_4^{2-}$ at pH 7.1 that exhibits an enhanced absorption tail in the 310–400 nm region. In general, ion pairs involving MV^{2+} and electron donors are photo-sensitive [3,67-72], yielding the stable radical cation MV^+ which exhibits a quite intense and characteristic absorption spectrum in the visible.

The conventional mechanism used to account for the observed behaviour of the ion pairs of MV^{2+} with electron donors is as follows. Light excitation of the $MV^{2+} \ldots D^-$ ion pair (D^- = electron donor) leads to a geminate pair $MV^+ \ldots D$ (8.18). The dissociation reaction (8.19) must compete with the highly exergonic back electron-transfer reaction within the solvent cage (8.20). If D is a stable species, the back electron-transfer process will occur anyway (8.21), and irradiation causes no permanent chemical change. If, however, the oxidized donor undergoes an irrever-sible transformation (sacrificial donor) within (8.22) and/or outside (8.23) the solvent cage, MV^+ can accumulate in the solution.

$$MV^{2+} \ldots D^- \xrightarrow{h\nu} MV^+ \ldots D \qquad\qquad (8.18)$$

$$MV^+ \ldots D \rightarrow MV^+ + D \tag{8.19}$$

$$MV^+ \ldots D \rightarrow MV^{2+} \ldots D^- \tag{8.20}$$

$$MV^+ + D \rightarrow MV^{2+} \ldots D^- \tag{8.21}$$

$$MV^+ \ldots D \rightarrow MV^+ + P \tag{8.22}$$

$$D \rightarrow P \tag{8.23}$$

The sacrificial nature of $C_2O_4^{2-}$ and EDTA as electron donors arises because, upon one electron oxidation, they undergo rapid irreversible reactions to form species that cannot be reduced by MV^+ (for $C_2O_4^-$, dissociation into $CO_2 + CO_2^-$ takes place). In fact, one of such secondary products (CO_2^- in the case of $C_2O_4^-$) is usually a *reducing* radical capable of reacting with MV^{2+} to yield a second equivalent of MV^+ (8.24).

$$MV^{2+} + P \rightarrow MV^+ + \text{final products} \tag{8.24}$$

For example, the reaction between MV^{2+} and CO_2^- to yield MV^+ and CO_2 is practically diffusion controlled [73]. According to the above model (8.18–8.24), fast kinetic measurements would be expected to show the generation of the first equivalent of MV^+ (8.18–8.19) within the pulse of a nanosecond laser equipment, followed by the slower formation of the second equivalent of MV^+ from reaction (8.24). In the absence of bulk D, reaction (8.21) would not occur and the ratio of the primary and secondary yields of MV^+ is expected to be unity if P quantitatively reduces the second equivalent of MV^{2+} in competition with its other degradative modes of decay (not shown). Detailed experimental results obtained for the $MV^{2+}/C_2O_4^{2-}$ [70,71] and MV^{2+}/EDTA [69] systems, however, have demonstrated that the conventional mechanism (8.18–8.24) cannot be valid. A new model has therefore been proposed [74] in which the ion pairs involve more than one MV^{2+} species and can be viewed as "pseudo-micelle" aggregates. Both the primary and secondary processes would occur inside such aggregates and the MV^+ species would be ultimately released into the bulk solution as the aggregate structure equilibrates.

Several other investigations have concerned ion pairs involving MV^{2+}. For the 1:1 complex with thiocyanate ion, back electron transfer in the solvent cage (8.20) occurs within 100 ps [75], while for the ion pairs with tetrakis[3,5-bis(trifluoromethyl)phenyl]borate in degassed organic solvents the MV^+ cation is produced with $\Phi = 5 \times 10^{-3}$ and exhibits a lifetime of about 1 h [76] (see also [77]). MV^{2+} ion pairs with tetraphenylborate and benzilate have been used in photoelectrochemical cells [78,79]. Ion pairs of viologens with metal dithiolenes exhibit charge transfer absorption bands in the near infrared and show semiconductor behaviour [80]. Detailed investigations involving complexes of MV^{2+} and neutral species have also been reported [3].

Ion pairs of organic dyes have also been investigated. Rose bengal (RB^{2-}), and other xanthene dyes form ion pairs with MV^{2+}. Excitation of the RB^{2-} moiety of the ion pair is followed by static quenching of the singlet excited state, via electron-transfer. The back electron-transfer reaction, however, is very fast and even in the presence of triethanolamine as a sacrificial electron donor the formation of MV^+ is quite inefficient ($\Phi = 8 \times 10^{-4}$) [81]. On addition of colloidal SiO_2, however, the ion pair is separated, and a different reaction mechanism leads to efficient ($\Phi = 0.1$) MV^+ formation. Photobleaching of cationic dyes like crystal violet in 2-propanol could be due to photolysis of their ion pairs with the halide counter ions [82].

8.4.4 Luminescence

As mentioned in section 8.2, the potential energy curves of the $A^{n+} \ldots B^{m-}$ and $A^{(n-1)+} \ldots B^{(m-1)-}$ species do not cross when the displacement along the nuclear coordinates is small and/or the energy difference is large (Fig. 8.3c). Under such conditions, if the IPCT level is the lowest excited state of the system, it can undergo radiative decay (ion-pair luminescence).

Such conditions are not likely to obtain for ion pairs of coordination compounds for the following reasons:

(a) the nuclear displacements are small when the solvent is not polar (so as to have little solvent repolarization) and the orbitals involved in the electronic transition are delocalized (so as to avoid large changes in bond distances): these conditions are encountered infrequently for coordination compounds;
(b) most coordination compounds have relatively low-lying excited states, so that when the energy gap between $A^{n+} \ldots B^{m-}$ and $A^{(n-1)+} \ldots B^{(m-1)-}$ is sufficiently large with respect to the reorganizational energy, localized levels are likely to lie below the lowest $A^{(n-1)+} \ldots B^{(m-1)-}$ level.

The situation, however, may be different for ion pairs between aromatic organic species, and luminescence from ion pairs involving MV^{2+} has been reported [67,68,83–87]. In most cases, however, it was demonstrated that such luminescence originates from impurities or from secondary photoproducts [67,84,85]. In the case of the ion pair with tetrakis[3,5-bis(trifluoromethyl)phenyl]borate anion in 1,2-dimethoxyethane, the observed luminescence seems to be a *bona fide* ion pair emission [87].

8.5   CONTROL AND TUNING OF EXCITED STATE PROPERTIES

In several cases ion pairing can be used to tune the excited-state properties of a species. For example, the emission energies and lifetimes of $[Os(phen)_3]X_2$ (X = PF_6^-, ClO_4^-, Cl^-, Br^-, or SCN^-) in a solvent of low dielectric constant like CH_2Cl_2 vary on changing the anion [88]. The microscopic origin of the anion effect is essentially electrostatic, and it is related to the localization of the excited electron into a single phenanthroline ligand. The emission lifetime (which is essentially controlled by the nonradiative rate constant) increases with increasing emission energy, as expected from the energy-gap law.

An interesting example of the tuning of excited state properties by ion-pair formation is given by the $[M \subset 2.2.1]^{3+}$ cryptates (M = Eu or Tb) [89]. On addition of F^- ions to aqueous solutions of the cryptates, the species $[M \subset 2.2.1]^{3+} \ldots F^-$ and

$[M \subset 2.2.1]^{3+} \dots 2F^-$ are formed, with profound changes in the absorption spectra, luminescence spectra, and excited state decay. The effects are particularly strong because the F^- ions replace water molecules in the holes of the cryptate structure and become more or less directly coordinated to M^{3+}. Ion pairing does not introduce low-lying energy levels, because the $F^- \rightarrow Eu^{3+}$ (and even more so the $F^- \rightarrow Tb^{3+}$) CT transitions occur at very high energies. However, association with F^- reduces the effective positive charge (or the oxidizing power) of the lanthanide ion, with a consequent increase in the energy of the LMCT transitions involving the nitrogen and oxygen donor atoms of the cryptand. For the Eu^{3+} cryptate such transitions give rise to bands in the near ultraviolet region; on addition of F^- ions, they are shifted to considerably higher energies. As regards the luminescence properties, the replacement of H_2O molecules in the coordination sphere of the lanthanide ion causes an increase in the luminescence lifetime and quantum yield because the main radiationless decay path for the emitting levels $^5D_0$ (for the Eu^{3+} cryptate) and $^5D_4$ (for the Tb^{3+} cryptate) is related to the coupling with the high-frequency O–H oscillators. Furthermore, in the Eu^{3+} cryptate the LMCT excited states, which also offer a way to the radiationless decay of $^5D_0$ at room temperature, become inaccessible because of the above-mentioned blue-shift on ion pairing. The splitting of the various composite transitions (for example, the $^5D_0 \rightarrow {}^7F_1$ for the Eu^{3+} cryptate) is also modified because of the changes in the symmetry of the Eu^{3+} coordination sphere.

Addition of increasing amounts of the antibiotic lasalocid to solutions of $Cr(bpy)_3^{3+}$ (up to a concentration ratio 1:1) causes noticeable changes in the absorption spectra, showing that ion pairs are formed. Strong effects are also observed in the luminescence intensity: on ion-pair formation, the luminescent excited state of lasalocid is quenched by $Cr(bpy)_3^{3+}$, whereas the luminescent excited state of $Cr(bpy)_3^{3+}$ is protected by lasalocid [90].

An extensive literature is available on the changes caused by metal ions on the excited-state properties of organic anions (see, for example [4]) and by halide ions (especially I^-) on the excited state properties of organic cations (see, for example [91]).

REFERENCES

[1] Haim, A. (1983) *Prog. Inorg. Chem.* **30** 273.

[2] Hogen-Esch, T. E. (1977) *Adv. Phys. Org. Chem.* **15** 153.

[3] Jones, G., II (1988) In Fox, M. A. and Chanon, M. (eds) *Photoinduced electron transfer*. Part A. Elsevier, p. 245.

[4] Soumillion, J. Ph., Vandereecken, P., Van der Auweraer, M., De Schryver, F. C., and Schanck, A. (1989) *J. Am. Chem. Soc.* **111** 2217.

[5] Curtis, J. C. and Meyer, T. J. (1982) *Inorg. Chem.* **21** 1562.

[6] Balzani, V., Sabbatini, N., and Scandola, F. (1986) *Chem. Rev.* **86** 319.

[7] Vogler, A. and Kunkely, H. (1990) *Topics Curr. Chem.*, in press.

[8] Vogler, A. and Kunkely, H. (1988) *Inorg. Chim. Acta* **144** 149.

[9] Vogler, A. and Kunkely, H. (1988) *Inorg. Chim. Acta* **150** 3.

[10] Vogler, A. and Kunkely, H. (1987) *Inorg. Chem.* **26** 1819.

[11] Creutz, C. (1983) *Progr. Inorg. Chem.* **30** 1.

[12] Balzani, V. and Scandola, F. (1988) In Fox, M. A. and Chanon, M. (eds) *Photoinduced electron transfer*. Part D. Elsevier, p. 148.

[13] Balzani, V., Ballardini, R., Gandolfi, M. T., and Prodi, L. (1990) In Schneider, H. J. and Dürr, H. (eds) *Frontiers in supramolecular organic chemistry and photochemistry*. Verlag Chemie, in press.

[14] Marcus, R. A. (1964) *Annu. Rev. Phys. Chem.* **15** 155.

[15] Hush, N. S. (1968) *Electrochim. Acta* **13** 1005.

[16] Haim, A. (1985) *Comments Inorg. Chem.* **4** 113.

[17] Hennig, H., Rehorek, D., and Billing, R. (1988) *Comments Inorg. Chem.* **8** 163.

[18] Hennig, H., Benedix, R., and Billing, R. (1986) *J. Prakt. Chem.* **328** 829.

[19] Hennig, H., Billing, R., and Rehorek, D. (1987) *J. Inf. Rec. Mater.* **15** 423.

[20] Sabbatini, N., Bonazzi, A., Ciano, M., and Balzani, V. (1984) *J. Am. Chem. Soc.* **106** 4055.

[21] Sabbatini, N. and Balzani, V. (1985) *J. Less Common Metals* **112** 381.

[22] Toma, H. E. (1980) *J. Chem. Soc., Dalton Trans.* 471.

[23] Scandola, F., Bignozzi, C. A., and Balzani, V. (1986) In Pelizzetti, E. and Serpone, N. (eds) *Homogeneous and heterogeneous photocatalysis*. Reidel, p. 29.

[24] Larson, R. (1967) *Acta Chem. Scand.* **21** 257.

[25] Vogler, A. and Kisslinger, J. (1982) *Angew. Chem. Int. Ed. Engl.* **21** 77.

[26] Vogler, A., Osman, A. H., and Kunkely, H. (1985) *Coord. Chem. Rev.* **64** 159.

[27] Pina, F., Mulazzani, Q. G., Venturi, M., Ciano, M., and Balzani, V. (1985) *Inorg. Chem.* **24** 848.

[28] Bockman, T. M. and Kochi, J. K. (1989) *J. Am. Chem. Soc.* **111** 4669.

[29] Vogler, A. and Kunkely, H. (1986) *J. Chem. Soc., Chem. Commun.* 1616.

[30] Curtis, J. C. and Meyer, T. J. (1978) *J. Am. Chem. Soc.* **100** 6284.

[31] Creutz, C., Kroger, P., Matsubara, T., Netzel, T. L., and Sutin, N. (1979) *J. Am. Chem. Soc.* **101** 5442.

[32] Vogler, A. and Kisslinger, J. (1982) *J. Am. Chem. Soc.* **104** 2311.

[33] Rybak, W., Haim, A., Netzel, T. L., and Sutin, N. (1981) *J. Phys. Chem.* **85** 2856.

[34] Demas, J. N. and Addington, J. W. (1976) *J. Am. Chem. Soc.* **98** 5800.

[35] Frank, R. and Rau, H. (1983) *J. Phys. Chem.* **87** 5181.

[36] Bolletta, F., Maestri, M., Moggi, L., and Balzani, V. (1974) *J. Phys. Chem.* **78** 1374.

[37] Ballardini, R., Gandolfi, M. T., and Balzani, V. (1987) *Inorg. Chem.* **26** 862.

[38] Ballardini, R., Gandolfi, M. T., and Balzani, V. (1985) *Chem. Phys. Lett.* **119** 459.

[39] Curtis, J. C., Sullivan, B. P., and Meyer, T. J. (1980) *Inorg. Chem.* **19** 3833.

[40] Fernandez, A., Görner, H., and Kisch, H. (1985) *Chem. Ber.* **118** 1936.

[41] Lahner, S., Wakatsuki, Y., and Kisch, H. (1987) *Chem. Ber.* **120** 1011.

[42] Megehee, E. G., Johnson, C. E., and Eisenberg, R. (1989) *Inorg. Chem.* **28** 2423.

[43] Barnett, J. R., Hopkins, A. S., and Ledwith, A. (1973) *J. Chem. Soc., Perkin Trans. 2* 80.

[44] Checchi, L., Chiorboli, C., Rampi Scandola, M. A., and Scandola, F. (1987) In

Yersin, H. and Vogler, A. (eds) *Photochemistry and photophysics of coordination compounds*. Springer-Verlag, p. 181.

[45] Mauzerall, D. (1982) *Annu. Rev. Phys. Chem.* **33** 377.

[46] Balzani, V., Juris, A., and Scandola, F. (1986) In Pelizzetti, E. and Serpone, N. (eds) *Homogeneous and heterogeneous photocatalysis*. Reidel, p. 1.

[47] Ballardini, R., Gandolfi, M. T., Balzani, V., and Scandola, F. (1987) *Gazz. Chim. Ital.* **117** 769.

[48] Sabbatini, N., Perathoner, S., Lattanzi, G., Dellonte, S., and Balzani, V. (1988) *Inorg. Chem.* **27** 1628.

[49] Balzani, V., Moggi, L., Manfrin, M. F., Bolletta, F., and Laurence G. S. (1975) *Coord. Chem. Rev.* **15** 321.

[50] White, H. S., Becker, W. G., and Bard, A. J. (1984) *J. Phys. Chem.* **88** 1840.

[51] Chiorboli, C., Indelli, M. T., Rampi, M. A., and Scandola, F. (1988) *J. Phys. Chem.* **92** 156.

[52] Lilie, J., Shinohara, N., and Simic, M. G. (1976) *J. Am. Chem. Soc.* **98** 6516.

[53] Curtis, J. C. and Meyer, T. J. (1978) *J. Am. Chem. Soc.* **100** 6284.

[54] Vogler, A. and Kunkely, H. (1988) *Organomet.* **7** 1449.

[55] Pina, F., Ciano, M., Moggi, L., and Balzani, V. (1985) *Inorg. Chem.* **24** 844.

[56] Pina, F., Maestri, M., Ballardini, R., Mulazzani, Q. G., D'Angelantonio, M., and Balzani, V. (1986) *Inorg. Chem.* **25** 4249.

[57] Sotomayor, J., Costa, J. C., Mulazzani, Q. G., and Pina, F. (1989) *J. Phochem. Photobiol. A: Chemistry* **49** 195.

[58] Sexton, D. A., Curtis, J. C., Cohen, H., and Ford, P. C. (1984) *Inorg. Chem.* **23** 49.

[59] Hennig, H., Walther, D., and Thomas Ph. (1983) *Z. Chem.* **23** 445.

[60] Hennig, H., Rehorek, D., and Archer, R. D. (1985) *Coord. Chem. Rev.* **61** 1.

[61] Hennig, H. and Rehorek, D. (1986) In Lever A. B. P. (ed.) *Excited states and reactive intermediates. ACS Symp. Ser.* **307** 104.

[62] Everly, R. M. and McMillin, D. R. (1989) *Photochem. Photobiol.* **50** 711.

[63] Stacy, E. M. and McMillin, D. R. (1990) *Inorg. Chem.* **29** 393.

[64] Serpone, N. (1988) In Fox, M. A. and Chanon, M. (eds) *Photoinduced electron transfer*. Part D. Elsevier, p. 47.

[65] Darwent, J. R. (1982) In Harriman, A. and West, M. A. (eds) *Photogeneration of hydrogen*. Academic, p. 23.

[66] White, B. G. (1969) *Trans. Faraday Soc.* 2000.

[67] Hoffman, M. Z., Prasad, D. R., Jones, G., II, and Malba, V. (1983) *J. Am. Chem. Soc.* **105** 6360.

[68] Kuczynski, J. P., Milosavljevic, B. H., Lappin, A. G., and Thomas, J. K. (1984) *Chem. Phys. Lett.* **104** 149.

[69] Kennelly, T., Strekas, T. C., and Gafney, H. D. (1986) *J. Phys. Chem.* **90** 5338.

[70] Prasad, D. R., Hoffman, M. Z., Mulazzani, Q. G., and Rodgers, M. A. J. (1986) *J. Am. Chem. Soc.* **108** 5135.

[71] Prasad, D. R. and Hoffman, M. Z. (1986) *J. Chem. Soc., Faraday Trans. 2* **82** 2275.

[72] Jones, G., II and Zisk, M. B. (1986) *J. Org. Chem.* **51** 947.

[73] Mulazzani, Q. G., D'Angelantonio, M., Venturi, M., Hoffman, M. Z., and Rodgers, M. A. J. (1986) *J. Phys. Chem.* **90** 5347.

[74] Hoffman, M. Z. and Prasad, D. R. (1987) In Balzani, V. (ed.) *Supramolecular photochemistry*. Reidel, p. 153.

[75] Ebbeson, T. W., Manring, L. E., and Peters, K. S. (1984) *J. Am. Chem. Soc.* **106** 7400.

[76] Nagamura, T. and Sakai, K. (1988) *J. Chem. Soc., Faraday Trans. 1* **84** 3529.

[77] Nagamura, T. and Sakai, K. (1989) *Ber. Bunsenges. Phys. Chem.* **93** 1432.

[78] Sullivan, B. P., Dressick, W. J., and Meyer, T. J. (1982) *J. Phys. Chem.* **86** 1473.

[79] Deronzier, A. and Esposito, F. (1983) *Nouv. J. Chim.* **7** 15.

[80] Nüsslein, F., Peter, R., and Kisch, H. (1989) *Chem. Ber.* **122** 1023.

[81] Willner, I. and Willner, B. (1990) In Schneider, H. J. and Dürr, H. (eds) *Supramolecular organic chemistry and photochemistry*. Verlag Chemie, in press.

[82] Jones, G., II and Goswami, K. (1986) *J. Phys. Chem.* **90** 5414.

[83] Poulous, A. T. and Kelly, C. K. (1983) *J. Chem. Soc., Faraday Trans. 1* **79** 55.

[84] Prasad, D. R. and Hoffman, M. Z. (1984) *J. Phys. Chem.* **88** 5660.

[85] Mau, A. W. H., Overbeek, J. M., Loder, J. W., and Sasse, W. H. F. (1986) *J. Chem. Soc. Faraday Trans. 2* **82** 869.

[86] Villemure, G., Detellier, C., and Szabo, A. G. (1986) *J. Am. Chem. Soc.* **108** 4658.

[87] Nagamura, T. and Sakai, K. (1987) *Chem. Phys. Lett.* **141** 553.

[88] Vining, W. J., Caspar, J. V., and Meyer, T. J. (1985) *J. Phys. Chem.* **89** 1095.

[89] Sabbatini, N., Perathoner, S., Lattanzi, G., Dellonte, S., and Balzani, V. (1987) *J. Phys. Chem.* **91** 6136.

[90] Moyá, L., unpublished results.

[91] Görner, H. (1987) *J. Phys. Chem.* **91** 1887.

9

Electron donor–acceptor complexes and exciplexes

9.1 INTRODUCTION

Molecular association can take place not only through covalent bonding (Chapters 5–7) and electrostatic interaction (Chapter 8), but also by weaker interaction types which include hydrogen bonding, charge transfer, polarization interaction, and higher-order coupling terms [1–9]. In several cases the result of these weaker interactions is the formation of well-defined supramolecular species which exhibit rather interesting photochemical and photophysical properties.

In supramolecular species structure-specific interactions (particularly, complementarity between concave and convex shapes) may or may not play an important role. This chapter deals with supramolecular systems based on intermolecular forces, but not characterized by structure-specific interactions. Supramolecular systems of the latter type are dealt with in Chapter 10.

Most of the supramolecular species originating from the intricate combination of weak interactions are traditionally classified as *electron donor–acceptor* (*EDA*) *complexes* (also called charge-transfer complexes) and *exciplexes*. The distinction between these two types of supramolecular species can be illustrated making reference to Figs 9.1 and 9.2.

An EDA complex results from a bonding interaction in the ground state and thus it corresponds to a minimum in the potential energy curve (Fig. 9.1) which describes the variation of the energy with the distance between the two components, D (donor) and A (acceptor). In terms of the Mulliken theory [1] the ground and excited state wavefunctions of the EDA complex, Ψ_g and Ψ_e, can be written to a first approximation in terms of a "no-bond" and a "charge transfer" limiting structures, A . . . D and A$^-$ . . . D$^+$, as in (9.1) and (9.2),

$$\Psi_g = a\Psi(\text{A} \ldots \text{D}) + b\Psi(\text{A}^- \ldots \text{D}^+) \tag{9.1}$$

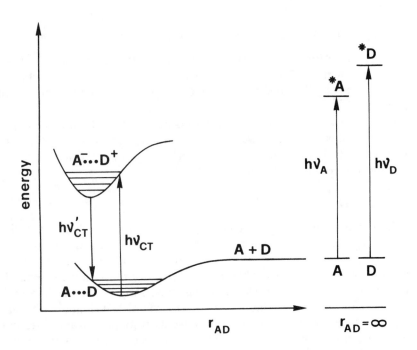

Fig. 9.1 — Potential energy curves for electron donor-acceptor complexes.

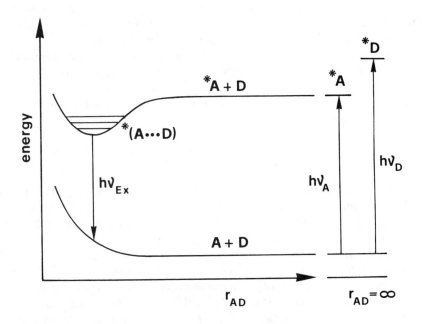

Fig. 9.2 — Potential energy curves for exciplexes.

$$\Psi_e = {}^*a\Psi(A\ldots D) + {}^*b\Psi(A^-\ldots D^+) \tag{9.2}$$

where a, b, $^*a$ and $^*b$ are coefficients which determine the relative contribution of the two limiting structures. For the ground state the weighting is $a \gg b$, while for the excited state $^*b \gg {}^*a$. This theoretical formalism can be extended to the case of a 2:1 complex [10].

Besides the (more or less perturbed) absorption bands of the two components, the EDA complex shows a new, low energy absorption band resulting from the charge-transfer (CT) transition $\Psi_g \to \Psi_e$, which corresponds virtually to an electron transfer from the donor to the acceptor (9.3).

$$A\ldots D \overset{h\nu}{\to} A^-\ldots D^+ \tag{9.3}$$

In the CT excited state there is an electrostatic interaction which increases the bond strength between donor and acceptor, thereby decreasing the intercomponent distance (Fig. 9.1). In a few cases, spectroscopic information concerning the association coordinate are available. For example, for tetracyanoethylene (TCNE) in mesitylene an IR band at $115\,\text{cm}^{-1}$ has been assigned to the intermolecular stretching frequency of the solute–solvent EDA complex [11], and for hexamethyl-benzene (HMB) and TCNE in various solvents resonance Raman studies have evidenced a vibration at $\sim 165\,\text{cm}^{-1}$ that has again been assigned to the intercomponent stretch [12]. Vibrational spectroscopic studies are also used for the analysis of exchange dynamics in EDA complexes (see, for example [13–15]).

In many cases fast radiationless decays (including chemical reactions) of the CT excited state prevent the observation of a CT luminescence. When luminescence occurs, it consists of a broad and structureless band because the nuclear configurations of the excited and ground states are quite different.

When the interaction between two molecular components in their ground states is weaker than the interaction of each component with the solvent molecules, the potential energy curve does not show a minimum on decreasing intermolecular distance (Fig. 9.2). However, the interaction may become much stronger when one of the two components is excited (9.4),

$$A + h\nu \to {}^*A \tag{9.4a}$$

$$^*A + D \to {}^*(A\ldots D) \tag{9.4b}$$

so that the curve describing the approach of $^*A$ and D (or A and $^*D$) may show a well-defined minimum at a certain intercomponent distance. Such an adduct is called an *exciplex*, which can thus be defined as an electronically excited complex which is dissociated in its ground state. If A and D are the same, the excited complex is termed an *excimer*. A simple description of exciplex formation (that is, of the enhanced stabilization of the $^*A\ldots D$ collision pair relative to a $A\ldots D$ collision pair) can be given on the basis of the molecular orbital theory. As schematized in Fig. 9.3a, the interaction of the HOMO and LUMO of A and D in the ground state does not lead to a net stabilization since two electrons are stabilized and two electrons are destabilized. The interaction of $^*A$ and D, however, leads to stabilization of three electrons and destabilization of one electron (Fig. 9.3b) accounting for the formation

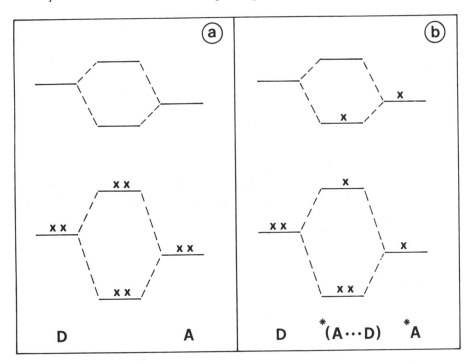

Fig. 9.3 — HOMO–HOMO and LUMO–LUMO interaction between D and A in their ground
states (a) and between ground state D and excited state *A (b).

of a weak chemical bond. Because of the absence of adduct formation in the ground
state, the absorption spectrum of a mixture of the two components in solution will be
exactly the superposition of the spectra of the separated components. Exciplex
formation, however, often implies the appearance of a broad, structureless emission,
red-shifted compared to those of the isolated components.

EDA complexes and exciplexes can occur not only between separated, but also
between covalently-linked D and A species [16,17]. In the latter case, of course, the
connecting linkage must be sufficiently short and/or flexible to allow some interaction between the two components.

The TICT species discussed in section 7.2.4 bear some similarities [18], but also
big differences, compared to the systems discussed here, as can be understood on
comparing Fig. 7.2 with Figs 9.1 and 9.2.

Extensive studies have been carried out on EDA complexes and exciplexes over
the past 20 years. Several books and review articles are therefore available on this
topic [1–4,6,16,17,19–33]. We shall only recall some general concepts and discuss
recent examples.

9.2 ABSORPTION SPECTRA

In general, a complex formed between an electron donor and an electron acceptor
retains the absorption bands of the components modified to a greater or lesser

extent, and shows one or more absorption bands of the complex as a whole. As mentioned before, such bands are due to $\Psi_g \rightarrow \Psi_e$ transitions and correspond to charge transfer from donor to acceptor. In several cases the CT bands are difficult to observe for the following reasons: (1) the concentration of the EDA complex may be very small because the equilibrium is strongly displaced towards the "free" components; (2) the CT band is weak because of the poor overlap of the orbitals involved; (3) the CT band may be covered by stronger absorptions due to locally excited states of A and/or D.

The CT bands are generally very broad and structureless because of the difference in the equilibrium nuclear geometry between the ground and excited states. Furthermore, the rather small binding energy of an EDA complex in the ground state may allow many different configurations to exist in equilibrium with one another, each one showing a slightly different CT maximum.

The approximate energy of the CT transition in the gas phase is given by (9.5),

$$h\nu_{CT} = IP_D - EA_A - W \qquad (9.5)$$

where IP_D is the ionization potential of the donor, EA_A the electron affinity of the acceptor, and W is the energy of electrostatic stabilization of the excited state, e^2/r, at the equilibrium distance r. In fluid solution, a linear correlation is expected between the transition energy and the redox potentials of the two components (9.6) [1,2,16].

$$h\nu_{CT} = E_0(D^+/D) - E_0(A/A^-) + \text{constant} \qquad (9.6)$$

Specific aspects of the theoretical treatment have been recently discussed [31–35]. Because of the change in the electronic structure indicated by (9.3), the CT absorption bands are sensitive to solvent and the energies of the band maxima in general correlate with solvent polarity parameters. The CT bands are also influenced by an external electric field and this effect can be related to variation of the dipole moment in going from the ground to the excited state [20].

For covalently-linked A–D species, the degree of interaction which dictates the position and intensity of the CT bands depends on several factors such as the nature

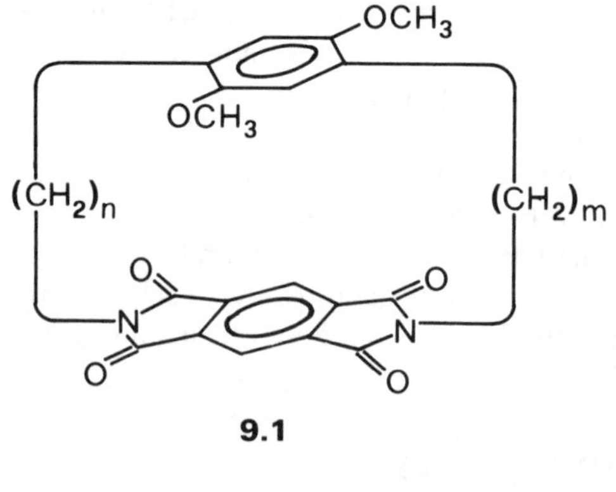

9.1

of the connecting linkage, separation distance, mutual orientation, nature of the solvent, and temperature [16,17,36–41]. Cyclophane (or capped) systems like **9.1** ($n = m = 5$; $n = 4$, $m = 8$), **9.2** ($n = m = 6$; $n = 3$, $m = 8$), and **9.3** ($n = m = 6$; $n = 4$,

$m = 8$) exhibit long-wavelength CT bands [42–44]. These bands decrease in intensity and shift to high energy as the chain length of the bridges increases. Moreover, their

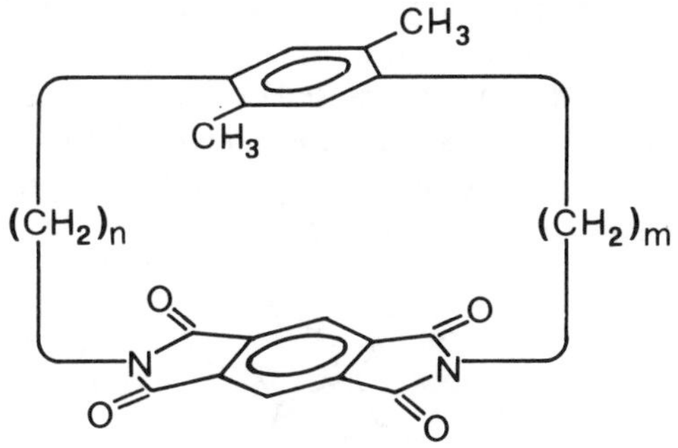

9.2

intensities are much stronger in asymmetrical than in symmetrical compounds, presumably because of the different relative orientation of the two moieties. CT absorption bands are also observed in systems where the interaction can only occur through rigid bonds. Very interesting investigations in this field were performed some years ago by Verhoeven *et al.* [45,46] which made a critical contribution to the advancement of electron-transfer theories [36,37], as discussed in detail in section 5.4.

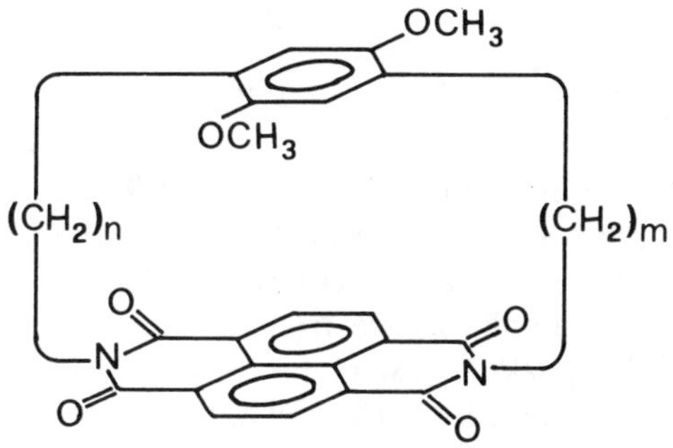

9.3

Several other cyclophane-type systems have been investigated. In compounds **9.4** and **9.5** [47], which contain strongly-bent naphthalene units, the characteristic naphthalene bands are shifted by about 20 nm to longer wavelengths compared with

9.4   **9.5**

those of 2,6-dimethylnaphthalene. This effect could arise from a combination of (through-bond and through-space) intercomponent interactions and deformations of the naphthalene moieties. Experimental and theoretical investigations on compounds **9.6**, **9.7**, and **9.8** have shown that through-bond and through-space interactions are both sizable but oppose each other causing small splitting in the π and π* manifolds [48]. The spectral changes caused by "face-to-face" interaction in dimeric phthalocyanines and porphyrins (see, for example [49]) and by donor–acceptor interactions in porphyrin-phenolphthalein [50] and porphyrin-quinone [51] systems have also been investigated (see section 5.6).

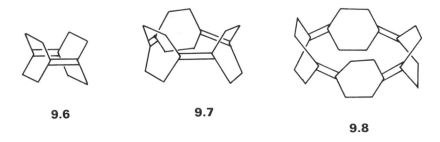

9.6   **9.7**

9.8

Oxygen dissolved in organic solvents gives rise to a CT band. Irradiation in this band leads to singlet oxygen and other transients [52]. Excitation of contact charge-transfer complexes of some styrene derivatives and oxygen leads to electron-transfer products [53].

9.3 LUMINESCENCE

As one can imagine by inspection of Figs 9.1 and 9.2, excited CT states of EDA complexes and exciplexes are somewhat similar species. From the experimental point of view, both CT states and exciplexes can give rise to a broad structure-less emission, although luminescence from CT states is less frequent than from exciplexes. Generally speaking, exciplexes are species of variable CT character (measured values [54], 0.6–0.9 of a full charge transfer), whereas CT states are virtually ion pairs. It has long been recognized [55,56] that one should consider a continuum of states (9.7),

$$^*A\ldots D \to {}^*(A\ldots D) \to (A^-\ldots D^+)_c \to (A^-)_s(D^+)_s \qquad (9.7)$$

which involves localized excitation, exciplex, contact ion pair (which can be obtained directly by excitation of an EDA complex in its CT band), and solvent-separated ion pair. Such a continuous range of states is likely to exist for the same donor–acceptor pair in solvents of different polarity. For solvents having a dielectric constant greater than 10, luminescence generally cannot be observed because of the formation of solvent-separated ions. The existence of conformational isomers of EDA complexes and the role of symmetry correlations in determining the rate of excited-state decay to radical-ion pairs have been analysed on theoretical grounds [57–60]. CT luminescence has been observed for both cyclophane-type systems like **9.2** [43] as well as for linear systems [36,37,46,61]. For compound **9.9** a dipole moment of 23 D can be calculated in the excited state from the red shift of its emission in polar media [61], and much higher values (up to 77 D) have been obtained for the linear systems shown in Fig. 5.6 (section 5.4).

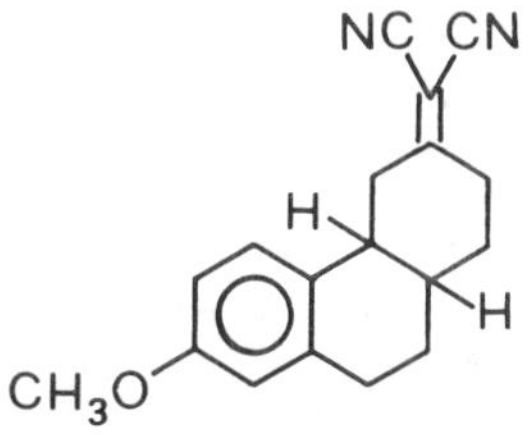

9.9

The complicated pathways for excimer and exciplex formation and the important role played by the solvent are well-illustrated by the detailed investigations carried out by Mataga and coworkers [62–65] on the behaviour of 1,2-di(1-anthryl)ethane **9.10**, 1,2-di(9-anthryl)ethane **9.11**, and other bichromophoric systems with identical halves. In addition to the fluorescence from the locally-excited state of the "monomeric" unit, an excimer fluorescence spectrum was observed in all cases. Picosecond time-resolved studies have shown that intramolecular excimer formation takes place through the following mechanisms. In a nonpolar solvent (*n*-hexane), excimers are formed by a slow process due to "intramolecular" diffusional rearrangement of the chromophores and, to a very small degree, by a rapid process due to the chromophores that are already close to each other in the ground state. In a slightly polar solvent (ethyl acetate), there is a slow process as above and a rapid process that is enhanced by photoinduced charge separation followed by a small conformational change prior to excimer formation. In a strongly polar solvent (acetonitrile), there are rapid and relatively slow photoinduced charge-separation processes which lead to large amounts of ion pairs, whose recombination causes excimer formation.

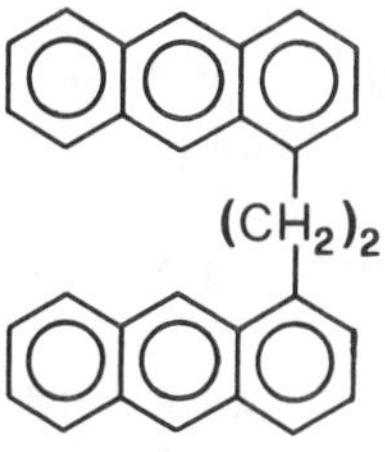

9.10

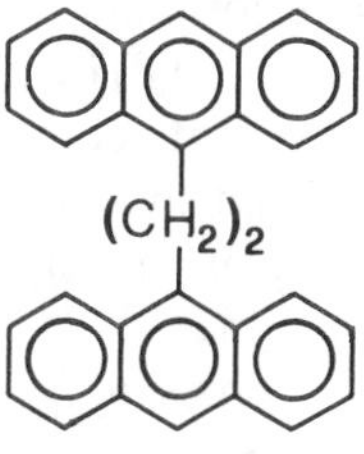

9.11

Several other investigations of the factors which control intramolecular exciplex or excimer formation have been reported (see, for example [41,66–70]). A specially interesting case is that of compound **9.12** which gives rise to a nonlinear triple exciplex where the two tertiary aliphatic amines are symmetrically involved in an interaction with anthracene [71]. The formation of exciplexes in polychromophoric systems is the object of extensive research (see, for example [72–75]). In particular, the interaction between arene-amine exciplexes and another amine in covalently-linked trichromophoric supramolecular structures has been investigated in detail by Yang *et al.* [74].

9.12

Experimental distinction between CT states and exciplexes can emerge from experiments carried out at different temperatures and from time-resolved fluorescence spectra [76,77]. For the TCNB-diethyl ether system in isopentane at ~ 120 K [77], the longer wavelength (~ 490 nm), longer-lived (~ 21 ns) luminescence component was assigned to exciplex emission, and the weaker, shorter-wavelength (~ 370 nm), short-lived (~ 2 ns) one to the CT state. At high temperatures, however, the exciplex emission is not observed. The processes that take place in this system can be rationalized on the basis of the schematic energy diagram shown in Fig. 9.4 [77]. Excitation of the EDA complex (process 1) leads to fluorescence from the equilibrated CT state (process 2). Excitation of the acceptor (process 3) leads to fluorescence of *A (process 4) and of the *A . . . D exciplex (process 5). Apart from radiative decay to the "repulsive" ground state (whose absorption is indicated by the dashed line), the exciplex can undergo an activated radiationless transition to the CT state (process 6).

High-resolution fluorescence emission spectra of EDA complexes [78,79] and exciplexes [30,78,80,81] can be obtained in supersonic nozzle beams. Under such conditions, different geometric isomers are "frozen" so that the role of the relative orientation of the two partners can be elucidated and a better control over the stereospecificity of ensuing reactions is possible by selective optical excitation of a given isomer.

Intersystem crossing from the singlet to the triplet CT state can take place under certain conditions [22,27]. Phosphorescence in a rigid matrix at 77 K has long been established [82]. For complexes between TCNB and methylated benzenes [83], phosphorescence occurs with high efficiencies and long lifetimes (second time scale). Heavy-atom enhancement of the relative yield of phosphorescence has also been observed for the complexes given by hexamethylbenzene with halophthalic anhydrides [84]. The presence of triplet CT levels in EDA complexes in the solid state is often studied by magnetic techniques (see, for example [85]).

Relatively little was known until a few years ago on triplet excimers and exciplexes because of experimental difficulties. With the advent of sensitive detec-

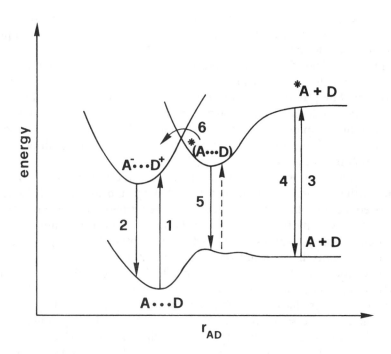

Fig. 9.4 — Schematic representation of the photophysical processes occurring for the tetracya-
nobenzene–diethyl ether EDA system [77].

tion techniques (for example, photon-counting luminescence equipment) rapid
progress has been made and triplet excimers of several aromatic molecules,
especially naphthalene [26,86], have been thoroughly investigated. Examples of
triplet exciplexes involving transition-metal complexes are known [87–94] and the
reasons why exciplexes are rare among transition-metal complexes have been
discussed [94].

Systematic studies on the photophysics of organic exciplexes have been reported
by several groups [95–98]. Exciplex luminescence is particularly useful in studying
conformations of polymers and of their model compounds [98–106]. Exciplex
luminescence has also been used extensively for probing the magnetic field effect on
radical ion-pair recombination processes [107–109].

9.4 PHOTOINDUCED PROCESSES

A predictable fate of the contact ion pair derived from an exciplex or an excited EDA
complex is dissociation into free, solvated ions (9.7). Electron-spin resonance
experiments have been used since the early sixties to demonstrate that this is indeed
the case [110,111]. Another earlier technique used to demonstrate the formation of
ions on excitation in the CT band was photoelectrochemistry [111]. Detailed
quantitative information has since been obtained by flash photolysis studies, first

with microsecond [112,113] and then with nano-, pico-, and femtosecond time resolution. In a series of classical experiments with ns lasers on the complexes of TCNB with toluene and other similar donors, Mataga and coworkers [114–118] showed that the yield of free ions is determined by the chemical properties of donor and acceptor and increases with increasing solvent polarity. On the basis of these and other results they proposed that the Franck–Condon (FC) excited state obtained upon excitation undergoes relaxation with respect to internal coordinates yielding a "transient ionic complex" which can then undergo relaxation with respect to solvent (with formation of solvent-shared ion pairs or free ions) in competition with other deactivation processes.

More detailed information on the dynamics of CT excited states has been obtained with ultrafast laser flash photolysis. Mataga and coworkers [119–121] have investigated the role played by the structure of the partners and solvent properties in further detail, with particular interest in the factors which determine the rates of forward and back electron-transfer reactions (Chapter 5). From pico- and femto-second experiments they concluded that the separation of contact ion pairs obtained by CT excitation of EDA complexes is much slower than the separation of the geminate ion pairs obtained *via* exciplex formation [122]. Furthermore, the energy gap dependence for charge recombination of ion pairs produced by excitation of EDA complexes is essentially different from that found for geminate ion pairs formed *via* fluorescence quenching [123]. In the chloranil (CA)–diphenylamine (DPA) system, which shows two well-separated CT absorption bands, they investi-gated the wavelength dependence of the mechanism and efficiency of ion-pair formation [124]. In cyclohexane solution irradiation in the lower energy CT band ($\lambda_{max} = 640$ nm) is followed by ultrafast ($\tau \sim 1.5$ ps) relaxation to the ground state, whereas irradiation in the higher energy band (which lies below 450 nm, overlapped by the free CA absorption) leads with $\sim 50\%$ efficiency to a transient that in 40 ps decays to the $^3(CA^- \ldots DPA^+)$ ion pair. This species decays directly to the ground state in cyclohexane ($\tau \sim 7$ ns), whereas in 1,2-dichloroethane it gives rise to free ions which undergo back electron-transfer on the microsecond time scale. The marked difference in behaviour of the S_1 and S_2 CT states was attributed to the presence of locally-excited triplet states of CA close to S_2 which speed up intersystem crossing of S_2 and allow it to compete effectively with internal conversion.

Rentzepis and coworkers [125–127] showed that on excitation of the TCNE complexes of 9-cyanoanthracene and indene a transient absorption composed of the superimposed spectra of A^- and D^+ is formed within the laser pulse (20 ps). For the complex with 9-cyanoanthracene the decay of the transient to the ground state occurred within 60 ps, whereas for the complex with indene the recovery of the ground state absorption required a time (500 ps) much longer than that of the decay of the spectrum of the ions.

In order to explain the results obtained with the fumaronitrile/*trans*-stilbene (FN/ TS) system, Goodman and Peters [128–130] used the scheme shown in Fig. 9.5. Direct irradiation in the CT band of the complex produces the singlet contact ion-pair. This may decay by one of the three pathways. The first is back electron transfer, k_{et}, to regenerate reactants. The second is by diffusional separation, k_{ips}, of the singlet contact ion pair to form a solvent-separated ion pair. The third process is singlet–triplet intersystem crossing, k_{isc}, to give a triplet contact ion-pair. The fate of

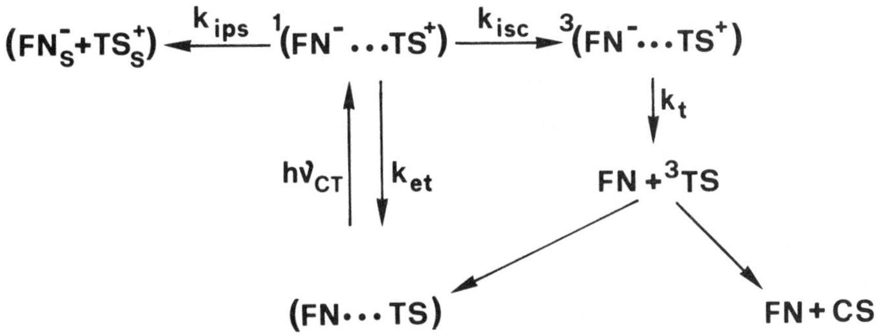

$$(FN_s^- + TS_s^+) \xleftarrow{\ k_{ips}\ } {}^1(FN^- \cdots TS^+) \xrightarrow{\ k_{isc}\ } {}^3(FN^- \cdots TS^+)$$

$$h\nu_{CT} \Big\updownarrow k_{et} \qquad\qquad \Big\downarrow k_t$$

$$FN + {}^3TS$$

$$(FN \cdots TS) \qquad\qquad\qquad FN + CS$$

Fig. 9.5 — Schematic representation of the processes occurring on CT excitation of the fumaronitrile(FN)/*trans*-stilbene(TS) system [130].

this intermediate involves an electron-transfer step, k_t, to form the triplet state of TS, which can isomerize to *cis*-stilbene, CS. The investigation showed that the observed changes in lifetime of ${}^1(FN^- \cdots TS^+)$ in various solvents are due primarily to changes in k_{ips} and k_{et} and not to k_{isc} which appears to be insensitive to solvent polarity. The temperature effect on the various rate constants in different solvents has also been measured.

Kochi and coworkers [32,131–133] studied the complexes of substituted anthracenes as donors (D) and tetranitromethane (TNO_2) as acceptor. This acceptor undergoes instantaneous fragmentation on electron transfer (9.8)

$$D \ldots TNO_2 \xrightarrow{h\nu} D^+ \ldots T^- \ldots NO_2 \rightarrow products \qquad (9.8)$$

leading to addition products at the 9,10-positions of anthracene with high quantum yield. The accurate study of such systems by time-resolved flash photolysis (from ps to μs) provided very detailed information on the intermediates and reaction mechanisms.

Some of the photochemical oxidations of organic substrates by polyoxometalates might proceed *via* excitation of EDA complexes. A picosecond study of the oxidation of 1,1,3,3-tetramethylurea and related dipolar aprotic organic media by α-$H_3PW_{12}O_{40}$ indicates that one transient is formed within the laser pulse (< 20 ps) and one or more other distinct species are formed on a time scale (< 80 ps) faster than bimolecular diffusion [134].

In the context of their investigations on electron-transfer reactions in the Marcus-inverted region [135–137], Gould, Farid, and coworkers performed a comparative study of the solvation and back electron-transfer reactions for contact ion-pairs obtained by CT excitation of EDA complexes and geminate ion-pairs obtained by photoinduced electron transfer [138]. The systems investigated comprised alkylbenzenes as donors and cyanoanthracenes as acceptors. The results obtained show that the back electron transfer in the contact ion-pair obeys the electron-transfer theories (in particular, it exhibits the inverted behaviour), and that electron-transfer quenching of the free excited acceptor does not proceed *via* the intermediacy of a contact ion-pair but results in the formation of the solvent-separated ion-pair directly.

Interesting studies have also been performed on electron donor–acceptor systems connected by methylene chains [139–141] or incorporated in polymers [142]. For *N*-[ω-(4-nitro-1-naphthoxyl)alkyl]anilines, the nature of the photoreaction switches from a photo-Smiles rearrangement to electron transfer when the number of methylene groups increases from 6 to 8 [140]. Strong effects of magnetic fields have also been observed [140,141].

CT excitation of EDA complexes may be followed by a variety of chemical processes including geometrical isomerization, valence isomerization, cyclization, and substitution processes [16]. Photocyclization reactions have been reported for several interesting species (see, for example [41,143–146]). The anthracene-crown ether **9.13** shows an absorption spectrum slightly different from that of the two independent chromophoric units. Its luminescence spectrum in degassed CH_2Cl_2 solution shows a structured band with $\lambda_{max} = 440$ nm and $\tau = 16$ ns characteristic of non-interacting anthracene-type chromophores as well as a broad, unstructured band with $\lambda_{max} = 520$ nm and $\tau = 85$ ns, typical of an excited state with interacting chromophores. This implies that (i) the species emitting at 440 nm belong to sets of conformers which do not lead to intramolecular excimers during their lifetimes, and (ii) the excimer-type emission is due to conformers with interacting chromophores in the ground state as well as to conformers in which the chromophores do not interact in the ground state but can interact in the excited state on a time scale shorter than the time resolution of the equipment used [145,146]. Upon 365 nm excitation, **9.13** gives rise to photocycloaddition both in CH_2Cl_2 and CH_3OH. The perturbations caused by guests on the photochemical and photophysical properties of **9.13** are discussed in section 10.2.

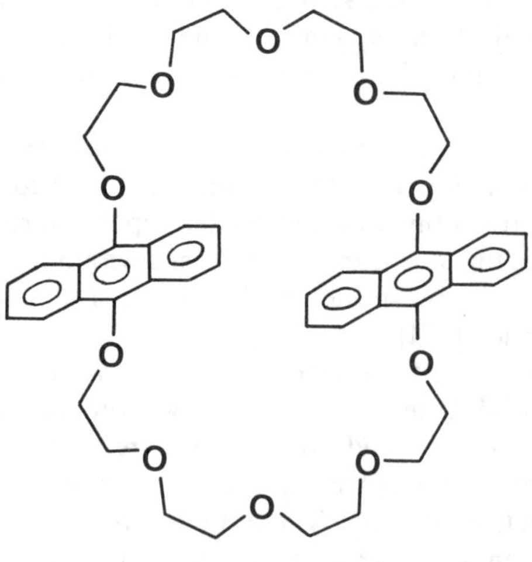

9.13

Exciplexes and excimers are intermediates in a very large number of organic photochemical reactions [147–152], and a discussion of these processes is outside the scope of this book.

REFERENCES

[1] Mulliken, R. S. and Person, W. B. (1969) *Molecular complexes.* Wiley-Interscience.

[2] Foster, R. (1969) *Organic charge-transfer complexes.* Academic.

[3] Birks, J. B. (1970) *Photophysics of aromatic molecules.* Wiley.

[4] Yarwood, J. (1973) *Spectroscopy and structure of molecular complexes.* Plenum.

[5] Morokuma, K. (1977) *Acc. Chem. Res.* **10** 294.

[6] Mataga N. and Ottolenghi M. (1979) In Foster, R. (ed.) *Molecular association.* Vol. 2. Academic, p. 1.

[7] Davidson, R. S. (1983) *Adv. Phys. Org. Chem.* **19** 1.

[8] Bender, C. J. (1986) *Chem. Soc. Rev.* **15** 475.

[9] Maitland, G. C., Rigby, M., Smith, E. B., and Wakeham, W. A. (1987) *Intermolecular forces.* Oxford Univ. Press.

[10] Merriam, M. J., Rodriguez, R., and McHale, J. L. (1987) *J. Phys. Chem.* **91** 1058.

[11] Larkindale, J. P. and Simkin, D. J. (1972) *J. Chem. Phys.* **56** 3730.

[12] Smith, M. L. and McHale, J. L. (1985) *J. Phys. Chem.* **89** 4002.

[13] Yarwood, J. and Catlow, B. (1987) *J. Chem. Soc., Faraday Trans. 2* **83** 1801.

[14] Eskola, S. M., Keskisaari, L., and Stenman, F. (1988) *J. Mol. Struct.* **175** 383.

[15] Besnard, M., Del Campo, N., Cavagnat, R. M., and Lascombe, J. (1989) *Chem. Phys. Lett.* **162** 132.

[16] Jones, G. II (1988) In Fox, M. A. and Chanon, M. (eds) *Photoinduced electron transfer.* Part A. Elsevier, p. 245.

[17] Connolly, J. S. and Bolton, J. R. (1988) In Fox, M. A. and Chanon, M. (eds) *Photoinduced electron transfer.* Part D. Elsevier, p. 301.

[18] Eckert, C., Heisel, F., Miehé, J. A., Lapouyade, R., and Ducasse, L. (1988) *Chem. Phys. Lett.* **153** 357.

[19] Lim, E. C. (ed.) (1969) *Molecular luminescence.* Benjamin.

[20] Birks, J. B. (1975) In Birks, J. B. (ed.) *Organic molecular photophysics.* Vol. 2. Wiley, p. 409.

[21] Stevens, B. (1971) *Adv. Photochem.* **8**, 161.

[22] Ottolenghi, M. (1973) *Acc. Chem. Res.* **6** 153.

[23] Gordon, M. and Ware, W. R. (1975) *The exciplex.* Academic.

[24] Yakhot, V., Cohen, M. D., and Ludmer, Z. (1979) *Adv. Photochem.* **11**, 489.

[25] Kuzmin, M. G. and Soboleva, I. V. (1986) *Progr. Reaction Kinetics* **14** 157.

[26] Lim, E. C. (1987) *Acc. Chem. Res.* **20** 8.

[27] Ferguson, J. (1986) *Chem. Rev.* **86** 957.

[28] Mattay, J. (1987) *Angew. Chem. Int. Ed. Engl.* **26** 825.

[29] Albini, A. and Sulpizio, A. (1988) In Fox, M. A. and Chanon, M. (eds) *Photoinduced electron transfer.* Part C. Elsevier, p. 88.

[30] Haas, Y. and Anner, O. (1988) In Fox, M. A. and Chanon, M. (eds) *Photoinduced electron transfer.* Part A. Elsevier, p. 305.

[31] Kochi, J. K. (1988) *Angew. Chem. Int. Ed. Engl.* **27** 1227.

[32] Yang, N. C., Minsek, D. W., Johnson, D. G., and Wasielewski, M. R. (1989) In Norris, J. R., Jr. and Meisel, D. (eds) *Photochemical energy conversion.* Elsevier, p. 111.

[33] Wittig, C., Sharpe, S., and Beaudet, R. A. (1988) *Acc. Chem. Res.* **21** 341.

[34] Egorochkin, A. N., Lopatin, M. A., and Razuvaev, G. A. (1988) *Doklady Phys. Chem.* **299** 306.

[35] Egorochkin, A. N., Lopatin, M. A., Skobeleva, S. E., and Razuvaev, G. A. (1988) *Doklady Phys. Chem.* **300** 473.

[36] Oevering, H., Verhoeven, J. W., Paddon-Row, M. N., and Warman, J. M. (1989) *Tetrahed.* **45** 4766.

[37] Hermant, R. M., Bakker, N. A. C., Scherer, T., Krijnen, B., and Verhoeven, J. W. (1990) *J. Am. Chem. Soc.* **112** 1214.

[38] Van Damme, M., Hofkens, J., and De Schryver, F.C. (1989) *Tetrahed.* **45** 4693 .

[39] Ferguson, J., Puza, M., Robbins, R. J., and Wilson, G. J. (1989) *Aust. J. Chem.* **42** 2215.

[40] Shin, D. M. and Whitten, D. G. (1988) *J. Phys. Chem.* **92** 2945.

[41] Wenska, G. (1989) *J. Photochem. Photobiol. A: Chemistry* **49** 167.

[42] Schroff, L. G., van der Weerdt, A. J. A., Staalman, D. J. H., Verhoeven, J. W., and de Boer, Th. J. (1973) *Tetrahed. Lett.* 1649.

[43] Borkent, J. H., Verhoeven, J. W., and de Boer, Th. J. (1976) *Chem. Phys. Lett.* **42** 50.

[44] Zom, R. L. J., Schroff, L. G., Bakker, C. J., Verhoeven, J. W., de Boer, Th. J., Wright, J. D., and Kuroda, H. (1978) *Tetrahed.* **34** 3225.

[45] Pasman, P., Verhoeven, J. W., and de Boer, Th. J. (1978) *Chem. Phys. Lett.* **59** 381.

[46] Pasman, P., Rob, F., and Verhoeven, J. W. (1982) *J. Am. Chem. Soc.* **104** 5127.

[47] Blank, N. E., Haenel, M. W., Krüger, C., Tsay, Y. H., and Wientges, H. (1988) *Angew. Chem. Int. Ed. Engl.* **27** 1064.

[48] Falcetta, M. F., Jordan, K. D., McMurry, J. E., and Paddon-Row, M. N. (1990) *J. Am. Chem. Soc.* **112** 579.

[49] Ohno, O., Ishikawa, N., Matsuzawa, H., Kaizu, Y., and Kobayashi, H. (1989) *J. Phys. Chem.* **93** 1713.

[50] D'Souza, F. and Krishnan, V. (1990) *Photochem. Photobiol.* **51** 285.

[51] Cormier, R. A., Posey, M. R., Bell, W. L., Fonda, H. N., and Connolly, J. S. (1989) *Tetrahed.* **45** 4831.

[52] Scurlock, R. D. and Ogilby, P. R. (1989) *J. Phys. Chem.* **93** 5493.

[53] Kojima, M., Sakuragi, H., and Tokumaru, K. (1989) *Bull. Chem. Soc. Japan* **62** 3863.

[54] Baumann, W., Fröhling, J. C., Brittinger, C., Okada, T., and Mataga, N. (1988) *Ber. Bunsenges. Phys. Chem.* **92** 700, and references therein.

[55] Knibbe, H., Röllig, K., Schäfer, F. P., and Weller, A. (1967) *J. Chem. Phys.* **47** 1184.

[56] Knibbe, H., Rehm, D., and Weller, A. (1969) *Ber. Bunsenges. Phys. Chem.* **73** 839.

[57] Stevens, B. (1984) *J. Phys. Chem.* **88** 702.

[58] Stevens, B. (1988) *Chem. Phys.* **122** 347.

[59] Stevens, B. (1988) *Photochem. Photobiol.* **47** 621.

[60] Glauser, W. A., Raber, D. J., and Stevens, B. (1989) *J. Phys. Chem.* **93** 1784.

[61] Pasman, P., Mes, G. F., Koper, N. W., and Verhoeven, J. W. (1985) *J. Am. Chem. Soc.* **107** 5839.

[62] Hayashi, T., Suzuki, T., Mataga, N., Sakata, Y., and Misumi, S. (1977) *J. Phys. Chem.* **81** 420.

[63] Hayashi, T., Mataga, N., Umemoto, T., Sakata, Y., and Misumi, S. (1977) *J. Phys. Chem.* **81** 424.

[64] Mataga, N. (1987) *Acta Phys. Polonica* **A71** 767.

[65] Yao, H., Okada, T., and Mataga, N. (1989) *J. Phys. Chem.* **93** 7388.

[66] Goedeweek, R., Van der Auweraer, M., and De Schryver, F. C. (1985) *J. Am. Chem. Soc.* **107** 2334.

[67] Zachariasse, K. A., Duveneck, G., and Kühnle, W. (1985) *Chem. Phys. Lett.* **113** 337.

[68] Zachariasse, K. A., Duveneck, G., Kühnle, W., Reynders, P., and Striker, G. (1987) *Chem. Phys. Lett.* **133** 390.

[69] Swinnen, A. M., van der Auweraer, M., De Schryver, F. C., Nakatani, K., Okada, T., and Mataga, N. (1987) *J. Am. Chem. Soc.* **109** 321.

[70] Imabayashi, S., Kitamura, N., and Tazuke, S. (1988) *Chem. Phys. Lett.* **153** 23.

[71] Fages, F., Desvergne, J. P., and Bouas-Laurent, H. (1989) *J. Am. Chem. Soc.* **111** 96.

[72] Beecroft, R. A., and Davidson, R. S. (1981) *Chem. Phys. Lett.* **77** 77.

[73] Lin, J. X., Yu, Q., Zhou, Q. F., and Xu, H. J. (1989) *J. Photochem. Photobiol. A: Chemistry* **49** 143.

[74] Yang, N. C., Minsek, D. W., Johnson, D. G., Larson, J. R., Petrich, J. W., Gerald III, R., and Wasielewski, M. R. (1989) *Tetrahed* **45** 4681.

[75] Turro, N. J. (1990) *J. Photochem. Photobiol. A: Chemistry* **51** 63.

[76] Itoh, M. (1974) *J. Am. Chem. Soc.* **96** 7390.

[77] Lim, B. T., Okajima, S., and Lim, E. C. (1986) *J. Chem. Phys.* **84** 1937.

[78] Amirav, A., Castella, M., Piuzzi, F., and Tramer, A. (1988) *J. Phys. Chem.* **92** 5500.

[79] Piuzzi, F. and Tramer, A. (1990) *Chem. Phys. Lett.* **166** 503.

[80] Anner, O. and Haas, Y. (1988) *J. Am. Chem. Soc.* **110** 1416.

[81] Itoh, M. and Hayashi, A. (1989) *J. Phys. Chem.* **93** 7789.

[82] Reid, C. (1952) *J. Chem. Phys.* **20** 1212.

[83] Iwata, S., Tanaka, J., and Nagakura, S. (1967) *J. Chem. Phys.* **47** 2203.

[84] Gronkiewicz, M. K., Kozankiewicz, B., and Prochorow, J. (1976) *Chem. Phys. Lett.* **38** 325.

[85] Bizzarro, G., Corvaja, C., Toffoletti, A., and Pasimeni, L. (1989) *J. Chem. Soc., Faraday Trans. 2* **85** 1913.

[86] Locke, R. J. and Lim, E. C. (1989) *Chem. Phys. Lett.* **160** 96.

[87]   Vogler, A., and Kunkely, H. (1980) *Inorg. Chim. Acta* **45** L265.

[88]   McMillin, D. R., Kirchhoff, J. R., and Goodwin, K. V. (1985) *Coord. Chem. Rev.* **64** 83.

[89]   Palmer, C. E. A., McMillin, D. R., Kirkmaier, C., and Holten, D. (1987) *Inorg. Chem.* **26** 3167.

[90]   Lever, A. B. P., Seymour, P., and Auburn, P. R. (1988) *Inorg. Chim. Acta* **145** 43.

[91]   Ayala, N. P., Demas, J. N., and DeGraff, B. A. (1988) *J. Am. Chem. Soc.* **110** 1523.

[92]   Nagle, J. K. and Brenman, B. A. (1988) *J. Am. Chem. Soc.* **110** 5931.

[93]   Ayala, N. P., Demas, J. N., and DeGraff, B. A. (1989) *J. Phys. Chem.* **93** 4104.

[94]   Stacy, E. M. and McMillin, D. R. (1990) *Inorg. Chem.* **29** 393.

[95]   Cheung, S. T. and Ware, W. R. (1983) *J. Phys. Chem.* **87** 466, and previous papers in this series.

[96]   Palmans, J. P., Van der Auweraer, M., Swinnen, A. M., and De Schryver, F. C. (1984) *J. Am. Chem. Soc.* **106** 7721, and references therein.

[97]   Elisei, F., Aloisi, G. G., and Masetti, F. (1989) *J. Chem. Soc., Faraday Trans. 2* **85** 789, and references therein.

[98]   Ikeda, N., Baba, H., Masuhara, H., Collart, P., De Schryver, F. C., and Mataga, N. (1989) *Chem. Phys. Lett.* **154** 207.

[99]   Gleria, M., Barigelletti, F., Dellonte, S., Lora, S., Minto, F., and Bortolus, P. (1981) *Chem. Phys. Lett.* **83** 559.

[100] Tazuke, S. and Yuan, H. L. (1982) *J. Phys. Chem.* **86** 1250.

[101] Tamai, Na., Masuhara, H., and Mataga, N. (1983) *J. Phys. Chem.* **87** 4461.

[102] Masuhara, H., Tamai, Na., Mataga, N., De Schryver, F. C., Vandendriessche, J., and Boens, N. (1983) *Chem. Phys. Lett.* **95** 471.

[103] Rabek, J. F., (1987) *Mechanisms of photophysical processes and photochemical reactions in polymers.* Wiley.

[104] Tazuke, S. (1985) *Makromol. Chem. Suppl.* **14** 145.

[105] Tazuke, S., Higuchi, Y., Tamai, No., Kitamura, N., Tamai, Na., and Yamazaki, I. (1986) *Macromolecules* **19** 603.

[106] De Schryver, F. C., Collart, P., Vanderdriessche, J., Goedeweeck, R., Swinnen, A., and Van der Auweraer, M. (1987) *Acc. Chem. Res.* **20** 159.

[107] Steiner, U. E. and Ulrich, T. (1989) *Chem. Rev.* **89** 51.

[108] Basu, S., Nath, D. N., and Chowdhury, M. (1989) *Chem. Phys. Lett.* **161** 449.

[109] Staerk, H., Busmanny, H. G., Kühnle, W., and Weller, A. (1989) *Chem. Phys. Lett.* **155** 603.

[110] Ward, R. L. (1963) *J. Chem. Phys.* **39** 852.

[111] Ilten, D. F. and Calvin, M. (1965) *J. Chem. Phys.* **42** 3760.

[112] Halpern, A. M. and Weiss, K. (1968) *J. Phys. Chem.* **72** 3863.

[113] Pilette, Y. P. and Weiss, K. (1971) *J. Phys. Chem.* **75** 3805.

[114] Masuhara, H. and Mataga, N. (1981) *Acc. Chem. Res.* **14** 312.

[115] Masuhara, H., Shimada, M., Tsujino, N., and Mataga, N. (1971) *Bull. Chem. Soc. Japan.* **44** 3310.

[116] Masuhara, H., Tsujino, N., and Mataga, N. (1973) *Bull. Chem. Soc. Japan.* **46** 1088.

[117] Masuhara, H., Hino, T., and Mataga, N. (1975) *J. Phys. Chem.* **79** 994.

[118] Hinatu, J., Yoshida, F., Masuhara, H., and Mataga, N. (1978) *Chem. Phys. Lett.* **59** 80.

[119] Mataga, N. (1984) *Pure Appl. Chem.* **56** 1255.

[120] Mataga, N., Karen, A., Okada, T., Nishitani, S., Sakata, Y., and Misumi, S. (1984) *J. Phys. Chem.* **88** 4650.

[121] Mataga, N. (1989). In Norris, J. R., Jr. and Meisel, D. (eds) *Photochemical energy conversion.* Elsevier, p. 32.

[122] Miyasaka, H., Ojima, S., and Mataga, N. (1989) *J. Phys. Chem.* **93** 3380.

[123] Asahi, T. and Mataga, N. (1989) *J. Phys. Chem.* **93** 6575.

[124] Kobashi, H., Funabashi, M., Shisuka, H., Okada, T., and Mataga, N. (1989) *Chem. Phys. Lett.* **160** 261.

[125] Hilinski, E. F., Masnovi, J. M., Amatore, C., Kochi, J. K., and Rentzepis, P. M. (1983) *J. Am. Chem. Soc.* **105** 6167.

[126] Hilinski, E. F., Masnovi, J. M., Kochi, J. K., and Rentzepis, P. M. (1984) *J. Am. Chem. Soc.* **106** 8071.

[127] Rentzepis, P. M., Steyert, D. W., Roth, H. D., and Abelt, C. J. (1985) *J. Phys. Chem.* **89** 3955.

[128] Goodman, J. L. and Peters, K. S. (1985) *J. Am. Chem. Soc.* **107** 1441.

[129] Goodman, J. L. and Peters, K. S. (1985) *J. Am. Chem. Soc.* **107** 6459.

[130] Goodman, J. L. and Peters, K. S. (1986) *J. Phys. Chem.* **90** 5506.

[131] Masnovi, J. M., Huffman, J. C., Kochi, J. K., Hilinski, E. F., and Rentzepis, P. M. (1984) *Chem. Phys. Lett.* **106** 20.

[132] Masnovi, J. M., Levine, A., and Kochi, J. K. (1985) *J. Am. Chem. Soc.* **107** 4356.

[133] Masnovi, J. M. and Kochi, J. K. (1985) *J. Am. Chem. Soc.* **107** 7880.

[134] Hill, C. L., Bouchard, D. A., Kadkhodayan, M., Williamson, M. M., Schmidt, J. A., and Hilinski, E. F. (1988) *J. Am. Chem. Soc.* **110** 5471, and references therein.

[135] Gould, I. R., Ege, D., Mattes, S. L., and Farid, S. (1987) *J. Am. Chem. Soc.* **109** 3794.

[136] Gould, I. R., Moser, J. E., Ege, D., and Farid, S. (1988) *J. Am. Chem. Soc.* **110** 1991.

[137] Gould, I. R., Moser, J. E., Armitage, B., and Farid, S. (1989) *J. Am. Chem. Soc.* **111** 1917.

[138] Gould, I. R., Moody, R., and Farid, S. (1988) *J. Am. Chem. Soc.* **110** 7242.

[139] Nakatani, K., Okada, T., Mataga, N., De Schryver, F. C., and Van der Auweraer, M. (1988) *Chem. Phys. Lett.* **145** 81.

[140] Nakagaki, R., Mutai, K., and Nagakura, S. (1989) *Chem. Phys. Lett.* **154** 581.

[141] Tanimoto, Y., Takashima, M., and Itoh, M. (1989) *Bull. Chem. Soc. Japan* **62** 3923.

[142] Tkachev, V. A., Mal'tsev, E. I., Vannikov, A. V., and Kryukov, A. Y. (1990) *Research Chem. Interm.* **13** 7.

[143] Yamashita, I., Fujii, M., Kaneda, T., Misumi, S., and Otsubo, T. (1980) *Tetrahed. Lett.* **21** 541.

[144] Eichner, M., and Herz, A. (1981) *Tetrahed. Lett.* **22** 1315.

[145] Bouas-Laurent, H., Castellan, A., Daney, N., Desvergne, J. P., Guinand, G., Marsau, P., and Riffaud, M. H. (1986) *J. Am. Chem. Soc.* **108** 315.
[146] Prodi, L., Ballardini, R., Gandolfi, M. T., Balzani, V., Desvergne, J. P., and Bouas-Laurent, H., *J. Phys. Chem.* in press.
[147] Lewis, F. D. (1979) *Acc. Chem. Res.* **12** 152.
[148] Mattes, S. L. and Farid, S. (1982) *Acc. Chem. Res.* **15** 80.
[149] Caldwell, R. A. and Creed, D. (1980) *Acc. Chem. Res.* **13** 45 .
[150] Mattes, S. L. and Farid, S. (1984) *Science* **226** 917.
[151] McCullough, J. J. (1987) *Chem. Rev.* **87** 811.
[152] Lewis, F. D. and Dykstra, R. E. (1989) *J. Photochem. Photobiol. A: Chemistry* **49** 109

10

Host–guest systems

10.1 INTRODUCTION

Organization in biological systems is quite often the result of molecular association phenomena based on noncovalent intermolecular forces (electrostatic interactions, hydrogen bonding, donor–acceptor interactions etc.). Enzymes, genes, antibodies, ionophores, and other biological systems possess receptor sites that can selectively bind suitable substrates, giving rise to highly specific molecular recognition, transformation, and translocation processes which form the chemical basis of life.

Natural receptors (and, sometimes, substrates) are extremely complicated molecules, but molecular recognition, transformation, and translocation processes can be performed, in principle, by smaller molecules that are synthetically accessible. This concept, introduced in 1967 by the fundamental discovery of the synthetic crown ethers [1], has strongly stimulated the imagination and ingenuity of chemists. In the last twenty years a "concave-oriented" chemistry [2] has thus begun to complement the almost exclusively "convex-oriented" chemistry previously developed, and a great variety of concave molecular species (*hosts*) have been prepared for selective complexation of convex molecular species (*guests*) [3–17]. The award of the 1987 Nobel Prize in Chemistry to Charles J. Pedersen [18], Donald J. Cram [19], and Jean-Marie Lehn [20] "for their development and use of molecules with structure-specific interactions of high selectivity" has given new impetus to the studies in this field. Investigations on the interactions between hosts and guests of known chemical composition and structure are indeed highly interesting and useful since they can help in elucidating the factors that control receptor–substrate interactions in biological systems, and can also lead to the creation of novel species and invention of novel chemical processes. Specific hosts for several metal ions, organic cations, organic and inorganic anions, and some classes of organic molecules and coordination compounds are already available for recognition and transportation purposes [3–14], and a variety of new hosts are currently appearing in the literature (see, for example [21–39]).

This chapter deals with processes that occur in host–guest systems, defined as supramolecular species where structure-specific interactions (particularly, complementarity between concave and convex shapes) play a key role in determining the photochemical and photophysical properties. Supramolecular systems based on intermolecular forces but not characterized by structure-specific interactions are dealt with in Chapters 8 and 9.

10.2 CROWN ETHERS AND RELATED SPECIES

10.2.1 Complexation of metal cations

Covalent linking of chromophores to crown ethers or related species (for example, cryptands) leads to *chromoionophores*, a class of compounds which is interesting for theoretical reasons (for example, for a better understanding of the process of charge separation) as well as for practical purposes [40–43]. Insertion of a cation into the ionophore cavity can, in fact, change the absorption spectrum, the luminescence properties, and the photochemical reactivity of the chromophore, opening the way to recognition and determination of metal cations and to a variety of other applications (for example, optical reading devices [20]).

Several families of chromoionophores have been investigated. They include neutral azacrown ether dyes [41], anionic crown ether dyes [42], anion-capped crown ether dyes [43], phenolic crown ether dyes [44,45], chromogenic spherands [46], anthraceno-cryptands [47], etc.

Most of the chromophores employed are electron donor–acceptor systems which exhibit strong charge-transfer bands in the visible. Both the energy and the intensity of these bands can be affected by metal coordination. The direction of the energy shift depends on the position of the coordination site relative to the electron donor and acceptor sites. Table 10.1 collates some data obtained by Vögtle and coworkers

Table 10.1 — Absorption maxima (nm) of crown ether dyes with and without added metal salts[a]

Metal salt	10.1	10.2	10.3	10.4	10.5
None	476	477	590	583	493, (317sh)
LiI	451	—	630	—	(430sh), 342
NaSCN	445[b]	480	612	554[b]	438, (330sh)
KSCN	464	490	608	579	487, (320sh)
Ca(SCN)$_2$	377	492	668	485	(412sh), 349
Ba(SCN)$_2$	394	560[c]	651	504	(415sh), 344

[a]acetonitrile solution, [metal salt]:[dye] ratio > 100 [41,48–50]; [b]NaClO$_4$; [c]Ba(ClO$_4$)$_2$

[41,48–50] for compounds **10.1** to **10.5**. As one can see, interaction of the metal ion with the donor amino-nitrogen of **10.4** causes a hypsochromic shift, whereas a bathochromic shift is observed in **10.3** since the coordinated metal cation interacts with the acceptor carbonyl site. Compound **10.5** is somewhat peculiar in that both the electron-donor and -acceptor sites can simultaneously interact with the coordinated metal ion. As a consequence, the long-wavelength absorption band at 493 nm, which

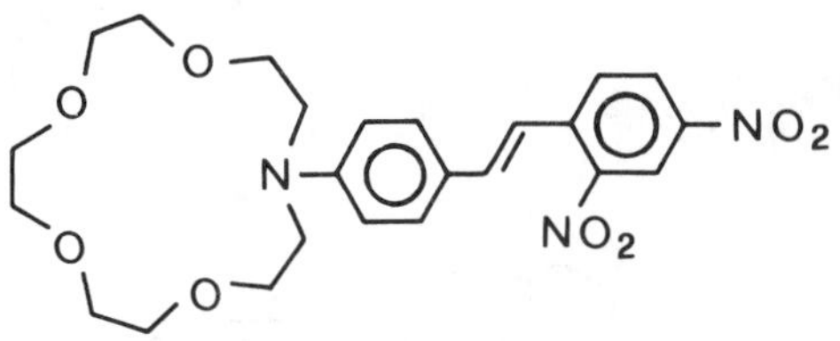

10.1 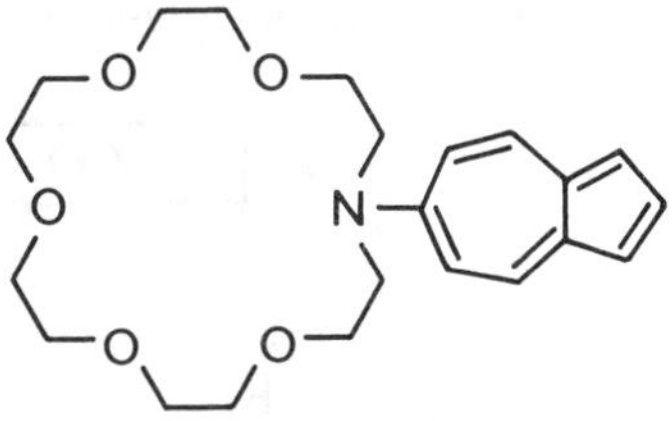   **10.2**

corresponds to a charge-transfer transition from the nitrogen atom to the aromatic system, is shifted hypsochromically, whereas the shoulder at 317 nm, which belongs to a quinoid transition, undergoes a bathocromic shift. Examination of the large number of available results shows that the extent of the band shift depends on (i) the fitting of the metal ion into the crown, (ii) the charge density of the metal cation, and (iii) the nature of the solvent [41,42].

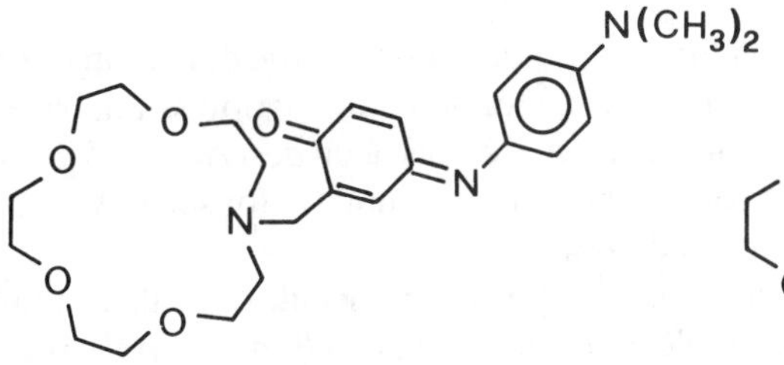

10.3 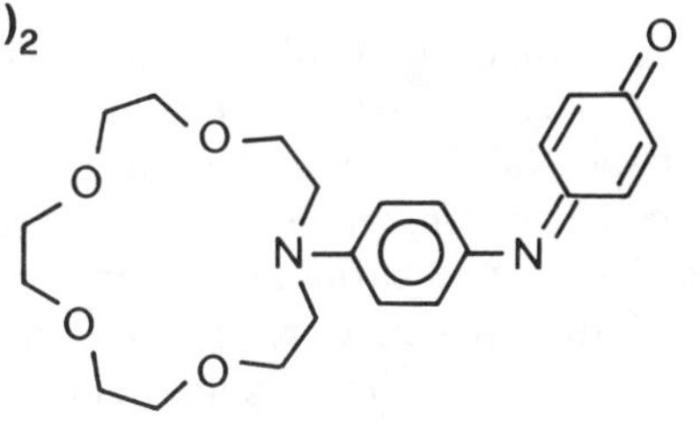   **10.4**

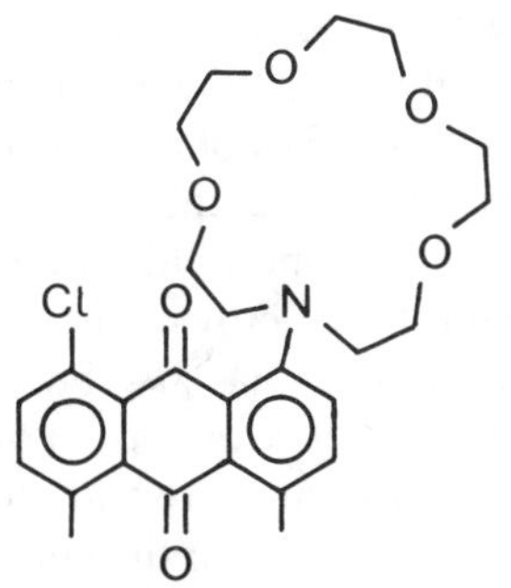

10.5

In specifically-designed molecules the colour change is steered by complexation [41,51,52]. A possible mechanism is schematized in Fig. 10.1 and a relevant example is given by the long-chain electron donor–acceptor podand **10.6** whose absorption band at ~400 nm shifts slightly to the red and becomes much more intense on addition of alkali metal salts [51]. This principle is also applicable to monocyclic ligands of the cyclophane type [53], and to the complexation of the diquat cation by a bpy biscrown ether [54] (*vide infra*). Chiral chromoionophores based on crown ether structures have also been synthesized [41,55].

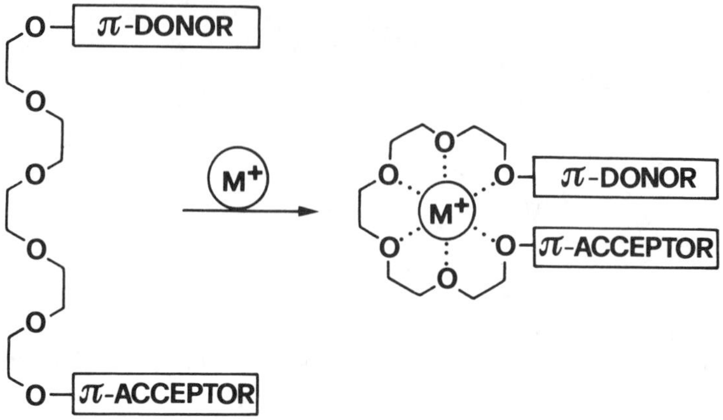

Fig. 10.1 — Cation-steered interaction [41].

An important field of application for chromoionophores is the detection of metal cations by optical fibre sensors [56–58]. A device for detection of potassium based on the spectral changes induced in a phenolic crown dye has been described [59], and an excellent chromogenic crown ether containing a merocyanin dye for selective sensing of calcium and barium has been reported [44].

Cation-selective crown ethers or other coordinating molecules can also be linked to fluorophores. The resulting supramolecular species, called *fluoroionophores* [41], are very interesting because their fluorescent properties (wavelength, intensity, lifetime) may be strongly and selectively affected by the coordinated metal. In one of the first investigations in this field, Sousa and Larson [60,61] showed that the fluorescence behaviour depends critically on the orientation of the fluorophore with respect to the coordination site. For 2,3-naphtho-20-crown-6, complexation with alkali metal chloride salts caused a decrease of the fluorescence quantum yield and an increase in the phosphorescence quantum yield, whereas the reverse effect was observed for the isomeric 2,6-naphtho-20-crown-6 compound. The dependence of the triplet-state properties on orientation has also been investigated using heavy-metal ions as triplet perturbers [62]. For dibenzo-18-crown-6 (**10.7**), metal-ion coordination enhances the fluorescence intensity and lifetime, presumably because of the rigidity imposed by coordination [63]. For the heavy cations Rb^+ and Cs^+, an increase in the $S_1 \rightarrow T_1$ and $T_1 \rightarrow S_0$ intersystem crossing rates was also evidenced in a rigid matrix at 77 K. For acetophenone-(crown ether) systems, complexation with

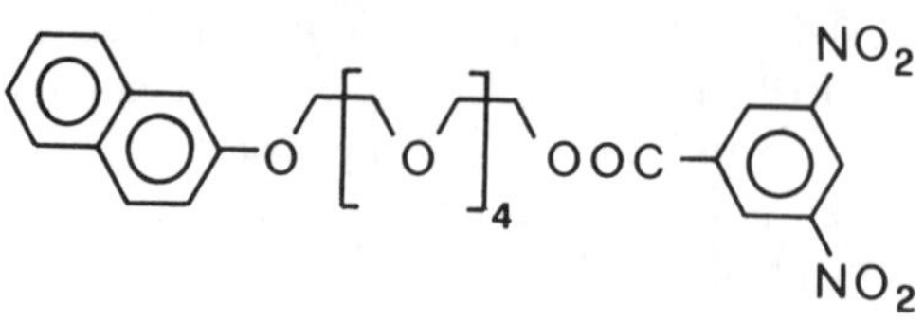

10.6

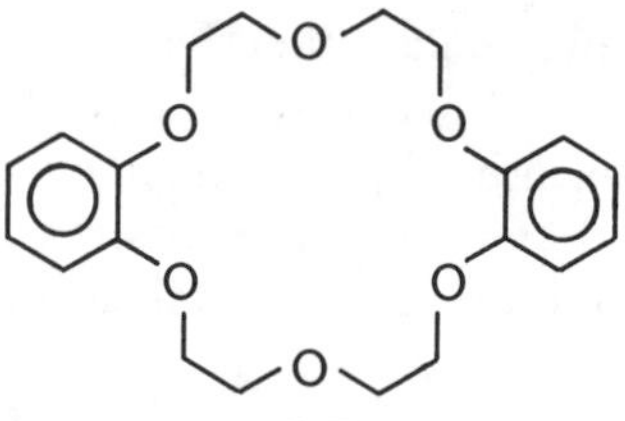

10.7

alkali and alkaline-earth metal cations decreases the phosphorescence quantum yield without affecting the lifetime [64]. A variety of selective fluoroionophores based on crown ethers or other coordinating units have been developed [65–69].

Linkage of monoaza-15-crown to 7-(dimethylamino)-3-(*p*-formyl-styryl)-1,4-benzoxazin-2-one leads to **10.8**. In the presence of Ca(ClO$_4$)$_2$ its fluorescence maximum shifts from 642 to 578 nm, and the fluorescence quantum yield increases

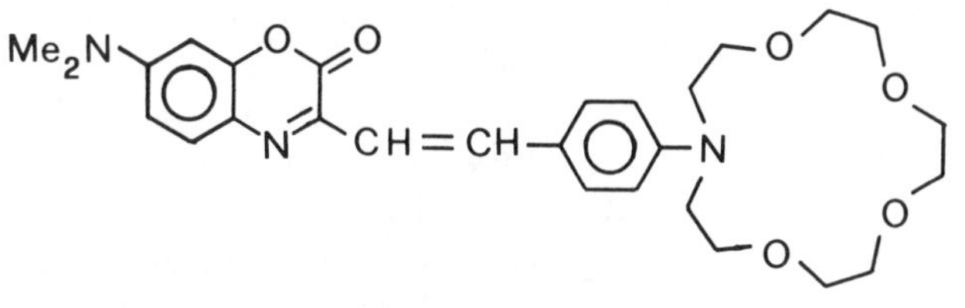

10.8

from 0.33 to 0.64 [70]. The effect of alkali metal ions is, as expected, much smaller [71]. The apparently similar compound **10.9** exhibits a quite different behaviour because of the lack of the dimethylamino substituent on the dye [52]. The quenching of the fluorescence intensity of chiral derivatives of crown ethers by Tb^{3+} is affected by chiral hosts, which allows the use of fluorescence to optimize the requirements for chiral recognition [72]. Anthracenoyl crowns (for example, **10.10**) have been used as fluorescence probes for the solid-phase transition of phosphatidylcholine [73].

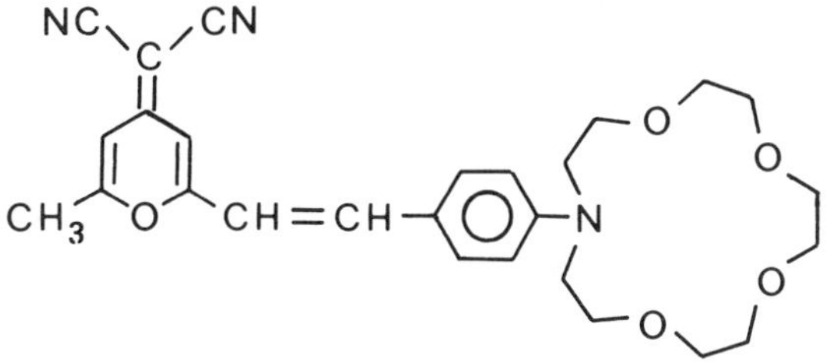

10.9

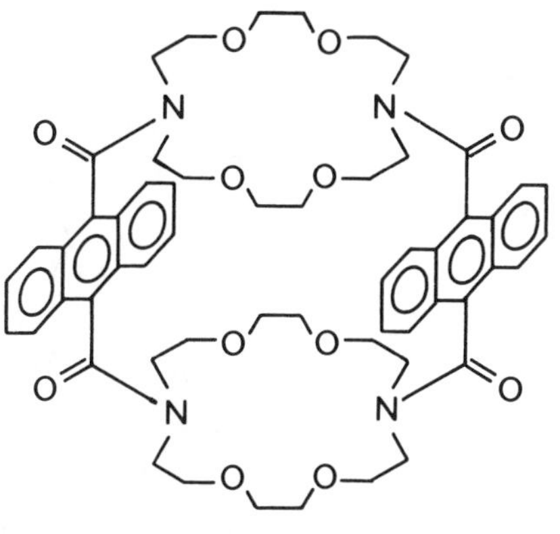

10.10

Complexation of benzo-15-crown-5 with Eu^{3+} and Tb^{3+} results in the quenching of the ligand fluorescence, accompanied by sensitized metal-ion emission in the case of Tb^{3+} [74]. This result is similar to those discussed in section 11.1 for the Eu^{3+} and Tb^{3+} cage-type complexes. An important difference is that in the benzo-15-crown-5 complex the metal ion is much more exposed to interaction with solvent molecules and counter ions. Large macrocyclic ligands can host two metal ions. Homo- and hetero-binuclear complexes of Eu^{3+} and Tb^{3+} have been prepared and their luminescence properties have been investigated [75]. In the heteronuclear complex, energy transfer takes place from Tb^{3+} to Eu^{3+}.

Complexation of metal cations can also induce or prevent photochemical reactions in suitably assembled chromophoric and coordinating moieties. This phenomenon is, in some way, the reverse of that discussed in Chapter 7, where a photochemical reaction induces or prevents metal coordination. Both effects can play an important role in the design of photochemical molecular devices (Chapter 12).

The anthraceno-(crown ether) **10.11** in methanol solution exhibits a structured (monomeric-type) and a broad (excimeric-type) emission [76]. On addition of $NaClO_4$, the intensity of the monomeric band decreases and the excimeric band is shifted to the red. X-ray structural analysis showed that coordination of Na^+ leads to compound **10.12**, where the two anthracene rings are aligned forming a slightly staggered sandwich. Such a change in geometry was also found to induce modification in the photoreactivity. On irradiation with 366-nm light a degassed methanol solution of **10.11** yields as a single product the photocyclomer with ring closure between the positions 9,1′ and 10,4′, whereas **10.12** undergoes the normal 9,9′-10,10′ photocycloaddition [76].

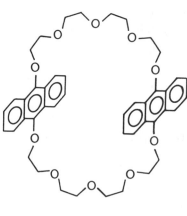

10.11   **10.12**

For a system composed of two anthracene units linked by a polyether chain, a photocycloisomerization was observed, which is rapidly reversed in the dark. In the presence of a Li$^+$ salt, however, the photoproduct is considerably stable, presumably because of the "locking effect" induced by complexation of Li$^+$ in the photogenerated crown ether structure [77].

In the anthraceno-cryptands **10.13**, exciplexes are formed between anthracene and nitrogen lone pairs [47,78]. On addition of an excess of K$^+$, Ag$^+$, and Tl$^+$ to fluid solutions of **10.13**, 1:1 cryptate-type complexes are formed [47,79]. Complexation causes drastic changes in the spectroscopic properties. Light cations (for example, Na$^+$) decrease the intensity of the exciplex emission and increase the intensity of the structured anthracene emission. Heavy-metal cations (Tl$^+$, Ag$^+$) interact strongly with the central ring of anthracene, as shown by an exciplex-type emission observed for the Ag$^+$ complex of **10.13** with n = 3.

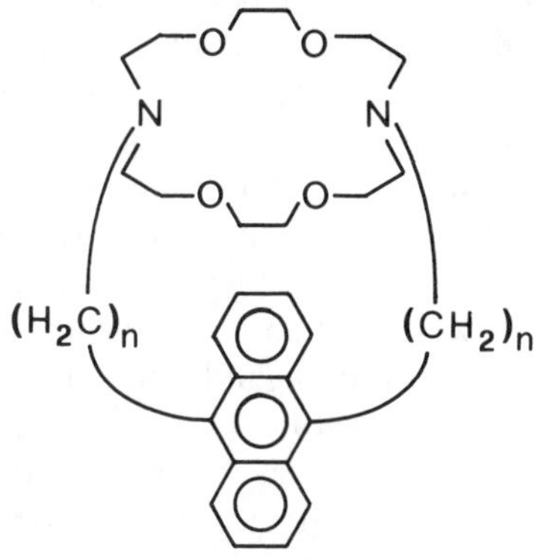

10.13

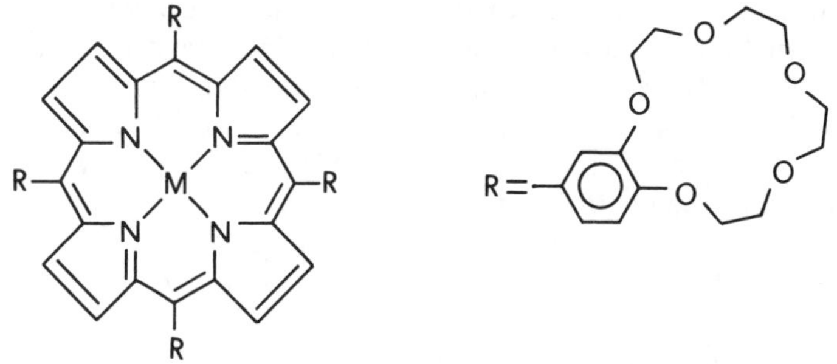

10.14

Porphyrins with attached benzo-(crown ether) groups (**10.14**) have been prepared and their photophysical and photochemical properties have been investigated [80,81]. Complexation of metal ions with the crown-ether moieties causes different effects, depending on the nature of the cation. Na$^+$ forms a stable complex, as expected for a crown-5-cavity, and does not affect the photophysical properties. Large cations such as K$^+$ and Ba^{2+}, that require two crown-ether cavities for complexation, cause face-to-face dimerization of the porphyrins. Transition-metal ions like Cu^{2+} and Ni^{2+} quench the excited states of the metalloporphyrin presumably *via* formation of weak complexes (or, alternatively, *via* encounters) by an energy-transfer mechanism [81]. Eu^{3+} gives rise to a more stable complex and quenches the excited states of the metalloporphyrin by electron transfer. The photogenerated Eu^{2+} ion, however, is too large to fit the crown ether cavity, and it is apparently displaced in competition with back electron transfer [81]. Photoinduced electron-transfer processes also occur when Ag^+ ions are coordinated to crown ethers of porphyrin (section 5.6.8, **5.19**) [82,83] or $Ru(bpy)_3^{2+}$ [82] macropolycyclic structures, and to crown ethers linked to pyrene [84].

10.2.2 Adducts with metal complexes

As first shown by Werner, coordination compounds, being made up of metal ions and ligands which are capable of separate existence, are already complex systems. However they are, at the same time, well-characterized molecular species which can interact with other molecules. Such interactions are promoted by the following properties of metal complexes: (a) electric charge; (b) residual interaction properties of the metal ion; (c) residual interaction properties of the ligands [85]. While the effect of electric charge is nondirectional, the residual interaction properties of metal and ligands are directional and therefore give rise to structure-specific interactions. Many supramolecular systems made of metal complexes hosted in suitable receptors are known. In such systems, which are sometimes called supercomplexes [82], the host plays the role of a second-sphere ligand [86]. This field has been extensively reviewed by Stoddart and coworkers [10,87].

The idea of exploiting crown ethers as second-sphere ligands for coordination compounds was inspired [88] by the structural analogy between organic ammonium ions $(R–\overset{+}{N}H_3)$ and transition-metal amine complexes $(L_nM–\overset{+}{N}H_3)$. For the former class of molecules a proposal had been advanced [89] (later confirmed by X-ray crystallography [90]) that the binding with crown ethers was due to a three-point hydrogen-bonding interaction, as schematized in **10.15** for an adduct of 18C6. The X-ray crystal structure of the 2:1 adduct of *trans*-[Pt(PMe$_3$)Cl$_2$(NH$_3$)] with 18C6 showed that the expected three-point hydrogen-bonding interaction was indeed present [91]. A large number of adducts between metal complexes and crown ethers have been reported in the last few years [10,87], but only a few photochemical investigations have been performed.

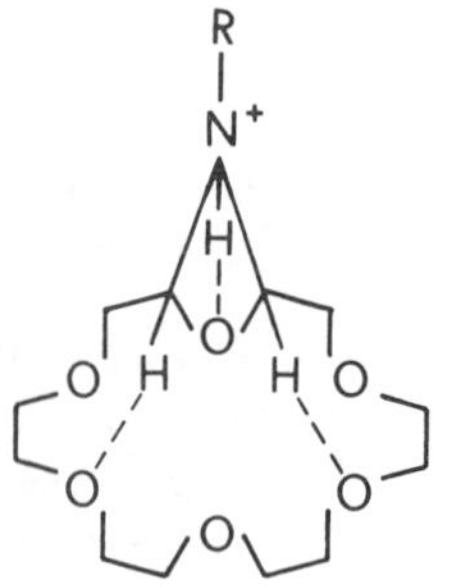

10.15 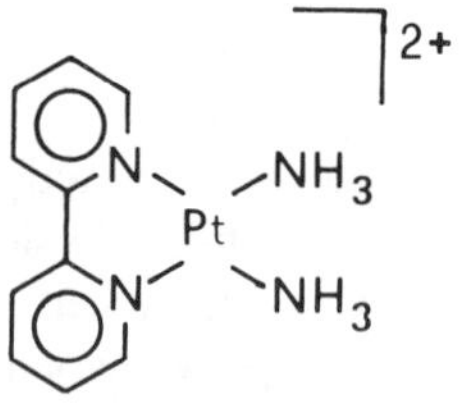   **10.16**

$Pt(bpy)(NH_3)_2^{2+}$ (**10.16**) is a square-planar complex which exhibits most of the properties needed to give stable adducts, namely electric charge, residual interaction ability on the metal ion, and residual interaction ability on both bpy and NH_3 ligands. Furthermore, this complex undergoes a clean (although not yet fully-characterized) photochemical reaction and exhibits luminescence at low and high temperature in a variety of solvents [92,93]. Because of these properties, $Pt(bpy)(NH_3)_2^{2+}$ is suitable for the design of adducts of photochemical and photophysical interest [85].

[$Pt(bpy)(NH_3)_2](PF_6)_2$ is insoluble in CH_2Cl_2, but it can be dissolved by addition of 18C6 to give a 1:1 adduct [94,95]. The absorption and emission spectra of the adduct in CH_2Cl_2 (Fig. 10.2) are practically identical to those of free

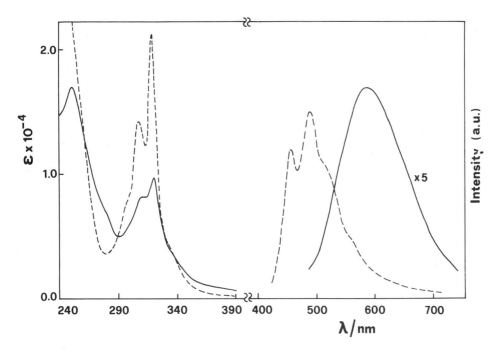

Fig. 10.2 — Absorption and emission spectra of the adducts of $Pt(bpy)(NH_3)_2^{2+}$ with 18C6 (– – –) and DB30C10 (———) [93].

$Pt(bpy)(NH_3)_2^{2+}$ in acetonitrile [76], showing that there is a negligible electronic interaction between $Pt(bpy)(NH_3)_2^{2+}$ and 18C6, which are kept together by hydrogen bonds (*vide supra*). The intense absorptions with maxima at 319 nm ($\varepsilon = 21\,000$ $M^{-1}cm^{-1}$) and 307 nm ($\varepsilon = 13\,300$ $M^{-1}cm^{-1}$) are due to spin-allowed $\pi\pi^*$ transitions of the bpy ligand. The structured luminescence band shows its maximum at 453 nm ($\tau = 25$ μs) at 77 K and at 488 nm ($\tau = 100$ ns) at room temperature. These data indicate that the luminescent excited state is a metal-perturbed ^3LC level [93]. When aromatic macrocyclic polyethers are used in place of 18C6, 1:1 adducts are also formed [10]. For the adduct with DB30C10 (**10.17**), single crystal X-ray diffraction studies showed [94,95] that the receptor assumes a U-shaped configuration around the metal complex (Fig. 10.3). This suggests that besides hydrogen bonding, a $\pi\pi$ interaction between the bpy ligand of the metal complex and the aromatic units of the macrocyclic polyether can take place. The spectroscopic, photophysical, and photochemical behaviour of $Pt(bpy)(NH_3)_2^{2+}$ adducts with the aromatic crown ethers have been investigated in detail [92,93,96]. Some of the data obtained are gathered in Table 10.2, while Fig. 10.2 shows the absorption and emission spectra of $Pt(bpy)(NH_3)_2^{2+}$.DB30C10 compared with those of the "free" metal complex. Adduct formation with aromatic crown ethers causes:

(a) a decrease of the crown ether absorption band;
(b) a strong decrease of the ligand (bpy)-centred absorption bands of $Pt(bpy)(NH_3)_2^{2+}$ in the 290–330 nm region;
(c) the appearance of a weak and broad absorption band in the 340–450 nm spectral region;
(d) the complete or partial quenching of the crown ether fluorescence and of the ligand-centred phosphorescence of $Pt(bpy)(NH_3)_2^{2+}$;
(e) the appearance of a new, broad, and short-lived luminescence band with a maximum in the 550–630 nm region;
(f) the quenching of the photoreaction of $Pt(bpy)(NH_3)_2^{2+}$ in CH_2Cl_2;
(g) a perturbation in the electrochemical reduction potentials of $Pt(bpy)(NH_3)_2^{2+}$.

These results clearly imply an electronic interaction, in the ground and excited states, between the bpy ligand of the Pt complex and the aromatic rings of the crowns. The intensity of such an electronic interaction depends on the size of the crown ring and on the nature and substitution positions of the aromatic rings present in the crown (Table 10.2).

Similar investigations have been carried out with the anthraceno-crown ether **10.11** [97], which is a rather interesting potential host [76]. In CH_2Cl_2 solution there is some interaction between the two anthracene units of **10.11** as shown by the absorption and luminescence spectra and by a photoreaction which leads to the formation of a cycle between the two aromatic units (section 9.4). In the 1:1 adduct of $Pt(bpy)(NH_3)_2^{2+}$ and **10.11**, neither the photocyclization of **10.11** nor the photoreaction of $Pt(bpy)(NH_3)_2^{2+}$ take place, which shows that in the adduct the two species protect each other from photoreaction [97]. Furthermore, both the dual fluorescence of **10.11** and the ^{3}LC emission of $Pt(bpy)(NH_3)_2^{2+}$ are completely quenched and no other luminescence band appears in CH_2Cl_2 solution at room temperature. In a rigid matrix at 77 K, however, a broad band is present with $\lambda_{max} = 565$ nm and $\tau = 30$ μs, which can be assigned to a CT transition between the two species. In this system, the lowest triplet-excited state localized on each anthracene moiety lies below the adduct CT level, so that the adduct emission can only be seen in the rigid matrix, where the deactivation of the CT level to the lower-lying anthracene-localized triplet can be slowed down by energy barriers related to the different nuclear configurations of the two excited states. The results obtained suggest that the most likely structure of the adduct between $Pt(bpy)(NH_3)_2^{2+}$ and **10.11** is a host–guest structure, with the flat Pt complex residing between the two anthracene units.

In conclusion, the assembly of a coordination compound into an appropriate supramolecular structure can protect the compound towards photoreaction and can profoundly change its spectroscopic and photophysical properties.

Table 10.2 — Absorption, emission, and lifetime data of $[Pt(bpy)(NH_3)_2](PF_6)_2$ and its adducts with crown ethers[a]

	Absorption 298 K		Emission			
			298 K		77 K	
	λ_{max} (nm)	ε (M^{-1}cm^{-1})	λ_{max} (nm)	τ (ns)	λ_{max} (nm)	τ (μs)
$[Pt(bpy)(NH_3)_2](PF_6)_2$	319[b] 307[b]	18 000 12 500	488[b]	< 10	453[c]	25[c]
$Pt(bpy)(NH_3)_2^{2+} \cdot 18C6$	319 308	21 000 13 300	488	100	486	25
$Pt(bpy)(NH_3)_2^{2+} \cdot DB18C6$	319 308	15 000 10 000	488	< 10	492	25
$Pt(bpy)(NH_3)_2^{2+} \cdot DB24C8$	319 308	13 200 10 500	585	< 10	487	26
$Pt(bpy)(NH_3)_2^{2+} \cdot DB30C10$	320 309	9 900 8 400	585	< 10	493	20
$Pt(bpy)(NH_3)_2^{2+} \cdot DB36C12$	320 309	14 400 10 800	585	< 10	488	23
$Pt(bpy)(NH_3)_2^{2+} \cdot MDB32C10$	319 312[d]	9 700 9 000	560	< 10	496	31
$Pt(bpy)(NH_3)_2^{2+} \cdot PDB34C10$	321	10 900	600	< 10	490	10
$Py(bpy)(NH_3)_2^{2+} \cdot DN30C10$	320 309	13 500 11 200	555	30	498	30
$Py(bpy)(NH_3)_2^{2+} \cdot ODN38C10$	319	12 900	608	< 10	550	22

[a]In CH_2Cl_2, unless otherwise noted; DB = *ortho*-di-benzo; MDB = *meta*-di-benzo; PDB = *para*-di-benzo; DN = 2,3-di-naphtho; ODN = 1,5-di-naphtho; from [93,96]; [b]AN solution; [c]butyronitrile solution; [d]shoulder.

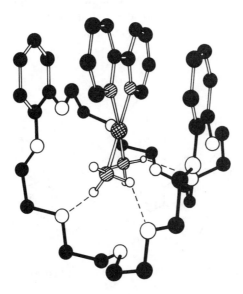

Fig. 10.3 — Skeletal representation of the X-ray structure of the adduct of $Pt(bpy)(NH_3)_2^{2+}$ with DB30C10 [93].

10.2.3 Other systems

As mentioned in the previous section, it is well known that crown ethers can form complexes with organic ammonium ions leading to a structure of the type schematized in **10.15** [89,90,98]. The excited-state proton-transfer reactions of the adducts between α- and β-naphthylammonium ions and 18-crown-6 in MeOH–H_2O (9:1) mixtures have been investigated [99–101]. The proton-dissociation rate in the singlet excited state decreases markedly in the adduct with a consequent increase of the fluorescence lifetime. The excited neutral complex, produced by proton dissociation, does not undergo back-protonation because the resulting amino group is structurally blocked by 18-crown-6 and by the naphthyl group of the adduct. The ground-state association constant, determined by fluorescence titration, is larger for the adduct of the β-naphthylammonium ion. The results are consistent with the model schematized in **10.15**.

The obvious constitutional similarities between $Pt(bpy)(NH_3)_2^{2+}$ (**10.16**) and the bipyridinium herbicide diquat, DQT^{2+} (**10.18**) has led to an investigation of the complexation of dibenzocrown ethers (for example **10.17**) and related receptors with DQT^{2+} and PQT^{2+} (paraquat, **10.19**, also known as methylviologen, MV^{2+}) [102–104]. DQT^{2+} is complexed both in the solid state and in acetone and acetonitrile solutions. Adduct formation leads to the appearance of an absorption band with a maximum at about 400 nm, which can be attributed to a charge transfer between the π-electron rich catechol-type units of the aromatic crown and the π-electron deficient bipyridinium ring system of the guest. For the most stable adduct, which involves DB30C10 (**10.17**), X-ray crystallography showed that the three aromatic rings exhibit a parallel alignment and that stabilization receives a contribution from electrostatic interactions between the phenolic oxygen atoms in DB30C10 and the nitrogen atoms of the bipyridinium ring system of DQT^{2+}. The DB3nCn ethers are highly selective in their complexation of DQT^{2+} relative to PQT^{2+}. The latter ion, however, can give adducts with the *bis*-metaphenylene-32-crown-10 derivative [104,105]. No photochemical or photophysical studies have yet been performed on these interesting host–guest systems.

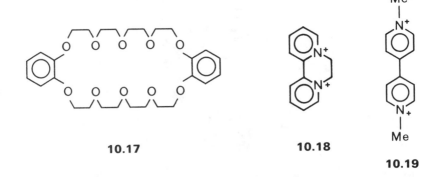

10.17   **10.18**

10.19

Aromatic crown ethers such as **10.7** give rise to 1:1 adducts with 2,3-dichloro-5,6-dicyano-1,4-benzoquinone [106] and 2-dicyanoethylene-1,3-indane dione [107] which show charge-transfer bands in the visible spectral region.

10.3   AZA MACROCYCLES AND RELATED SPECIES

Macrocyclic polyamines, some of which are shown in Fig. 10.4, are analogous to

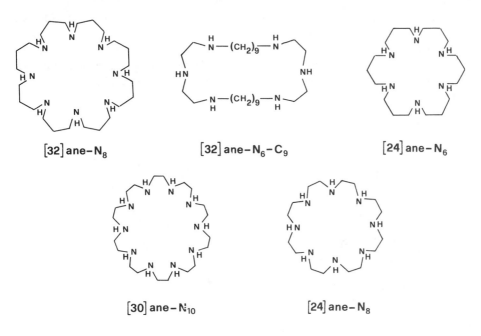

$[32]$ ane–N_8 $[32]$ ane–N_6–C_9 $[24]$ ane–N_6

$[30]$ ane–N_{10} $[24]$ ane–N_8

Fig. 10.4 — Structures of some polyaza macrocycles.

macrocyclic polyethers. Small aza macrocycles can coordinate a metal ion to give simple coordination compounds [20,108–114]. Large macrocycles can enclose two or more metal ions [20,114–121]. Ditopic macrocyclic polyamine receptors have also been designed and used for catalytic purpose [122]. Practically no photochemical or photophysical investigations have been carried out on complexes or adducts of macrocyclic polyamines, except in the case of cage compounds like sepulchrates, discussed in section 11.1. There is no doubt, however, that several polyaza ligands (for example, aza macrocycles with pendant arms [123], N,N-linked bis(cyclam) ligands [124], cryptate-type structures [125], and aza macrocycles capable of selective transport of electrons across liquid membranes [126]) will be used for the design of supramolecular systems of photochemical and photophysical interest.

   A most interesting property of macrocyclic polyamines is their strong tendency to protonate even in neutral solution [127], giving rise to macrocyclic polyammonium species which can coordinate simple [20,125] and complex [20,128–132] anions. The adducts obtained in the latter cases are called supercomplexes [20] or second-sphere coordination compounds [10,86,87]. Some detailed investigations on such species have been performed.

10.3.1 Complexation of metal cations

Complexation of a metal cation by a macrocyclic polyamine linked to a chromophore can cause changes in the spectroscopic properties as in the case of crown ethers (section 10.2.1). Most of the crown ethers described in that section are, in fact, aza-crown ethers, with the nitrogen atom playing an important spectroscopic role.

A case of a fully aza macrocyclic chromoionophore has been reported by Ogawa *et al.* [133]. They found that the tetra-aza macrocycle **10.20** selectively binds Li$^+$ giving rise to a remarkable spectral change (from red to colourless). This is explained by a decrease in the extension of the conjugation caused by the displacement of hydrogens from nitrogens to the carbons bearing the aliphatic chains, which yields two bpy units bridged by saturated carbons.

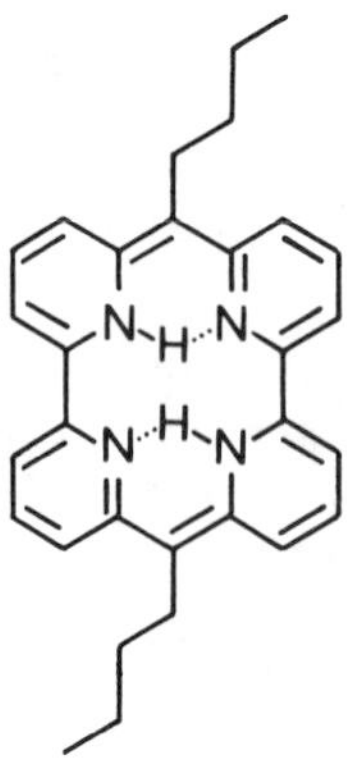

10.20

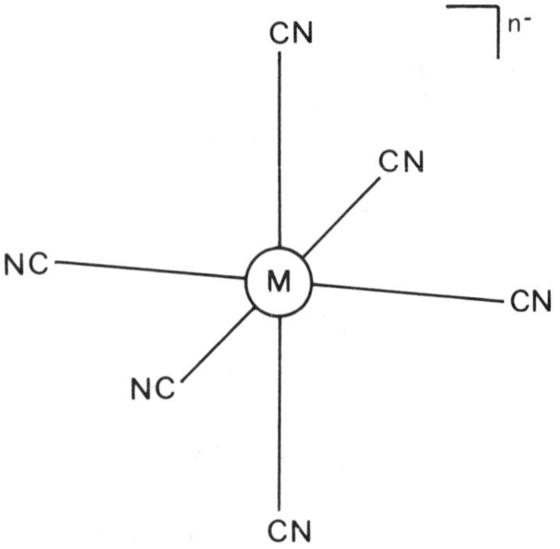

10.21

Kimura *et al.* [134] have prepared a rather interesting supramolecular species made of a Ni(II)-cyclam complex appended to a Ru(bpy)$_3^{2+}$ unit. Preliminary investigations show that the luminescence of the Ru(bpy)$_3^{2+}$ unit is quenched by the Ni complex.

Porphyrins, of course, can also be considered a family of polyazamacrocycles. Their absorption and luminescence spectra are strongly affected on coordination of metal ions [135].

10.3.2 Adducts of metal complexes with polyammonium macrocyclic receptors

Because of their high electric charge and ability to give hydrogen bonds, cyanide complexes (**10.21**) are very suitable species for adduct formation with the protonated forms of aza macrocycles [85,86,127,128]. Unfortunately, no cyanide complex is luminescent in fluid solution at room temperature, with the exception of Cr(CN)$_6^{3-}$ in non-hydroxylic solvents [136]. As far as photochemical behaviour is concerned, most of the cyanide complexes undergo efficient ligand substitution reactions. However in all cases, with the remarkable exception of $Co(CN)_6^{3-}$ (10.1),

$$Co(CN)_6^{3-} + H_2O \xrightarrow{h\nu} Co(CN)_5(H_2O)^{2-} + CN^- \qquad (10.1)$$

the primary photoreaction, which involves the replacement of a CN^- ligand by a solvent molecule in the coordination sphere, is accompanied by secondary thermal and/or photochemical processes which make difficult, if not impossible, an exact determination of the primary quantum yield [136].

$Co(CN)_6^{3-}$ gives rise to very stable 1:1 adducts with the protonated forms of the polyazamacrocyclic receptors shown in Fig. 10.4 [127,128]. The absorption spectrum above 265 nm is essentially the same as that of free $Co(CN)_6^{3-}$ [137], as expected for adducts between partners which cannot give rise to low-energy charge-transfer transitions. The relatively high values of the association constants (10^3–10^6 M^{-1}) reflect an ionic-type interaction, probably with a large contribution from hydrogen bonds between the peripheral nitrogen atoms of the complex anion and the hydrogen atoms of the polyammonium cation. When solutions containing the adducts are excited in the lowest-energy metal-centred band of $Co(CN)_6^{3-}$ ($\lambda_{exc} = 313$ nm), the same photoaquation reaction exhibited by $Co(CN)_6^{3-}$ alone is observed, but with a noticeably smaller quantum yield (Table 10.3) [137,138]. Reduction of the photo-

Table 10.3 — Photochemical quantum yields for $Co(CN)_6^{3-}$ (Φ_0) and its adducts with polyazamacrocycles (Φ)[a]

	pH	Φ_0	Φ	Φ_0/Φ
$Co(CN)_6^{3-}$	1.0	0.33		
	1.2–5.0	0.30		
([24]ane-N_6H_6)$^{6+}$.$Co(CN)_6^{3-}$	2.5		0.15	2
([24]ane-N_8H_7)$^{7+}$.$Co(CN)_6^{3-}$	2.5		0.15	2
([24]ane-N_8H_8)$^{8+}$.$Co(CN)_6^{3-}$	1.0		0.16	2
([30]ane-$N_{10}H_{10}$)$^{10+}$.$Co(CN)_6^{3-}$	2.3		0.14	2
([32]ane-N_6H_6-C_9)$^{6+}$.$Co(CN)_6^{3-}$	1.2		0.11	3
([32]ane-N_8H_8)$^{8+}$.$Co(CN)_6^{3-}$	2.5		0.10	3

[a]Aqueous solution, room temperature [137,138].

aquation quantum yield is also observed in the presence of protonated polyethylenei- mines [139].

Since adduct formation does not introduce low-energy levels and does not cause any appreciable spectral change in the low-energy metal-centred levels of $Co(CN)_6^{3-}$, the reactive excited state of the adducts has to be the same as that of free $Co(CN)_6^{3-}$. Experimental and theoretical evidence indicate that the $Co(CN)_6^{3-}$

photoreaction takes place from the lowest excited level ($^3T_{1g}$) *via* a dissociative mechanism [140]. Therefore, each one of the six CN$^-$ ligands of Co(CN)$_6^{3-}$ involved in the adducts should have the same probability of undergoing photodissociation as in the free complex. A striking result is that adduct formation with the poly-ammonium macrocycles reduces the quantum yield Φ to values substantially equal to a multiple of $(1/6)\Phi_0$ (Table 10.3), which suggests that in the adduct only a discrete number of the six CN$^-$ groups are allowed to undergo photodissociation. Although the structures of the adducts are unknown, space-filling models suggest that the results obtained can be rationalized in the following way [137,138]. In the adducts of the large 32-atom-ring receptors, the octahedral cobalt complex can be enclosed within the macrocycle with four CN$^-$ ligands involved in hydrogen bonding (Fig. 10.5a), so that only 2 out of the 6 CN$^-$ ligands are allowed to escape. For the smaller

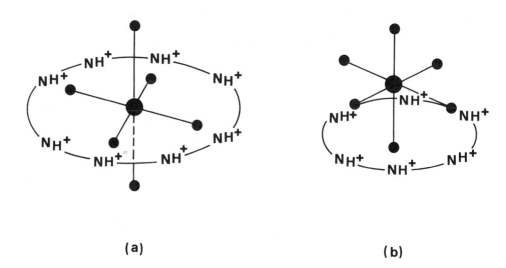

(a)   (b)

Fig. 10.5 — Proposed supramolecular structures of the adducts formed in solution between Co(CN)$_6^{3-}$ and polyammonium macrocyclic receptors of different size [138].

macrocycles, the ring cannot encircle Co(CN)$_6^{3-}$ and the most likely structure is that shown in Fig. 10.5b, with 3 CN$^-$ ligands free to escape. In terms of potential energy curves, the above interpretation of the results obtained implies that on adduct formation there is practically no perturbation of the Co(CN)$_6^{3-}$ levels in the spectroscopic region, whereas there is a strong perturbation in the non-spectroscopic region (Fig. 4.5) for those CN$^-$ ligands which are involved in hydrogen bonding with the macrocyclic structure.

These results show that the photoreactivity of transition-metal complexes can be controlled by adduct formation. Besides offering a generic protection against photodissociation, adduct formation might also find interesting applications in the case of mixed-ligand complexes by orienting photosubstitution reactions towards specific products. Finally, the results suggest that in favourable cases photochemistry may be a probe for the structures of the adducts.

10.4 CYCLOPHANES AND RELATED SPECIES

Cyclophanes are macrocyclic compounds containing aromatic rings, bridged by nonadjacent positions, as components of the macrocyclic structure. Because of their hydrophobic cavity, water-soluble cyclophanes may host apolar molecules in aqueous solution. Research on synthetic cyclophane hosts has grown rapidly in recent years together with the increased interest in catalysis and molecular recognition. An excellent review on complexation of neutral molecules by cyclophane hosts is available [13].

Very few photochemical and photophysical investigations on host–guest adducts of cyclophane species have been reported. Some pieces of information on the excited-state behaviour of such systems come from the use of absorption and luminescence spectra as analytical techniques to demonstrate that inclusion has occurred and to determine the host–guest association constants.

The first stoichiometric inclusion of an apolar guest in the cavity of a synthetic cyclophane host in aqueous solution and in the solid state was reported by Koga *et al.* [141,142]. They designed the host **10.22**, which is soluble in aqueous solution below

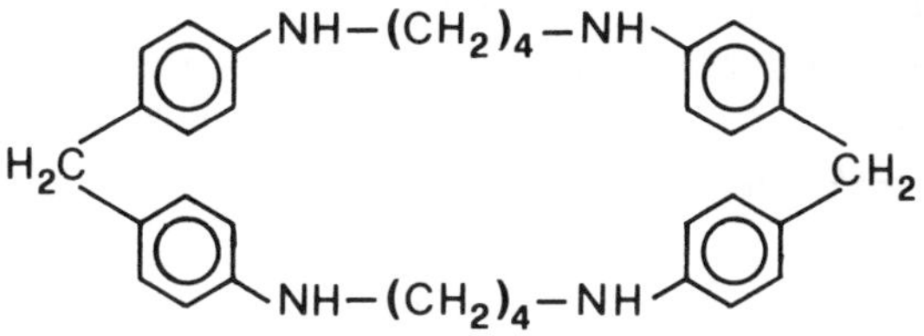

10.22

pH 2, and found that the fluorescence intensity of 1-anilinonaphthalene-8-sulpho-nate (ANS) was markedly enhanced in the presence of **10.22**. Since similar fluorescent changes had been previously observed for inclusion of ANS in α-cyclodextrin [143] (see also section 10.5), they concluded that ANS must be enclosed in the cyclophane cavity. Further and definitive evidence for the formation of inclusion compounds came from NMR and from the crystallization of a 1:1 complex of **10.22** and durene [141]. X-ray analysis of such crystals showed that the four benzene rings of **10.22** are perpendicular to the mean plane of the macroring ("face" conformation), and that the guest molecule is located exactly at the middle of the cavity, with the benzene ring parallel to the inner wall and the methyl substituents protruding from the cavity. The cyclophane design introduced by Koga was developed by Diederich and coworkers [144,145] with cyclophane **10.23**. The presence of the piperidinium groups remote from the cavity increases the water solubility and allows the cavity to be quite apolar, and the two spiro-units confer rigidity to the molecular structure. The four benzene rings are again in the "face-to-face" conformation, oriented perpendicular to the mean molecular plane of **10.23**, and the cavity is large enough to host polycyclic aromatic hydrocarbons. The electronic absorption spectra of pyrene, fluoranthene, and perylene hosted in **10.23** exhibit a bathochromic shift and a significant reduction in intensity compared to the spectra obtained in organic solvents [144]. These changes result from interactions between the host and the guest rather than from changes of the medium, since the spectra are known to be barely affected by solvent polarity. In the luminescence spectra, no excimer emission was observed, indicating that host–guest complexes with 1:2 and 2:2 stoichiometry are not formed. A large blue shift and a strong increase in intensity was observed for the fluorescence of aqueous ANS solutions on addition of **10.23**, indicating again the formation of an inclusion complex, with an association constant $3.2 \times 10^6 \, M^{-1}$. Such a high association constant and the exclusive 1:1 complexation can be understood by inspecting molecular models which show that **10.23** provides a binding site for both the naphthalene and benzene moieties of ANS simultaneously. Competitive fluorescence-inhibition experiments were used to estimate the association constants of **10.23** with other guests, and acceleration of the transport of aromatic hydrocarbons through an aqueous phase using **10.23** as a molecular carrier was achieved [144]. Spherical versions of **10.23** were later prepared [145] following the cryptand concept of Lehn [3], that is by introducing a third diphenylmethane unit into the cyclophane structure **10.23**. Such new hosts exhibit a highly preorganized structure prior to complexation and offer a binding cavity with a suitably-sized entrance for large arenes. They are able to enclose perylene, pyrene, and fluoranthene, causing spectral changes like those described above. One of the macrobicyclic hosts prepared was able to give complexes with perylene and pyrene even in benzene solution [145]. Addition of the host to a methanol solution of 1,3-bis-(1-pyrenyl)propane caused the disappearance of the excimeric emission of the latter, indicating that only one of the two pyrene moieties is encapsulated in the host.

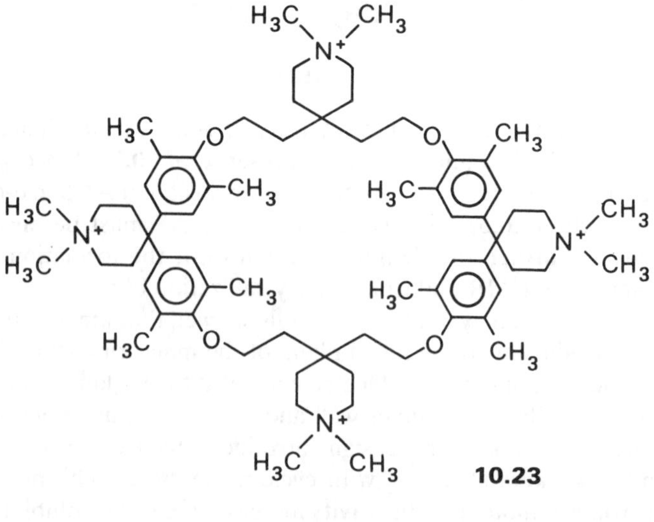

10.23

The macrobicyclic host **10.24** designed by Vögtle *et al.* [146] has a large disc-shaped cavity and shows remarkable guest selectivities towards polycyclic arenes. The binding of guest molecules by **10.24** in dilute HCl solutions was estimated by fluorescence spectroscopy, using perylene as a fluorescent guest. As expected from examination of space-filling models, **10.24** preferentially accommodates in its cavity condensed arenes with flat and discoid shape such as triphenylene, pyrene and perylene.

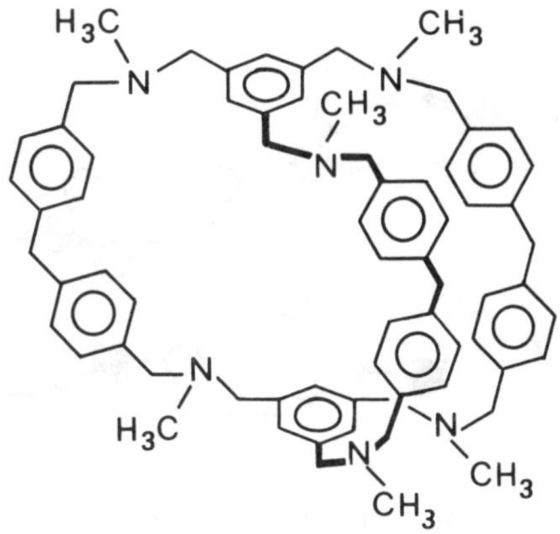

10.24

A bis(4-pyridyl)-porphyrin has been shown to bind within the cavity of a host metalloporphyrin-pyromellitimide dimer [147]. The quenching of the enclosed guest indicates a photoinduced electron transfer to the host.

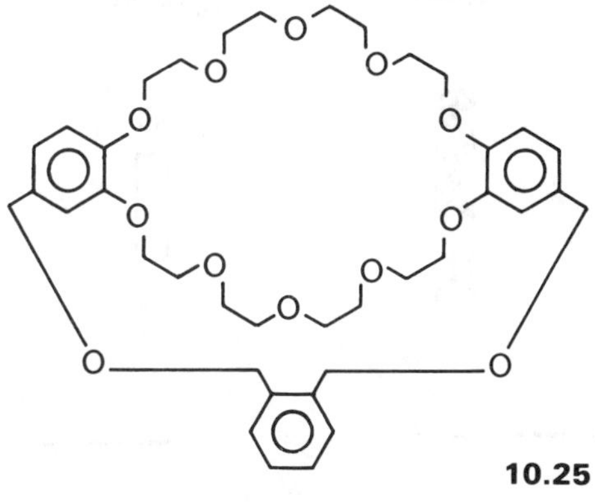

10.25

Very interesting cyclophane-like hosts have been synthesized by Stoddart and coworkers. The macrobicyclic receptor **10.25** [148] and another similar receptor [149] are able to enclose diquat (**10.18**). A series of box-shaped receptors for diquat (**10.18**) and paraquat (**10.19**) have been designed [105,150–153]. The host–guest adduct of a bis-paraphenylene-34-crown-10 derivative (**10.26**) with paraquat (**10.19**) shown in Fig. 10.6, exhibits a band with $\lambda_{max} = 435$ nm in acetone solution, assigned to a CT transition from the electron-rich hydroquinol units of the host to the electron-deficient paraquat guest [151]. A further step [154] on this development was to reverse the roles of the receptor and the substrate, using the tetracation cyclobis(paraquat-p-phenylene) **10.27** as a host for the 1,4-dimethoxybenzene guest **10.28** (see also [155]). This strategy, illustrated in Fig. 10.7, constitutes the basis for the synthesis of catenanes (section 11.2). Complexation of dimethoxybenzenes (for

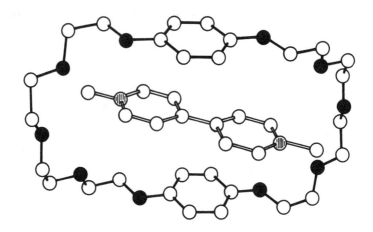

Fig. 10.6 — Host–guest adduct of **10.26** with **10.19** [151].

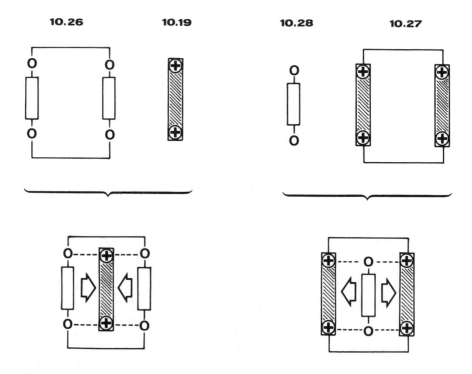

Fig. 10.7 — Host–guest adducts based on paraquat and dimethoxybenzene units [154].

example, **10.28**) by **10.27** results in the appearance of moderately intense ($\varepsilon =$ 10^2–10^3 M^{-1} cm^{-1}) CT bands in the 450–500 nm region. The PF$_6^-$ salt of host **10.27** has a strong propensity to form a channel structure in crystalline state and its 1:1

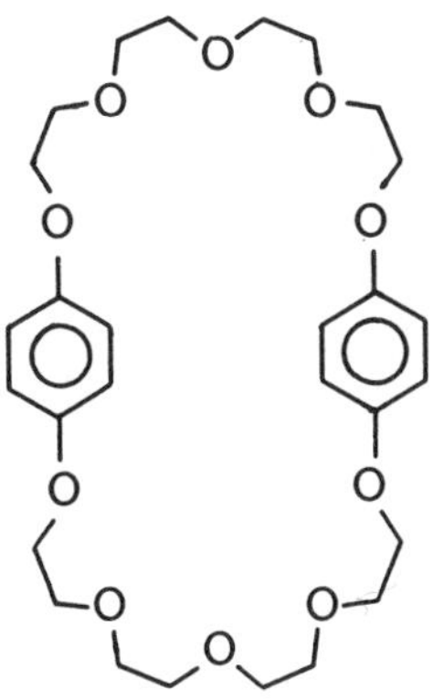

10.26 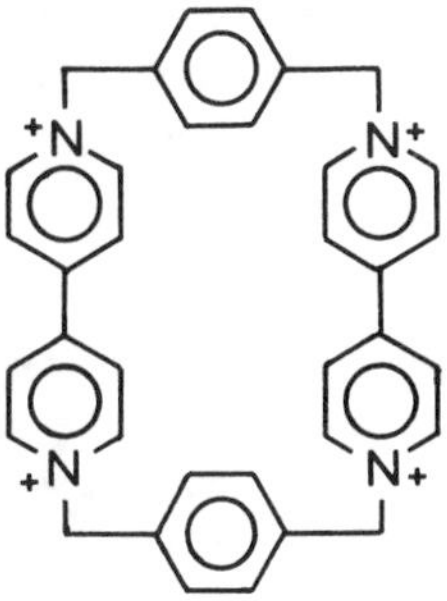   **10.27**

inclusion complexes with hydroquinone dimethyl ether give rise to alternately-charged receptor stacks in the solid state [156]. A polymolecular donor–acceptor stack made of paraquat and a 1,5-dihydroxynaphthalene-derived crown ether has been described [157] as a last step towards the design of a template-directed synthesis of a catenane. Paraquat (**10.19**) can also form strong complexes with a macrocyclic receptor made of four carboxylate groups and two diphenylmethane units [158]. Irradiation leads to the rupture of the macrocyclic ring, presumably *via* photooxidation, and to liberation of the substrate ("photosuicide" process) [82].

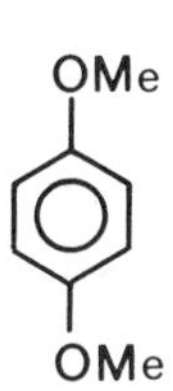

10.28

The great number of cyclophane-type hosts already available, the continuing development in synthetic strategies (see, for example [27,30,159–161]) and the variety of structures and interactions (see, for example [162,163]) that can be found in these systems allow us to forecast that the study of host–guest systems based on cyclophane-type receptors will soon become a major theme in photochemical research.

10.5   CYCLODEXTRIN INCLUSION COMPOUNDS

10.5.1   Introduction

Cyclodextrins (CDs) are cyclic oligosaccharides composed of D(+)-glucopyranose units linked by an α-(1,4) glycosidic linkage [164,165]. Commonly available cyclo-dextrins are those with 6, 7 and 8 glucose units and are called α-, β- and γ-CD, respectively. CDs are shaped like truncated cones, with a smaller and a larger diameter opening at, respectively, the primary hydroxyl and the secondary hydroxyl faces of the cyclic sugar network (Fig. 10.8). The exterior of the molecule is,

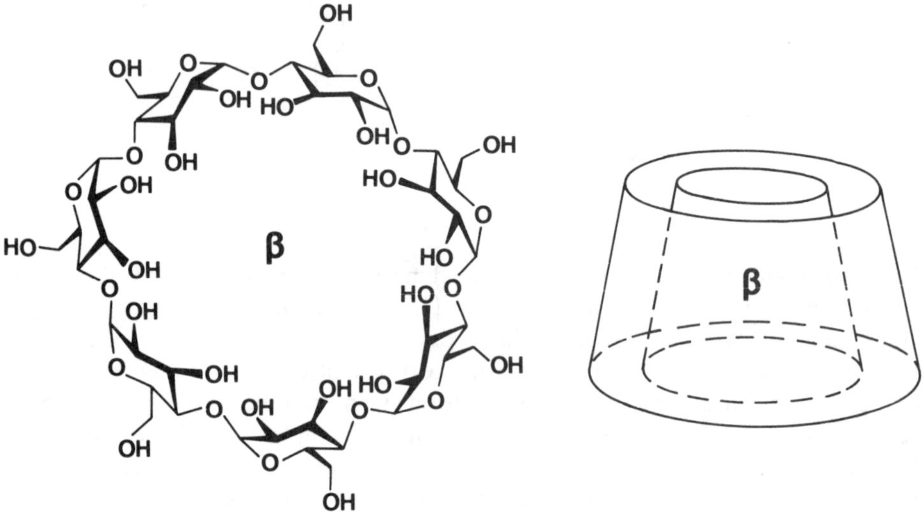

Fig. 10.8 — Schematic representation of β-cyclodextrin.

therefore, relatively hydrophilic and CDs are soluble in water. The internal diameter of the cavity is approximately 5 Å (α-CD), 6.5 Å (β-CD), and 8.5 Å (γ-CD) and its depth is approximately 8 Å. The interior of the cavity consists of a ring of C–H groups, a ring of glucosidic oxygens, and another ring of C–H groups; thus, the interior of CDs is relatively hydrophobic.

   The most important property of CDs is their ability to admit a variety of molecules of appropriate size into the cavity, with formation of inclusion complexes. Several types of interaction have been proposed as being responsible for the complexation: hydrophobic interactions, van der Waals interactions, hydrogen bonding, and stabilization due to the displacement of "high-energy water" from the hydrophobic CD cavity. Modification of the CD molecule (for example, replacement of OH by charged groups or regiospecific functionalization) can strongly modify the affinity for potential guest species [164–169]. Moreover, CDs and their derivatives are interesting models for enzyme-active sites [167,170–173].

The restricted space and reduced polarity of the CD cavity can markedly influence a number of molecular properties. For coordination compounds, CDs may also play the role of second-sphere ligands [10,86] since hydrophilic ligands such as amines can form hydrogen bonds with the hydroxyl groups of cyclodextrins, and hydrophobic ligands or ligand substituents can be enclosed in the cavity of a CD [170,174]. Involvement of a CD as a third-sphere ligand is also conceivable [174] since several complexes can be second-sphere coordinated by crown ethers [10,93,174] which, in turn, can be coordinated by γ-CD [175–178]. Direct coordination of β-CD to Co(III) ions through two deprotonated secondary hydroxyl groups of the same glucose unit has also been reported [179] (see also [174]).

Photoprocesses occurring in cyclodextrin supramolecular systems have been reviewed recently [180–182]. We shall mainly discuss recent results to illustrate some principles and applications.

10.5.2 Photophysics

As mentioned above, CDs are very soluble in water but their cavity is hydrophobic. Therefore, hydrophobic species can be "extracted" from water and enclosed in the CDs, with a consequent change in the luminescence properties [180–182]. Even diastereomeric discrimination can be achieved [183]. Fluorescence techniques can thus be used to determine the association constants, the extent of accessibility of the enclosed probe to other molecules, and the stereochemical mode of inclusion of the probe. The freedom of motion experienced by the probe and the micro-viscosity within the CD cavity can also be investigated by fluorescence depolarization measurements. The changes in the luminescence properties derive from a variety of factors such as a change in the radiative rate constant, limitations in rotational freedom, lack of coupling with water protons, shielding from external quenchers, and association with another species inside the cavity.

The experimental results obtained for supramolecular systems involving a CD are often quite complicated. For example, in the case of xanthene dyes the effect of β-CD on fluorescence and lasing depends on the dye concentration [184]. For concentrated solution both fluorescence and lasing are enhanced, due to dissociation of the dye aggregates. For dilute solutions, different effects are observed depending on the nature and concentration of the dye, presumably because different inclusion complexes can be formed (see also [185,186]). For pyrene, conflicting results have been found concerning fluorescence enhancement and quenching, and association constants which range from 44 to 675 M^{-1} for the complex with β-CD have been reported [187].

Aromatic alcohols are known to become stronger acids on photoexcitation. In aqueous media of appropriate pH, excitation of β-naphthol is followed by adiabatic deprotonation in competition with luminescence and radiationless decay. The excited alcoholate anion formed is also luminescent so that a dual emission is observed. When β-naphthol is enclosed in the β-CD cavity, the photoinduced adiabatic deprotonation becomes too slow to occur within the excited-state lifetime, and a single emission is observed [181,182]. This is taken as an indication that the proton-acceptor ability of the cavity is low, comparable with that of ethanol. Deprotonation is also blocked for the excited state of enclosed 1,6-naphthalenediol.

Four pH-dependent fluorescence bands can thus be observed for the complexes of 1,6-naphthalenediol with α- and β-CD, which are attributed to a molecular form, two different monoionic forms, and a diionic form [188].

In several cases, inclusion in cyclodextrin has a relatively small or no effect, but the co-inclusion of another species may drastically change the luminescence properties [189,190]. For α-naphthyloxyacetic acid [191], inclusion in γ-CD only causes a slight increase in the fluorescence intensity, but a marked enhancement is observed on addition of cyclohexanol. The role of cyclohexanol is presumably that of narrowing the cavity of γ-CD to allow the inclusion of the fluorophore (space regulator). Time-resolved emission spectroscopy is of considerable help in these investigations since the fluorescence spectra of complexed and uncomplexed species may overlap [192]. For aromatic hydrocarbons, nitrogen heterocycles and other organic molecules, room-temperature phosphorescence can easily be observed upon formation of supramolecular species consisting of the luminophore and a heavy-atom containing species (for example, brominated alcohols) enclosed in the CD cavity [193–196]. The effect of an external heavy atom in inducing phosphorescence can also be observed on inclusion of aromatic hydrocarbons in heavy-atom substituted β-CD [197]. Since inclusion of molecules in the CD cavities is shape-selective, heavy-atom enhanced phosphorescence can have important applications in analytical chemistry.

A systematic study on the luminescence properties of Ru(II)-polypyridine complexes showed that only complexes possessing phenyl-substituted ligands exhibit significant interaction with β-CD [198]. This indicates that the interaction occurs *via* inclusion of the phenyl group in the CD cavity. Multiple binding was observed for those complexes which contain more than one phenyl substituent. In these systems, CDs clearly play the role of second-sphere ligands. An important effect of such second-sphere coordination is protection from oxygen quenching [198]. Apparently, second-sphere coordination also occurs in the case of Eu^{3+} complexes having naphthyl-, phenyl-, and thienyl-substituted 1,3-diketones as ligands, but this offers no real advantage in the use of such complexes as fluorescent probes [199].

Surfactant-active chains are ideal guests for CD cavities [180,200]. A rather interesting investigation in this area concerns the surfactant-active complexes of the form $[(bpy)Re(CO)_3NC(CH_2)_nCH_3]^+$ ($n = 0$–17) studied by Demas, De Graff, and co-workers [201], which exhibit luminescence from a $Re \rightarrow bpy$ CT excited state. Intramolecular fold-back of the alkyl chain (Fig. 10.9a) alters the solvent environment around the excited portion of the molecule, with noticeable effects on the excited-state energies and decay paths. In the presence of α- or β-CD, the luminescence lifetime of the complexes with a sufficiently long chain ($n = 9$) decreases in deoxygenated solution and increases in oxygen-bubbled solutions. These effects have been attributed to binding of CD to the alkyl chain (Fig. 10.9b), with more exposure of bpy to the deactivating action of solvent, and to an increased shielding of the excited state towards oxygen by the bulky CD *vs* the more penetrable alkyl group. For a longer alkyl chain ($n = 13$), α-CD has no effect. This is explained on the assumption that, as suggested by molecular models, α-CD can only cap the end of the alkyl chain and cannot efficiently slide down over the chain, so that the chain can continue to protect the excited portion of the molecule. With β-CD, however, there

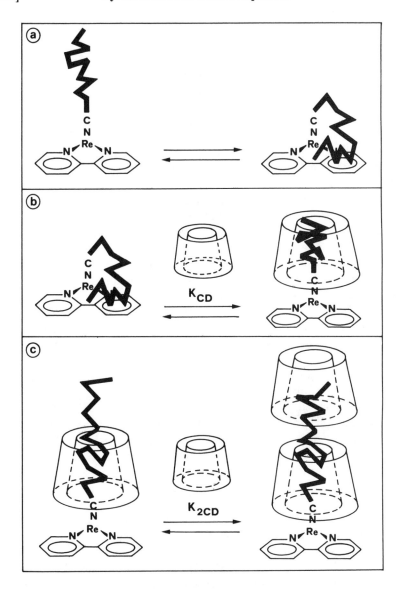

Fig. 10.9 — (a) Intramolecular fold-back of the alkyl chain of
$[(bpy)Re(CO)_3NC(CH_2)_9CH_3]^+$; (b) binding of CD to the alkyl chain; (c) binding of
a second CD [201].

is a large increase in the association constant with increasing n. For $n = 13$, the plot of
τ *vs* β-CD concentration shows biphasic behaviour which has been interpreted as
indicating a second binding of β-CD to the long alkyl chain (Fig. 10.9c). The values of

K_{CD} and K_{2CD} are estimated to be 9300 and 100 M^{-1}, respectively. In this system, the two β-CDs could be considered as second- and third-sphere ligands.

The protection offered by CDs to the luminescent levels of the coordination compounds discussed above towards oxygen quenching is indirect. For chromophoric species that can be entirely enclosed into the CD cavity, direct protection from quenchers can take place as long as the probe remains inside the cavity [202]. The rate of formation and dissociation of the inclusion complex can thus be studied from an analysis of the dependence of phosphorescence lifetimes on the quencher concentration [195,203]. Oxygen indicators based on the ratio of two luminescence signals (one quenched and one unquenched by O$_2$) in systems containing β- or γ-CD have also been proposed [204].

For molecules giving rise to intramolecular interactions such as methyl salicylate, fluorescence measurements show that complexation into α- or β-CD requires different geometrical arrangements [205]. When the cavity size of the CD matches the size of the potential guest, a 1:1 complexation takes place. For example, binding of pyrene to β-CD results in pyrene monomer fluorescence only, indicating that a 1:1 complex is formed [206]. However, when the cavity is much larger than the molecule, 1:2 (host:guest) species can be obtained, but the results are often controversial. With the larger γ-CD, pyrene excimer-like fluorescence is observed [207,208]. In time-resolved studies the fluorescence of pyrene in γ-CD shows two rise-components [209,210]: a strong and very fast component ($\tau < 1.5$ ns), attributed to a pyrene-dimer species which gives rise to the excimer-like excited state immediately after excitation, and a slower component ($\tau = 24$ ns) assigned to pyrene couples which must change their configuration after excitation. Interestingly, circular dichroism and circular polarized fluorescence spectra indicate that the dimers (and the excimers) are chiral, presumably because the formation of the achiral face-to-face configuration, which takes place in fluid media, is prohibited by steric factors [209,210]. More recent investigations [211] on the inclusion compounds of pyrene with γ-CD seem to indicate that the pyrene excimer fluorescence is due to a 2:2 (pyrene: γ-CD) complex, and that 1:1 and 1:2 complexes are also present (see also [212]).

Intramolecular excimers are formed in the case of covalently-linked aromatic moieties when the eclipsed conformation of the bichromophoric system can be hosted in CD cavities [181,182,213–215]. An interesting example is given by 1,1'-bis(α-methylnaphthyl)-1,3-dithiane, **10.29** [181,182], which in hydrocarbon solution shows emission from naphthyl-localized chromophores (~ 320 nm) and excimers (~ 400 nm). Furthermore, interaction of the naphthyl-localized excited states with the sulphur atoms can lead to charge-transfer photochemistry. In contrast, the complex of **10.29** with γ-CD is photoinert and does not show naphthyl-localized emission, but emits only in the 400–480 nm region from an excimer-type level. These results indicate that both naphthyl groups of **10.29** are enclosed in the γ-CD cavity. With the smaller β-CD, complex formation leads to the disappearance of the excimeric emission, but the naphthyl-localized emission decay is biexponential, indicating that the two chromophoric units are no longer degenerate as in the free molecule. This suggests that only one naphthyl unit can be enclosed in the β-CD cavity.

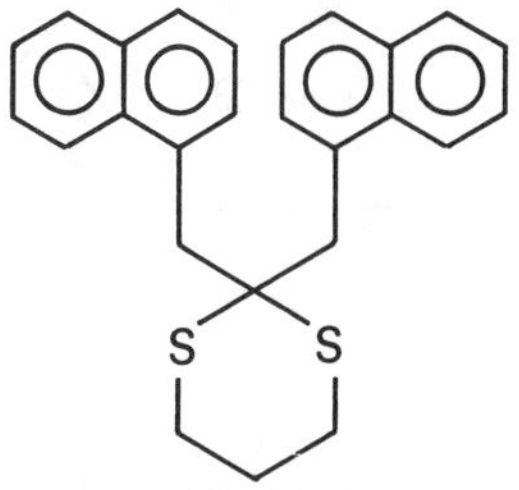

10.29

For *p*-methoxy-β-phenylpropiophenone, inclusion occurs both in β- and γ-CD, but the efficient intramolecular quenching of the carbonyl triplet by the aromatic ring observed in free solution is only prevented by γ-CD [216]. This is thought to derive from a different mode of inclusion of the molecule in the two different CD cavities.

The behaviour of CD derivatives bearing one or two aromatic moieties has been thoroughly investigated by Ueno, Osa, and their coworkers [212,217–225]. The mono-anthracene derivative of γ-CD **10.30** gives rise to an association dimer at high concentration. No excimer-type fluorescence is observed, presumably because the

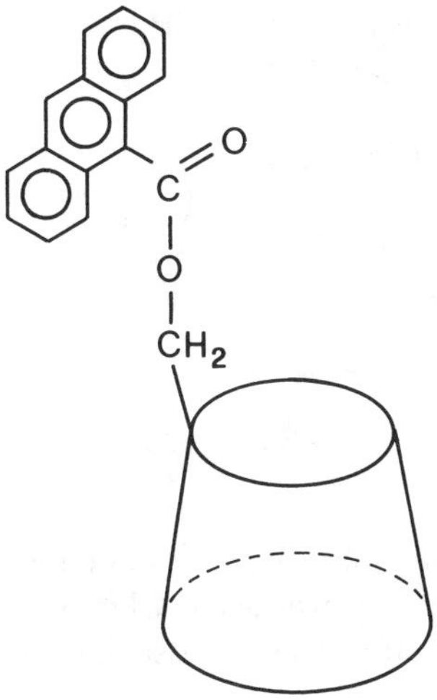

10.30

excimeric structures are rapidly converted into the observed photodimer. On addition of a guest (*l*-borneol), guest-induced dissociation of the dimer takes place (Fig. 10.10) and photodimerization is suppressed. In the dianthracene derivative **10.31** the two chromophoric groups are intramolecularly associated and also exhibit (intramolecular) photodimerization. In the presence of *l*-borneol, the two anthracene moieties are extruded from the cavity [217]. The behaviour of γ-CD bearing two naphthyl moieties has also been studied [223]. The changes caused by reactions of the

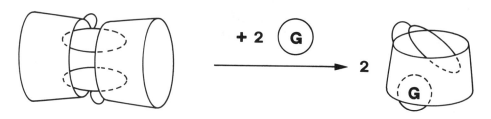

Fig. 10.10 — Guest-induced dissociation of the dimer of **10.30** [217].

type shown in Fig. 10.10 on the photochemical and photophysical properties of CD derivatives can be used to detect a variety of organic compounds [224].

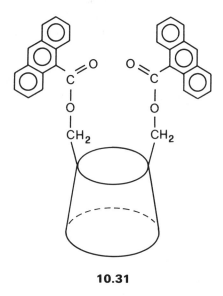

10.31

Fluorescence measurements have shown [218] that naphthyl groups appended to γ-CD prefer to be included in the cavity of added β-CD rather than in the cavity of the linked γ-CD. This size-preference effect leads to association of γ- and β-CD using the naphthyl moiety as a connector.

Besides inter- and intramolecular excimers, inter- and intramolecular exciplexes have also been observed in CD-inclusion systems [180,226]. For example, 1-(1-naphthyl)-3-(dimethylamino)propane **10.32** exhibits only naphthalene monomer fluorescence in aqueous solution, but an exciplex-type emission is observed on addition of β-CD [180]. The exciplex emission, however, is not observed with α- or γ-CD, presumably because the cavity is either too small (α-CD) to allow the inclusion of the entire probe or too large (γ-CD) to give a stable complex. Comparison of the exciplex emission maxima and yields for β-CD and organic solvents indicates that the environment of the probe within the cavity is similar to that of an ethanol solution.

Recent studies on the 1:1 inclusion complexes of β-CD with several amino-naphthalene sulphonate probes have shown that the probe is not in a single discrete environment within the CD cavity but in an array of environments in equilibrium with one another [227].

Electron and energy transfer have also been found to occur between guests within the CD cavity [228–230] or between CD-substituents and guests [231] (see also [232,233]). Finally β-CD inclusion complexes of *p*-nitroaniline and other organic molecules have been proposed as materials for second-harmonic generation [181,234,235].

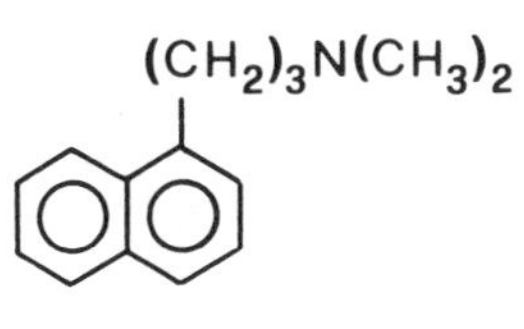

10.32

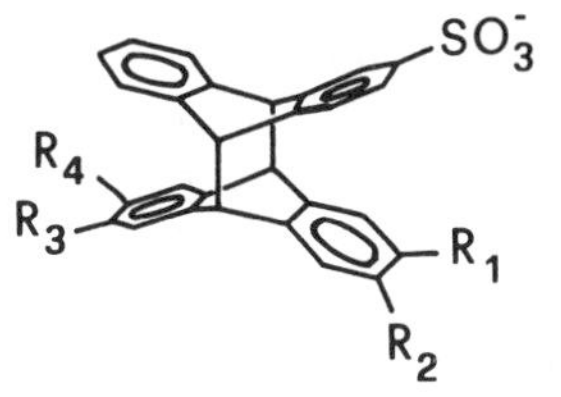

10.33

10.5.3 Photochemistry

The restricted shape and size and the hydrophobic nature of the cavities of CDs offer the opportunity to carry out selective phototransformations and to investigate specific aspects of photochemical mechanisms. The nature of the CD microvessel can impose constraints to nuclear motions (Fig.4.5) and stabilize conformations which are less favoured in free solution. The finite dimension of the cavity may also affect the recombination of radical fragments generated upon photoexcitation. The hydrophobic nature of the cage can affect photoprocesses which are sensitive to solvent polarity or dielectric properties.

Conformational control of photoreactions by a CD cavity has been reported in several cases. An example is given by the $4+4$ dimerization of water-soluble anthracenes [236]. The quantum yield of photodimerization increases dramatically in the presence of CDs for those systems capable of forming 2:1 (guest:host) complexes. Effects were also seen upon the regiochemistry of dimerization. For example, in the absence of CDs, four photodimers of anthracene-2-sulphonate are formed, corresponding to structure 10.33 where any one of R_1, R_2, R_3 and R_4 can be a second sulphonate group instead of hydrogen. When the reaction is carried out in the presence of β-CD, the isomer 10.33 where R_3 is the second sulphonate is almost exclusively formed. When the compound is irradiated in the presence of γ-CD, the four isomers are again obtained, as in the absence of added CD [236]. The face-selectivity of the photocycloaddition of derivatives of adamantan-2-one with fumaro-nitrile is reversed by β-CD, whereas it is unaffected by α- and γ-CD [237]. Formation of inclusion complexes of CDs can also be used for asymmetric induction purposes [238].

10.34

10.35

Restriction on rotational motion can affect photoisomerization reactions. Geometrical photoisomerization of *trans* (**10.34**) to *cis* (**10.35**) stilbene is known to take place by a twisting around the central double bond [239]. Steady-state studies in aqueous solutions in the presence of CD have shown that the *trans* → *cis* photoisomerization is restricted by the CD cavity, whereas the reverse process is apparently unaffected [240]. Picosecond laser studies on the photoisomerization of *trans*-stilbene have also been reported [241]. In the presence of α-CD the excited-state fluorescence decay, which is related to the twisting around the double bond, is monoexponential and considerably longer (137 ps) than the fluorescence decay in a homogeneous non-viscous medium (34 ps), but quite similar to that observed in *n*-alcohols of relatively high viscosity (*n*-octyl alcohol and *n*-decyl alcohol). According to the authors [241] this result suggests that *trans*-stilbene gives rise to only one type of inclusion complex with α-CD and that in the α-CD cavity there is strong hindrance to the rotation of one phenyl group about the double bond. In β-CD, a two-component decay was observed. The value of the short-lived component was 35 ps, that is the same as observed for free *trans*-stilbene. However, the presence of free *trans*-stilbene molecules in solution was ruled out. The long-lived component was of the order of 450 ps. The formation of 1:2 (β-CD:*trans*-stilbene) complexes could also be ruled out by experiments at different β-CD concentrations as well as by inspection of molecular models. The two components were therefore attributed to "loose" and "tight" forms of complexes (Fig.10.11) which should interconvert slowly on the time

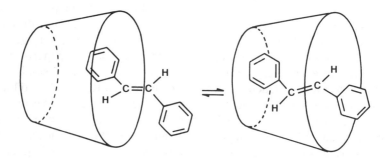

Fig. 10.11 — "Loose" and "tight" forms of the complexes between β-CD and *trans*-stilbene [241].

scale of the photoisomerization. According to the authors [241], the slower decay time for the "tight" complex with β-CD compared with that of α-CD might reflect the fact that the stilbene molecule is embedded more deeply into the β-CD cavity.

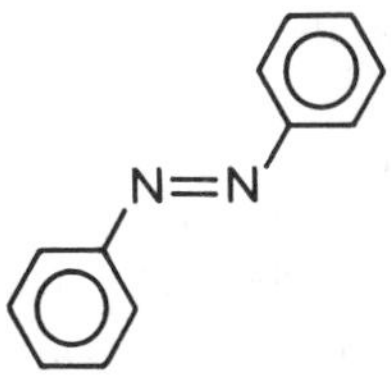

10.36 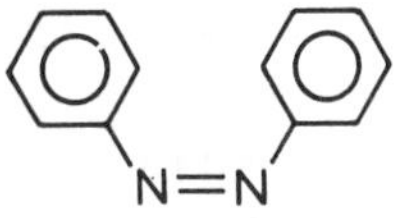   **10.37**

Restrictions are also imposed on azobenzene (**10.36**, **10.37**) photoisomerization [242], which can take place by twisting around the $-N=N-$ double bond and/or by in-plane inversion at one of the two nitrogen atoms (section 7.2.2). In homogeneous solution the *trans* → *cis* photoisomerization occurs with a much lower quantum yield on excitation at 313 nm in the $S_2(\pi\pi^*)$ band compared with excitation at 436 nm in the $S_1(n\pi^*)$ band. The quantum yield of the *cis* → *trans* photoisomerization, however, is wavelength-independent (Table 10.4). In the presence of CD, the quantum yield of

Table 10.4 — Quantum yields for the photoisomerization of azobenzene

	$\Phi(trans \rightarrow cis)$		$\Phi(cis \rightarrow trans)$	
	313 nm	436 nm	313 nm	436 nm
H_2O/CH_3OH (80/20 v/v)	0.20	0.31	0.40	0.42
n-hexane	0.10	0.25	0.44	0.51
tetrahydrofuran	0.08	0.19	0.40	0.48
α-CD (10^{-2} M)	0.11	0.15	0.40	0.41
β-CD (10^{-2} M)	0.13	0.13	0.40	0.48
γ-CD (10^{-2} M)	—	—	0.40	0.46

From [242].

the *cis* → *trans* isomerization is practically unaffected, the quantum yield of *trans* → *cis* isomerization upon excitation in S_1 decreases, and the wavelength dependence of the *trans* → *cis* quantum yield tends to vanish. The absence of an excitation wavelength effect in the *trans* → *cis* photoisomerization is also observed when the rotation about the $-N=N-$ double bond is blocked by a cyclophane structure [243]. This shows that inclusion of *trans*-azobenzene in the CD cavity prevents torsional motion around the double bond and that isomerization from S_1 occurs *via* inversion at one of the two nitrogen atoms [242].

The photoisomerization of azobenzene is used extensively as a photosensitive switch mechanism to regulate chemical and physical phenomena (Chapter 7). Azobenzene-appended γ-CDs are photoresponsive since the *cis* and *trans* isomers of azobenzene interact differently with the CD cavities and therefore affect the ability of CDs to complex other guests [244]. Azobenzene-capped CDs are also photoresponsive [245,246] and offer a means to photocontrol catalytic activity [246] (section 7.3.2).

The photochemistry of aqueous solutions of benzophenone in the presence of CDs offers an interesting example of the role played by the dimension of the CD cavity in controlling the decay of photogenerated radical pairs [247]. The benzophenone phosphorescence is quenched because the triplet-excited states are enclosed in the CD cavity and quantitatively abstract hydrogen from the interior wall of the CD. This leads to the formation of triplet radical-pairs whose fate depends critically on the dimension of the cavity: escape processes dominate in α- and γ-CD; intersystem crossing prevails in β-CD. The fate of the escaped radicals also depends on the cavity dimensions, presumably because in the case of β- and γ-CD they may re-enter into the cavity. Restrictions imposed by the size of the cavity on the reorientational process of geminate radical pairs are also evident in other cases [248].

An example of the effect played by the low polarity of the CD cavity is given by TICT compounds. Enclosure of *p-N,N*-dimethylaminobenzonitrile **10.38** in the hydrophobic cavity of β-CD causes an enhancement of the nonpolar emission [249,250], as expected because the energy barrier for the TICT process increases with decreasing polarity of the medium [251] (section 7.2.4).

10.38

The effects of CDs on the competition between Norrish type I and Type II photoreactions, photo-Fries rearrangement, and photo-Claisen rearrangement have been recently reviewed [182]. The photo-Smiles rearrangement can also be affected [252]. The influence of CDs on the photoisomerization of 7-substituted norbornadiene [253] and on the efficiency and selectivity of aromatic nucleophilic substitution of 2-fluoroanisole [254] has been reported. The photodimerization of anthracene moieties linked to CDs [217] has been discussed in the previous section.

CDs play important roles in other processes of photochemical interest. Zn(II)-*meso*-tetra(*N*-propylsulphonato)pyridinium porphyrin, Zn-TPSPyP, forms a ground-state complex with anthraquinone-2-sulphonate, AQS⁻ [255]. Excitation of this complex results in internal electron transfer followed by the rapid recombination of the photoproducts even in the presence of cysteine as a potential sacrificial donor. However, when β-CD is added to the solution, the protonated semianthraquinone radical anion, AQHS⁻, accumulates in solution. The proposed reaction mechanism involves selective association of AQS⁻ with the receptor, with separation of the components of the electron donor–acceptor complex, and protection of the primary AQS⁻ product against the back electron-transfer reaction. Other examples of separation of ground-state complexes and protection from back electron-transfer by CDs have been reported [256]. Colloidal particles can also be stabilized by cyclodextrins [257] and photosensitization of semiconductors may be improved [258]. External magnetic field effects on photoinduced electron-transfer reactions in phenothiazine-viologen linked systems complexed with α- or β-CD have been investigated [259].

The modifications of photochemical and photophysical properties by inclusion of molecules in zeolites (for a recent paper, see [260]) are in some ways related to those discussed for CD inclusion complexes.

10.6 ADDUCTS OF DNA

10.6.1 Introduction

Many molecular species, particularly metal complexes, bind to the DNA helix. DNA is a polyanion with a very complex structure and a rich selection of coordinating sites which range from the hard phosphate oxygen atoms to the soft heterocyclic base nitrogen positions. Both electrostatic interaction and covalent binding can therefore lead to formation of DNA complexes. Other types of interactions, such as van der Waals forces or hydrogen bonds, can also play an important role. Another special and rather effective binding mode of small molecules to double-stranded DNA is *intercalation* [261], a noncovalent interaction resulting from the insertion of a planar heterocyclic aromatic moiety between the base pairs of the DNA helix. Most of the DNA adducts of photochemical and photophysical interest involve cationic species and therefore they are ion pairs. In these ion pairs, however, structure-specific interactions play a very important role. This is the reason why the DNA adducts are discussed in this chapter rather than in Chapter 8.

The DNA molecule contains all the genetic information necessary for cellular function. The control and regulation of genetic information at the chemical level depend upon binding proteins or small molecules to specific sites along the DNA strand. The design of molecules that target specific sites along a DNA helix is therefore a subject of considerable interest. The most important aims of these studies are:

(1) the development of selective DNA-cleaving agents for mapping or footprinting experiments [262];
(2) the recognition of local structures along the DNA strand [263,264];
(3) the rational design of drugs [265].

10.6.2 Intercalation

Intercalation is a general and, at the same time, a very specific binding mode of small molecules to DNA. Therefore, intercalating agents are extensively used as reporters of specific local DNA conformations and as site-specific reagents (for example, for conformation-specific cleavage of the DNA helix). The photophysical and photochemical properties of the intercalating agents may prove most useful for both the above purposes, as shown by numerous investigations carried out by Barton, Turro, and coworkers [263,264,266–282], Kelly and coworkers [283–289], and others [290–301].

A typical group of molecules that display a high affinity for DNA and other double-stranded polynucleotides consists of organic cations, such as the dyes ethidium (**10.39**) and acridine (**10.40**) [302], which bear an extended heteroaromatic system. In addition to non-specific ionic interactions, these molecules bind DNA by intercalation since the flat dye slides between the hydrophobic base pairs and is stabilized by a π-stacking interaction [303]. Metal complexes also may bind to DNA by intercalation, as was first demonstrated by Lippard and coworkers [304–308] for square-planar Pt(II) complex cations containing a heterocyclic aromatic ligand, such as phen or terpy. Later it was realized that full insertion of the metal complex into the helix is not a necessary condition for intercalation since only a planar aromatic ligand is sufficient [266]. Advantage can also be taken of additional non-intercalated ligands in designing complexes that bind to the helix with some degree of specificity. For example, discrimination can be based upon stereochemistry. Chiral discrimination was first observed for intercalation of M(phen)$_3^{2+}$ complexes in DNA helices [266] and this effect has then been extensively used for recognition and selective reactivity purposes [276]. Chiral metal complexes such as those shown in Fig. 10.12 are indeed

10.39   **10.40**

unique molecular probes for the DNA helix, which is also an asymmetric molecule. In the following, we shall mainly deal with the interaction of DNA with metal complexes.

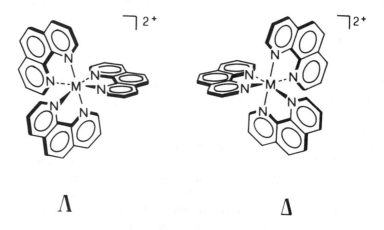

Λ Δ

Fig. 10.12 — Λ-and Δ-forms of chiral M(phen)$_3^{2+}$ complexes.

Among the various octahedral complexes that can be used in these studies, those containing Ru(II) and polypyridine ligands are the most suitable for the following reasons [309]:

(1) they are thermally inert towards both ligand substitution and racemization;
(2) they have intense metal-to-ligand charge-transfer bands which are quite sensitive to the local environment;
(3) their luminescence properties are very sensitive to the local environment as well as to dynamic (quenching) interactions;
(4) upon photoexcitation these complexes can exhibit strong oxidant and/or reductant properties;
(5) several hundreds of different ligands can be used, and both homoleptic and heteroleptic complexes can be synthesized rather easily;
(6) by systematic variation of the ligands, it is possible to tune both the factors which contribute to DNA binding affinity (geometry, size, hydrophobicity, hydrogen-bonding ability, π-stacking interaction, etc.) and the excited-state properties which constitute the basis for the probing and cleaving effects (absorption and emission bands, luminescence lifetimes, excited-state redox potentials, etc.);
(7) an increasing number of oligonuclear Ru-polypyridine complexes are currently synthesized and characterized [310]: because of their high ionic charge, large dimensions, and complex stereochemistry, such complexes may become useful in mapping larger portions of the DNA structure.

Voltammetric studies are also used to investigate the interaction of metal chelates with DNA [311].

10.6.3 Spectroscopy and photophysics

When Ru(II)-polypyridine complexes interact with DNA, more-or-less noticeable changes occur in their spectroscopic and photophysical properties. Typical data are collated in Table 10.5 [280,287]. Fig.10.13 shows the structural formulae of the ligands. In general, the metal-to-ligand charge-transfer bands in the visible region decrease in intensity and their maxima undergo blue or red shifts. The degree of hypochromism and the spectral shift generally correlate well with the overall binding strength. For $Ru(bpy)_2(biq)^{2+}$ and $Ru(bpy)(biq)_2^{2+}$ the Ru$\rightarrow$bpy CT bands are hardly affected, while the bands corresponding to the Ru$\rightarrow$biq transitions are subject to hypochromicity and a red shift, showing that intercalation involves the larger biq ligand [287]. The hypochromicity data for the complexes $Ru(bpy)_2(LL)^{2+}$ show that the interaction decreases along the ligand series HAT > biq = DIP > phen > bpy, indicating again that the dimensions of the ligand play an important role. No effect is observed for the neutral $Ru(phen)_2(CN)_2$ complex. This shows that ion–pair interaction is necessary for the stability of the adducts.

The luminescence of Ru(II)-polypyridine complexes is affected by interaction with DNA. In most cases, the emission intensity increases (even by a factor of 2 or 3), the bandwidth decreases, and the band maximum is shifted. Furthermore, the luminescence decay is strongly affected and becomes quite complicated. The results obtained depend on the experimental conditions and particularly on the presence of added salts since other cations can shield the DNA phosphate groups, thereby reducing the role played by electrostatic interactions in the binding [284,286]. The

Table 10.5 — Spectroscopic and photophysical properties of Ru(II)-polypyridine complexes and their adducts with DNA

Complex	Absorption λ_{max} (nm)		Emission λ_{max} (nm)		Emission $\tau(\mu s)$	
	Free	Bound	Free	Bound	Free	Bound
Ru(bpy)$_3^{2+}$ [a]	453	453	610	614	0.37	0.58(70%)
Ru(bpy)$_2$(biq)$^{2+}$ [a]	440,524	440,538	—	—	—	—
Ru(bpy)(biq)$_2^{2+}$ [a]	480,545	480,553	—	—	—	—
Ru(bpy)$_2$(HAT)$^{2+}$ [a]	430,480sh	408,450,510sh	716	710	0.10	0.33(100%)
Ru(phen)$_3^{2+}$ [b]	443	445	591	593	0.53	0.63, 2.3
Ru(phen)$_2$(DIP)$^{2+}$ [b]	427	432	614	606	0.97	1.2, 5.3
Ru(phen)(DIP)$_2^{2+}$ [b]	433	439	616	621	0.99	1.2, 5.1
Ru(phen)(bpy)$_2^{2+}$	450	453	610	615	0.42	1.1(75%)
Ru(phen)$_2$(bpy)$^{2+}$ [b]	446	448	608	604	0.56	0.53, 2.1
Ru(phen)$_2$(CN)$_2$ [a]	430	430	620	620	—	—
Ru(TAP)$_3^{2+}$ [a]	408	410	590	593	0.22	0.045(40%)[c]

[a]Air-saturated buffer solution (1 mM phosphate buffer, pH 7), 22°C [287]; [b]air-saturated buffer solution (50 mM NaCl, 5mM Tris, pH 7.5), 25°C [280]; [c] a very short-lived component (< 5 ns (50%)) is present in unbuffered solutions.

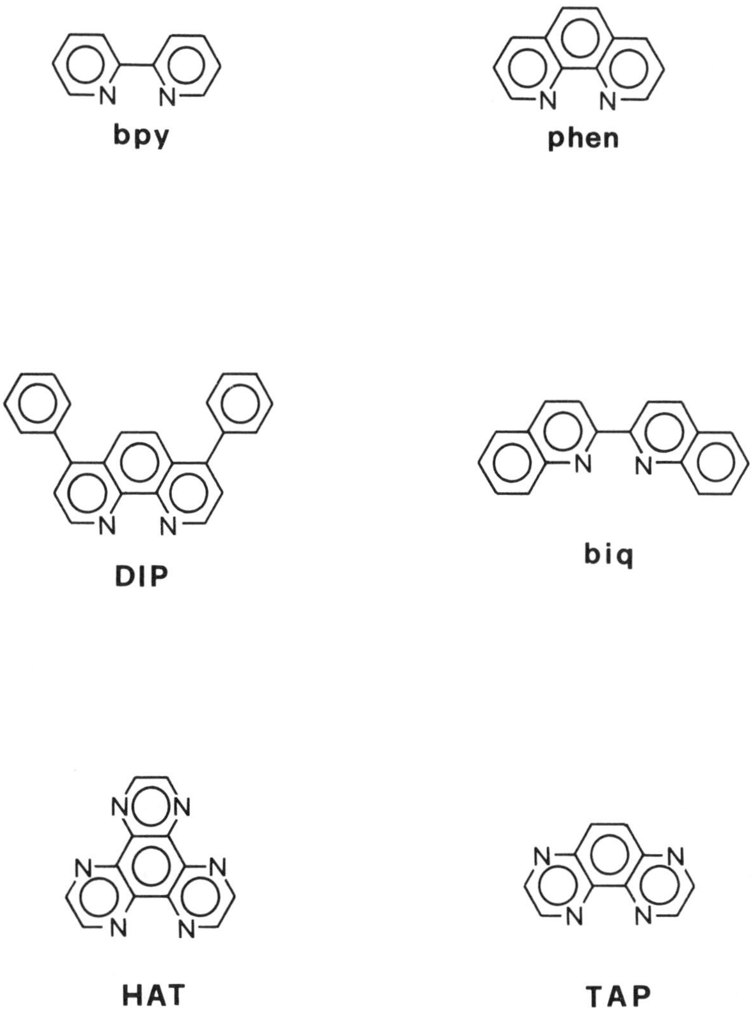

Fig. 10.13 — Structures of some polypyridine-type ligands.

red or blue shift of the emission band in the mixed-ligand complexes depends on whether or not the lowest (luminescent) excited state involves CT to the intercalated ligand [280]. Time-resolved emission experiments show that at least two emitting species are present [280,287]. At low ionic strength, a component which is longer-lived than the unbound excited state is responsible for the greater part of the emission intensity [287]. In most cases such a long-lived decay is thought to be due to species which are intercalated [280,287]. A fast component with a lifetime similar to that of the free complex, observed for Ru(phen)$_3^{2+}$ and Ru(DIP)$_3^{2+}$ in solutions containing some NaCl, has been attributed to surface-bound species [271]. The reduction in the luminescence intensity and the very short-lived (<20 ns) component observed for $Ru(TAP)_3^{2+}$ in the presence of DNA are related to the excited-state quenching by a redox process (*vide infra*). A minor decay component with lifetime shorter than the unbound complex, observed for other complexes, was attributed to a fraction of partially-quenched excited states [287,296]. Protection from oxygen quenching plays a major role in increasing the lifetime of the intercalated complexes. A similar protection from ferrocyanide, but not from Co^{2+}, is effective upon intercalation [274]. Intercalation has also been evidenced by emission polarization measurements [271], excited-state resonance Raman spectra [280], and electric dichroism spectra [292].

Luminescence and quenching experiments have demonstrated that the enantiomeric forms (Fig.10.12) of $Ru(phen)_3^{2+}$, and even more so those of $Ru(DIP)_3^{2+}$, possess a quite different affinity for the right-handed B-form of DNA [268,270,271,274]. Specifically, interaction with the Δ-$Ru(DIP)_3^{2+}$ enantiomer is much stronger than the interaction with the Λ-enantiomer, while a small enantioselectivity for the Λ-form has been observed for the left-handed Z-DNA. This enantiomeric specificity of binding may be understood in terms of the intercalation model and on the basis of its steric requirements [276]. Other interpretations, however, have also been advanced [301].

Detailed studies on the flexibility of DNA have been performed, making use of the time-resolved polarization technique and aromatic luminescent intercalators [299].

10.6.4 Photochemistry

Drug-dependent damage of DNA is a sensitive marker of binding, not only of the number of bound sites but also of their location along the strand. Since electron-transfer reactions offer a pathway for oxidative cleavage of the DNA strand, redox-active metal complexes are very useful in the design of specific cleaving agents [262,312–315]. Dervan and coworkers have cleverly used redox-induced cleavage to determine the sequence-specific binding sites of antitumour antibiotics [262,313–315]. Their technique is based on a supramolecular species, MPE.Fe(II) (Fig. 10.14), which consists of an intercalator (methidium ion) and a redox centre ($Fe^{II}(EDTA)$) connected by a short hydrocarbon tether [313]. The intercalator binds tightly to the duplex and delivers high concentrations of $Fe^{II}(EDTA)$ to the helix. Ferrous ions can then activate O_2 to yield radicals or iron-bound oxygen species which can cleave the sugar phosphate backbone. This metal reagent is a nonspecific DNA cleaver. However, at those points along the DNA strand occupied by a given drug or protein, cleavage is blocked and the sequence-specific binding sites can be determined ("footprinting" technique [262]). As a further step, reactivity has been coupled to recognition so as to cleave DNA in the vicinity of the binding site [262].

It was then realized that *light-induced* redox reactions of metal complexes could be used to cause DNA cleavage [290], and that one can take advantage of the conformation-specific recognition for metal-polypyridine complexes to photoinduce conformation-specific cleavage of DNA [269,276]. It was shown that the DNA intercalator $Co(phen)_3^{3+}$ is indeed a DNA cleaver, and that cleavage by $Co(DIP)_3^{3+}$ enantiomers is enantiospecific [269]. Thus, the chiral metal complex Λ-$Co(DIP)_3^{3+}$, which binds to the left-handed DNA-Z conformation, may be used to map specific cleavage sites and hence to learn something about the frequency of occurrence of the

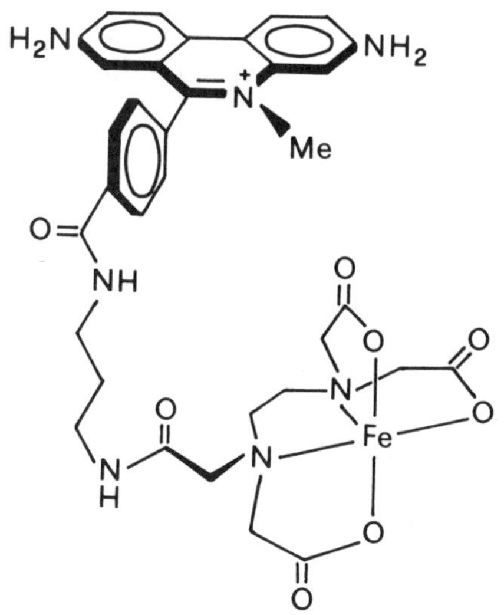

Fig. 10.14 Structure of the synthetic footprint tool MPE.Fe(II) [313].

conformationally distinct sites that it recognizes. A probe which is specific for the cleavage of the A-form helices, namely $Ru(TMP)_3^{2+}$, has been designed [272], and shape-selective targeting of DNA by Rh complexes containing the phenanthrenequinone diimine ligand has been achieved [281]. Sequence-directed photochemical damage has also been obtained using a $Ru(phen)_3^{2+}$ moiety linked to a short oligonucleotide [288].

The photoinduced cleavage of DNA by metal complexes is not limited to complexes which intercalate. In fact the quantum efficiency for cleavage is higher for $Co(NH_3)_6^{3+}$ than for $Co(phen)_3^{3+}$ [279]. $Co(en)_3^{3+}$ [279], $Rh(phen)_3^{3+}$ [279], and a variety of Ru(II)-polypyridine complexes [268,272,273,283,284,287–289] are able to cause light-induced cleavage, whereas $Cr(phen)_3^{3+}$ [279] is inactive. Organic aromatic cations (for example, 2,7-diazopyreniumdications [316]) are also efficient photocleavers of DNA.

The variety of photoactive metal complexes and of excited states involved (MC, CT, LC) point to the presence of different reaction mechanisms. A series of experiments based on the effect of D_2O and quenchers indicate that one possible mechanism is the sensitized production of singlet oxygen by the bound complexes which have a long-lived excited state [273,287,297]. However, it cannot be excluded that oxygen plays other roles such as generation of superoxide or hydroxyl radical, or merely trapping of DNA-centred radicals [280]. Cleavage is also very efficient for complexes like $Ru(bpy)_3^{2+}$ in the presence of the radical precursor $S_2O_8^{2-}$ [287,297].

Another mechanism is direct photooxidation of a DNA base by the photoexcited complex. This is the case for $Ru(TAP)_3^{2+}$, which is the most efficient cleaver among the Ru(II) complexes examined by Kelly and coworkers [284,287]. Evidence for this mechanism comes from the following:

(1) the potential of the $^*Ru(TAP)_3^{2+}/Ru(TAP)_3^+$ couple is positive enough to oxidize guanine;

(2) binding to DNA leads to a decrease in the luminescence intensity of $Ru(TAP)_3^{2+}$, while the opposite behaviour is observed for the other Ru(II) complexes;

(3) the excited-state lifetime for the bound complex is shorter than that for the free complex (Table 10.5);

(4) luminescence intensity and lifetime quenching is observed also for interaction with poly[d(G-C)] but not with poly[d(A-T)]. Direct demonstration of the excited-state electron-transfer mechanism has recently been obtained by flash photolysis experiments [289].

A further line of research in this field concerns the study of electron-transfer processes between donor and acceptor species interacting with DNA [275,279,293,317]. Along this line, Netzel and coworkers have synthesized DNA oligomers and duplexes containing a covalently-attached derivative of $Ru(bpy)_3^{2+}$ and have shown that DNA duplexes can be used as molecular scaffolds to construct macromolecules with specifically located, redox-active subunits [318]. Other luminescent labels have been covalently attached to DNA oligomers and duplexes [319,320], and label/label interactions have also been investigated.

10.6.5 Conclusions

Luminescence investigations performed with probes that exhibit specific structures are useful in recognizing local sites along the DNA strand. Luminescent and redox-active probes can combine selective recognition and selective photocleavage. The picture which emerges from the photocleavage experiments is certainly complicated, and much work will be needed to arrive at a complete rationalization. However, the phenomenon seems to be quite general, which indicates the possibility of tuning and selecting conditions for conformation-specific cleavage.

The great advantage of the Ru(II)-polypyridine complexes and related families lies in the large number of already available homo- and heteroleptic complexes which are luminescent and redox active [309], the relatively easy synthesis of other desired complexes, and recent progress in the design of polymetallic species [310] which have high positive charges and interesting structures (see, for example [321,322]). Future development, in fact, will probably concern the design of molecular species capable of giving rise to bis and tris intercalators. This would not only increase the binding strength, but also the specificity. The naturally-occurring quinoxaline antibiotic echinomycin is a bis intercalator and its modification has led to a highly A-T specific tandem [323]. Artificial specific cleavers based on bis intercalation have been designed [262] and photoactive bis intercalators have already been used [316]. The design of specific photocleavers based on the assembly of Ru(II)-polypyridine species, with or without other organic or inorganic moieties, is a promising research field.

REFERENCES

[1] Pedersen, C. J. (1967) *J. Am. Chem. Soc.* **89** 2495.

[2] Vögtle, F. (1982) *Topics Curr. Chem.* **101** 201.

[3] Lehn, J. M. (1973) *Struct. Bonding* **16** 1.

[4] Cram, D. J. and Cram, J. M. (1974) *Science* **183** 803.

[5] de Jong, F. and Reinhoudt, D. N. (1980) *Adv. Phys. Org. Chem.* **17** 279.

[6] Vögtle, F. (ed.) (1981) *Host–guest complex chemistry I. Topics Curr. Chem.* **98**.

[7] Vögtle, F. (ed.) (1982) *Host–guest complex chemistry II. Topics Curr. Chem.* **101**.

[8] Vögtle, F. and Weber, E. (eds) (1984) *Host–guest complex chemistry III. Topics Curr. Chem.* **121**.

[9] Meade, T. J. and Busch, D. H. (1985) *Prog. Inorg. Chem.* **33** 59.

[10] Colquhoun, H. M., Stoddart, J. F., and Williams, D. J. (1986) *Angew. Chem. Int. Ed. Engl.* **25** 487.

[11] Rebek, J., Jr. (1987) *Science* **235** 1478.

[12] Izatt, R. M. and Christensen, J. J. (eds) (1987) *Progress in macrocyclic chemistry*. Vol.3. Wiley.

[13] Diederich, F. (1988) *Angew. Chem. Int. Ed. Engl.* **27** 362.

[14] Lindoy, L. F. (1989) *The chemistry of macrocycles*. Cambridge University Press.

[15] Hancock, R. D. and Martell, A. E. (1989) *Chem. Rev.* **89** 1875.

[16] Gutsche, C. D. (1989) *Calixarenes*. Royal Society of Chemistry, Cambridge.

[17] Rebek, J., Jr. (1990) *Angew. Chem. Int. Ed. Engl.* **29** 245.

[18] Pedersen, C. J. (1988) *Angew. Chem. Int. Ed. Engl.* **27** 1021.

[19] Cram, D. J. (1988) *Angew. Chem. Int. Ed. Engl.* **27** 1009.

[20] Lehn, J. M. (1988) *Angew. Chem. Int. Ed. Engl.* **27** 89.

[21] Zimmerman, S. C., Vanzyl, C. M., and Hamilton, G. S. (1989) *J. Am. Chem. Soc.* **111** 1373.

[22] Gutsche, C. D., Alam, I., Iqbal, M., Mangiafico, T., Nam, K. C., Rogers, J., and See, K. A. (1989) *J. Inclusion Phenom.* **7** 61.

[23] Dijkstra, P. J., Brunink, J. A. J., Bugge, K. E., Reinhoudt, D. N., Harkema, S., Ungaro, R., Ugozzoli, F., and Ghidini, E. (1989) *J. Am. Chem. Soc.* **111** 7567.

[24] Menif, R., Chen, D., and Martell, A. E. (1989) *Inorg. Chem.* **28** 4633.

[25] van Veggel, F. C. J. M., Harkema, S., Bos, M., Verboom, W., van Staveren, C. J., Gerritsma, G. J., and Reinhoudt, D. N. (1989) *Inorg. Chem.* **28** 1133.

[26] Diederich, F. and Lutter, H. D. (1989) *J. Am. Chem. Soc.* **111** 8438.

[27] Vögtle, F., Wallon, A., Müller, W. M., Werner, U., and Nieger, M. (1990) *J. Chem. Soc., Chem. Commun.* 158.

[28] Schneider, H. J., Kramer, R., Theis, I., and Zhou, M. Q. (1990) *J. Chem. Soc., Chem. Commun.* 276.

[29] McKervey, M. A., Owens, M., Schulten, H. R., Vogt, W., and Böhmer, V. (1990) *Angew. Chem. Int. Ed. Engl.* **29** 280.

[30] Seward, E. M., Hopkins, R. B., Sauerer, W., Tam, S. W., and Diederich, F. (1990) *J. Am. Chem. Soc.* **112** 1783.

[31] Shinkai, S., Nishi, T., Ikeda, A., Matsuda, T., Shimamoto, K., and Manabe, O. (1990) *J. Chem. Soc., Chem. Commun.* 303.

[32] Beer, P. D. (1989) *Chem. Soc. Rev.* **18** 409.

[33] Beer, P. D., Blackburn, C., McAleer, J. F., and Sikanyta, H. (1990) *Inorg. Chem.* **29** 378.

[34] Bryant, J. A., Knobler, C. B., and Cram, D. J. (1990) *J. Am. Chem. Soc.* **112** 1254.

[35] Bryant, J. A., Ericson, J. L., and Cram, D. J. (1990) *J. Am. Chem. Soc.* **112** 1255.

[36] Saigo, K., Kihara, N., Hashimoto, Y., Lin, R. J., Fujimura, H., Suzuki, Y., and Hasegawa, M. (1990) *J. Am. Chem. Soc.* **112** 1144.

[37] Teixidor, F., Viñas, C., Rius, J., Miravitlles, C., and Casabó, J. (1990) *Inorg. Chem.* **29** 149.

[38] Vincenti, M., Dalcanale, E., Soncini, P., and Guglielmetti, G. (1990) *J. Am. Chem. Soc.* **112** 445.

[39] Wienk, M. M., Stolwijk, T. B., Sudhölter, E. J. R., and Reinhoudt, D. N. (1990) *J. Am. Chem. Soc.* **112** 777.

[40] Bajaj, A. V. and Poonia, N. S. (1988) *Coord. Chem. Rev.* **87** 55.

[41] Löhr, H. G. and Vögtle, F. (1985) *Acc. Chem. Res.* **18** 65.

[42] Takagi, M. and Ueno, K. (1984) *Topics Curr. Chem.* **121** 39.

[43] Shinkai, S. and Manabe, O. (1984) *Topics Curr. Chem.* **121** 67.

[44] van Gent, J., Sudhölter, E. J. R., Lambeck, P. V., Popma, T. J. A., Gerritsma, G. J., and Reinhoudt, D. N. (1988) *J. Chem. Soc., Chem. Commun.* 893.

[45] Misumi, S. and Kaneda, T. (1989) *J. Inclusion Phenom.* **7** 83.

[46] Cram, D. J., Carmack, R. A., and Helgeson, R. C. (1988) *J. Am. Chem. Soc.* **110** 571.

[47] Fages, F., Desvergne, J. P., Bouas-Laurent, H., Marsau, P., Lehn, J. M., Kotzyba-Hibert, F., Albrecht-Gary, A. M., and Al-Joubbeh, M. (1989) *J. Am. Chem. Soc.* **111** 8672.

[48] Dix, J. P. and Vögtle, F. (1980) *Chem. Ber.* **113** 457.

[49] Dix, J. P. and Vögtle, F. (1981) *Chem. Ber.* **114** 638.

[50] Löhr, H. G. and Vögtle, F. (1985) *Chem. Ber.* **118** 905.

[51] Löhr, H. G. and Vögtle, F. (1985) *Chem. Ber.* **118** 915.

[52] Bourson, J. and Valeur, B. (1989) *J. Phys. Chem.* **93** 3871.

[53] Bauer, H., Briaire, J., and Staab, H. A. (1983) *Angew. Chem. Int. Ed. Engl.* **22** 334.

[54] Beer, P. D. and Rothin, A. S. (1988) *J. Chem. Soc., Chem. Commun.* 52.

[55] Hollmann, G. and Vögtle, F. (1984) *Chem. Ber.* **117** 1355.

[56] Narayanaswamy, R. and Sevilla, F. (1986) *Analyst* **111** 1085, and references therein.

[57] Narayanaswamy, R. and Sevilla, F. (1986) *J. Opt. Sensors* **1** 403, and references therein.

[58] Janata, J. (1989) *Principles of chemical sensors.* Plenum.

[59] Alder, J. F., Ashworth, D. C., Narayanaswamy, R., Moss, R. E., and Sutherland, I. O. (1987) *Analyst* **112** 1191.

[60] Sousa, L. R. and Larson, J. M. (1977) *J. Am. Chem. Soc.* **99** 307.

[61] Sousa, L. R. and Larson, J. M. (1978) *J. Am. Chem. Soc.* **100** 1943.

[62] Ghosh, S., Petrin, M., Maki, A. H., and Sousa, L. R. (1987) *J. Chem. Phys.* **87** 4315.

[63] Shizuka, H., Takada, K., and Morita, T. (1980) *J. Phys. Chem.* **84** 994.

[64] Shirai, M. and Tanaka, M. (1988) *J. Chem. Soc., Chem. Commun.* 381.

[65] Street, K. W., Jr. and Krause, S. A. (1986) *Anal. Lett.* **19** 735.

[66] Nakashima, K., Nagaoka, Y., Nakatsuji, S., Kaneda, T., Tanigawa, I., Hirose, K., Misumi, S., and Akiyama, S. (1987) *Bull. Chem. Soc. Japan* **60** 3219.

[67] de Silva, A. P. and Sandanayake, K. R. A. S. (1989) *J. Chem. Soc., Chem. Commun.* 1183.

[68] Huston, M. E., Akkaya, E. U., and Czarnik, A. W. (1989) *J. Am. Chem. Soc.* **111** 8735.

[69] de Silva, A. P. and Gunaratne, H. Q. N. (1990) *J. Chem. Soc., Chem. Commun.* 186.

[70] Fery-Forgues, S., Le Bris, M. T., Guetté, J. P., and Valeur, B. (1988) *J. Chem. Soc., Chem. Commun.* 384.

[71] Fery-Forgues, S., Le Bris, M. T., Guetté, J. P., and Valeur, B. (1988) *J. Phys. Chem.* **92** 6233.

[72] Tundo, P. and Fendler, J. H. (1980) *J. Am. Chem. Soc.* **102** 1760.

[73] Herrmann, U., Tummler, B., Maass, G., Koo Tze Mew, P., and Vögtle, F. (1984) *Biochemistry* **23** 4059.

[74] de B. Costa, S. M., Queimado, M. M., and da Silva, J. J. R. F. (1980) *J. Photochem.* **12** 31.

[75] Guerriero, P., Vigato, P. A., Bünzli, J. C., and Moret, E. (1990) *J. Chem. Soc., Dalton Trans.* 647.

[76] Bouas-Laurent, H., Castellan, A., Daney, M., Desvergne, J. P., Guinand, G., Marsau, P., and Riffaud, M. H. (1986) *J. Am. Chem. Soc.* **108** 315.

[77] Desvergne, J. P. and Bouas-Laurent, H. (1978) *J. Chem. Soc., Chem. Commun.* 403.

[78] Fages, F., Desvergne, J. P., and Bouas-Laurent, H. (1989) *J. Am. Chem. Soc.* **111** 96.

[79] Fages, F., Desvergne, J. P., Bouas-Laurent, H., Hinschberger, J., and Marsau, P. (1988) *New J. Chem.* **12** 95.

[80] Thanabal, V. and Krishnan, V. (1982) *J. Am. Chem. Soc.* **104** 3643.

[81] Blondeel, G., Harriman, A., Porter, G., and Wilowska, A. (1984) *J. Chem. Soc., Faraday Trans. 2* **80** 867.

[82] Lehn, J. M., (1987). In Balzani, V. (ed.) *Supramolecular photochemistry.* Reidel, p. 29.

[83] Gubelmann, M., Harriman, A., Lehn, J. M., and Sessler, J. L. (1990) *J. Phys. Chem.* **94** 308.

[84] Jao, T. C., Beddard, G. S., Tundo, P., and Fendler, J. H. (1981) *J. Phys. Chem.* **85** 1963.

[85] Balzani, V., Ballardini, R., Gandolfi, M. T., and Prodi, L. (1990) In Schneider, H. J. and Dürr, H. (eds) *Frontiers in supramolecular organic chemistry and photochemistry.* Verlag Chemie, in press.

[86] Balzani, V., Sabbatini, N., and Scandola, F. (1986) *Chem. Rev.* **86** 319.

[87] Stoddart, J. F. and Zarzycki, R. (1990) In Inoue, Y. and Gokel, G. W. (eds) *Cation binding by macrocycles: complexation of cationic species by crown ethers*, in press.

[88] Colquhoun, H. M. and Stoddart, J. F. (1981) *J. Chem. Soc., Chem. Commun.* 612.

[89] Cram, D. J. and Cram, M. J. (1978) *Acc. Chem. Res.* **11** 8.

[90]   Trueblood, K. N., Knobler, C. B., Lawrence, D. S., and Stevens, R. V. (1982) *J. Am. Chem. Soc.* **104** 1355.

[91]   Colquhoun, H. M., Stoddart, J. F., and Williams, D. J. (1981) *J. Chem. Soc., Chem. Commun.* 847.

[92]   Ballardini, R., Gandolfi, M. T., Balzani, V., Kohnke, F. H., and Stoddart, J.F. (1988) *Angew. Chem. Int. Ed. Engl.* **27** 692.

[93]   Ballardini, R., Gandolfi, M. T., Prodi, L., Ciano, M., Balzani, V., Kohnke, F. H., Shahriari-Zavareh, H., Spencer, N., and Stoddart, J. F. (1989) *J. Am. Chem. Soc.* **111** 7072.

[94]   Colquhoun, H. M., Stoddart, J. F., Williams, D. J., Wolstenholme, J. B., and Zarzycki, R. (1981) *Angew. Chem. Int. Ed. Engl.* **20** 1051.

[95]   Colquhoun, H. M., Doughty, S. M., Maud, J. M., Stoddart, J. F., Williams, D. J., and Wolstenholme, J. B. (1985) *Isr. J. Chem.* **25** 5.

[96]   Ballardini, R., Gandolfi, M. T., Prodi, L., Zappi, T., Balzani, V., Spencer, N., and Stoddart, J. F., to be published.

[97]   Prodi, L., Ballardini, R., Gandolfi, M. T., Balzani, V., Desvergne, J. P., and Bouas-Laurent, H., *J. Phys. Chem.* in press.

[98]   Izatt, R. M., Lam, J. D., Swain, C. S., Christensen, J. J., and Haymore, B. L. (1980) *J. Am. Chem. Soc.* **102** 3032.

[99]   Shizuka, H., Kameta, K., and Shinozaki, T. (1985) *J. Am. Chem. Soc.* **107** 3956.

[100] Shizuka, H. (1985) *Acc. Chem. Res.* **18** 141.

[101] Shizuka, H. and Serizawa, M. (1986) *J. Phys. Chem.* **90** 4573.

[102] Colquhoun, H. M., Goodings, E. P., Maud, J. M., Stoddart, J. F., Wolstenholme, J. B., and Williams, D. J. (1985) *J. Chem. Soc., Perkin Trans. 2* 607.

[103] Allwood, B. L., Colquhoun, H. M., Doughty, S. M., Kohnke, F. H., Slawin, A. M. Z., Stoddart, J. F., Williams, D. J., and Zarzycki, R. (1987) *J. Chem. Soc., Chem. Commun.* 1054.

[104] Stoddart, J. F. (1988) *Pure Appl. Chem.* **60** 467.

[105] Allwood, B. L., Shahriari-Zavareh, H., Stoddart, J. F., and Williams, D. J. (1987) *J. Chem. Soc., Chem. Commun.* 1058.

[106] Malini, R. and Krishnan, V. (1980) *J. Phys. Chem.* **84** 551.

[107] Malini, R. and Krishnan, V. (1984) *Spectrochim. Acta* **40A** 323.

[108] Fabbrizzi, L. (1985) *Comments Inorg. Chem.* **4** 33.

[109] Henrick, K. and Tasker, P. A. (1985) *Prog. Inorg. Chem.* **33** 1.

[110] Chaudhuri, P. and Wieghardt, K. (1987) *Prog. Inorg. Chem.* **35** 329.

[111] Lappin, A. G. and McAuley, A. (1988) *Adv. Inorg. Chem.* **32** 241.

[112] Anelli, P. L., Montanari, F., and Quici, S. (1988) *J. Org. Chem.* **53** 5292.

[113] Hancock, R. D., Dobson, S. M., Evers, A., Wade, P. W., Ngwenya, M. P., Boeyens, J. C. A., and Wainwright, K. P. (1988) *J. Am. Chem. Soc.* **110** 2788.

[114] Krakowiak, K. E., Bradshaw, J. S., and Zamecka-Krakowiak, D. J. (1989) *Chem. Rev.* **89** 929.

[115] Jazwinski, J., Lehn, J. M., Lilienbaum, D., Ziessel, R., Guilhem, J., and Pascard, C. (1987) *J. Chem. Soc., Chem. Commun.* 1691.

[116] Bencini, A., Bianchi, A., Garcia-España, E., Micheloni, M., and Paoletti, P. (1988) *Inorg. Chem.* **27** 176.

[117] Fenton, D. E. and Vigato, P. A. (1988) *Chem. Soc. Rev.* **17** 69.

[118] Micheloni, M. (1988) *Comments Inorg. Chem.* **8** 79.

[119] Motekaitis, R. J., Martell, A. E., Murase, I., Lehn, J. M., and Hosseini, M. W. (1988) *Inorg. Chem.* **27** 3630.

[120] Bencini, A., Bianchi, A., Castello, M., Dapporto, P., Faus, J., Garcia-España, E., Micheloni, M., Paoletti, P., and Poli, P. (1989) *Inorg. Chem.* **28** 3175.

[121] Christodoulou, D., Kanatzidis, M. G., and Coucouvanis, D. (1990) *Inorg. Chem.* **29** 191.

[122] Hosseini, M. W. and Lehn, J. M. (1987) *J. Am. Chem. Soc.* **109** 7047.

[123] Kaden, T. A. (1984) *Topics Curr. Chem.* **121** 157.

[124] Ciampolini, M., Fabbrizzi, L., Perotti, A., Poggi, A., Seghi, B., and Zanobini, F. (1987) *Inorg. Chem.* **26** 3527.

[125] Hosseini, M. W. and Lehn, J. M. (1988) *Helv. Chim. Acta.* **71** 749.

[126] De Santis, G., Di Casa, M., Mariani, M., Seghi, B., and Fabbrizzi, L. (1989) *J. Am. Chem. Soc.* **111** 2422.

[127] Dietrich, B., Hosseini, M. W., Lehn, J. M., and Sessions, R. B. (1983) *Helv. Chim. Acta.* **66** 1262.

[128] Dietrich, B., Hosseini, M. W., Lehn, J. M., and Sessions, R. B. (1981) *J. Am. Chem. Soc.* **103** 1282.

[129] Peter, F., Gross, M., Hosseini, M. W., and Lehn, J. M. (1983) *J. Electroanal. Chem.* **144** 279.

[130] Bianchi, A., Micheloni, M., Orioli, P., Paoletti, P., and Mangani, S. (1988) *Inorg. Chim. Acta* **146** 153.

[131] Bencini, A., Bianchi, A., Garcia-España, E., Giusti, M., Mangani, S., Micheloni, M., Orioli, P., and Paoletti, P. (1987) *Inorg. Chem.* **26** 3902.

[132] Bernhardt, P. V., Lawrence, G., Skelton, B. W., and White, A. H. (1989) *Aust. J. Chem.* **42** 1035.

[133] Ogawa, S., Narushima, R., and Arai, Y. (1984) *J. Am. Chem. Soc.* **106** 5760.

[134] Kimura, E., Wada, S., Shionoya, M., Takahashi, T., and Iitaka, Y. (1990) *J. Chem. Soc., Chem. Commun.* 397.

[135] Gouterman, M. (1978) In Dolphin, D. (ed.) *The porphyrins*. Vol. 3. Academic, p. 1.

[136] Balzani, V. and Carassiti, V. (1970) *Photochemistry of coordination compounds.* Academic.

[137] Manfrin, M. F., Moggi, L., Castelvetro, V., Balzani, V., Hosseini, M. W., and Lehn, J. M. (1985) *J. Am. Chem. Soc.* **107** 6888.

[138] Pina, F., Moggi, L., Manfrin, M. F., Balzani, V., Hosseini, M. W., and Lehn, J. M. (1989) *Gazz. Chim. Ital.* **119** 65.

[139] Schlaepfer, C. W. and von Zelewsky, A. (1990) *Comments Inorg. Chem.* **9** 181.

[140] Vanquickenborne, L. G., Hendrickx, M., Hyla-Kryspin, I., and Haspeslagh, L. (1986) *Inorg. Chem.* **25** 885, and references therein.

[141] Odashima, K., Itai, A., Iitaka, Y., and Koga, K. (1980) *J. Am. Chem. Soc.* **102** 2504.

[142] Soga, T., Odashima, K., and Koga, K. (1980) *Tetrahed. Lett.* 4351.

[143] Cramer, F., Saenger, W., and Spatz, H. Ch. (1967) *J. Am. Chem. Soc.* **89** 14.

[144] Diederich, F. and Dick, K. (1984) *J. Am. Chem. Soc.* **106** 8024.
[145] Diederich, F., Dick, K., and Griebel, D. (1986) *J. Am. Chem. Soc.* **108** 2273.
[146] Vögtle, F., Müller, W. M., Werner, U., and Losensky, H. W. (1987) *Angew. Chem. Int. Ed. Engl.* **26** 901.
[147] Anderson, H. L., Hunter, C. A., and Sanders, J. K. M. (1989) *J. Chem. Soc., Chem. Commun.* 226.
[148] Allwood, B. L., Kohnke, F. H., Slawin, A. M. Z., Stoddart, J. F., and Williams, D. J. (1985) *J. Chem. Soc., Chem. Commun.* 311.
[149] Allwood, B. L., Kohnke, F. H., Stoddart, J. F., and Williams, D. J. (1985) *Angew. Chem. Int. Ed. Engl.* **24** 581.
[150] Allwood, B. L., Spencer, N., Shahriari-Zavareh, H., Stoddart, J. F., and Williams, D. J. (1987) *J. Chem. Soc., Chem. Commun.* 1061.
[151] Allwood, B. L., Spencer, N., Shahriari-Zavareh, H., Stoddart, J. F., and Williams, D. J. (1987) *J. Chem. Soc., Chem. Commun.* 1064.
[152] Ashton, P. R., Slawin, A. M. Z., Spencer, N., Stoddart, J. F., and Williams, D. J. (1987) *J. Chem. Soc., Chem. Commun.* 1066.
[153] Slawin, A. M. Z., Spencer, N., Stoddart, J. F., and Williams, D. J. (1987) *J. Chem. Soc., Chem. Commun.* 1070.
[154] Odell, B., Reddington, M. V., Slawin, A. M. Z., Spencer, N., Stoddart, J. F., and Williams, D. J. (1988) *Angew. Chem. Int. Ed. Engl.* **27** 1547.
[155] Buhner, M., Geuder, W., Gries, W. K., Hünig, S., Koch, M., and Poll, T. (1988) *Angew. Chem. Int. Ed. Engl.* **27** 1553.
[156] Ashton, P. R., Odell, B., Reddington, M. V., Slawin, A. M. Z., Stoddart, J. F., and Williams, D. J. (1988) *Angew. Chem. Int. Ed. Engl.* **27** 1550.
[157] Ortholand, J. Y., Slawin, A. M. Z., Spencer, N., Stoddart, J. F., and Williams, D. J. (1989) *Angew. Chem. Int. Ed. Engl.* **28** 1394.
[158] Dahenens, M., Lacombe, L., Lehn, J. M., and Vigneron, J. P. (1984) *J. Chem. Soc., Chem. Commun.* 1097.
[159] Bottino, F., Di Grazia, M., Finocchiaro, P., Fronczek, F.R., Mamo, A., and Pappalardo, S. (1988) *J. Org. Chem.* **53** 3521.
[160] Vögtle, F., Ostrowicki, A., Knops, P., Fischer, P., Reuter, H., and Jansen, M. (1989) *J. Chem. Soc., Chem. Commun.* 1757.
[161] Tunstad, L. M., Tucker, J. A., Dalcanale, E., Weiser, J., Bryant, J. A., Sherman, J. C., Hegelson, R. C., Knobler, C. B., and Cram, D. J. (1989) *J. Org. Chem.* **54** 1305.
[162] Canceill, J., Cesario, M., Collet, A., Guilhem, J., Lacombe, L., Lozach, B., and Pascard, C. (1989) *Angew. Chem. Int. Ed. Engl.* **28** 1246.
[163] Benson, D. R., Valentekovich, R., and Diederich, F. (1990) *Angew. Chem. Int. Ed. Engl.* **29** 191.
[164] Saenger, W. (1980) *Angew. Chem. Int. Ed. Engl.* **19** 344.
[165] Szejtli, J. (1982) *Cyclodextrins and their inclusion complexes.* Akademiai Kiado, Budapest.
[166] Matsui, Y., Nishioka, T., and Fujita, T. (1985) *Topics Curr. Chem.* **128** 61.
[167] Tabushi, I. (1986) *Pure Appl. Chem.* **58** 1529.
[168] Fornasier, R., Reniero, F., Scrimin, P., and Tonellato, U. (1987) *J. Chem. Soc., Perkin Trans. 2* 1121.

[169] Matsui, Y., Fujie, M., and Hanaoka, K. (1989) *Bull. Chem. Soc. Japan* **62** 1451.

[170] Breslow, R. (1983) *Chem. Brit.* **19** 126.

[171] Tabushi, I. (1982) *Acc. Chem. Res.* **15** 66.

[172] D'Souza, V. T. and Bender, M. L. (1987) *Acc. Chem. Res.* **20** 146.

[173] Rao, R. R., Srinivasan, T. N., Bhanumathi, N., and Sattur, P. B. (1990) *J. Chem. Soc. Chem., Commun.* 10.

[174] Stoddart, J. F. and Zarzycki, R. (1988) *Rec. Trav. Chim. Pays-Bas* **107** 515, and references therein.

[175] Vögtle, F. and Müller, W. M. (1979) *Angew. Chem. Int. Ed. Engl.* **18** 623.

[176] Vögtle, F., Sieger, H., and Müller, W. M. (1981) *Topics Curr. Chem.* **98** 107.

[177] Kamitori, S., Hirotsu, K., and Higuchi, T. (1987) *J. Am. Chem. Soc.* **109** 2409.

[178] Kamitori, S., Hirotsu, K., and Higuchi, T. (1988) *Bull. Chem. Soc. Japan*, **61** 3825.

[179] Yamanari, K., Nakamichi, M., and Shimura, Y. (1989) *Inorg. Chem.* **28** 248.

[180] Kalianasundaram, K. (1987) *Photochemistry in microheterogeneous systems.* Academic, Chapter 9.

[181] Eaton, D. F. (1987) *Tetrahed.* **43** 1551.

[182] Ramamurthy, V. and Eaton, D. F. (1988) *Acc. Chem. Res.* **21** 300.

[183] Tran, C. D. and Fendler, J. H. (1984) *J. Phys. Chem.* **88** 2167.

[184] Politzer, I. R., Crago, K. T., Hampton, T., Joseph, J., Boyer, J. H., and Shah, M. (1989) *Chem. Phys. Lett.* **159** 258.

[185] Agbaria, R. A. and Gill, D. (1988) *J. Phys. Chem.* **92** 1052.

[186] Scypinski, S. and Drake, J. M. (1985) *J. Phys. Chem.* **89** 2432.

[187] Kusumoto, Y. (1987) *Chem. Phys. Lett.* **136** 535.

[188] Agbaria, R. A., Uzan, B., and Gill, D. (1989) *J. Phys. Chem.* **93** 3855.

[189] Hamai, S. (1989) *J. Phys. Chem.* **93** 2074.

[190] Nelson, G. and Warner, I. M. (1990) *J. Phys. Chem.* **94** 576.

[191] Ueno, A., Takahashi, K., Hino, Y., and Osa, T. (1981) *J. Chem. Soc., Chem. Commun.* 194.

[192] Nelson, G., Patonay, G., and Warner, I. M. (1989) *Talanta* **36** 199.

[193] Scypinski, S. and Cline Love, L. J. (1984) *Anal. Chem.* **56** 322, 331.

[194] Alak, A. M. and Vo-Dinh, T. (1988) *Anal. Chem.* **60** 596.

[195] Hamai, S. (1989) *J. Am. Chem. Soc.* **111** 3954.

[196] Alak, A. M., Contolini, N., and Vo-Dinh, T. (1989) *Anal. Chim. Acta* **217** 171.

[197] Femia, R. A. and Cline Love, L. J. (1985) *J. Phys. Chem.* **89** 1897.

[198] Cline III, J. I., Dressick, W. J., Demas, J. N., and DeGraff, B. A. (1985) *J. Phys. Chem.* **89** 94.

[199] Himanen, J. P. and Korpela, T. (1986) *J. Inclusion Phenom.* **4** 177.

[200] Turro, N. J., Okubo, T., and Chung, C. J. (1982) *J. Am. Chem. Soc.* **104** 1789.

[201] Reitz, G. A., Demas, J. N., DeGraff, B. A., and Stephens, E. M. (1988) *J. Am. Chem. Soc.* **110** 5051.

[202] Turro, N. J., Cox, G. S., and Li, X. (1983) *Photochem. Photobiol.* **37** 149.

[203] Turro, N. J., Bolt, J. D., Kuroda, Y., and Tabushi, I. (1982) *Photochem. Photobiol.* **35** 69.

[204] Lee, E. D., Werner, T. C., and Seitz, W. R. (1987) *Anal. Chem.* **59** 279.

[205] Cox, G. S. and Turro, N. J. (1984) *Photochem. Photobiol.* **40** 185.

[206] Patonay, G., Shapira, A., Diamond, P., and Warner, I. M. (1986) *J. Phys. Chem.* **90** 1963, and references therein.

[207] Kobayashi, N., Saito, R., Hino, H., Hino, Y., Ueno, A., and Osa, T. (1983) *J. Chem. Soc., Perkin Trans. 2* 1031, and references therein.

[208] Herkstroeter, W. G., Martic, P. A., Evans, T. R., and Farid, S. (1986) *J. Am. Chem. Soc.* **108** 3275.

[209] Kano, K., Matsumoto, H., Hashimoto, S., Sisido, M., and Imanishi, Y. (1985) *J. Am. Chem. Soc.* **107** 6117.

[210] Kano, K., Matsumoto, H., Yoshimura, Y., and Hashimoto, S. (1988) *J. Am. Chem. Soc.* **110** 204.

[211] Hamai, S. (1989) *J. Phys. Chem.* **93** 6527.

[212] Ueno, A., Suzuki, I., and Osa, T. (1989) *J. Am. Chem. Soc.* **111** 6391.

[213] Turro, N. J., Okubo, T., and Weed, G. C. (1982) *Photochem. Photobiol.* **35** 325.

[214] Itoh, M. and Fujiwara, Y. (1984) *Bull. Chem. Soc. Japan* **57** 2261.

[215] Arad-Yellin, R. and Eaton, D. F. (1983) *J. Phys. Chem.* **87** 5051.

[216] Netto-Ferreira, J. C. and Scaiano, J. C. (1988) *J. Photochem. Photobiol. A: Chemistry.* **45** 109.

[217] Ueno, A., Moriwaki, F., Osa, T., Hamada, F., and Murai, K. (1988) *J. Am. Chem. Soc.* **110** 4323.

[218] Ueno, A., Tomita, Y., and Osa, T. (1983) *J. Chem. Soc., Chem. Commun.* 1515.

[219] Ueno, A., Moriwaki, F., Osa, T., Hamada, F., and Murai, K. (1986) *Bull. Chem. Soc. Japan* **59** 465.

[220] Moriwaki, F., Ueno, A., Osa, T., Hamada, F., and Murai, K. (1986) *Chem. Lett.* 1865.

[221] Moriwaki, F., Kaneko, H., Ueno, A., Osa, T., Hamada, F., and Murai, K. (1987) *Bull. Chem. Soc. Japan* **60** 3619.

[222] Ueno, A., Suzuki, I., and Osa, T. (1988) *J. Chem. Soc., Chem. Commun.* 1373.

[223] Hamada, F., Murai, K., Ueno, A., Suzuki, I., and Osa, T. (1988) *Bull. Chem. Soc. Japan* **61** 3758.

[224] Ueno, A., Suzuki, I., and Osa, T. (1989) *Chem. Lett.* 1059.

[225] Suzuki, I., Ueno, A., and Osa, T. (1989) *Chem. Lett.* 2013.

[226] Kano, K., Hashimoto, S., Imai, A., and Ogawa, T. (1984) *J. Inclusion Phenom.* **2** 737.

[227] Bright, F. V., Catena, G. C., and Huang, J. (1990) *J. Am. Chem. Soc.* **112** 1343.

[228] Cox, G. S., Turro, N. J., Yang, N. C., and Chen, M. J. (1984) *J. Am. Chem. Soc.* **106** 422.

[229] DeLuccia, F. J. and Cline Love, L. J. (1984) *Anal. Chem.* **56** 2811.

[230] DeLuccia, F. J. and Cline Love, L. J. (1985) *Talanta* **32** 665.

[231] Gonzales, M. C. and Weedon, A. C. (1985) *Canad. J. Chem.* **63** 602.

[232] Neckers, D. C. and Paczkowski, J. (1986) *J. Am. Chem. Soc.* **108** 291.

[233] Ueno, A., Moriwaki, F., Tomita, Y., and Osa, T. (1985) *Chem. Lett.* 493.

[234] Wang, Y. and Eaton, D. F. (1985) *Chem. Phys. Lett.* **120** 441.
[235] Eaton, D. F., Anderson, A. G., Tam, W., and Wang, Y. (1987) *J. Am. Chem. Soc.* **109** 1886.
[236] Tamaki, T., Kokubu, T., and Ichimura, K. (1987) *Tetrahed.* **43** 1485.
[237] Chung, W. S., Turro, N. J., Silver, J., and le Noble, W. J. (1990) *J. Am. Chem. Soc.* **112** 1202.
[238] Rao, V. P. and Turro, N. J. (1989) *Tetrahed. Lett.* **30** 4641.
[239] Courtney, S. H. and Fleming, G. R. (1985) *J. Chem. Phys.* **83** 215.
[240] Syamala, M. S., Devanathan, S., and Ramamurthy, V. (1986) *J. Photochem.* **34** 219.
[241] Duveneck, G. L., Sitzmann, E. V., Eisenthal, K. B., and Turro, N. J. (1989) *J. Phys. Chem.* **93** 7166.
[242] Bortolus, P. and Monti, S. (1987) *J. Phys. Chem.* **91** 5046.
[243] Rau, H. and Lüddecke, E. (1982) *J. Am. Chem. Soc.* **104** 1616.
[244] Ueno, A., Tomita, Y., and Osa, T. (1983) *Tetrahed. Lett.* **24** 5245.
[245] Ueno, A., Yoshimura, H., Saka, R., and Osa, T. (1979) *J. Am. Chem. Soc.* **101** 2779.
[246] Ueno, A., Takahashi, K., and Osa, T. (1981) *J. Chem. Soc., Chem. Commun.* 94.
[247] Monti, S., Flamigni, L., Martelli, A., and Bortolus, P. (1988) *J. Phys. Chem.* **92** 4447.
[248] Rao, B. N., Turro, N. J., and Ramamurthy, V. (1986) *J. Org. Chem.* **51** 460.
[249] Nag, A., Dutta, R., Chattopadhyay, N., and Bhattacharyya, K. (1989) *Chem. Phys. Lett.* **157** 83.
[250] Nag, A. and Bhattacharyya, K. (1990) *J. Chem. Soc., Faraday Trans.* **86** 53.
[251] Hicks, J. M., Vandersall, M. T., Sitzmann, E. V., and Eisenthal, K. B. (1987) *Chem. Phys. Lett.* **135** 413.
[252] Wubbels, G. G., Stevetson, B. R., and Kaganove, S. N. (1986) *Tetrahed. Lett.* **27** 3103.
[253] Yumoto, T., Hayakawa, K., Kowase, K., Yamakita, H., and Taoda, H. (1985) *Chem. Lett.* 1021.
[254] Liu, J. H. and Weiss, R. G. (1985) *J. Photochem.* **30** 303.
[255] Adar, E., Degani, Y., Goren, Z., and Willner, I. (1986) *J. Am. Chem. Soc.* **108** 4696.
[256] Willner, I., Adar, E., Goren, Z., and Steinberger, B. (1987) *New J. Chem.* **11** 769.
[257] Willner, I. and Mandler, D. (1989) *J. Am. Chem. Soc.* **111** 1330.
[258] Willner, I., Eichen, Y., and Frank, A. J. (1989) *J. Am. Chem. Soc.* **111** 1884.
[259] Yonemura, H., Nakamura, H., and Matsuo, T. (1989) *Chem. Phys. Lett.* **155** 157.
[260] Ramamurthy, V., Caspar, J. V., Corbin, D. R., and Eaton, D. F. (1990) *J. Photochem. Photobiol. A: Chemistry* **51** 259.
[261] Lerman, L. S. (1961) *J. Mol. Biol.* **3** 18.
[262] Dervan, P. B. (1986) *Science* **232** 464.
[263] Barton, J. K. (1988) *Chem. Eng. News.* (Sept. 26), 30.
[264] Barton, J. K. (1985) *Comments Inorg. Chem.* **3** 321.
[265] Remers, W. A. (1984) *Antineoplastic agents.* Wiley.

[266] Barton, J. K., Dannenberg, J. J., and Raphael, A. L. (1982) *J. Am. Chem. Soc.* **104** 4967.
[267] Barton, J. K. (1983) *J. Biol. Struct. Dyn.* **1** 621.
[268] Barton, J. K., Danishefsky, A. T., and Goldberg, J. M. (1984) *J. Am. Chem. Soc.* **106** 2172.
[269] Barton, J. K. and Raphael, A. L. (1984) *J. Am. Chem. Soc.* **106** 2466.
[270] Barton, J. K., Basile, L. A., Danishefsky, A. T., and Alexandrescu, A. (1984) *Proc. Natl. Acad. Sci. USA* **81** 1961.
[271] Kumar, C. V., Barton, J. K., and Turro, N. J. (1985) *J. Am. Chem. Soc.* **107** 5518.
[272] Mei, H. Y. and Barton, J. K. (1986) *J. Am. Chem. Soc.* **108** 7414.
[273] Fleisher, M. B., Waterman, K. C., Turro, N. J., and Barton, J. K. (1986) *Inorg. Chem.* **25** 3549.
[274] Barton, J. K., Goldberg, J. M., Kumar, C. V., and Turro, N. J. (1986) *J. Am. Chem. Soc.* **108** 2081.
[275] Barton, J. K., Kumar, C. V., and Turro, N. J. (1986) *J. Am. Chem. Soc.* **108** 6391.
[276] Barton, J. K. (1986) *Science* **233** 727.
[277] Basile, L. A. and Barton, J. K. (1987) *J. Am. Chem. Soc.* **109** 7548.
[278] Basile, L. A., Raphael, A., and Barton, J. K. (1987) *J. Am. Chem. Soc.* **109** 7550.
[279] Purugganan, M. D., Kumar, C. V., Turro N. J., and Barton, J. K. (1988) *Science* **241** 1645.
[280] Pyle, A. M., Rehmann, J. P., Meshoyer, R., Kumar, C. V., Turro, N. J., and Barton, J. K. (1989) *J. Am. Chem. Soc.* **111** 3051.
[281] Pyle, A. M., Long, E. C., and Barton, J. K. (1989) *J. Am. Chem. Soc.* **111** 4520.
[282] Kirsch-De Mesmaeker, A., Orellana, G., Barton, J. K., and Turro, N. J. (1990) *Photochem. Photobiol.* **52** 461.
[283] Kelly, J. M., Tossi, A. B., McConnell, D. J., and OhUigin, C. (1985) *Nucl. Acids Res.* **13** 6017.
[284] Kelly, J. M., McConnell, D. J., OhUigin, C., Tossi, A. B., Kirsch-De Mesmaeker, A., Masschelein, A., and Nasielski, J.(1987) *J. Chem. Soc., Chem. Commun.* 1821.
[285] Blau, W. J., Croke, D. T., Kelly, J. M., McConnell, D. J., OhUigin, C., and van der Putten, W. J. M. (1987) *J. Chem. Soc., Chem. Commun.* 751.
[286] Görner H., Tossi, A. B., Stradowsky, C., and Schulte-Frohlinde, D. (1988) *J. Photochem. Photobiol. B: Biol.* **2** 67.
[287] Tossi, A. B. and Kelly, J. M. (1989) *Photochem. Photobiol.* **49** 545.
[288] Kelly, J. M., Tossi, A. B., McConnell, D. J., OhUigin, C., Helene, C., and Le Doan, T. (1989) In Beaumont, P. C., Deeble, D. J., Parsons, B. J., and Rice-Evans, C. (eds) *Free radicals, metal ions and biopolymers*. Richelieu Press, London, p. 143.
[289] Kelly, J. M., Feeney, M. M., Tossi, A. B., Lecomte, J. P., and Kirsch-De Mesmaeker, A. (1990) *Anti-Cancer Drug Design* **5** 69.
[290] Chang, C. H. and Meares, C. F. (1982) *Biochemistry* **21** 6332.
[291] Yamagishi, A. (1983) *J. Chem. Soc., Chem. Commun.* 572.

[292] Yamagishi, A. (1984) *J. Phys. Chem.* **88** 5709.

[293] Subramanian, R. and Meares, C. F. (1986) *J. Am. Chem. Soc.* **108** 6429.

[294] Lamos, M. L., Lobenstine, E. W., and Turner, D. H. (1986) *J. Am. Chem. Soc.* **108** 4278.

[295] Fujimoto, B. S. and Schurr, J. M. (1987) *J. Phys. Chem.* **91** 1947.

[296] Stradowski, C., Görner, H., Currel, L. J., and Schulte-Frohlinde, D. (1987) *Biopolymers* **26** 189.

[297] Aboul-Enein, A. and Schulte-Frohlinde, D. (1988) *Photochem. Photobiol.* **48** 27.

[298] Tamilarasan, R., Ropartz, S., and McMillin, D. R. (1988) *Inorg. Chem.* **27** 4082.

[299] Härd, T., Fan, P., Madge, D., and Kearns, D. R. (1989) *J. Phys. Chem.* **93** 4338.

[300] Farrow, S. J., Mohammad, T., Baird, W., and Morrison, H. (1990) *Photochem. Photobiol.* **51** 263.

[301] Hiort, C., Nordén, B., and Rodger, A. (1990) *J. Am. Chem. Soc.* **112** 1971.

[302] Zimmerman, H.W. (1986) *Angew. Chem. Int. Ed. Engl.* **25** 115.

[303] Gale, E. F., Cundliffe, E., Reynolds, P. E., Richmond, M. H., and Waring, M. (1972) *The molecular basis of antibiotic action.* Wiley.

[304] Lippard, S. J., Bond, P. J., Wu, K. C., and Bauer, W. R. (1976) *Science* **194** 726.

[305] Lippard, S. J. (1978) *Acc. Chem. Res.* **11** 211.

[306] Lippard, S. J. (1982) *Science* **218** 1075.

[307] Sherman, S. E. and Lippard, S. J. (1987) *Chem. Rev.* **87** 1153.

[308] Sundquist, W. I., Bancroft, D. P., and Lippard, S. J. (1990) *J. Am. Chem. Soc.* **112** 1590.

[309] Juris, A., Balzani, V., Barigelletti, F., Campagna, S., Belser, P., and von Zelewsky, A. (1988) *Coord. Chem. Rev.* **84** 85.

[310] Scandola, F., Indelli, M.T., Chiorboli, C., and Bignozzi, C. A. *Topics Curr. Chem.,* **158** in press.

[311] Carter, M. T., Rodriguez, M., and Bard, A. J. (1989) *J. Am. Chem. Soc.* **111** 8901.

[312] Reich, K. A., Marshall, L. E., Graham, D. R., and Sigman, D. S. (1981) *J. Am. Chem. Soc.* **103** 3582.

[313] Hertzberg, R. P. and Dervan, P. B. (1982) *J. Am. Chem. Soc.* **104** 313.

[314] Van Dyke, M. W., Hertzberg, R. P., and Dervan, P. B. (1982) *Proc. Natl. Acad. Sci. USA* **79** 5470.

[315] Van Dyke, M. W. and Dervan, P. P. (1983) *Biochemistry* **22** 2372.

[316] Blacker, A. J., Jazwinski, J., Lehn, J. M., and Wilhelm, F. X. (1986) *J. Chem. Soc., Chem. Commun.* 1035.

[317] Fromherz, P. and Rieger, B. (1986) *J. Am. Chem. Soc.* **108** 5361.

[318] Telser, J., Cruickshank, K. A., Schanze, K. S., and Netzel, T. L. (1989) *J. Am. Chem. Soc.* **111** 7221.

[319] Telser, J., Cruickshank, K. A., Morrison, L. E., Netzel, T. L., and Chan, C. K. (1989) *J. Am. Chem. Soc.* **111** 7226.

[320] Telser, J., Cruickshank, K. A., Morrison, L. E., and Netzel, T. L. (1989) *J. Am. Chem. Soc.* **111** 6966.

[321] Denti, G., Campagna, S., Sabatino, L., Serroni, S., Ciano, M., and Balzani, V. (1990) *Inorg. Chim. Acta* **176** 175.

[322] De Cola, L., Belser, P., Ebmeyer, F., Barigelletti, F., Vögtle, F., von Zelewsky, A., and Balzani, V. (1990) *Inorg. Chem.* **29** 495.

[323] Atwell, G. A., Leupin, W., Twigden, S. J., and Denny, W. A. (1983) *J. Am. Chem. Soc.* **105** 2913.

11

Other systems

11.1 CAGED METAL IONS

11.1.1 Introduction

For a variety of fundamental studies and practical applications there is a need for molecules that, when excited, are able to give luminescence (*luminophores*) and/or to transfer energy or electrons (*photosensitizers*) without undergoing photo-decomposition. From a spectroscopic and photophysical point of view, several coordination compounds would be quite suitable to play these roles [1–9]. In most cases, however, such compounds do not exhibit sufficient chemical stability.

A way to remedy this drawback is to link the ligands together so as to make a cage around the metal ion. As discussed in section 4.4.1, this effect may be considered a supramolecular perturbation that does not change the composition and symmetry of the first coordination sphere of the metal ion (and thus the spectroscopic properties), but prevents processes that require extensive nuclear motions such as ligand dissociation or radiationless decay *via* strongly distorted structures [10].

11.1.2 Cobalt complexes

It is well known that Co(III) complexes are kinetically inert while Co(II) complexes are very labile because of the presence of electrons in the $\sigma_M^*(e_g)$ antibonding orbitals [11]. As a consequence, reduction of the metal ion of cobalt(III) amine complexes in aqueous solution causes a fast and complete decomposition of the complex. For $Co(NH_3)_6^{3+}$, pulse radiolysis experiments [12] show that ligand dissociation occurs in successive steps and is completed in the microsecond time scale:

$$Co(NH_3)_6^{3+} + e_{aq}^- \rightarrow Co(NH_3)_6^{2+} \tag{11.1a}$$

$$Co(NH_3)_6^{2+} \xrightarrow{H_3O^+} Co_{aq}^{2+} + 6NH_4^+ \qquad k > 10^3 \, s^{-1} \tag{11.1b}$$

Simple chelation does not substantially change the situation since $Co(en)_3^{3+}$ (en = ethylenediamine) after pulse radiolysis reduction loses its ligands with $k \geqslant 25\,s^{-1}$ [12]. The extreme lability of the Co(II) species accounts for the efficient photodecomposition of $Co(NH_3)_6^{3+}$ upon intramolecular ligand-to-metal charge transfer (LMCT) photoexcitation ($\Phi = 0.16$) [13]

$$Co^{III}(NH_3)_6^{3+} \overset{h\nu}{\rightleftarrows} Co^{II}(NH_3)_5(NH_3^+)^{3+} \qquad (11.2a)$$

$$Co^{II}(NH_3)_5(NH_3^+)^{3+} \overset{H_3O^+}{\rightarrow} Co_{aq}^{2+} + 5NH_4^+ + products \qquad (11.2b)$$

and ion-pair charge transfer (IPCT) photoexcitation ($\Phi = 0.2$) [13]:

$$Co^{III}(NH_3)_6^{3+}\ldots I^- \overset{h\nu}{\rightleftarrows} Co^{II}(NH_3)_6^{2+}\ldots I \qquad (11.3a)$$

$$Co^{II}(NH_3)_6^{2+}\ldots I \overset{H_3O^+}{\rightarrow} Co_{aq}^{2+} + 6NH_4^+ + 1/2I_2 \qquad (11.3b)$$

The photodecomposition of $Co(en)_3^{3+}$ upon LMCT excitation occurs with comparable quantum yield [13].

In 1977 Sargeson and coworkers [14] reported the capping reaction of $Co(en)_3^{3+}$ along its C_3 axis with NH_3 and CH_2O (Fig. 11.1). The encapsulated metal ion so

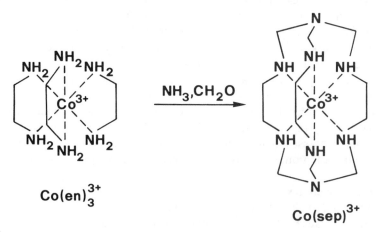

Fig. 11.1 — Capping reaction of $Co(en)_3^{3+}$ to obtain $Co(sep)^{3+}$ [14].

obtained, (1,3,6,8,10,13,16,19-octaazabicyclo-[6,6,6]-icosane)-cobalt(III) ion, was called cobalt(III) sepulchrate, $Co(sep)^{3+}$. Since then, a great number of similar cage-type complexes of a variety of metal ions have been synthesized and studied by Sargeson's group [15–24].

Co(sep)$^{3+}$ and analogous cage-type complexes exhibit several quite interesting properties. From the photochemical point of view, the most important consequence of encapsulation is the kinetic inertness of the reduced species. In fact, formation of Co$_{aq}^{2+}$ cannot be observed on one-electron reduction of Co(sep)$^{3+}$ in pulse radiolysis experiments (except in strong acid solution) [25]:

$$\text{Co(sep)}^{3+} + e_{aq}^- \rightarrow \text{Co(sep)}^{2+} \tag{11.4a}$$

$$\text{Co(sep)}^{2+} \not\rightarrow \text{Co}_{aq}^{2+} + \text{products} \qquad k < 10^{-6}\,\text{s}^{-1} \tag{11.4b}$$

Such a kinetic inertness prevents photodecomposition ($\Phi < 10^{-6}$) on intramolecular LMCT excitation (11.5) and also on IPCT excitation (11.6) [26]:

$$\text{Co}^{III}(\text{sep})^{3+} \overset{h\nu}{\rightleftarrows} \text{Co}^{II}(\text{sep}^+)^{3+} \tag{11.5a}$$

$$\text{Co}^{II}(\text{sep}^+)^{3+} \not\rightarrow \text{Co}_{aq}^{2+} + \text{products} \tag{11.5b}$$

$$\text{Co}^{III}(\text{sep}^+)^{3+} \dots \text{I}^- \overset{h\nu}{\rightleftarrows} \text{Co}^{II}(\text{sep})^{2+} \dots \text{I} \tag{11.6a}$$

$$\text{Co}^{II}(\text{sep})^{2+} \dots \text{I} \not\rightarrow \text{Co}_{aq}^{2+} + \text{products} \tag{11.6b}$$

In spite of their completely different photochemical behaviour, Co(NH$_3$)$_6^{3+}$ (or Co(en)$_3^{3+}$) and Co(sep)$^{3+}$ have quite similar absorption spectra. These differences and similarities between caged and uncaged hexaamine complexes can be explained on the basis of the potential energy curves shown in Fig. 11.2 (see also section 4.4.1 and Fig. 4.5). Encapsulation does not substantially modify the composition and symmetry of the first coordination sphere which is made up of the ligand atoms surrounding the metal ion. Thus, the potential energy curves for the caged and uncaged complexes in the spectroscopic region are nearly the same, which accounts for the strong similarity in the absorption spectra. The photodecomposition reaction of Co(NH$_3$)$_6^{3+}$ on LMCT (or IPCT) excitation is due to dissociation along the LMCT (or IPCT) potential energy curve in competition with radiationless decay to the ground state (Fig. 11.2). Encapsulation, however, introduces nuclear constraints to large-amplitude nuclear motions and particularly to the dissociation of the coordinated units, which are kept in almost fixed positions by the pattern of covalent bonds. Therefore, for the sepulchrate complex the dissociative branch of the LMCT (or IPCT) potential energy curve is replaced by an upward steep curve (Fig. 11.2) so that ligand detachment can no longer compete with radiationless decay to the ground state [10].

Because of the inertness of their one-electron reduction products, Co(sep)$^{3+}$ and other Co(III) cage complexes can be used as relays for hydrogen generation from water [27–33]. From a supramolecular point of view, the most interesting cases are those of the species **11.1** and **11.2** in which a cobalt cage complex is covalently linked

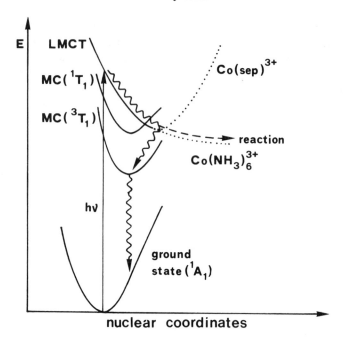

Fig. 11.2 — Schematic potential energy curves evidencing some differences and similarities between caged and uncaged Co(III) complexes [10].

to an anthracene moiety [32,33]. The proposed mechanism [33] for hydrogen

11.1

11.2

production in the presence of EDTA and colloidal Pt involves the following steps: (1) excitation of the anthracene chromophore; (2) reductive quenching of the singlet excited state of anthracene by EDTA in competition with radiative and radiationless decay, with formation of the reduced cage complex (possibly, *via* electron transfer from the anthracene anion to the cobalt complex); (3) reduction of H^+ to H_2 by the Co(II) complex, mediated by the Pt catalyst. It was also shown that the efficiency of the nonradiative decay (presumably *via* energy transfer from the anthracene singlet excited state to the cobalt complex) decreases with increasing length of the bridge [33]. The stability of the reduced forms of the cage cobalt complexes may also be exploited to generate a cage radical and thus to obtain dimeric cage complexes [17]. Interesting results have been obtained on irradiation of the ion pairs given by $Co(sep)^{3+}$ with a variety of anions [26,34–36] (see also section 8.4.3.1). For the $Co(sep)^{3+}\ldots I^-$ ion pair, excitation in the IPCT bands leads to no net reaction (11.6) [26]. However, when aerated aqueous solutions of the $Co(sep)^{3+}\ldots I^-$ ion pair are irradiated in the presence of methyl isobutyl ketone (which is a good solvent for I_2 and does not mix with water) a net photoreaction (11.7) is observed with $\Phi = 10^{-3}–10^{-2}$, depending on the experimental conditions [26]. As is better shown

$$4I^- + O_2 + 4H^+ \quad \xrightarrow[\text{Co(sep)}^{3+}]{h\nu} \quad 2I_2 + 2H_2O \tag{11.7}$$

shown in Fig. 11.3, in this system $Co(sep)^{3+}$ plays the role of an electron-transfer photosensitizer.

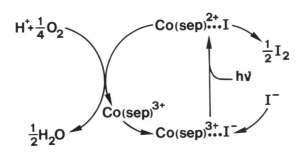

Fig. 11.3 — Photosensitized oxidation of I^- by O_2 mediated by $Co(sep)^{3+}$ [26].

A similar but perhaps more interesting case is that of the $Co^{III}(sep)^{3+}\ldots C_2O_4H^-$ ion pair [34], which exhibits an IPCT band with a maximum at 275 nm. Excitation in the IPCT spectral region in deoxygenated solutions causes the reduction of $Co^{III}(sep)^{3+}$ to $Co^{II}(sep)^{2+}$ and the oxidation of oxalate ions ($\Phi = 0.1–0.3$). This is due to the fast decomposition of the primary oxidation product of $C_2O_4H^-$, which competes with the back electron-transfer reaction. When colloidal platinum is present in the solution, no net reduction of Co(sep)$^{3+}$ takes place and, besides carbon dioxide, dihydrogen also evolves from the solution (11.8).

$$C_2O_4H^- + H^+ \xrightarrow[\text{Co(sep)}^{3+}]{hv} 2CO_2 + H_2 \tag{11.8}$$

The Co(sep)$^{3+}$ complex plays again the role of an electron-transfer photosensitizer (Fig. 11.4). The turnover number of Co(sep)$^{3+}$ is higher than 700 at pH 3 and the quantum yield for H$_2$ evolution is around 0.1 [34].

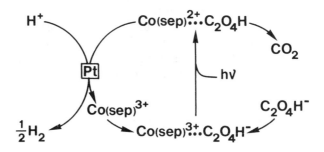

Fig. 11.4 — Photosensitized evolution of H$_2$ from water based on the excitation of Co(sep)$^{3+}$ ... C$_2$O$_4$H$^-$ ion pairs [34].

11.1.3 Chromium complexes

Cr(III) complexes are known to undergo ligand substitution and luminescence emission when excited in the $^4T_2$ and $^4T_1$ MC bands [13]. While emission originates from the lowest spin-forbidden ^2E level, there is controversy regarding the pathway which leads to the ligand-substitution reaction [37–40]. The ^2E level and the ground state belong to the same electronic configuration and thus they exhibit nearly the same equilibrium nuclear coordinates and vibrational frequencies. In other words, there is no nuclear reorganization in passing from the $^4A_2$ ground state to the ^2E excited state. By contrast, the $^4T_2$ and $^4T_1$ excited states are expected to be strongly distorted with respect to the ground-state geometry because of the presence of a σ^* antibonding electron. The relative positions of the $^4A_2$, ^2E, and $^4T_2$ potential energy curves along a metal-ligand stretching coordinate is shown in Fig. 11.5.

As discussed above, encapsulation is not expected to affect the spectroscopic properties. This expectation is substantially confirmed by the data reported in Table 11.1, where the spectroscopic and photophysical properties of the uncaged Cr(NH$_3$)$_6^{3+}$ and Cr(en)$_3^{3+}$ complexes are compared with those of the caged complexes **11.3** (Y=H, Cr(sar)$^{3+}$; Y=NH$_2$, Cr(diamsar)$^{3+}$) [41,42]. As one can see, Cr(sar)$^{3+}$ and Cr(diamsar)$^{3+}$ are structurally very similar to Cr(sep)$^{3+}$, whose reported synthesis [43] has been questioned [41,42]. The most noticeable differences

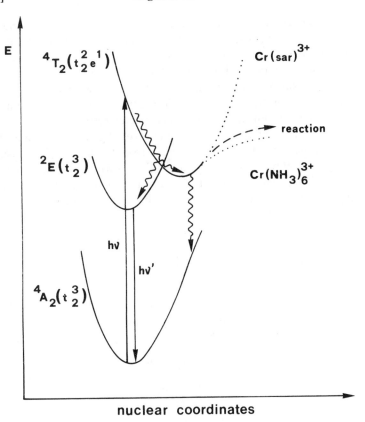

Fig. 11.5 — Schematic potential energy curves evidencing some differences and similarities between caged and uncaged Cr(III) complexes [10].

Table 11.1 — Comparison between uncaged and caged Cr(III) complexes[a]

	Absorption, 298 K E^{max}, cm^{-1} (ε, M^{-1} cm^{-1})		Em. 77 K $E^{(0)}$, cm^{-1}	$\tau(^2E)$ μs		Photodec. 298 K
	$^4T_2$	$^4T_1$	^2E	77 K	298 K	Φ_d
$Cr(NH_3)_6^{3+}$	21 790(45)	28 490(72)	15 200[b]	—	17[b,c]	0.45
$Cr(en)_3^{3+}$	21 930(38)	28 330(63)	15 000[b]	22	1.5	0.37
$Cr(sar)^{3+}$	21 860(153) 22 250(156)	28 820(91)	14 590	60	<0.01	2×10^{-5}
$Cr(diamsar)^{3+}$	21 950(203) 22 350(208)	28 940(109)	14 520	65	<0.01	2×10^{-5}

[a] From [41] and references therein; aqueous solutions unless otherwise noted; [b] DMF solution; [c] 273 K.

in the absorption spectra of the uncaged and caged compounds are the small splitting of the $^4T_2$ band and the higher extinction coefficients observed for the cage compounds, caused by their lower symmetry. Encapsulation is also seen to have little effect on the low-temperature lifetime of the ^2E excited state, which is again a "vertical" property since it is determined essentially by a weak coupling mechanism with high frequency vibrations. Encapsulation, however, has a very strong effect on the photoaquation quantum yields. This is an expected result because the constraints to nuclear motions imposed by encapsulation prevent ligand detachment, no matter whether the photoreaction takes place directly from the distorted $^4T_2$ excited state (Fig. 11.5), or from vibrationally-activated ^2E via surface crossing to $^4T_2$ or to a chemical intermediate. The strong temperature-dependence of the lifetime of ^2E for the cage-type complexes does not seem to have a straightforward explanation [41,42].

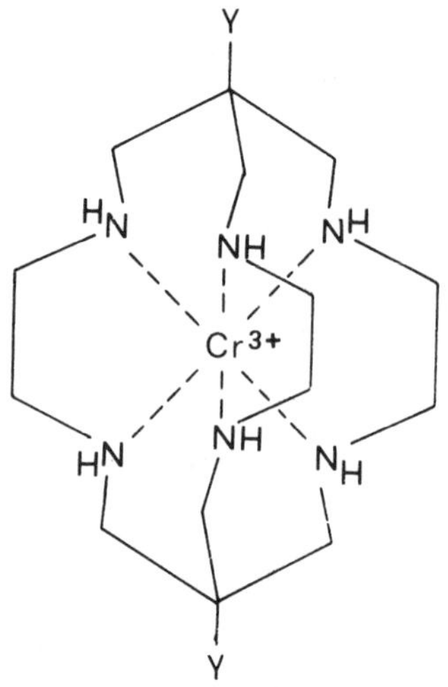

11.3

11.1.4 Lanthanide complexes

Some lanthanide ions, particularly Eu^{3+} and Tb^{3+}, possess strongly luminescent and long-lived excited states. For example, the lowest excited state of Eu^{3+}_{aq}, $^5D_0$, lives 3.2 ms and emits with efficiency 0.8 in D_2O solutions [44]. These ions however, are very poor light absorbers. The strongest absorption band of Eu^{3+}_{aq} in the near ultraviolet and visible region occurs at 393 nm with a molar absorption coefficient lower than $3 M^{-1} cm^{-1}$. To make use of the excellent emitting properties of these ions one has to overcome the difficulty in populating their excited states, i.e. to compensate for the lack of intense absorption bands. Complexation of the luminescent lanthanide ion with suitable ligands is, of course, a possibility. Complexation would also shield the lanthanide ion from interaction with water molecules, thereby decreasing the rate of radiationless decay of the luminescent excited state, which takes place mainly via coupling with the high energy O–H vibrations [45]. Unfortunately, however, lanthanide ions do not exhibit strong coordinating ability [11], and to give rise to well-defined and stable complexes they must be enclosed into cage-type ligands like the 2.2.1 [46] and the bpy.bpy.bpy [47] cryptands. The [Eu ⊂ 2.2.1]$^{3+}$ complex **11.4** exhibits LMCT absorption bands with $\varepsilon \sim 100\,M^{-1}cm^{-1}$ in the ultraviolet region (Fig. 11.6). The greatly improved absorption capacity, how-

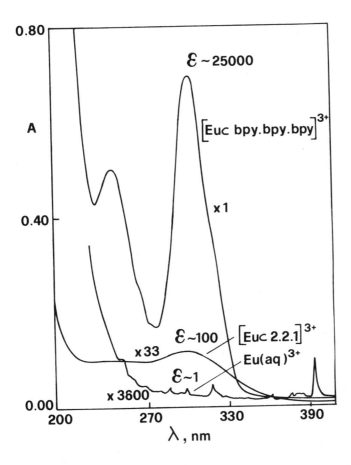

Fig. 11.6 — Absorption spectra of the Eu^{3+} aquo ion and of the Eu^{3+} cryptates **11.4** and **11.5** [50].

ever, is counterbalanced by the small efficiency of conversion (0.6%) of the originally populated CT states to the luminescent *f–f* level [48]. A much better result is obtained when Eu^{3+} is enclosed in the bpy.bpy.bpy cryptand (**11.5**). This cryptand, in fact, is made by bpy (2,2'-bipyridine) units that exhibit very strong absorption bands in the near ultraviolet region (Fig. 11.6). Excitation of [Eu ⊂ bpy.bpy.bpy]$^{3+}$ to the spin-allowed $^1\pi\pi^*$ LC level (Fig. 11.7a) leads to population of

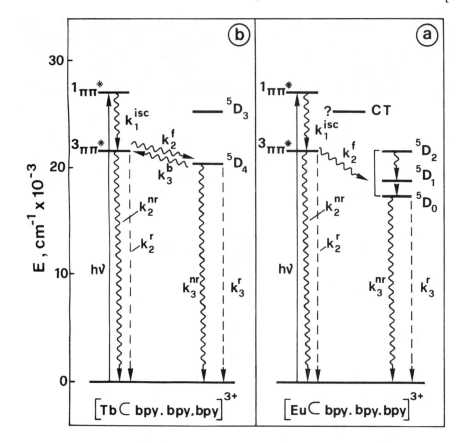

Fig. 11.7 — Schematic energy level diagrams for the Eu^{3+} cryptate **11.5** and for the analogous Tb^{3+} compound [52].

the luminescent metal-centred (MC) $^5D_0$ *f–f* level of Eu^{3+} with reasonably high

11.4   **11.5**

efficiency (10%), with the result that such a complex is an efficient molecular device for the conversion of ultraviolet light absorbed by the ligands into visible light emitted by the metal ion (antenna effect, section 12.5.2). Even in very dilute aqueous solution (10^{-5}M), this complex is able to convert about 1% of the incident ultraviolet photons into emitted visible photons, which is an interesting result for several applications, in particular for labelling biological materials [49,50]. The less-than-unity conversion efficiency is attributed to the presence of CT levels which offer to the $^1\pi\pi^*$ level a deactivation path which goes directly to the ground state [51,52]. In the analogous Tb^{3+} complex (Fig. 11.7b), such CT levels are not present because Tb^{3+} is very difficult to reduce. As a consequence, the efficiency of population of the luminescent $^5D_4$ level upon excitation to $^1\pi\pi^*$ is much higher ($\sim 100\%$), but the $^5D_4$ level (which lies $2250\,cm^{-1}$ higher than the $^5D_0$ Eu^{3+} level) can undergo radiationless decay *via* thermal activation back to $^3\pi\pi^*$, so that the luminescence yield upon $^1\pi\pi^*$ excitation is smaller than in the case of the Eu^{3+} complex [52]. Both the Eu^{3+} and Tb^{3+} complexes have also been investigated in the solid state [53,54].

Preliminary data on Eu^{3+} cryptates containing heterocyclic *N*-oxides have been reported. The complex of the ligand **11.6** is quite stable in aqueous solution and its luminescence quantum yield is 0.016 [55]. The complex of a cryptand incorporating a 3,3'-biisoquinoline 2,2'-dioxide has $\Phi_{em} = 0.04$ and $\tau = 640\,\mu s$ in acetonitrile solution [56].

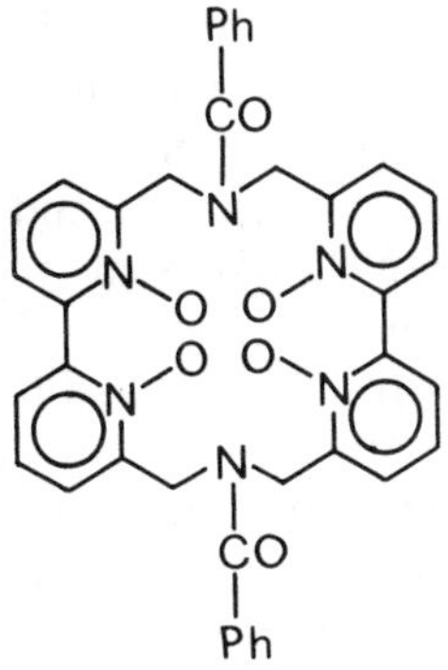

11.6 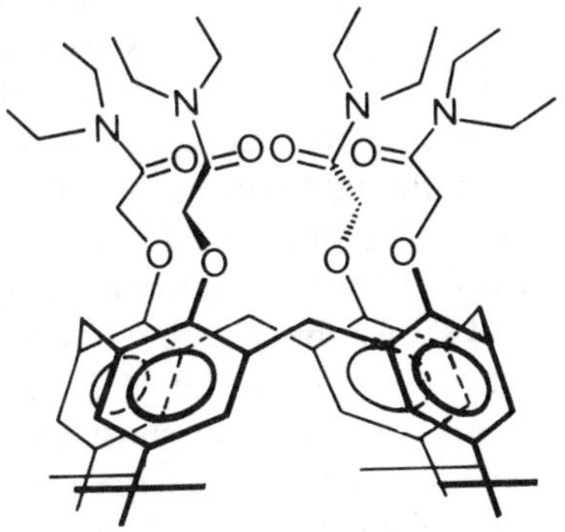   **11.7**

Another family of potentially encapsulating ligands is that of calixarenes (cyclic oligomers made up of benzene units) [57,58]. Following Cram's principle [59] that preorganization of the binding sites leads to much stronger complexes, ether or amide derivatives of calix[4]arene with fixed "cone" structures and convergent binding chains have been synthesized, where cations of suitable size can be encapsulated in a cavity of eight oxygen atoms [60–62]. The Eu^{3+}, Tb^{3+} and Gd^{3+} complexes of *p-t*-butyl-calix[4]arene tetracetamide **11.7** have been investigated [63]. The free ligand shows a phosphorescence band with a maximum at 420 nm and $\tau = 1.8$ s in a rigid matrix at 77 K. Such a band is also observable in the Gd^{3+} complex ($\lambda_{max} = 410$ nm, $\tau = 69$ ms), but not in the Eu^{3+} and Tb^{3+} complexes, because of ligand-to-metal energy transfer. In fact, for the last two complexes excitation in the ligand-centred band at ~ 280 nm in aqueous solution at room temperature causes the characteristic Eu^{3+} and Tb^{3+} emissions. The luminescence quantum yield of the Eu^{3+} complex is very low (2×10^{-4}), presumably because of deactivation of the $^1\pi\pi^*$ ligand level and/or of the luminescent $^5D_0$ metal level *via* CT levels (the tail of a CT band is observable in the absorption spectrum above 300 nm). For the Tb^{3+} complex the luminescence quantum yield is very high (0.20) because low-energy CT levels are not present and the large energy gap between $^3\pi\pi^*$ and $^5D_4$ prevents radiationless decay of $^5D_4$ *via* thermal activation back to $^3\pi\pi^*$. Luminescence lifetime measurements in H_2O and D_2O indicate that only one water molecule is coordinated to the metal ion in these calixarene complexes. Because of its high luminescence quantum yield, long excited-state lifetime (1.5 ms in H_2O at room temperature), and high molar absorption coefficient in the near ultraviolet region, the Tb^{3+} complex is an interesting label for time-resolved fluoroimmunoassay.

The europium complexes of *p-t*-butyl-calix[6]arene and *p-t*-butyl-calix[8]arene prepared by Harrowfield *et al.* [64] were characterized in the solid state as dimethyl-formamide (DMF) solvates. They contain two metal ions, two ligands, and several DMF molecules. Some preliminary results concerning their luminescence behaviour have been reported.

11.1.5 Ruthenium complexes

In the last ten years Ru(II)-polypyridine complexes have attracted the attention of several research groups because of a unique combination of ground- and excited-state properties [4–7,65]. The prototype of these complexes is the well-known $Ru(bpy)_3^{2+}$ that is extensively used as (1) a photoluminescent compound, (2) an excited-state reactant in energy and electron-transfer processes, (3) an excited-state product in chemiluminescent and electrochemiluminescent reactions, and (4) a mediator in the interconversion of light and chemical energy [7,65]. Some relevant data concerning this complex are summarized in the scheme of Fig. 5.13 [7,8]. Comparison with the requirements needed for photosensitizers [8,66] and lumino-phores [9] shows that the main drawbacks of $Ru(bpy)_3^{2+}$ are (i) the relatively fast radiationless decay of the ^{3}CT excited state to the ground state (with, as a consequence, a relatively short excited-state lifetime and a small luminescence efficiency), and (ii) the occurrence of a ligand photosubstitution reaction whose efficiency is strongly dependent on the experimental conditions (from $\sim 10^{-5}$ in water at room temperature to 10^{-1} in CH_2Cl_2 solutions containing Cl^- ions). It is generally agreed [7,67,68] that the ligand photosubstitution reaction proceeds (Fig. 11.8) *via* a thermally-activated radiationless transition from the luminescent ^{3}CT level to a distorted ^{3}MC level (a), with subsequent competition between radiation-less decay to the ground state (b), and cleavage of a Ru–N bond with formation of an intermediate containing a monodentate bpy ligand (c). Such an intermediate can undergo either loss of bpy (d), or chelate ring closure with reformation of $Ru(bpy)_3^{2+}$ (e).

If the bpy ligands are linked together to make a cage around the ruthenium ion, ligand photodissociation can be prevented [10]. Furthermore, a cage ligand can also confer more rigidity to the molecule, slowing down radiationless decays and thereby making stronger the luminescence emission and lengthening the excited-state lifetime. The cage ligand, however, should allow Ru to attain octahedral coordination and a suitable Ru–N bond distance, otherwise the lowest ^{3}MC level would drop

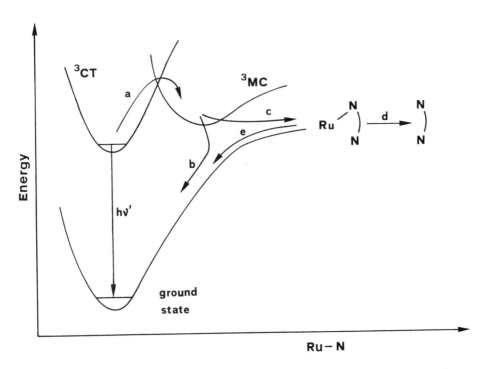

Fig. 11.8 — Schematic representation of the ligand photodissociation mechanism of Ru(bpy)$_3^{2+}$
[68].

below the lowest ^{3}CT level and most of the valuable properties of the complex would disappear. Molecular models show that this would indeed be the case for the bpy.bpy.bpy cryptand of **11.5**. Such a ligand, in fact, is appropriate for the larger, not symmetry-demanding Eu^{3+} ion, but it is clearly too rigid to create the octahedral coordination required by Ru^{2+}. This is apparently confirmed by the lack of luminescence at room temperature from a complex claimed to be the [Ru ⊂ bpy.bpy.bpy]$^{3+}$ cryptate [69].

As pointed out by Vögtle and coworkers [70,71], the dimensions of the cage and the symmetry of the coordination sphere may be "tuned" using spacers of different dimensions (Fig. 11.9). Molecular models show that with a larger spacer and three bpy coordinating groups it is possible to obtain a cage-type ligand more flexible than that shown in **11.5**, and capable of offering a suitable coordination site for Ru^{2+}. The cage complex **11.8**, where R is a benzyl group, was therefore designed and prepared *via* a template reaction [72]. As expected, it shows spectroscopic properties ($\lambda_{max}^{abs} = 455$ nm, $\lambda_{max}^{em} = 612$ nm) quite similar to those of Ru(bpy)$_3^{2+}$, but a longer excited-state lifetime at room temperature (1.7 *vs* 0.8 μs), and a much larger stability towards ligand photosubstitution ($\Phi_r < 10^{-6}$ compared to $\Phi_r = 0.017$ for Ru(bpy)$_3^{2+}$, in CH$_2$Cl$_2$ solution containing 0.01 M Cl$^-$) [73,74]. This should assure a reasonably high turnover number when the Ru-cage complex is used as a photosensitizer.

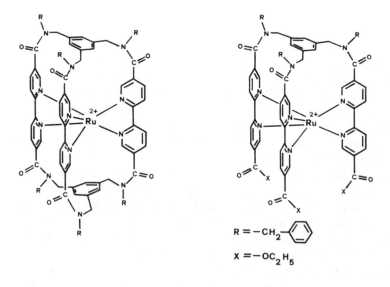

Fig. 11.9 — Tuning of the cavity of a cage ligand by using spacers of different sizes [71].

Interesting properties are also exhibited by hemicaged Ru^{2+} complexes obtained from tripod-type *tris*-bipyridine ligands. The hemicaged Ru(II) complex **11.9** was prepared [74] from a tripod ligand (**6.10**) which can also be used to obtain dinuclear and trinuclear polypyridine complexes of great photochemical interest [75] (section 6.3.2). Other similar tripod ligands [70,71,76–78], cage-type complexes [70,71,76,78,79], and hemicaged complexes [77,80] have also been characterized.

11.8   **11.9**

Ru(sar)$^{2+}$ [81] and a closed cage Ru(II) diimine complex [82] have also been prepared, but their photochemical properties have not yet been investigated.

11.1.6 Conclusions

Suitably designed cage-type complexes exhibit peculiar and quite interesting photo-chemical and photophysical properties and it is easy to predict that investigations concerning such species will rapidly grow in the next few years. Of course, chelating ligands different from amines or bipyridines can also be used such as, for example, catechol binding units [71,83–86]. A great variety of spacers can also be designed [71]. Furthermore, several synthetic approaches are available to synthesize non-collapsible molecular cells (carcerands [87,88], cyanospherands [89]), ditopic co-receptors [90,91], dinuclear cage-type complexes [91,92], receptors for neutral molecules [91,93–96] and anions [91,96], and other supramolecular cage structures (see, for example [97]) that might reveal quite interesting photochemical and photophysical properties.

Encapsulation–de-encapsulation equilibria can also be photocontrolled in suitable structures, as shown by an investigation on $[AgIr_2(dimen)_4(DMSO)_2]^{3+}$ (dimen = 1,8-diisocyanomenthane) [98].

11.2 CATENANES, ROTAXANES, AND RELATED SPECIES

11.2.1 Introduction

Catenanes are supramolecular species composed of interlocked rings, the simplest case being that of [2]-catenane shown in Fig. 11.10a. *Rotaxanes* are formed by a ring

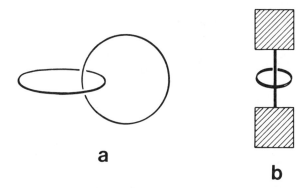

a

b

Fig. 11.10 — Schematic representation of catenanes (a) and rotaxanes (b).

which is threaded by a linear fragment with bulky groups on either end (Fig. 11.10b). Catenanes, rotaxanes and related species [99] are supramolecular architectures very attractive from an aesthetical viewpoint. The design of such sophisticated systems has long received much attention, but only recent achievements in synthetic and analytical methods have made possible their synthesis [100]. Nowadays several catenanes, rotaxanes, knots, helicates, etc. have been prepared, but few photo-chemical and photophysical investigations have been performed on such species.

11.2.2 Catenanes and rotaxanes

Catenanes and rotaxanes were first prepared by the so-called statistical threading method, which relies on the probability that a molecular thread, functionalized on both ends, may enter a macrocycle of adequate size: subsequent cyclization of the thread or insertion of bulky groups at both ends leads to catenanes and rotaxanes, respectively [101,102]. To increase the probability of catenane or rotaxane forma-tion, the threading process must be favoured by some type of interaction. This strategy has been followed, for example, to prepare the cyclodextrin rotaxane **11.10**, taking advantage of the affinity between the hydrophobic interior of α- or β-CD and the alkyl chain linked to a Co complex [103]. Incidentally, one can expect that ultraviolet light, causing the photodecomposition of the bulky Co complex, destroys such a rotaxane structure.

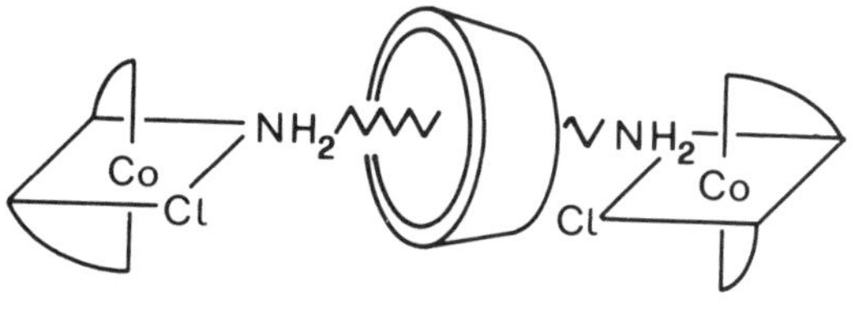

11.10

Electron donor–acceptor interactions provide the basis for the highly efficient template-directed synthesis of the [2]-catenane **11.11** [104] and rotaxane **11.12** [105] performed by Stoddart and coworkers. As mentioned in section 10.4, the possibility of obtaining **11.11** and **11.12** was suggested by the observation that **11.13** and **11.14** give EDA complexes with **11.16** (*p*-dimethoxybenzene) and **11.15** (paraquat), respectively (Fig. 10.7).

The photochemical behaviour of the catenane **11.11**, the rotaxane **11.12**, and of their parent compounds have been examined [106]. A mixture of equimolar, diluted CH_2Cl_2 solutions of **11.13** and **11.14** shows the absorption spectrum (a) on Fig. 11.11, which is exactly that expected from the summation of the two spectra of

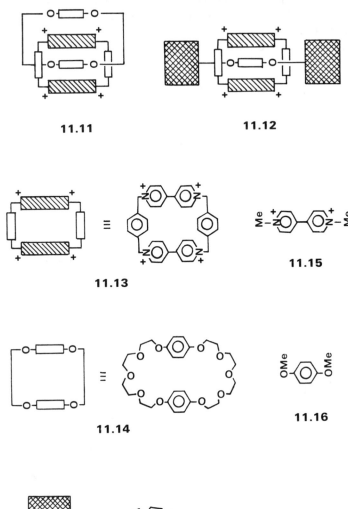

11.11

11.12

11.13

11.15

11.14

11.16

11.17

11.13 and **11.14**. This indicates that under the experimental conditions used (that is, very dilute solutions) there is practically no adduct formation, in agreement with the determined values of the association constants between **11.13** and **11.16** ($16\,\mathrm{M}^{-1}$) and **11.14** and **11.15** ($730\,\mathrm{M}^{-1}$). The absorption spectrum of the catenane **11.11**, in which **11.13** and **11.14** are interlocked, is shown by curve (b) of Fig. 11.12. This

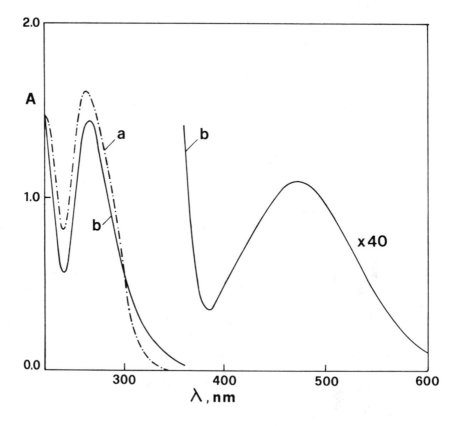

Fig. 11.11 — Absorption spectra of an equimolar mixture of **11.13** and **11.14** (a) and of their catenane **11.11** (b) [106].

spectrum is noticeably different from that of curve (a) in the same figure. The intense band of the paraquat units with $\lambda_{max} = 260$ nm undergoes a small red shift ($\lambda_{max} = 263$ nm) and a noticeable decrease in intensity. Furthermore, a tail appears in the 300–380 nm region and a broad, weak band is present in the visible region with $\lambda_{max} = 478$ nm ($\varepsilon = 700\,M^{-1}\,cm^{-1}$).

Of the two components of the catenane, **11.13** is not luminescent whereas **11.14** exhibits a fluorescence emission at room temperature with $\lambda_{max} = 320$ nm and $\tau \sim 1$ ns. In a rigid matrix at 77 K the fluorescence is accompanied by a very weak phosphorescence with $\lambda_{max} = 420$ nm and $\tau \sim 1$ sec. The equimolecular mixture of **11.13** and **11.14** shows the same luminescence properties of **11.14** alone. However, when **11.13** and **11.14** are interlocked in **11.11**, the fluorescence at room temperature and both the fluorescence and phosphorescence at 77 K of **11.14** are almost completely quenched.

The behaviour of the rotaxane **11.12**, which is made by **11.13** and **11.17** (a derivative of **11.16**), is quite similar to that of the catenane **11.11**.

The above results show that in the catenane **11.11** and in the rotaxane **11.12** an electronic interaction occurs between the π-electron-deficient paraquat residues of **11.13** and the π-electron-rich hydroquinol units of **11.14** or **11.17**. The maxima of the visible bands of **11.11** and **11.12** coincide, as expected because of the identical nature of the interacting chromophores in the two cases. Furthermore the molar absorption coefficient (at $\lambda_{max} = 478$ nm) of **11.11** is practically twice that of **11.12**, suggesting that in **11.11** *both* the hydroquinol-type units of **11.14** interact with a paraquat residue. This result is consistent with the almost complete quenching of the fluorescence of the hydroquinol units in both **11.11** and **11.12**.

11.2.3 Catenands and catenates

A very effective route to obtain interlocked species is the template synthesis around a metal complex, cleverly developed by Sauvage and coworkers [100]. This approach takes advantage of the affinity of diimine-type ligands for Cu(I) and of the tetrahedral coordination of this metal ion. For example, two ligands **11.18** fit together and form the very stable complex **11.19**. When the latter is reacted with two molecules of the diiodo derivative of pentaethylene glycol, the *catenate* **11.20** is obtained in excellent yield [107,108]. The topologically more complex [3]-catenates can be prepared following the strategy schematized in Fig. 11.12a [109]. A more efficient method of synthesis of [3]-catenates, however, is based on acetylenic oxidative coupling, schematized in Fig. 11.12b [110]. By this strategy, gram-scale preparation of the dinuclear [3]-catenate **11.21** is possible.

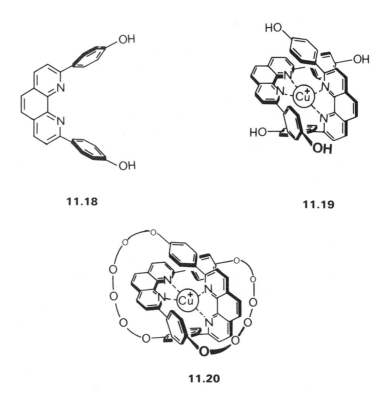

11.18   **11.19**

11.20

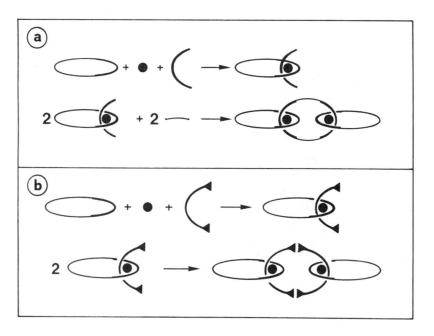

Fig. 11.12 — Strategies for the synthesis of [3]-catenates [100].

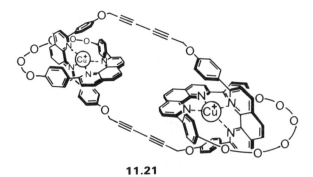

11.21

Demetalation of the catenates leads to the free ligands, *catenands*, where the coordinating subunits are disentangled (see, for example, **11.22**) [111,112]. Starting from the free catenands, a number of catenates of other metal ions were obtained [113], including heteronuclear catenates of catenand **11.22** [114].

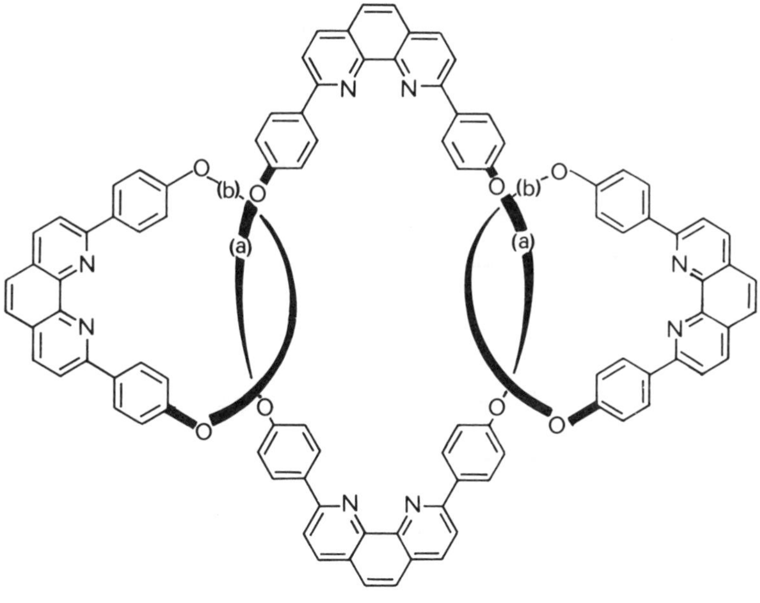

11.22

$(a) = -CH_2-C\equiv C-C\equiv C-CH_2-;$ (b) $= -CH_2CH_2\left(O-CH_2CH_2\right)_4$

The synthesis of molecular knots has also been achieved by Dietrich-Buchecker and Sauvage [115] *via* a three-dimensional template effect which leads two Cu(I) metal centres to interlace two molecular threads. Subsequent cyclization of the precursor dinuclear double helix leads to a knotted complex **11.23**. Demetalation of the complex leads to the free ligand, the trefoil knot **11.24**.

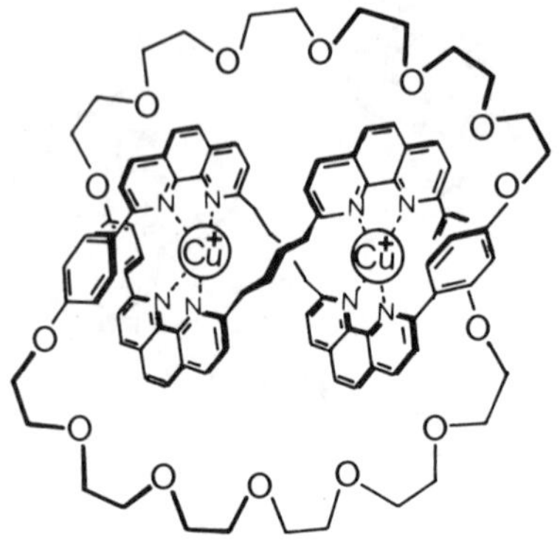

11.23

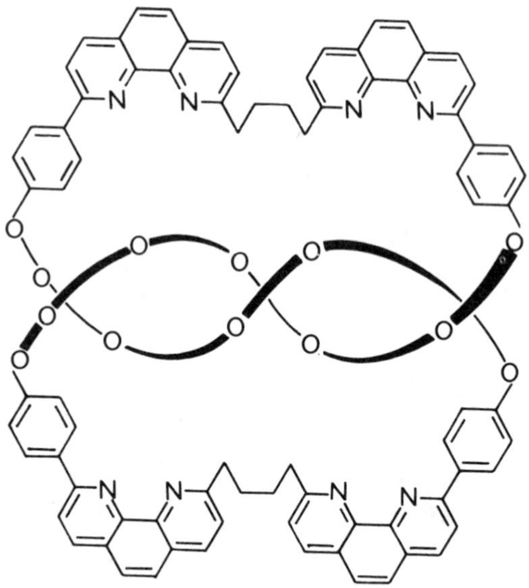

11.24

The catenates prepared by Sauvage and coworkers contain the well-known $Cu(DPP)_2^{2+}$ (DPP = 2,9-diphenyl-1,10-phenanthroline) chromophore [116,117], which shows broad and intense metal-to-ligand charge transfer bands in the visible region and a relatively intense luminescence ($\lambda_{max} = 710$ nm, $\Phi = 1 \times 10^{-3}$, $\tau = 250$ ns in CH_2Cl_2 solution at room temperature) [118]. $Cu(cat-30)^+$ (**11.20**) exhibits quite similar absorption and emission ($\lambda_{max} = 730$ nm, $\Phi = 8 \times 10^{-4} = 190$ ns, same experimental conditions) properties to those of $Cu(DPP)_2^{2+}$ [119]. Interestingly, for both compounds the absorption spectra become noticeably more resolved [119] and the luminescence intensity and lifetime decrease [120] in an alcoholic rigid matrix at 90 K, suggesting the occurrence of considerable structural changes. The strong tendency of MLCT excited states of Cu(I)-polypyridine complexes to undergo quenching *via* exciplex formation, thoroughly investigated by McMillin and co-workers [121], should also be recalled. The free cat-30 shows a fluorescence band at 400 nm ($\tau \sim 3$ ns) in fluid solution at room temperature and a phosphorescence band at 510 nm ($\tau \sim 1$ s) in a rigid matrix at 77 K [113,120]. For the complexes of cat-30 with several metals (for example, Li^+, Zn^{2+}, Co^{2+}) the ligand-centred fluorescence appears to be shifted [113], presumably because of $\pi\pi$ interaction between a phenanthroline ligand and a phenoxy substituent of the other phenanthroline ligand.

Preliminary investigations have also been performed on the luminescence properties of [3]-catenand and [3]-catenates [122]. The free catenand **11.22** shows behaviour quite similar to the free cat-30. The dinuclear Cu^+ complex **11.21** shows a MLCT luminescence ($\lambda_{max} = 700$ nm, $\tau = 91$ ns) in fluid solution at room temperature, the intensity of which decreases in passing to a rigid matrix at 77 K. The mononuclear Cu^+ complex of **11.22** shows, as expected, the luminescence properties of **11.21** and **11.22**.

11.2.4 Helicates

Cu^+ and poly(bipyridine) strands **11.25** ($n = 2,3$) can be used to build up double helix complexes (helicates), schematically shown by **11.26** and **11.27** [123,124]. No luminescence could be observed from such supramolecular species [125], presumably because the ethereal oxygen can quench the excited state *via* intramolecular exciplex formation.

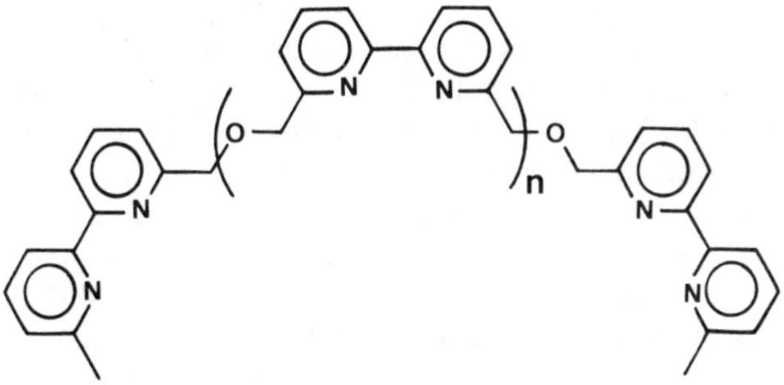

11.25

11.26

11.27

Double-helical binuclear complexes have also been obtained with quinquepyridine and sexipyridine (**11.28**) ligands [126,127]. The X-ray structure of the Cd^{2+} complex of two **11.28** ligands shows that the metal occupies an irregular six-coordinate N_6 environment and that each terpyridyl moiety is approximately planar, with stacking interactions between terpyridyl fragments of the two ligands [127]. The preparation of the analogous complexes of Cu^{2+}, Mn^{2+}, Fe^{2+}, Pd^{2+}, and Ru^{2+} has also been announced [127].

Helicates are promising subjects for supramolecular photochemical and photophysical investigation. It should also be recalled that they represent interesting examples of *spontaneous* organization of components into supramolecular structures (section 12.8).

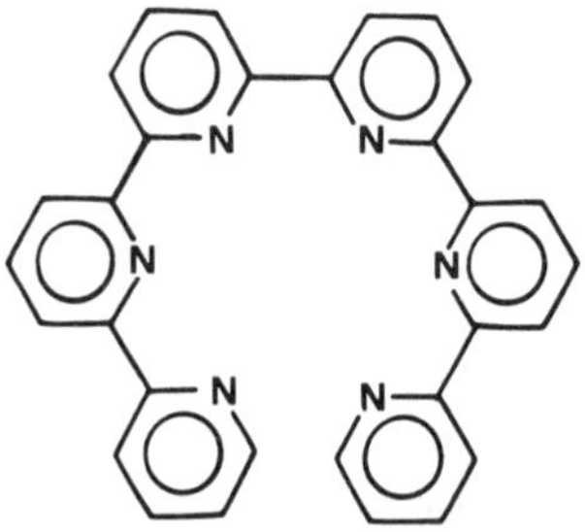

11.28

11.3   PROTON-TRANSFER PROCESSES

11.3.1   Introduction

The hydrogen bond plays an important role in molecular interactions. Several supramolecular systems involving intercomponent hydrogen bonds have been discussed in Chapter 10. The existence of a hydrogen bond is a prerequisite for the occurrence of excited-state proton transfer. This is due to the fact that proton transfer must compete with the other decay processes of the excited state. When a hydrogen bond is already present in the ground state, the required displacement of the proton in the excited-state reaction is relatively small and the reaction can be efficient. In this section we shall briefly recall the principles and review some recent examples of photoinduced proton transfer in hydrogen-bonded species.

About 40 years ago Förster [128] and Weller [129] showed that the acidity constant in the excited state can be significantly different from that in the ground state. Since then, proton transfer in electronically excited states has been extensively investigated [130–138]. Most of the work has dealt with singlet and triplet excited states of aromatic molecules in aqueous solution, for which p^*K data have been obtained by using the spectroscopic Förster cycle, fluorescence titration, or titration of the excited-state absorption. The acid-base equilibrium in the excited state may or may not be established, depending on the excited-state lifetime, p^*K, and acid concentration.

Excited-state proton transfer can be illustrated making use of double-well potential energy curves. Those schematized in Fig. 11.13 refer to a system where proton transfer corresponds to a structural change; if the initial and final structures are identical, the double-well potential energy curves are symmetric.

For the systems represented by Fig. 11.13, excitation of the normal ground-state

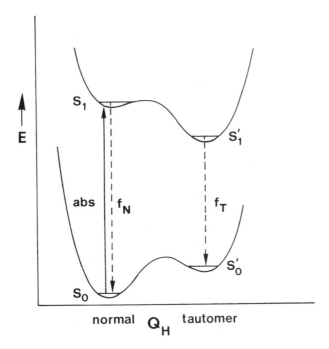

Fig. 11.13 — Double-well potential energy curves along the proton transfer coordinate Q_H.

molecular or supramolecular configuration S_0 is followed by adiabatic proton transfer with formation of the excited state tautomer S_1'. Emission from S_1' to the ground state tautomer S_0' is usually strongly red-shifted from the normal S_1 emission since proton transfer is caused by a massive electronic structural change in the system involved. As is apparent from Fig. 11.13, the energy-level diagram of a proton-transfer system of this type is like that of a typical 4-level laser [139]. In effect, proton-transfer lasers have been found to be extraordinarily efficient [136]. Formation of S_1' occurs on the picosecond time scale, and complete population inversion is possible because S_0' has zero population initially and is rapidly depopulated thereafter. As mentioned above, if the initial and final structures are identical (see, for example, Fig. 11.14a), the two wells of each curve are also identical and they are separated by a potential barrier. In such cases proton transfer takes place *via* a tunnelling mechanism [134].

A more complete potential energy diagram should take into account triplet as well as singlet excited states. In fact, Kasha and coworkers have shown that for some substituted aminosalicylates *normal molecule* phosphorescence at 77 K is greatly enhanced *via* tunnelling through proton-transfer potentials [140]. It should also be noticed that for *p*-dimethylaminosalicylic acid methyl ester **11.29** (a molecule of the TICT type, section 7.2.4), *three* competitive types of fluorescence have been observed: (i) normal local excited-state emission, (ii) proton-transfer tautomer emission, and (iii) twisted-intramolecular-charge-transfer emission [141].

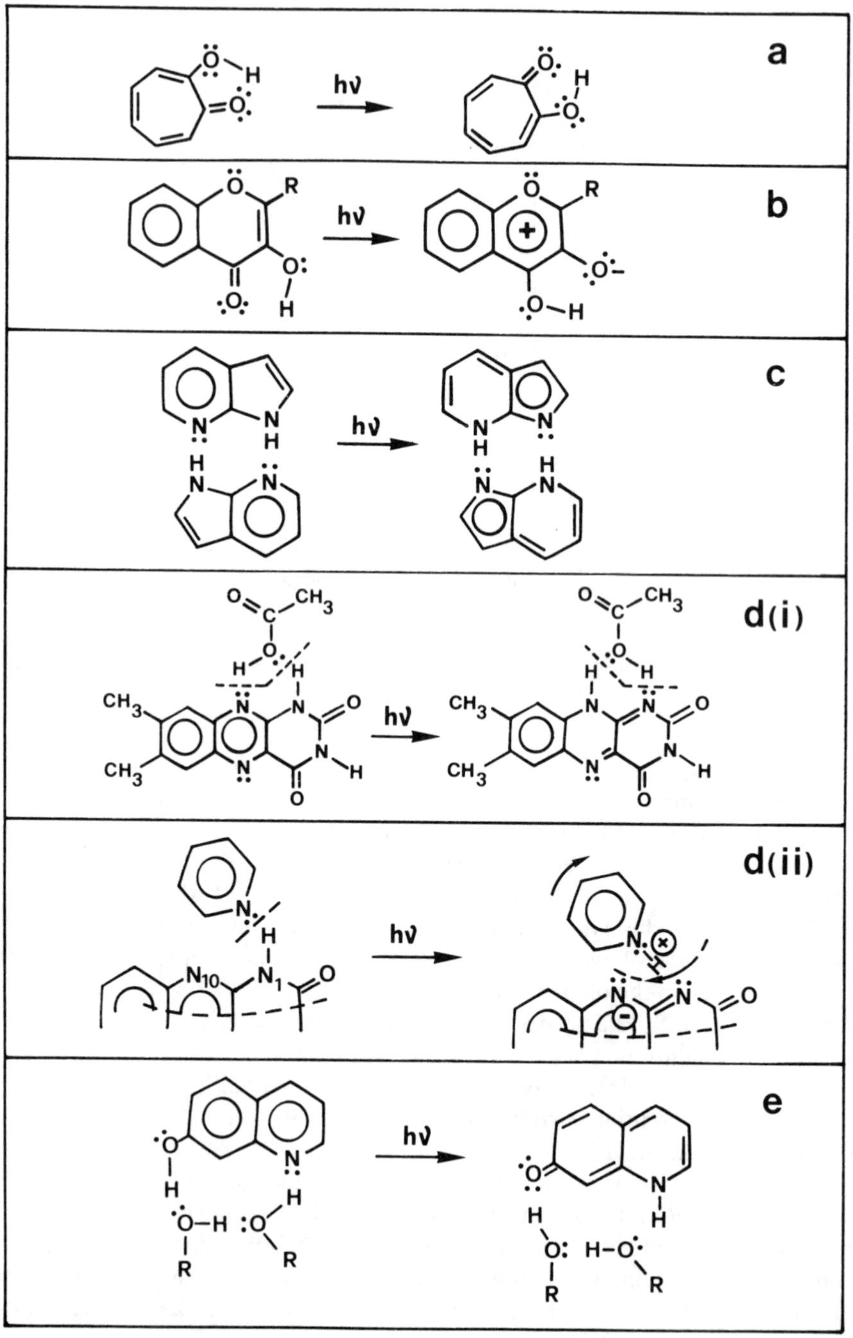

Fig. 11.14 — Types of photoinduced proton-transfer reactions [136].

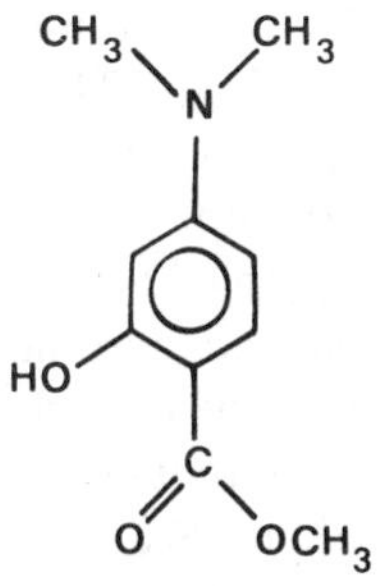

CH$_3$, CH$_3$

N

HO

C

O   OCH$_3$

11.29

We would also like to recall that commercial picosecond lasers can produce a photon concentration greater than 10^{-3} einsteins in a sample, so that in a solution containing a suitable species one can photoinduce a pH jump of several units, which can be used for a variety of applications [134]. The use of photoinduced proton transfer for a triggering action in molecular electronic devices has also been proposed [136].

The excited-state acid-base equilibria of organic molecules are described in photochemistry textbooks (see, for example [142]). We shall briefly discuss in the following (i) photoinduced tautomerization reactions and (ii) excited-state proton-transfer in metal complexes.

11.3.2 Photoinduced tautomerization

Following Kasha [133,136], five fundamental cases of proton transfer can be distinguished (Fig. 11.14):

(a) symmetric intramolecular proton transfer, involving identical tautomers which exhibit an intramolecular vicinal H-bond (for example, 2-hydroxytropolone [136], Fig. 11.14a);

(b) intrinsic intramolecular proton transfer, involving different tautomers which exhibit an intramolecular vicinal H-bond (for example, 3-hydroxyflavone [143], Fig. 11.14b);

(c) concerted biprotonic transfer, involving a doubly H-bonded dimer (for example, 7-azaindole dimer [144], Fig. 11.14c);

(d) strong-catalysis proton transfer between distal proton–donor and proton–acceptor groups. This may occur by (i) a static mechanism involving a doubly H-bonded complex (for example, luminochrome and acetic acid [133,145], Fig. 11.14d(i), or (ii) a dynamic mechanism involving a H-bonded complex (for example, luminochrome and pyridine [133,146], Fig. 11.14d(ii);

(e) proton relay transfer between very distant groups, involving solvent H- bonded bridges (for example, 7-hydroxyquinoline with two methanol molecules [133,147], Fig. 11.14e).

Cases (a) and (b) are truly intramolecular proton transfer processes. In cases (c), (d), and (e) the proton–donor and proton–acceptor groups of the molecule are too far apart to form an intramolecular H-bond. In such cases, the intramolecular tautomerization can be mediated by other components in an organized supramolecular structure.

The dynamic studies of excited-state proton transfer usually show complexity in the rise-time curves for tautomers [133,134,138,140]. In general this may be attribuited to the presence of multiple species and/or of multiple reaction pathways.

Several investigations on intercomponent proton tranfer in supramolecular systems have appeared (see, for example [148–154]). To illustrate some details, we shall make reference to the photoinduced tautomerization of 2-aminopyridine which occurs in 1:1 complexes with carboxylic acids [154]. In the ground state, 2-aminopyridine is by far more stable than its tautomer 2-imino-1,2-dihydropyridine (the structural formulae of the two species are shown in Fig. 11.15). On photoexci-

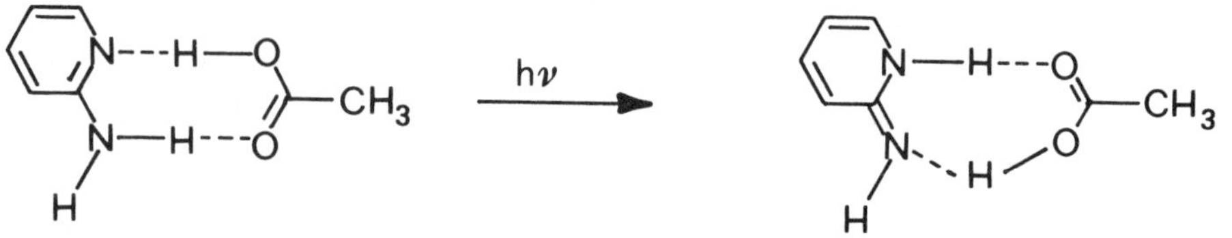

Fig. 11.15 — Tautomeric structures of complexes of 2-aminopyridine with acetic acid.

tation, partial charge transfer takes place from the amino group to the pyridine ring, with a consequent increase in acidity of the former and in basicity of the latter. In the excited state the tautomeric equilibrium is therefore displaced towards the imino-structure. In the isolated molecule, however, there is no suitable reaction coordinate for fast proton transfer since an intramolecular H-bond is lacking. In fact, phototautomerization does not occur in apolar solvents, where only the 2-aminopyridine fluorescence band with maximum at 330 nm can be observed. In alcohol solution the fluorescence band shows a slight bathochromic shift, but there is no evidence of phototautomerization. In 0.1 M HCl, only a fluorescence band with a maximum at 367 nm is observed, which is due to the singlet excited state of the protonated 2-aminopyridine cation. When acetic acid is added to an alkane solution of 2-aminopyridine, dramatic changes in the fluorescence spectrum are observed. The intensity of the 330 nm band is substantially reduced and the spectrum is dominated by a structured band around 420 nm [153,154]. For N,N-dimethyl-2-aminopyridine, where the two amine protons are replaced by methyl groups, addition of acetic acid causes no spectral change. Excitation spectra and experiments carried out at different acetic acid concentrations show that 1:1 complexes of 2-aminopyridine/acetic acid are formed in the ground state and that proton transfer takes place within these species (Fig. 11.15). Crystal structure analysis showed that the complex in the solid state is a hydrogen-bonded ion pair, consisting of a 2-aminopyridinium cation and a carboxylate anion. The fluorescence spectrum of the solid, in fact, shows only the 370 nm band which is found in HCl solution. When the crystalline complex is dissolved in alkane, the 370 nm band disappears and the dual fluorescence discussed above is present. Time-resolved fluorescence experiments showed that a lower limit for the rate constant of the adiabatic excited-state double proton-transfer process is

$10^{11} s^{-1}$ [154]. This suggests that the excited-state potential energy surface presents probably only a single minimum along the reaction coordinate, which is located at the equilibrium nuclear configuration of the excited tautomer complex. Relaxation of the electronically-excited tautomer leads to the ground state of the tautomer (S_0' in Fig. 11.13). From there, the thermal reverse proton transfer takes place faster than the risetime (a few nanoseconds) of the detection system used for transient absorption spectroscopy studies [154]. This suggests that the ground-state potential energy surface presents a single minimum located at the original equilibrium nuclear configuration.

11.3.3 Excited-state proton-transfer in metal complexes

Excited-state proton-transfer processes of inorganic coordination compounds [10] have been less commonly investigated than those of organic molecules. Notable exceptions are a number of complexes of Ru(II) with polypyridine-type ligands containing non-coordinated proton-transfer sites. These systems have low-lying d-π^* metal-to-ligand charge-transfer (MLCT) excited states [7]. If the proton-transfer site resides on the polypyridine ligands (for example, on carboxyl [155–157], Fig. 11.16a,

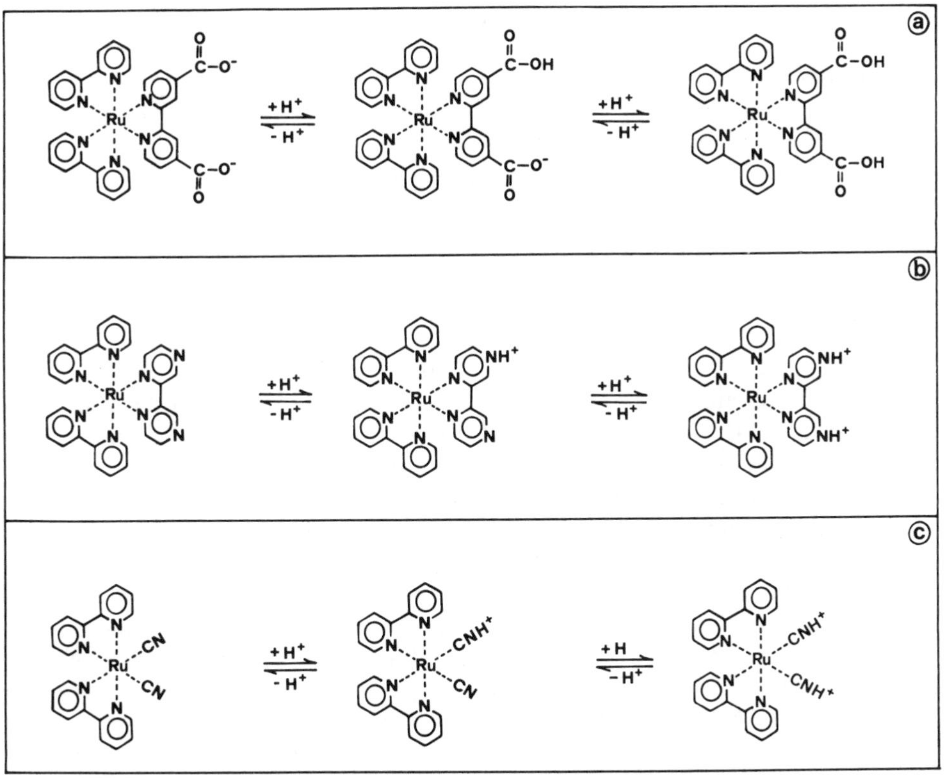

Fig. 11.16 — Examples of acid-base equilibria in Ru^{2+} metal complexes; the charges of the metal ions and of the CN$^-$ ligands have been omitted for simplicity.

or imino [158–161], Fig. 11.16b, groups), the increased electron density on the ligand makes the site less acidic in the excited state than in the ground state ($p^*K > pK$). The opposite result is obtained when in a mixed-ligand Ru(II) polypyridine complex the non-coordinated proton-transfer sites reside on ancillary ligands as, for example, in the mixed-ligand cyano bipyridine complexes of Ru(II) (Fig. 11.16c). In this case, the MLCT excitation still involves the polypyridine ligands, but the diminished electron density at the metal is efficiently sensed by the ancillary ligands, which are more acidic in the excited state than in the ground state ($p^*K < pK$) [162–164].

The first example of the last type of behaviour has been reported several years ago by Demas [162,163], who investigated the excited-state equilibria of cis-Ru(NN)$_2$(CN)$_2$ ($NN = 2,2'$-bipyridine, bpy, or 1,10-phenanthroline, phen) in water/sulphuric acid mixtures. In these systems, two basic sites are available at the nitrogen ends of the cyanide ligands, so that two successive proton-transfer steps must be considered (Fig. 11.16c). Demas' work gave clear evidence that (i) the protonated forms are considerably more acidic in the excited state than in the ground state, (ii) deprotonation efficiently competes with unimolecular deactivation of the excited state, and (iii) deprotonation of the excited states proceeds without excited-state deactivation. Quenching of the emission of the unprotonated Ru(bpy)$_2$(CN)$_2$ complex was observed in highly concentrated sulphuric acid solutions, but unfortunately no emission could be detected at room temperature from the mono and diprotonated forms.

More recently, the excited-state acid-base behaviour of Ru(bpy)$_2$(CN)$_2$ has been studied in acetonitrile and acetonitrile/water solvent mixtures, with HClO$_4$ as the proton source [165]. With respect to water, the use of these solvent systems has several practical advantages (smaller acid concentrations required, better separation and experimental accessibility of successive steps, observation in fluid solution of emission from the monoprotonated form). The ground-state equilibria shift toward lower pK values as the water content of the mixed solvent is increased. Depending on the concentration of water, three distinct kinetic regimes for the excited-state proton transfer can be observed. In neat acetonitrile, excited-state deprotonation does not compete with excited-state deactivation, so that emission follows the distribution of variously protonated species present in the ground state. At large (>0.1 M) water concentrations, the proton-transfer steps are fast, and excited-state equilibration is established prior to deactivation, allowing determination of excited-state p^*K values from emission titration. At low water concentrations (0.01–0.04 M), a hybrid regime is obtained, in which no equilibrium is established in spite of the efficient deprotonation observed. In the excited-state equilibrium regime, the p^*K values depend linearly on the third power of the water concentration. This result suggests that in the acetonitrile/water mixtures the proton has an average hydration number of 3. Clusters of water molecules of various sizes acting as proton carriers in pure water or water/organic solvent mixtures have been suggested by several authors. For example, an average hydration number of 4 for protons in methanol/water mixtures has been suggested by Robinson and coworkers [166] on the basis of a Markow random-walk analysis of solvent composition effects on the lifetime of β-naphthol.

Ru(bpy)(CN)$_4^{2-}$ exhibits a MLCT luminescence as Ru(bpy)$_2$(CN)$_2$. The excited-state proton transfer equilibria of Ru(bpy)(CN)$_4^{2-}$ in sulphuric acid aqueous solutions have also been investigated [164]. Here again excited-state protonation starts at considerably higher acidities than ground-state protonation, consistent with the predicted acidity changes. Emission spectra obtained at various acid concentrations again exhibit the blue shift in the energy of the MLCT state upon protonation. The four protonation steps, however, are not separable in this system and no definite emission spectra for the various protonated forms can be obtained. In concentrated sulphuric acid, where $Ru(bpy)(CNH)_4^{2+}$ is presumably the dominant species, the complex exhibits a structured emission typical of the LC $\pi\pi^*$ bipyridine phosphorescence. This suggests that the upward shift of the MLCT state caused by protonation ultimately leads to an inversion in the nature of the lowest excited state of the complex.

REFERENCES

[1] Fleischauer, P. D. and Fleischauer, P. (1970) *Chem. Rev.* **70** 199.

[2] Crosby, G. A. (1975) *Acc. Chem. Res.* **8** 231.

[3] Sutin, N. and Creutz, C. (1978) *Adv. Chem. Ser.* **168** 1.

[4] Kalyanasundaram, K. (1982) *Coord. Chem. Rev.* **46** 159.

[5] Meyer, T. J. (1986) *Pure Appl. Chem.* **58** 1193.

[6] Krause, R. A. (1987) *Structure and Bonding* **67** 1.

[7] Juris, A., Balzani, V., Barigelletti, F., Campagna, S., Belser, P., and von Zelewsky, A. (1988) *Coord. Chem. Rev.* **84** 85.

[8] Balzani, V. (1989) *Gazz. Chim. Ital.* **119** 311.

[9] Balzani, V. and Ballardini, R. (1990) *Photochem. Photobiol.* **52** 409.

[10] Balzani, V., Sabbatini, N., and Scandola, F. (1986) *Chem. Rev.* **86** 319.

[11] Cotton, F. A. and Wilkinson, G. (1988) *Advanced inorganic chemistry.* Interscience.

[12] Lilie, J., Shinohara, N., and Simic, M. G. (1976) *J. Am. Chem. Soc.* **98** 6516.

[13] Balzani, V. and Carassiti, V. (1970) *Photochemistry of coordination compounds.* Academic.

[14] Creaser, I. I., Harrowfield, J. M., Herlt, A. J., Sargeson, A. M., Springborg, J., Geue, R. J., and Snow, M. R. (1977) *J. Am. Chem. Soc.* **99** 3181.

[15] Sargeson, A. M. (1979) *Chem. Brit.* **15** 23.

[16] Boucher, H. A., Lawrance, G. A., Lay, P. A., Sargeson, A. M., Bond, A. M., Sangster, D. F., and Sullivan, J. C. (1983) *J. Am. Chem. Soc.* **105** 4652.

[17] Sargeson, A. M. (1984) *Pure Appl. Chem.* **56** 1603.

[18] Sargeson, A. M. (1986) *Pure Appl. Chem.* **58** 1511.

[19] Bernhard, P. and Sargeson, A. M. (1987) *Inorg. Chem.* **26** 4122.

[20] Hagen, K. S., Lay, P. A., and Sargeson, A. M. (1988) *Inorg. Chem.* **27** 3424.

[21] Bernhard, P., Sargeson, A. M., and Anson, F. C. (1988) *Inorg. Chem.* **27** 2754.

[22] Bernhard, P. and Sargeson, A. M. (1989) *J. Am. Chem. Soc.* **111** 597.

[23] Bernhard, P., Burgi, H. B., Raselli, A., and Sargeson, A. M. (1989) *Inorg. Chem.* **28** 3234.

[24] Martin, L. L., Martin, R. L., Murray, K. S., and Sargeson, A. M. (1990) *Inorg. Chem.* **29** 1387.

[25] Creaser, I. I., Geue, R. J., Harrowfield, J. M., Herlt, A. J., Sargeson, A. M., Snow, M. R., and Springborg, J. (1982) *J. Am. Chem. Soc.* **104** 6016.

[26] Pina, F., Ciano, M., Moggi, L., and Balzani, V. (1985) *Inorg. Chem.* **24** 844.

[27] Houlding, V., Geiger, T., Kolle, V., and Graetzel, M., (1982) *J. Chem. Soc., Chem. Commun.* 681.

[28] Scandola, M., Scandola, F., Indelli, A., and Balzani, V. (1983) *Inorg. Chim. Acta* **76** L67.

[29] Lay, P. A., Mau, A. W. H., Sasse, W. H. F., Creaser, I. I., Gahan, L. R., and Sargeson, A. M. (1983) *Inorg. Chem.* **22** 2347.

[30] Launikonis A., Lay, P. A., Mau, A. W. H., Sargeson, A. M., and Sasse, W. H. F. (1984) *Sci. Papers Inst. Phys. Chem. Res.* **78** 198.

[31] Creaser, I. I., Gahan, L. R., Geue, R. J., Launikonis, A., Lay, P. A., Lydon, J. D., McCarthy, M. G., Mau, A. W. H., Sargeson, A. M., and Sasse, W. H. F. (1985) *Inorg. Chem.* **24** 2671.

[32] Mau, A. W. H., Sasse, W. H. F., Creaser, I. I., and Sargeson, A. M. (1986) *Nouv. J. Chim.* **10** 589.

[33] Creaser, I. I., Hammershoi, A., Launikonis, A., Mau, A. W. H., Sargeson, A. M., and Sasse, W. H. F. (1989) *Photochem. Photobiol.* **49** 19.

[34] Pina, F., Mulazzani, Q. G., Venturi, M., Ciano, M., and Balzani, V. (1985) *Inorg. Chem.* **24** 848.

[35] Pina, F., Ciano, M., Mulazzani, Q. G., Venturi, M., Balzani, V., and Moggi, L. (1984) *Sci. Papers Inst. Phys. Chem. Res.* **78** 166.

[36] Sotomayor, J., Costa, J. C., Mulazzani, Q. G., and Pina, F. (1989) *J. Photochem. Photobiol. A: Chemistry* **49** 195.

[37] Kirk, A. D. (1981) *Coord. Chem. Rev.* **39** 225.

[38] Endicott, J. F. (1983) *J. Chem. Educ.* **60** 824.

[39] Kirk, A. D. (1983) *J. Chem. Educ.* **60** 843.

[40] Kane-Maguire, N. A. P., Wallace, K. C., and Miller D. B. (1985) *Inorg. Chem.* **24** 597.

[41] Comba, P., Mau, A. W. H., and Sargeson, A. M. (1985) *J. Phys. Chem.* **89** 394.

[42] Comba, P., Creaser, I. I., Gahan, L. R., Harrowfield, J. M., Lawrance, G. A., Martin, L. L., Mau, A. W. H., Sargeson, A. M., Sasse, W. H. F., and Snow, M. R. (1986) *Inorg. Chem.* **25** 384.

[43] Ramasami, T., Endicott, J. F., and Brubaker, G. R. (1983) *J. Phys. Chem.* **87** 5057.

[44] Haas, Y. and Stein, G. (1971) *J. Phys. Chem.* **75** 3668.

[45] Horrocks, W. D. and Albin, M. (1984) *Progr. Inorg. Chem.* **31** 1.

[46] Lehn, J. M. and Sauvage, J. P. (1975) *J. Am. Chem. Soc.* **97** 6700.

[47] Rodriguez-Ubis, J. C., Alpha, B., Plancherel, D., and Lehn, J. M. (1984) *Helv. Chim. Acta* **67** 2264.

[48] Sabbatini, N., Dellonte, S., Ciano, M., Bonazzi, A., and Balzani, V. (1984) *Chem. Phys. Lett.* **107** 212.

[49] Alpha, B., Lehn, J.M., and Mathis, G. (1987) *Angew. Chem. Int. Ed. Engl.* **26** 266.

[50] Sabbatini, N., Perathoner, S., Balzani, V., Alpha, B., and Lehn, J. M. (1987) In Balzani, V. (ed.) *Supramolecular photochemistry*. Reidel, p. 187.

[51] Alpha, B., Balzani, V., Lehn, J. M., Perathoner, S., and Sabbatini, N. (1987) *Angew. Chem. Int. Ed. Engl.* **26** 1266.

[52] Alpha, B., Ballardini, R., Balzani, V., Lehn, J. M., Perathoner, S., and Sabbatini, N. (1990) *Photochem. Photobiol.* **52** 229.

[53] Blasse, G., Dirksen, G. J., Van der Voort, D., Sabbatini, N., Perathoner, S., Lehn, J. M., and Alpha, B. (1988) *Chem. Phys. Lett.* **146** 347.

[54] Blasse, G., Dirksen, G. J., Sabbatini, N., Perathoner, S., Lehn, J. M., and Alpha, B. (1988) *J. Phys. Chem.* **92** 2419.

[55] Pietraszkiewicz, M., Pappalardo, S., Finocchiaro, P., Mamo, A., and Karpiuk, J. (1989) *J. Chem. Soc., Chem. Commun.* 1907.

[56] Lehn, J. M., Pietraszkiewicz, M., and Karpiuk, J. (1990) *Helv. Chim. Acta* **73** 106.

[57] Gutsche, C. D. (1989) *Calixarenes*. The Royal Society of Chemistry. Cambridge.

[58] Ungaro, R. and Pochini, A. (1990) In Vicens, J. and Bohmer, V. (eds) *Calixarenes: a versatile class of macrocyclic compounds*. Kluwer, p. 133.

[59] Cram, D. J. (1986) *Angew. Chem. Int. Ed. Engl.* **25** 1039.

[60] Calestani, G., Ugozzoli, F., Arduini, A., Ghidini, E., and Ungaro, R. (1985) *J. Chem. Soc., Chem. Commun.* 344.

[61] Arduini, A., Pochini, A., Reverberi, S., Ungaro, R., Andreetti, G. D., and Ugozzoli, F. (1986) *Tetrahed.* **42** 2089.

[62] Arduini, A., Ghidini, E., Pochini, A., Ungaro, R., Andreetti, G. D., Calestani, G., and Ugozzoli, F. (1988) *J. Incl. Phenom.* **6** 119.

[63] Sabbatini, N., Guardigli, M., Mecati, A., Balzani, V., Ungaro, R., Casnati, A., and Pochini, A. (1990) *J. Chem. Soc., Chem. Commun.* 878.

[64] Harrowfield, J. M., Odgen, M. I., White, A. H., and Wilner, F. R. (1989) *Aust. J. Chem.* **42** 949, and references therein.

[65] Balzani, V., Barigelletti, F., and De Cola, L. (1990) *Topics Curr. Chem.* **158**, 31.

[66] Balzani, V., Juris, A., and Scandola, F. (1986) In Pelizzetti, E. and Serpone, N. (eds) *Homogeneous and heterogeneous photocatalysis*. Reidel, p. 1.

[67] Van Houten, J. and Watts, R. J. (1978) *Inorg. Chem.* **17** 3381.

[68] Durham, B., Caspar, J. V., Nagle, J. K., and Meyer, T. J. (1982) *J. Am. Chem. Soc.* **104** 4803.

[69] Diirr, H., Zengerle, K., and Trierweiler, H. P. (1988) *Z. Naturforsch. B* **43** 361.

[70] Grammenudi, S. and Vögtle, F. (1986) *Angew. Chem. Int. Ed. Engl.* **25** 1122.

[71] Stutte, P., Kiggen, W., and Vögtle, F. (1987) *Tetrahed.* **43** 2065.

[72] Belser, P., De Cola, L., and von Zelewsky, A. (1988) *J. Chem. Soc., Chem. Commun.* 1057.

[73] De Cola, L., Barigelletti, F., Balzani, V., Belser, P., von Zelewsky, A., Vögtle, F., Ebmeyer, F., and Grammenudi, S. (1988) *J. Am. Chem. Soc.* **110** 7210.

[74] Barigelletti, F., De Cola, L., Balzani, V., Belser, P., von Zelewsky, A., Vögtle, F., Ebmeyer, F., and Grammenudi, S. (1989) *J. Am. Chem. Soc.* **111** 4662.

[75] De Cola, L., Belser, P., Ebmeyer, F., Barigelletti, F., Vögtle, F., von Zelewsky, A., and Balzani, V. (1990) *Inorg. Chem.* **29** 495.

[76] Ebmeyer, F. and Vögtle, F. (1989) *Angew. Chem. Int. Ed. Engl.* **28** 79.

[77] Beeston, R. F., Larson, S. L., and Fitzgerald, M. C. (1989) *Inorg. Chem.* **28** 4189.

[78] Vögtle, F., private communication.

[79] Ebmeyer, F. and Vögtle, F. (1989) *Chem. Ber.* **122** 1725.

[80] Belser, P. and von Zelewsky, A., private communication.

[81] Bernhard, P. and Sargeson, A. M., (1985) *J. Chem. Soc., Chem. Commun.* 1516.

[82] Muller, J. G., Takeuchi, K. J., and Grzybowski, J. J. (1989) *Polyhedron* **8** 1391.

[83] Kiggen, W., Vögtle, F., Franken, S., and Puff, H. (1986) *Tetrahed.* **42** 1859.

[84] McMurry, T. J., Hosseini, M. W., Garret, T. M., Hahn, F. E., Reyes, Z. E., and Raymond, K. N. (1987) *J. Am. Chem. Soc.* **109** 7196, and references therein.

[85] McMurry, T. J., Raymond, K. N., and Smith, P. H. (1989) *Science* **244** 938.

[86] Raymond, K. N. (1989) *J. Inclusion Phenom.* **7** 169.

[87] Cram, D. J. (1988) *Angew. Chem. Int. Ed. Engl.* **27** 1009.

[88] Sherman, J. C. and Cram, D. J. (1989) *J. Am. Chem. Soc.* **111** 4527.

[89] Paek, K., Knobler, C. B., Maverick, E. F., and Cram, D. J. (1989) *J. Am. Chem. Soc.* **111** 8662.

[90] McKervey, M. A., Owens, M., Schulten, H. R., Vogt, W., and Böhmer, V. (1990) *Angew. Chem. Int. Ed. Engl.* **29** 280.

[91] Lehn, J. M. (1988) *Angew. Chem. Int. Ed. Engl.* **27** 89.

[92] Jazwinski, J., Lehn, J. M., Libienbaum, D., Ziessel, R., Guilhem, J., and Pascard, C. (1987) *J. Chem. Soc., Chem. Commun.* 1691.

[93] Diederich, F. (1988) *Angew. Chem. Int. Ed. Engl.* **27** 362.

[94] Canceill, J., Cesario, M., Collet, A., Guilhem, J., Lacombe, L., Lozach, B., and Pascard, C. (1989) *Angew. Chem. Int. Ed. Engl.* **28** 1246.

[95] Seward, E. M., Hopkins, R. B., Sauerer, W., Tam, S. W., and Diederich, F. (1990) *J. Am. Chem. Soc.* **112** 1783.

[96] Vögtle, F., Wallon, A., Müller, W. M., Werner, U., and Nieger, M. (1990) *J. Chem. Soc., Chem. Commun.* 158.

[97] Dijkstra, P. J., Brunink, J. A. J., Bugge, K. E., Reinhoudt, D. N., Harkema, S., Ungaro, R., Ugozzoli, F., and Ghidini, E. (1989) *J. Am. Chem. Soc.* **111** 7567.

[98] Sykes, A. and Mann, K. R. (1988) *J. Am. Chem. Soc.* **110** 8252; (1990) *ibid.* **112** 1297.

[99] Schill, G. (1971) *Catenanes, rotaxanes, and knots.* Academic.

[100] Dietrich-Buchecker, C. O. and Sauvage, J. P. (1987) *Chem. Rev.* **87** 795.

[101] Agam, G., Graiver, D., and Zilkha, A. (1976) *J. Am. Chem. Soc.* **98** 5206.

[102] Agam, G. and Zilkha, A. (1976) *J. Am. Chem. Soc.* **98** 5214.

[103] Ogino, H. and Ohata, K. (1984) *Inorg. Chem.* **23** 3312.

[104] Ashton, P. R., Goodnow, T. T., Kaifer, A. E., Reddington, M. V., Slawin, A. M. Z., Spencer, N., Stoddart, J. F., Vicent, C., and Williams, D. J. (1989) *Angew. Chem. Int. Ed. Engl.* **28** 1396.

[105] Anelli, P. L., Reddington, M. V., Spencer, N., Stoddart, J. F., and Vicent, C. (1989) In *Proc. Workshop on "Frontiers in supramolecular organic chemistry and photochemistry"*, Saarbrücken (FRG) August 27–September 1, 1989, p. P4.

[106] Anelli, P. L., Ashton, P. R., Ballardini, R., Balzani, V., Gandolfi, M. T., Goodnow, T. T., Kaifer, A. E., Pietraszkiewicz, M., Prodi, L., Reddington, M. V., Slawin, A. M. Z., Spencer, N., Stoddart, J. F., Vicent, C., and Williams, D. J. *J. Am. Chem. Soc.*, submitted.

[107] Dietrich-Buchecker, C. O., Sauvage, J. P., and Kern, J. M. (1984) *J. Am. Chem. Soc.* **106** 3043.

[108] Albrecht-Gary, A. M., Saad, Z., Dietrich-Buchecker, C. O., and Sauvage, J. P. (1985) *J. Am. Chem. Soc.* **107** 3205.

[109] Sauvage, J. P. and Weiss, J. (1985) *J. Am. Chem. Soc.* **107** 6108.

[110] Dietrich-Buchecker, C. O., Khemiss, A. K., and Sauvage, J. P. (1986) *J. Chem. Soc., Chem. Commun.* 1376.

[111] Cesario, M., Dietrich-Buchecker, C. O., Guilhem, J., Pascard, C., and Sauvage, J. P. (1985) *J. Chem. Soc., Chem. Commun.* 244.

[112] Guilhem, J., Pascard, C., Sauvage, J. P., and Weiss, J. (1988) *J. Am. Chem. Soc.* **110** 8711.

[113] Dietrich-Buchecker, C. O., Sauvage, J. P., and Kern, J. M. (1989) *J. Am. Chem. Soc.* **111** 7791.

[114] Sauvage, J. P., private communication.

[115] Dietrich-Buchecker, C. O. and Sauvage, J. P. (1989) *Angew. Chem. Int. Ed. Engl.* **28** 189.

[116] McMillin, D. R., Buckner, M. T., and Ahn, B. T. (1977) *Inorg. Chem.* **16** 943.

[117] Kirchhoff, J. R., Gamache, R. E., Jr., Blaskie, M. W., Del Paggio, A. A., Lengel, R. K., and McMillin, D. R. (1983) *Inorg. Chem.* **22** 2380.

[118] Ichinaga, A. K., Kirchhoff, J. R., McMillin, D. R., Dietrich-Buchecker, C. O., Marnot, P. A., and Sauvage, J. P. (1987) *Inorg. Chem.* **26** 4290.

[119] Gushurst, A. K. I., McMillin, D. R., Dietrich-Buchecker, C. O., and Sauvage, J. P. (1989) *Inorg. Chem.* **28** 4070.

[120] Barigelletti, F. and De Cola, L., unpublished observations.

[121] Stacy, E. M. and McMillin, D. R. (1990) *Inorg. Chem.* **29** 393, and references therein.

[122] Armaroli, N., Balzani, V., Barigelletti, F., De Cola, L., Sauvage, J. P., and Hemmert, C., work in progress.

[123] Lehn, J. M. and Rigault, A. (1988) *Angew. Chem. Int. Ed. Engl.* **27** 1095.

[124] Lehn, J. M., Rigault, A., Siegel, J., Harrowfield, J., Chevrier, B., and Moras, D. (1987) *Proc. Natl. Acad. Sci. USA* **84** 2565.

[125] Monti, S., Balzani, V., and Lehn, J. M., unpublished results.

[126] Constable, E. C., Ward, M. D., Drew, M. G. B., and Forsyth, G. A. (1989) *Polyhedron* **8** 2551.

[127] Constable, E. C., Ward, M. D., and Tocher, D. A. (1990) *J. Am. Chem. Soc.* **112** 1256.

[128] Förster, Th. (1950) *Z. Elektrochem.* **54** 531.

[129] Weller, A. (1952) *Z. Elektrochem.* **56** 662.

[130] Weller, A. (1961) *Prog. React. Kinet.* **1** 187.

[131] Vander Donckt, E. (1970) *Prog. React. Kinet.* **5** 273.

[132] Klöpffer, W. (1977) *Adv. Photochem.* **10** 311.

[133] Kasha, M. (1986) *J. Chem. Soc., Faraday Trans. 2* **82** 2379.

[134] Kosower, E. M. and Huppert, D. (1986) *Annu. Rev. Phys. Chem.* **37** 127.

[135] Scandola, F. and Indelli, M. T. (1988) *Pure Appl. Chem.* **60** 973.

[136] Kasha, M. (1988). In Carter, F. L., Siatkowski, R. E., and Wohltjen, H. (eds) *Molecular electronic devices*. North-Holland, p. 107.
[137] *Chem. Phys.* (1989) **136** No. 2, pp. 153–360 (special issue on "Spectroscopy and dynamics of elementary proton transfer in polyatomic systems").
[138] Barbara, P. F., Walsh, P. K., and Brus, L. E. (1989) *J. Phys. Chem.* **93** 29.
[139] Chou, P., McMorrow, D., Aartsma, T. J., and Kasha, M. (1984) *J. Phys. Chem.* **88** 4596.
[140] Gormin, D., Heldt, J., and Kasha, M., (1990) *J. Phys. Chem.* **94** 1185.
[141] Heldt, J., Gormin, D., and Kasha, M. (1989) *Chem. Phys.* **136** 321.
[142] Turro, N. J. (1978) *Modern molecular photochemistry*. Benjamin.
[143] McMorrow, D. and Kasha, M. (1984) *J. Phys. Chem.* **88** 2235.
[144] Ingham, K. C., Abu-Elgheit, M., and El-Bayoumi, M. A. (1971) *J. Am. Chem. Soc.* **93** 5023.
[145] Choi, J. D., Fugate, R. D., and Song, P. S. (1980) *J. Am. Chem. Soc.* **102** 5293.
[146] Koziolawa, A. (1979) *Photochem. Photobiol.* **29** 459.
[147] Itoh, M., Adachi, T., and Tokumura, K. (1983) *J. Am. Chem. Soc.* **105** 4828.
[148] Ikeda, N., Miyasaka, H., Okada, T., and Mataga, N. (1983) *J. Am. Chem. Soc.* **105** 5206.
[149] Waluk, J., Komorowski, S. J., and Herbich, J. (1986) *J. Phys. Chem.* **90** 3868.
[150] Inuzuka, K. and Fujimoto, A. (1986) *Spectrochim. Acta, Part A* **42** 929.
[151] Hadjoudis, E., Vittorakis, M., and Moustakali-Mavridis, I. (1987) *Tetrahed.* **43** 1345.
[152] Brucker, G. A. and Kelley, D. F. (1989) *Chem. Phys.* **136** 213.
[153] Brucker, G. A. and Kelley, D. F. (1989) *J. Phys. Chem.* **93** 5179.
[154] Konijnenberg. J., Huizer, A. H., and Varma, C. A. G. O. (1989) *J. Chem. Soc., Faraday Trans. 2* **85** 1539.
[155] Giordano, P. J., Bock, C. R., Wrighton, M. S., Interrante, L. V., and Williams, R. F. X. (1977) *J. Am. Chem. Soc.* **99** 3187.
[156] Lay, P. A. and Sasse, W. H. F. (1984) *Inorg. Chem.* **23** 4123.
[157] Shimidzu, T., Iyoda, T., and Izaki, K. (1985) *J. Phys. Chem.* **89** 642.
[158] Rillema, D. P., Allen, G., Meyer, T. J., and Conrad, D. (1983) *Inorg. Chem.* **22** 1617.
[159] Crutchley, R. J., Kress, N., and Lever, A. B. P. (1983) *J. Am. Chem. Soc.* **105** 1170.
[160] Braunstein, C. H., Baker, A. D., Strekas, T. C., and Gafney, H. D. (1984) *Inorg. Chem.* **23** 857.
[161] Hosek, W., Tysoe, S. A., Gafney, H. D., Baker, A. D., and Strekas, T. C. (1989) *Inorg. Chem.* **28** 1228.
[162] Peterson, S. H. and Demas, J. N. (1976) *J. Am. Chem. Soc.* **98** 7880.
[163] Peterson, S. H. and Demas, J. N. (1979) *J. Am. Chem. Soc.* **101** 6571.
[164] Indelli, M. T., Bignozzi, C. A., Marconi, A., and Scandola, F. (1987). In Yersin, H. and Vogler, A. (eds) *Photochemistry and photophysics of coordination compounds*. Springer-Verlag, p. 159.
[165] Davila, J., Bignozzi, C. A., and Scandola, F. (1989) *J. Phys. Chem.* **93** 1373.
[166] Lee, J., Griffin, R. D., and Robinson, G. W. (1985) *J. Chem. Phys.* **82** 4920.

12

Photochemical molecular devices

12.1 INTRODUCTION

In everyday life we make extensive use of *macroscopic devices*, that is of assemblies of components designed to achieve specific functions. Each component of the device is involved in, and/or performs, one or more single acts, while the entire device performs one or more complex functions, characteristic of the assembly. For example, the function performed by a hairdryer (production of hot wind) is the result of acts performed by a switch, a heater, and a fan suitably connected by electric wires and assembled in an appropriate framework.

The concept of a device can be extended to the molecular level [1–9]. We can define a *molecular device* as an assembly of molecular components (that is, a supramolecular structure) designed to achieve specific functions. Each molecular component performs one or more single *acts*, while the entire (supramolecular) device performs one or more complex *functions*, characteristic of the assembly. Molecular devices obviously operate chemically, that is *via* electronic and/or nuclear rearrangements. The extension of the *device* concept to the molecular level is currently the object of lively discussions in the field of microelectronics where a *small upward* approach is thought to offer substantial advantages compared to the conventional *large downward* approach of lithographic miniaturization techniques [1–6,9–18] (12.7.1).

In molecular photochemistry, a single molecule performs simple intramolecular and/or intermolecular photoinduced acts such as bond breaking, light emission, electron transfer, etc. (Chapter 2). These simple acts may find useful applications in the field of photochemical synthesis, photodecomposition, photochromism, photoluminescence, etc. However, more complex light-induced functions, such as vectorial electron transfer, migration of electronic energy, and switch on/off of receptor ability, cannot be performed by single molecules but need the cooperation of several components. An assembly of molecular components capable of performing light-induced functions can be called a *photochemical molecular device* (**PMD**)

[7]. The aim of this chapter is to illustrate some valuable functions that can or could be performed by **PMD**s, to review their possible applications, to single out their molecular components, and to examine the requirements needed for these components and for the entire device.

12.2 ARTIFICIAL *VS* NATURAL DEVICES

PMDs are present, of course, in nature where they perform functions essential to life such as photosynthesis and vision. Important progress towards the understanding of such natural **PMD**s has been made in recent years [19–25]. For example, the structure of the reaction centre of bacterial photosynthesis is known and its function is reasonably well understood. As discussed in Chapter 5, the very efficient photo-induced charge separation achieved in this system arises from the successful competition of forward over back electron-transfer reactions, which is made possible by the *very specific* supramolecular organization reached as the result of evolution.

Examination of this and other natural **PMD**s teaches us the following lesson: valuable photochemical functions can only be obtained upon a complex elaboration of the absorbed light energy input in the dimensions of *space*, *energy*, and *time* by means of a suitably organized supramolecular system. It should be realized, however, that natural systems are extremely complicated and that any synthetic effort aimed at their exact duplication would be hopeless. Such a complexity is related to their *living* nature, that requires the interconnection among many different functions. To obtain a *single*, valuable photoinduced function, simpler strategies can be followed. For example, to convert solar energy into fuel we do not need to design an artificial chloroplast that mimics the natural photosynthetic process; rather, it would be sufficient to construct a **PMD**, presumably much simpler than a chloroplast, capable of performing, for example, the photoinduced splitting of water into hydrogen and oxygen [26]. The history of science shows that man has indeed been extremely successful in achieving valuable functions by using systems different from, and simpler than, the natural ones. For example, the wheel works better than the knee for locomotion along a highway, and a computer works better than the brain for mathematical calculations. Thus, there are good reasons to believe that natural photochemical functions can be successfully duplicated by artificial **PMD**s. It is also conceivable that suitably designed **PMD**s will be able to perform new useful functions not found in nature.

12.3 MACHINERY OF PHOTOCHEMICAL MOLECULAR DEVICES

To perform a particular function and to be useful for a specific application, a **PMD** needs to be constructed of suitable molecular components, each having a specific role. We may distinguish three fundamental types of molecular component:

(i) *active components*, which perform an elementary act or a sequence of elementary acts directly related to the desired fuction;

(ii) *perturbing components*, which can be used to modify the properties of the active components;

(iii) *connecting components*, which can be used to link together the other components in the desired spatial arrangement.

The first requisite of any **PMD** is chemical stability. Only stable devices, in fact, are able to process a large number of photons and can thus be useful for practical applications. For this reason, the elementary acts that take place in **PMDs** are only those which may be reversed. The commonest ones are listed in Table 12.1, where the abbreviations used are also shown [7].

Table 12.1 — Some elementary acts occurring in **PMDs** [7].

Al	light acceptance (absorption)	**Dl**	light donation (emission)
Ael	electron acceptance	**Del**	electron donation
Aen	energy acceptance	**Den**	energy donation
	I	isomerization	
	W	nonradiative decay	

Each active molecular component of a **PMD** must perform a specific elementary act or a sequence of elementary acts. For example, an energy-transfer photosensitizer must first absorb light (elementary act **Al**) and then donate electronic energy to another component (elementary act **Den**): its role can thus be defined as **Al–Den**. Some important active components, their roles, and the abbreviations used are shown in Table 12.2. The main requirements needed for the active components will

Table 12.2 — Active components of **PMDs** [7]

Symbol	Role	Elementary acts[a]
Pel	Electron-transfer photosensitizer	**Al(or Aen)–Ael–Del** **Al(or Aen)–Del–Ael**
Pen	Energy-transfer photosensitizer	**Al–Den**
Pi	Photoisomerizable component	**Al(or Aen)–I**
L	Luminophore	**Al(or Aen)–Dl**
Rel	Electron relay	**Ael–Del; Del–Ael**
Ren	Energy relay	**Aen–Den**
Sel	Electron store	**Ael–Ael–2Del**
U	Energy up-converter	**Aen–Aen–Dl**
H	Holder	**specific binding**

[a]For the abbreviations used to indicate elementary acts, see Table 12.1.

be examined when describing specific types of **PMD**s. The characteristics of perturbing components are discussed in Chapter 4. The connectors are chemical bonds or bridging groups which should have (a) high chemical and photochemical stability, (b) no low-energy excited state to prevent energy trapping, (c) no low-energy redox level to prevent electron or hole trapping, and, usually, (d) a rigid structure to assure vectorial energy or electron migration (for details on specific connectors, see Chapters 5–7).

Several types of **PMD**s have been encountered in previous chapters. Most of the functions performed are based on three main mechanisms: (a) photoinduced electron transfer; (b) electronic energy transfer; (c) photoinduced structural change. The principles underlying these mechanisms are discussed in Chapters 5–7.

12.4 PMDs BASED ON PHOTOINDUCED ELECTRON TRANSFER

As we have seen in section 2.7.2, the electronically excited states of molecules are usually good oxidants and/or good reductants. One can take advantage of this property to photoinduce charge separation and vectorial transport of electric charge. This function can only be performed by means of a suitable elaboration of the light energy input in the dimensions of energy, space, and time on an appropriate sequential assembly of molecular components. Necessary acts are light absorption and electron transfer. The **PMD** must thus comprise an electron transfer photosensitizer (**Pel**), which plays the role of interface towards light, assembled with electron relays (**Rel**) by suitable connectors (**C**).

To make practical use of the light input (Fig.12.1a), a two-component **Pel–Rel** system (diad, Fig. 12.1b) is usually unsuitable because of the occurrence of the fast back electron-transfer reaction. A three-component system (triad, Fig. 12.1c) is expected to be more efficient because a fast secondary electron-transfer step can compete with the back reaction, resulting in charge separation over a larger distance. It should also be noted that, in such systems, light can play the role of either an electron pump or an electron switch (Fig. 12.1c). In the former case, part of the absorbed light energy is converted into chemical energy since the electron is transferred to an upper energy level, resulting in the thermodynamically-forbidden reduction of Rel by Rel'. In the latter case, light plays a kinetic role since excitation is used to overcome the kinetic barrier of the thermodynamically-allowed reduction of Rel by Rel'. It should be noted that the actual synthesis of a triad system of the latter type can be severely hampered by the occurrence of relatively fast intercomponent or intertriad thermal redox processes.

As mentioned above, the interface towards light of a **PMD** performing the photoinduced electron-transfer function has to be an electron-transfer photosensitizer (**Pel**), which is a molecular component that (a) can be excited (directly by light absorption, or indirectly by energy transfer), (b) can undergo oxidation (or reduction) in the excited state, and (c) can be subsequently reduced (or oxidized) back to the ground state. Specific requirements for **Pel**s to be used in practical systems are [27]: (1) stability towards thermal and photochemical decomposition reactions; (2) fully reversible redox behaviour; (3) suitable ground- and excited-state redox

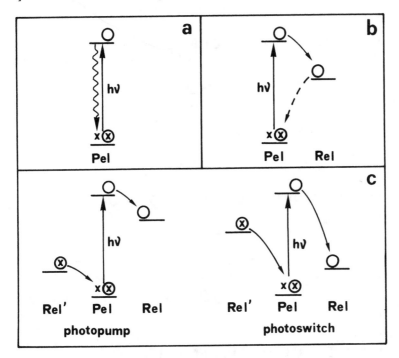

Fig. 12.1 — Light excitation in a single molecular component (a), and photoinduced charge separation in a diad (b), and a triad (c) [7].

potentials; (4) high efficiency of population of the reactive excited state; (5) sufficiently long lifetime of the reactive excited state; (6) appropriate kinetic factors for ground- and excited-state electron-transfer reactions. $Ru(bpy)_3^{2+}$ (section 5.7, Fig. 5.13) is a typical example of a good **Pel**.

A second fundamental component is **Rel**, a molecular species capable of undergoing a one-electron reversible redox reaction. Specific requirements for **Rels** to be utilized as components of practical devices include (1) suitable redox potentials, (2) thermal and photochemical stability in both the oxidation states, (3) lack of quenching ability *via* mechanisms not involving electron transfer, and (4) suitable kinetic factors for electron-transfer reactions [27].

Connections between the active components are often present in **PMDs** based on photoinduced electron transfer. The connectors (**C**) should meet the general requirements given in section 12.2 and should also provide (or allow to occur) an appropriate weak coupling between the connected components (section 5.2).

Photoinduced electron transfer in supramolecular systems may be useful for several applications which include (Fig. 12.2) (a) light energy conversion, (b) photoinduced electron collection, (c) remote photosensitization, and (d) photoswitching.

12.4.1 Conversion of light into chemical or electrical energy
Conversion of light into chemical energy (artificial photosynthesis) can be performed by **PMDs** like those shown in Fig. 12.3. The principal components of the device must

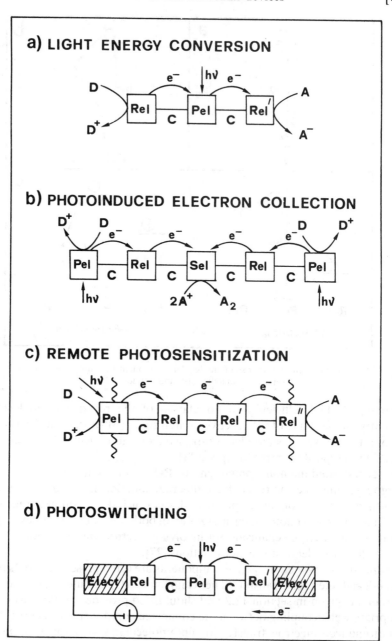

Fig. 12.2 — Block diagrams illustrating the operation of some **PMDs** based on photoinduced
electron transfer [7].

have appropriate relative (ground- and/or excited-state) redox potentials and must
be assembled in a correct energy sequence. Photoinduced charge separation can
occur either between two relay components as in Fig. 12.3a, or between the

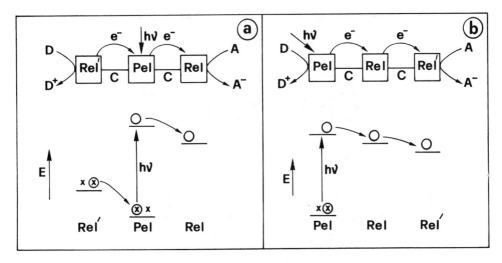

Fig. 12.3 — Photoinduced charge separation between (a) two relay components and (b) a photosensitizer and a relay [7].

photosensitizer and a relay as in Fig. 12.3b. When interfaced with electrodes, these devices can convert light into electrical energy and constitute the fundamental elements of molecule-based photovoltaic cells [28–34].

In the last few years photoinduced charge separation in covalently-linked systems has been extensively investigated [35–38] (Chapter 5). For illustration of these systems as **PMDs** we have chosen the triad developed by Wasielewski *et al.* [39] (Fig. 12.4), which consists of a porphyrin (**Pel**), a quinone (**Rel**), and an aromatic amine (**Rel'**) linked together by rigid triptycene connectors (**C**). As shown in Fig. 12.4, the photoinduced charge separation involves the following sequence of events (rate constants for butyronitrile solution):

— light absorption:

$$\text{Rel'–Pel–Rel} + h\nu \rightarrow \text{Rel'–}^*\text{Pel–Rel} \qquad (12.1)$$

— primary electron transfer:

$$\text{Rel'–}^*\text{Pel–Rel} \xrightarrow{1.1 \times 10^{11}\,\text{s}^{-1}} \text{Rel'–Pel}^+\text{–Rel}^- \qquad (12.2)$$

— secondary electron transfer

$$\text{Rel'–Pel}^+\text{–Rel}^- \xrightarrow{1.4 \times 10^{10}\,\text{s}^{-1}} \text{Rel'}^+\text{–Pel–Rel}^- \qquad (12.3)$$

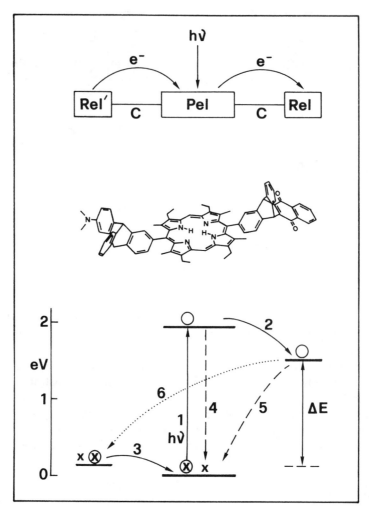

Fig. 12.4 — Block diagram, molecular structure, and one-electron energy diagram for an aniline(**Rel′**)-porphyrin(**Pel**)-quinone(**Rel**) triad [39]. Processes 1–6 in the energy diagram correspond to eqns (12.1–12.6) in the text.

Such a charge separation takes place because the primary electron-transfer step (12.2) competes favourably with the porphyrin excited-state decay (12.4), and the secondary electron-transfer step (12.3) competes favourably with the primary

$$\text{Rel′--*Pel--Rel} \xrightarrow{10^8 \text{s}^{-1}} \text{Rel′--Pel--Rel} \tag{12.4}$$

charge-recombination process (12.5), the rate of which is known from the study of the corresponding Pel–Rel diad [40].

$$\text{Rel}'-\text{Pel}^+-\text{Rel}^- \quad \overset{5.6 \times 10^9 \text{s}^{-1}}{\rightarrow} \quad \text{Rel}'-\text{Pel}-\text{Rel} \tag{12.5}$$

In the absence of other species (such as A and D in Fig. 12.3a) that can react with the Rel'^+ and/or Rel^- components, a relatively slow charge recombination reaction (12.6) takes place:

$$\text{Rel}'^+-\text{Pel}-\text{Rel}^- \quad \overset{4.1 \times 10^5 \text{s}^{-1}}{\rightarrow} \quad \text{Rel}'-\text{Pel}-\text{Rel} \tag{12.6}$$

This **PMD** performs photoinduced charge separation over a long distance (centre-to-centre aniline-quinone distance, 25 Å) with a high efficiency (71%) and stores a considerable amount of energy (1.39 eV) for a relatively long time (2.5 μs). This result, obtained with a simple artificial system, should be compared with the performance of the reaction centre of the photosynthetic bacteria where ~ 0.6 eV, produced with 100% efficiency, are stored for ~ 0.1 s [25].

A possible strategy to increase the efficiency of long-lived charge separation is that of increasing the number of molecular components. Such a strategy is illustrated by the work performed by Gust, Moore, *et al.* [41,42]. By adding a terminal benzo-quinone moiety (**Rel''**) to a carotene(**Rel'**)–porphyrin(**Pel**)–naphthoquinone(**Rel**) triad they have obtained the tetrad shown in Fig. 12.5 where the porphyrin is linked to the attached units by amide bonds, and a saturated bicyclic bridge links the two quinones. In such a **PMD** light excitation is followed by electron-transfer from the excited porphyrin to the naphthoquinone ($k \geqslant 2 \times 10^{10}$ s^{-1}, step (2) in Fig. 12.5). In competition with the primary back electron-transfer reaction (step (7)), two path-ways (step (3) followed by step (4), or *vice versa*) lead to the final charge-separated state which stores ~ 1.1 eV for 460 ns with 23% efficiency in CH_2Cl_2. For the parent triad, which lacks the terminal benzoquinone unit, step (3) cannot occur and the yield of charge separation is only 4%. It should be emphasized that the design of triads and tetrads has been made possible by the large amount of theoretical and experimental information accumulated in the field of electron-transfer kinetics, as discussed in detail in Chapter 5. The design of **PMDs** to obtain long-lived photo-induced charge separation lies now on a relatively firm basis. This rapidly moving field promises to lead to outstanding developments in the near future.

Charge separation processes in Langmuir–Blodgett films (for example, [28–34]), polymer films (for example [43,44]), and vesicular systems (for example [45,46]) are currently the object of extensive investigation (see also section 12.7.3).

12.4.2 Photoinduced electron collection

Fig. 12.2b shows the schematic structure of a **PMD** featuring photoinduced electron collection. In the example shown, the device consists essentially of two "fused" triad systems in which two photons pump two electrons toward a common electron store **Sel**, capable of effecting concerted two-electron redox processes. Such devices can be considered as molecular analogues of metallized and nonmetallized semiconductor

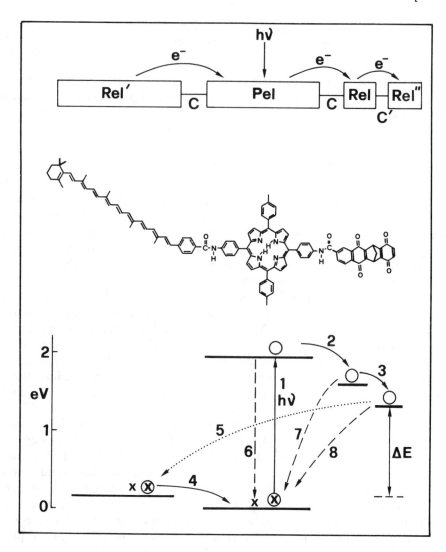

Fig. 12.5 — Photoinduced charge separation in a four-component (tetrad) system [41]. For details, see text.

particles, where a variety of photoinduced multielectron-transfer processes can take place [47–50]. A hybrid device for photoinduced electron collection, composed of porphyrin molecules linked to RuO_2 microcolloids, has been described [51].

In principle, any molecular species capable of undergoing multiple redox processes (for example [4,52–54]) could play the role of **Sel**. In practice, however, accumulation of redox equivalents *via* sequential photochemical processes where one photon produces one redox equivalent poses several problems. The great development in the field of redox-reactive polynuclear metal complexes (for example [55–57]) and of multiple redox sites bound to polymers (for example [58–60]), might lead to important progress in this area. Manganese complexes, which are involved in plant photosynthesis [61,62], are also promising multielectron catalysts [63] and have already been used to catalyse O_2 evolution in phospholipid vesicles using $Ru(bpy)_3^{2+}$ as a photosensitizer [64].

12.4.3   Remote electron-transfer photosensitization

Photosensitization of an electron-transfer process between remote reactants (for example, reactants separated by a membrane) is, in principle, another important application of **PMDs** based on photoinduced electron transfer (Fig.12.2c). This process may become useful both in the case of a spontaneous overall electron-transfer process that is prevented by an intermediate step having a high energy barrier (photoswitching mode), as well as in the case of an endergonic process (photopumping mode) if product mixing and back reaction have to be avoided.

An example of photodriven electron transfer across a bilayer membrane is that concerning the exergonic oxidation of ascorbic acid by ferricyanide ions in aqueous solutions separated by a phospholipid membrane containing a carotenoid–porphyrin–quinone triad (C–P–Q) [65,66]. As shown schematically in Fig. 12.6a,

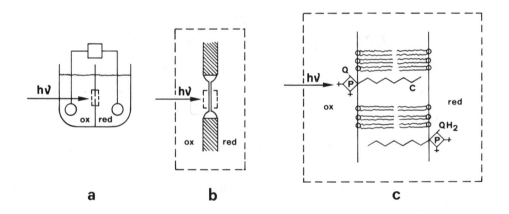

Fig. 12.6 — Photodriven electron transfer across a bilayer membrane [66]. For details, see text.

the half cells containing the oxidant and the reductant are separated by a teflon divider where a hole of 1 mm diameter was drilled. Application of a phospholipid solution containing the triad to this hole results in the formation of a bilayer membrane across the hole (Fig. 12.6b). Since such a membrane is a good insulator, only a very small current was observed in the dark. However, upon illumination of the membrane with 600 nm light from a laser, a photocurrent was detected. The explanation given [65,66] is that the triad molecules span the organic interior of the membrane, with the amphipathic porphyrin (and therefore the attached quinone) lying near the membrane surface (Fig. 12.6c). Excitation of the porphyrin generates the C^+–P–Q^- charge-separated state. In those triad molecules where the porphyrin-quinone moiety faces the aqueous solution containing the oxidant, the reduced quinone transfers an electron to a ferricyanide ion regenerating the neutral quinone, and the carotenoid radical cation accepts an electron from the ascorbic acid to regenerate the neutral carotenoid. The net result is electron transfer from one side of the membrane to the other side, with a consequent electron flow through the external circuit.

12.4.4   Switching electric signals
Figure 12.7 shows schematically the potential application of photoinduced electron

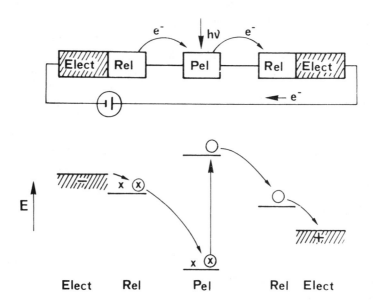

Fig. 12.7 — Schematic representation of a **PMD** for switching electric signals [7].

transfer to the switching of electric signals [7]. The relay components are connected to electrodes and separated by a **Pel**. Application of a potential difference to the electrodes does not allow electronic migration when the **Pel** is kept in the dark because its lowest unoccupied and highest occupied orbitals lie at too high an energy with respect to the Fermi levels of the electrodes. On light excitation, however, an electron is transferred from the highest occupied to the lowest unoccupied orbital of the photosensitizer, making possible the transfer of an electron between the two relays and, therefore, between the two electrodes. Because of the possibility of modifying its electrical resistance with light, this **PMD** can be considered as a phototransistor.

The molecular shift register described in section 12.7.3 is based essentially on this function. The idea of using single molecules for switching and rectification purposes was first advanced by Aviram and Ratner [1] and constitutes one of the basic concepts of molecular electronic devices [3,5,12,14,67–71].

12.5 PMDs BASED ON ELECTRONIC ENERGY TRANSFER

Light absorption by a component of a **PMD** generates "localized" electronic energy. For several practical purposes an important function is represented by the possibility of transmitting this electronic energy (or at least part of it) to another component of the **PMD** over a more or less long distance, where the energy will be used for chemical purposes or reconverted into light. Once again this function can only be obtained upon elaboration of the light energy input in the dimensions of energy, space, and time by means of an appropriate sequence of suitable components. Some interesting applications of **PMD**s based on energy transfer (Fig. 12.8) [7] concern (a) spectral sensitization, (b) antenna effect, (c) remote photosensitization, and (d) light-energy up-conversion.

For **PMD**s performing this function, the interface toward light must be an energy-transfer photosensitizer (**Pen**, Table 12.2), that is a molecular species capable of absorbing light and donating electronic energy to another component of the device. A **Pen** must satisfy several requirements [72]. A very important one is, of course, (1) stability towards photochemical decomposition reactions. Furthermore, the **Pen** excited state involved as a donor in the energy-transfer process should (2) be populated with high efficiency from the **Pen** excited state directly obtained by light absorption, (3) have a sufficiently high energy content, (4) be reasonably long-lived, and (5) have appropriate kinetic factors for energy-transfer processes.

In a two-component **PMD** based on electronic energy transfer, the **Pen** is connected to either an energy-transfer relay, **Ren**, or a luminophore, **L**. When energy migration over a long distance is required, intermediate **Ren** components are needed. Specific requirements for **Ren**s are (1) photochemical stability, (2) presence of suitable energy levels, and (3) appropriate kinetic factors to make energy-transfer processes competitive with radiationless decay of the excited states. Specific requirements for **L** are (1) photochemical stability and (2) high luminescence efficiency. The chemical bonds or groups which act as connectors (**C**) should exhibit the general properties listed in section 12.2. Furthermore, they should provide a suitable coupling to enable the occurrence of energy transfer between the connected components (Chapter 6).

12.5.1 Spectral sensitization

Spectral sensitization may be important when the light absorption properties of a potentially luminescent or photoreactive species do not permit excitation in a desired wavelength range. The device needed may be a simple diad as shown in Fig. 12.8a. Such a device can also be useful as a bathochromic luminescent shifter.

A simple example of spectral sensitization and bathochromic luminescent shift is given by the binuclear $[NC–Ru(bpy)_2–NC–Cr(CN)_5]^{2-}$ complex where *visible light* absorption by the $Ru(bpy)_2^{2+}$ chromophoric unit causes luminescence from the *colourless* $Cr(CN)_6^{3-}$ luminophore [73]. As shown in Fig. 12.9, the steps involved are:

a) SPECTRAL SENSITIZATION

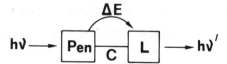

b) ANTENNA EFFECT

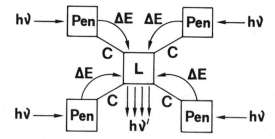

c) REMOTE PHOTOSENSITIZATION

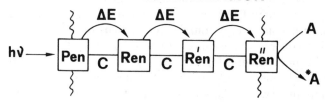

d) LIGHT ENERGY UP-CONVERSION

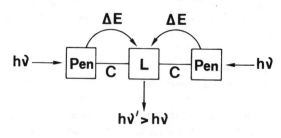

Fig. 12.8 — Block diagrams illustrating the operation of some **PMDs** based on electronic energy transfer [7].

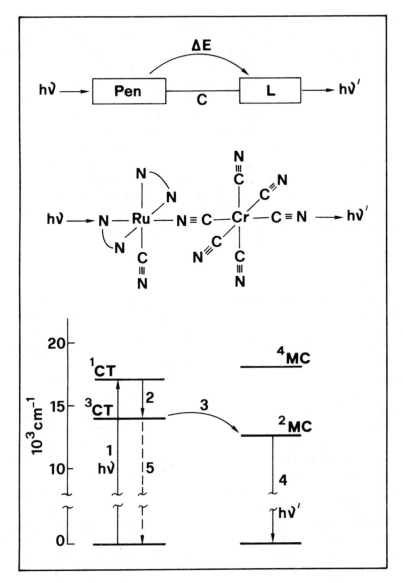

Fig. 12.9 — Block diagram, molecular structure (schematic), and energy level diagram for the [Ru(bpy)₂(CN)–NC–Cr(CN)₅]²⁻ diad performing spectral sensitization [73].

(1) light excitation in the intense spin-allowed charge-transfer (^{1}CT) bands of the Ru-based chromophore;
(2) complete conversion to the lowest spin-forbidden charge-transfer (^{3}CT) level;
(3) energy transfer to the metal-centred doublet (^{2}MC) state of the Cr-based luminophore;
(4) phosphorescence from the luminophore.

The energy-transfer process, which occurs by an exchange mechanism, has unit efficiency and is fast enough ($k > 10^9$ s^{-1}) to prevent radiative and radiationless deactivation within the Ru-based chromophore (step (5) in Fig. 12.9). It should also be noted that the efficiency of ^{2}MC population in the **Pen-L** device is twice that obtainable upon direct excitation of the spin-allowed quartet metal-centred (^{4}MC) level in the free **L** because the ^{4}MC → ^{2}MC intersystem-crossing process is only 50% efficient. Furthermore, the device is photostable whereas the free **L** is highly photolabile because of ^{4}MC reactivity. In conclusion, a comparison of the properties of the **Pen-L** devices with those of the isolated **Pen** and **L** components provides not only examples of spectral sensitization and bathochromic luminescence shift, but also of the antenna effect (section 12.5.2) and photoprotection.

Several examples of energy transfer and spectral sensitization in monolayers and other organized molecular assemblies have been reported (for example [74–80]). Spectral sensitization of semiconductor electrodes by supramolecular species [81] will be discussed in section 12.7.2.

12.5.2   Antenna effect
This effect consists of an enhanced light-sensitivity obtained by an increase in the overall cross-section for light absorption. To achieve this result, energy should be conveyed from several **Pen**s to a common component that represents the interface toward use. In the scheme shown in Fig. 12.8b such a component is a luminophore, but components playing other roles (for example, that of electron-transfer photo-sensitizer as in the natural photosynthetic process [19,20]) could also be used (section 12.7.3).

Many **PMD**s performing the antenna effect have been discussed in Chapter 6 and in section 11.1.4. As an illustrative example, we report here the case of the tetrametallic Os[(2,3-dpp)Ru(bpy)$_2$]$_3^{8+}$ complex shown in Fig. 12.10, where the photon energy collected by the peripheral Ru-containing chromophores is conveyed to the central Os-containing luminescent core [82]. In this system the distinction between active components and connectors is not clearcut since the 2,3-dpp bridging ligands are also involved in low-energy metal-to-ligand CT excited states. For the sake of simplicity, in the energy-level diagram of Fig. 12.10 only one singlet and triplet CT levels are shown for each metal-containing unit. The energy-transfer process is fast enough to prevent the radiative and radiationless decay of ^{3}CT within each **Pen** unit and is 100% efficient. This device mimics some of the features of natural photosynthetic antennas, with the Os-based chromophore playing the role of the special pair as an energy trap. Interestingly, when the central Os^{2+} ion is replaced by Ru^{2+}, the energy levels of the central unit become higher than the corresponding levels of the peripheral units, so that energy transfer takes place in the reverse direction (that is, from the centre to the periphery) [82,83]. A homometallic heptanuclear complex of the same type has also been synthesized [84].

Tetrad and pentad systems in which light harvesting *via* energy transfer is followed by charge separation have been reported [85]. Light harvesting by func-tionalized polymers (for example [58,59,86]) and by mono- or multilayers of dyes deposited on semiconductor electrodes (for example [33,74,81]) has been widely investigated.

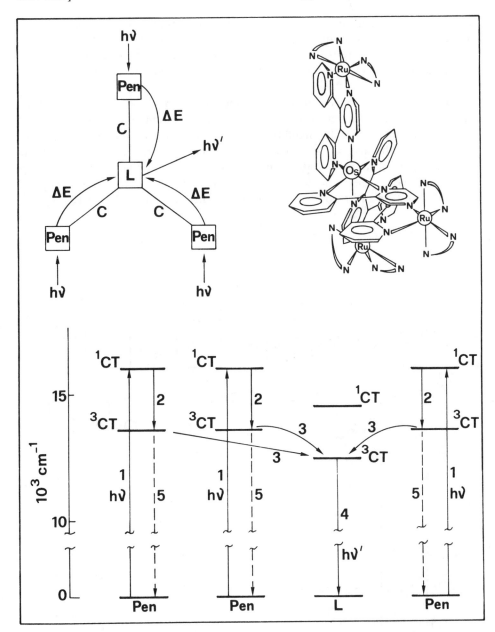

Fig. 12.10 — Block diagram, schematic molecular structure, and schematic energy level diagram for the $Os[(2,3\text{-dpp})Ru(bpy)_2]_3^{8+}$ tetrad performing the antenna effect [82].

12.5.3 Remote photosensitization

For remote energy-transfer photosensitization, the component interfacing the **PMD** towards use may be either an energy-transfer relay (remote energy-transfer sensitization of a photochemical reaction, Fig. 12.8c) or a luminophore (remote generation of optical signals). To obtain migration over long distances, the **PMD** should involve a sequence of energy-transfer processes along a vectorial array of components. The initial and final energy-transfer steps must be exergonic, but the other steps can also be isoergonic.

Energy transfer between molecular components across more or less long bridges has been investigated extensively (Chapter 6), but no example is known concerning multicomponent covalently-linked rigid systems. To illustrate remote photosensitization, we cite the case of a soluble polymer (which, of course, is not rigid) with the active components appended [86]. The polymer is a 1:1 copolymer of styrene:m,p-(chloromethyl)styrene. The active components are $Ru(bpy)_3^{2+}$ and $Os(bpy)_3^{2+}$ derivatives and an anthryl derivative (Fig.12.11), hereafter indicated by Ru^{2+}, Os^{2+}

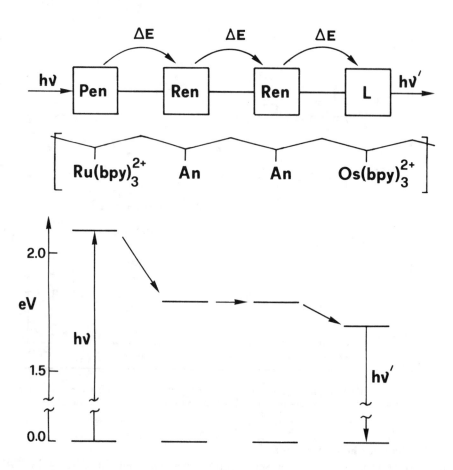

Fig. 12.11 — Remote luminescence sensitization in a functionalized polymer [86].

and An. The An component has its lowest triplet-excited state at 1.8 eV, which is lower than the luminescent state of Ru^{2+} (2.1 eV) and higher than the luminescent level of Os^{2+} (1.7 eV). The three components are located randomly along the ~ 30 available functionalized sites in the backbone of the polymer, in the ratio $12(An):3(Ru^{2+}):3(Os^{2+})$. Polymers containing only An and Ru^{2+} (12:3) and Ru^{2+} and Os^{2+} (3:3) were also investigated. In the latter system, separate emissions from the Ru^{2+} and Os^{2+} chromophores were obtained, showing that no energy transfer takes place. In the Ru^{2+}–An system, the luminescence of Ru^{2+} is quenched by An. Finally, in the three-component system Ru^{2+} luminescence is quenched and Os^{2+} luminescence is sensitized, apparently *via* energy transfer mediated by the An components as schematized in Fig.12.11.

Another example of remote photosensitization comes from solid state materials. For the system $NaY_{0.98-x}Gd_xCe_{0.01}Tb_{0.01}F_4$ (Fig. 12.12), absorption of uv light by

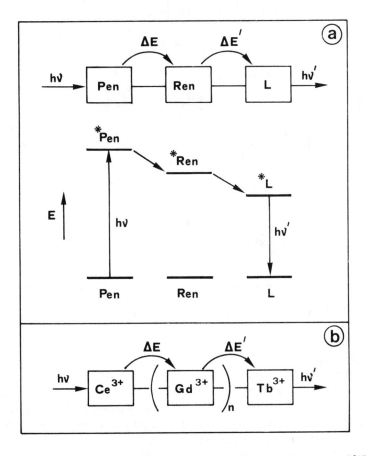

Fig. 12.12 — Remote photosensitization in a solid state lanthanide compound [87].

Ce^{3+} ions is followed by energy transfer to the Gd^{3+} sublattice where energy migrates until it is trapped by Tb^{3+} ions which give a green emission [87]. The number of transfer steps is of the order of 10^5 and the quantum yield can reach 0.95. These systems can be very efficient phosphors for energy-saving lamps if suitable sensitizers and activators are used [88]. One-dimensional energy migration can also occur in suitable solid structures [89].

12.5.4 Light-energy up-conversion

Another potential utilization of electronic energy transfer is that shown in Fig. 12.8d and, in more detail, in Fig. 12.13 which represents a **PMD** capable of performing

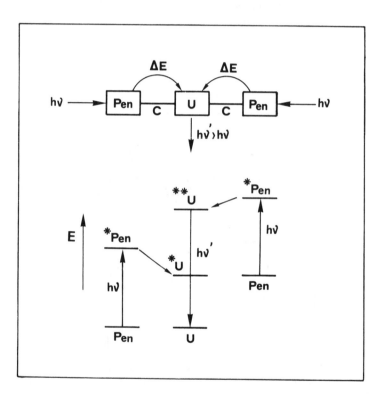

Fig. 12.13 — Schematic representation of a **PMD** for light-energy up-conversion [7].

light-energy up-conversion so as to obtain, for example, anti-Stokes luminescence. To reach such a result one needs a very peculiar component, **U**, which has to be raised to a suitable energy level *U by a first energy-transfer step and, before undergoing deactivation, has to be further raised to an upper excited level, **U, by a second energy-transfer step. The two exciting photons can be absorbed by two different **Pen** components or, sequentially, by the same **Pen**. The requirements needed to obtain this result are not easy to meet. They include suitably-spaced energy levels in **U**, a long lifetime of the first excited level *U, high density of excited **Pens** around U, and luminescent emission from the upper excited state **U. Light-energy up-conversion by means of this mechanism has been obtained in the solid state with a system containing Er^{3+} and Yb^{3+} ions [90,91] (Fig. 12.14). The Yb^{3+} ion absorbs 970 nm

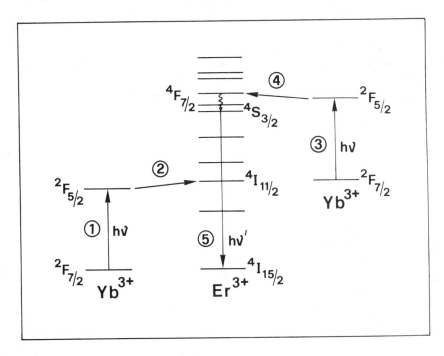

Fig. 12.14 — Infrared-to-green light-energy up-conversion by rare earth ions [90].

photons, and in phonon-assisted energy transfer an erbium ion is excited to its $^4I_{11/2}$ state. Then, a second photon is absorbed by another ytterbium ion and produces $^4I_{11/2} \rightarrow {}^4F_{7/2}$ in the excited erbium by resonance energy transfer. This successive double energy transfer accounts for the fact that the excitation spectrum of such a system agrees with the reflection spectrum of Yb^{3+}, and that the intensity of the resulting green emission $^4S_{3/2} \rightarrow {}^4I_{15/2}$ of erbium is proportional to the square of the flux of infrared photons. Up-conversion can also take place by energy transfer between two Er^{3+} ions [92] or by stepwise photon absorption [93].

12.6 PMDs BASED ON PHOTOINDUCED STRUCTURAL CHANGES

Photoinduced structural changes can constitute the basis of a number of interesting devices, particularly for switch on/off applications. The characteristic active components of **PMDs** based on this function are photoisomerizable components (**Pi**), discussed in section 7.2.

Important requirements for **Pi** are (1) capacity of being excited either by light absorption or by energy transfer, (2) chemical and photochemical stability towards decomposition, and (3) possibility of reverting the photoinduced structural change by a thermal or photochemical reaction. The nature of the other components of a **PMD** performing this function depend on its specific utilizations, which include (a) switching electrical signals, (b) switching receptor (coordinating) ability, (c) modification of cavity size, and (d) activation of coreceptor catalysis.

12.6.1 Switching electric signals

Any photoinduced configurational change which modifies the electronic conduction in a sequence of components can switch on/off electrical signals. This is shown schematically in Fig. 12.15a where a photoisomerizable component is supposed to be

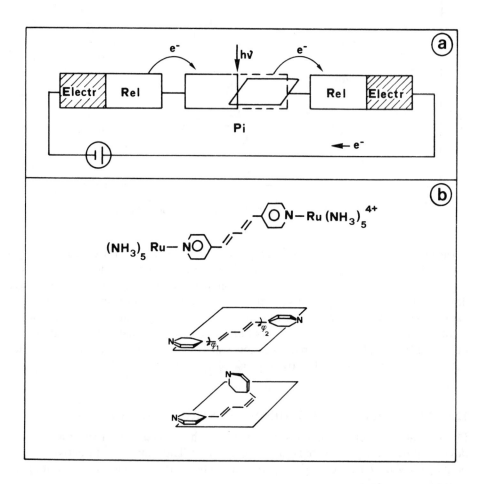

Fig. 12.15 — (a) Schematic representation of a **PMD** for switching electrical signals; (b) a photoisomerizable bridging ligand [97].

connected to two relay components deposited on the electrodes of a battery. The electric charge can flow when the two **Rels** are conjugated through **Pi**. Photoisomerization of **Pi**, however, may destroy the conjugation, thereby switching off the charge flow. Such a device can be considered as another type of molecule-based phototransistor. Scaling-up and interfacing with electrodes are, of course, necessary for practical applications (section 12.7.1).

In looking for suitable photoisomerizable compounds the attention of several research groups has been focused on molecules with twisted-intramolecular-charge-transfer states (the so called TICT compounds [94–96], section 7.2.4). Particularly interesting are also model systems made up of binuclear metal complexes linked by photoisomerizable bridging ligands (section 7.3.3). In the example of Fig.12.15b [97,98], two $Ru(NH_3)_5^{2+}$ units are connected by a bipyridylbutadiene bridge. Such a bridge is expected to undergo EE to ZE isomerization upon light excitation, with a remarkable change of geometry. The EE isomer should be nearly planar, whereas in the ZE isomer steric crowding is expected to cause a considerable twist angle between one of the pyridine rings and the central butadiene skeleton. Preliminary results based on the properties of the intervalence transfer band in the mixed valence Ru^{II}–Ru^{III} system indicate some change in the metal–metal coupling upon isomerization [98].

12.6.2 Switching receptor ability

Switch on/off of receptor (coordinating) ability is another rather interesting possibility for **PMDs** based on conformational changes. Fig. 12.16a shows schematically a mode of operation of such a device. Upon light excitation, a photoisomerizable component **Pi** undergoes a structural change which brings two potential ligands (holders, **H**) to a suitable distance to enclose a metal ion. A subsequent thermal or photochemical reverse structural change can also be envisaged to move a holder away from the other, with release of the metal ion. Such **PMDs** can be used as photoresponsive ionophores (phototweezer) [99].

Suitable structural changes can be obtained by *cis–trans* photoisomerization, and appropriate holder components to bind metal ions are crown ethers. A classical example of a photoresponsive crown ether is shown in Fig. 12.16b [100]. In the *trans* configuration of the azobenzene chromophore the supermolecule has a weak coordinating ability for large cations. Light excitation causes the *trans* → *cis* isomerization yielding a supramolecular configuration suitable to enclose large metal ions between the two crown ethers, with a strong increase in the coordination ability. The *trans* form extracts Na^+ 5.6 times more efficiently than the *cis* form, whereas the *cis* form extracts K^+ 42.5 times more efficiently than the *trans* form. Such devices offer a means of controlling the rate of ion transport through membranes [99].

12.6.3 Modification of cavity size

Photoinduced structural changes can also offer the opportunity of modifying the size of a cavity. Such a device, schematized in Fig. 12.17a, is composed of a **Pi** component linked to an appropriate sequence of flexible connectors in a cyclic-type structure. A photoinduced structural change of **Pi** can cause profound modifications in the size of the cyclic structure. This may lead to a change in the coordination ability and

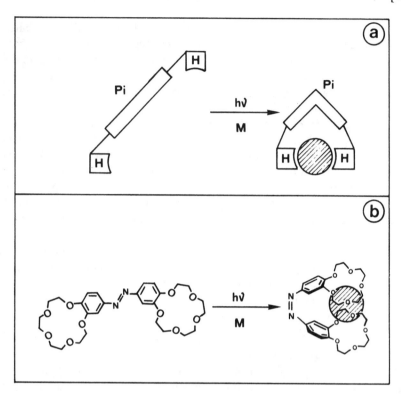

Fig. 12.16 — (a) Schematic representation of a **PMD** for switching receptor ability; (b) switching coordinating ability of a photoresponsive crown ether by light excitation [100].

may also play the role of a chemical gate which could be used, for example, to change the permeability of a membrane.

Photoresponsive crown ethers can also be used for this purpose, as shown in Fig. 12.17b. The *EE* isomer of the compound shown in the figure has a cylindrical shape, whereas the *ZZ* isomer, obtained by light excitation, exhibits a cavity of different geometry [101]. An even more interesting, three-dimensional example is given by compound **7.41** [102] (section 7.3.4).

12.6.4   Activation of coreceptor catalysis

Another potential application of photoinduced structural changes is the activation of coreceptor catalysis. As schematized in Fig. 12.18, a change in the geometry of a photoisomerizable component linked to two suitable host-type sites **H** and **H'** could be used to bring two guest species close together, thereby facilitating their interaction. Examples of this type of PMD are not yet known. Complex enzyme-type catalytic processes [103–105] which involve recognition of the reactants, transformation, and release of the products, could be strongly facilitated by such a type of photoprocess.

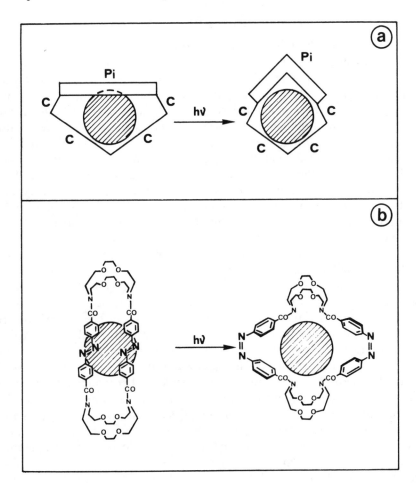

Fig. 12.17 — (a) Schematic representation of a **PMD** for modification of cavity size; (b) change of the cavity size in a photoresponsive crown ether [101].

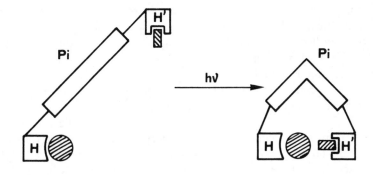

Fig. 12.18 — Schematic representation of a PMD for activation of coreceptor catalysis [7].

12.7 FROM PMDs TO PRACTICAL MICRODEVICES

12.7.1 Interfacing with the macroscopic world

Some **PMDs** illustrated in the previous sections can be used directly. **PMDs** exhibiting the antenna effect (Fig. 12.8b), such as the Eu^{3+} cage-type complexes described in section 11.1, have been patented as luminescence probes for biological applications, and other similar devices absorbing in the visible be used as fluorescent concentrators [106]. Photochromic molecules can be directly used as elements for optical storage memories (see, for example [107–109]), and dyes can be employed as optical modulators (see, for example [110]). Photoresponsive ionophores, such as the azocrown ethers discussed in sections 7.3.1 and 12.6.2, can photoinduce ion transport through liquid membranes. **PMDs** based on configurational changes can be used to photocontrol micellar catalysis (section 7.4.1), to photoregulate the properties of membranes, and to cause photoviscosity, photomechanic, and other photoeffects in polymers (section 7.4.2).

In most cases, however, practical utilization of **PMDs** requires the solution of a complicated problem, namely that of a suitable interfacing of the supramolecular species with structures of the macroscopic world. A quite similar situation is encountered in the related field of molecular electronic devices [2–6,9–18]. It is envisioned, in fact, that the next generation of electronic devices will be constructed by assembling individual molecular components into arrays, thereby engineering from *small upward* rather than from *large downward* as do current lithographic techniques. Such arrays would be characterized by two revolutionary aspects: (i) the functional elements would be individual molecules rather than macroscopic ensembles, and (ii) the operation would be based on molecular properties rather than on bulk properties. The realm of the molecular has indeed extraordinary potentialities for problems concerning memory, logic, and amplification. Using molecules, information bits can be associated to oxidation states, excited levels, geometrical conformations, etc. A chip based on molecular computing elements would have a memory density thousands of times larger than conventional very large scale integrated (VLSI) chips and would require much less energy. The complex problem of the "communication" between the molecular size world and the macroscopic world for the full utilization of these potentialities is currently the object of much interest [2–6,9–14,67–71].

Interfacing with the macroscopic world can require the assembly of a great number of identical or different supramolecular species on a "terminal" that can in some way be handled with available technologies and used to collect the energy and/ or signals generated by light absorption. Obvious candidates to play the role of macroscopic terminals for **PMDs** are electrodes (metallic or semiconductor), which can process both redox energy and electric signals, and semiconductor powders (metallized or non-metallized), which can use charge-separation energy for chemical purposes. While other solutions can also be envisaged, in the last few years an extensive literature has accumulated both on derivatized electrodes (for some recent papers, see [58–60,81,111–120]), and on semiconductor powders [47–51,121,122]. Detailed discussion of these topics is outside the scope of this book. In the next sections, however, we shall illustrate some proposed devices involving supramolecular species interfaced with electrodes.

12.7.2 Spectral sensitization of semiconductor electrodes

Dye sensitization, that is charge injection from an electronically-excited adsorbed dye, is a well-established technique [123,124] that enables the driving of photoelectrochemical and photocatalytic processes on wide-bandgap semiconductors using sub-bandgap excitation. This feature is of obvious relevance to the use of semiconductors in solar energy conversion [122]. The main drawback of this technique is that, at monolayer coverage, light absorption by the dye is often inefficient. On the other hand, multilayer adsorption does not help, as the inner layers tend to act as insulators with respect to the outer ones [125]. Thus, the only successful strategy to obtain good light-harvesting efficiency of sensitized semiconductors has so far been that of increasing the surface area. In photocatalytic systems this can be achieved by using colloidal semiconductor particles [126,127]. In photoelectrochemical systems, substantial advances have been made with the use of electrodes of extremely high surface area, such as the "fractal" TiO_2 electrodes developed by Graetzel [128].

A conceptually different (but complementary) strategy for improving the light-absorption efficiency of a sensitized semiconductor is to replace the sensitizer molecule at the semiconductor-solution interphase with a *sensitizer-antenna* molecular device made of an electron-transfer photosensitizer, **Pel**, covalently linked to energy-transfer photosensitizers, **Pen**s (Fig.12.19b) [81,129]. In this way, both the light directly absorbed by the **Pel** unit and that absorbed by the **Pen** antenna components can be used to effect charge injection, with an increase in the overall cross-section for light absorption compared to the case of a simple molecular sensitizer. How this increase is actually distributed over the action spectrum depends on the spectral characteristics of the antenna and sensitizer chromophores (subject to the obvious condition that $h\nu' \leq h\nu$). In principle, antenna-sensitizer molecular devices with higher light-harvesting efficiency could be designed making use of several antenna components in parallel (**Pen–Pel–Pen**) or in series (**Pel–Pen–Pen**), although for the latter case efficient **Pen**-to-**Pen** energy transfer would be an additional requisite. Polynuclear Ru(II) polypyridine complexes are particularly suitable candidates to play the role of antenna-sensitizer devices since they exhibit very intense absorption bands and appropriate redox properties [38] (section 6.3).

In an attempt to demonstrate the use of antenna-sensitizer molecular devices in the sensitization of semiconductor electrodes, the trinuclear complex $[Ru(bpy)_2(CN)_2]_2Ru(bpyCOO)_2]^{2-}$, schematized in Fig. 12.19a, has been designed [81]. The role of the carboxylate groups is that of anchoring the central component to the semiconductor surface by electrostatic interaction. From previous studies on the parent $[Ru(bpy)_2(CN)_2]_2Ru(bpy)_2^{2+}$ complex [130] it was already known that the lowest excited state of the trinuclear complex is a MLCT level of the central component (which is hardly affected by carboxyl substitution), so that the light energy absorbed by the peripheral units can be efficiently transferred to the central unit. In experiments carried out using TiO_2-coated electrodes (aqueous solution, pH 3.5, NaI as electron donor) photocurrents were obtained on irradiation with visible light [81]. The photocurrent spectrum reproduces closely the absorption spectrum of the complex, indicating that the efficiency of conversion of absorbed light to electrons is constant throughout the spectrum, regardless of whether the incident light is absorbed by the central unit or by the terminal ones. This indicates that the energy absorbed by the peripheral units is efficiently transferred to the

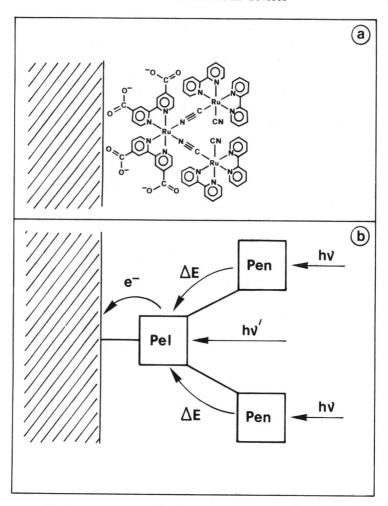

Fig. 12.19 — Spectral sensitization of a semiconductor electrode by a supramolecular species [81].

central one, where it is used for electron injection. Thus, the trinuclear complex $[Ru(bpy)_2(CN)_2]_2Ru(bpyCOO)_2)_2^{2-}$ appears indeed to perform as an antenna-sensitizer molecular device on the surface of TiO_2 as schematized in Fig. 12.19b.

The use of polynuclear complexes exhibiting even better absorption characteristics (a heptanuclear Ru(II) complex with $\varepsilon = 76\,000\ M^{-1}cm^{-1}$ at $\lambda = 547$ nm has been recently reported [84]) could bring about a further gain in the already remarkable conversion efficiencies reached by state-of-the-art regenerative photoelectrochemical cells [113] for solar energy conversion.

12.7.3 Sensitization of charge separation in Langmuir–Blodgett films

The device discussed in the previous section works on the basis of cooperation of **Pen** and **Pel** components, that is of energy- and electron-transfer processes. Other devices employing **Pen** and **Pel** components have been described (see, for example [74,131]. Fujihira *et al.*, after several investigations on photoinduced processes in covalently linked components and Langmuir–Blodgett films (see, for example [132–135]), have designed a complex supramolecular system, schematized in Fig. 12.20, which simulates the photosynthetic process [136]. The system is made of

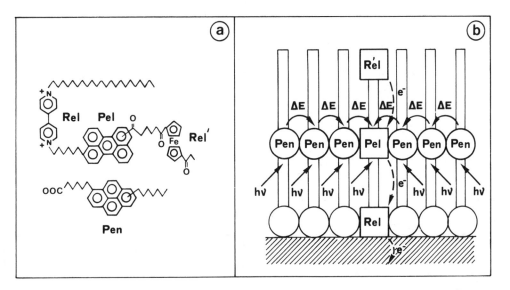

Fig. 12.20 — Sensitization of charge separation in a Langmuir–Blodgett film [136].

amphiphilic pyrene derivatives (**Pen**) and amphiphilic triads (**Rel–Pel–Rel′**) which consist of a central perylene component (**Pel**) covalently linked to a viologen (electron acceptor, **Rel**) and a ferrocene moiety (electron donor, **Rel′**). A mixture of the **Pen** and **Rel–Pel–Rel′** species (5:1) gives a mixed monolayer that can be transferred onto a gold optical transparent electrode with the orientation shown in Fig. 12.20b. Upon light excitation, a photocurrent is observed whose excitation spectrum shows both the pyrene and perylene absorption bands. Quenching of the pyrene (**Pen**) luminescence is also observed in the mixed monolayers. As schematized in Fig.12.20b, these results suggest that light absorption by the antenna **Pen** array is followed by energy transfer to the **Pel** component of the covalent triad where the excited **Pel** reduces the electron acceptor **Rel** (which, in turn, transfers an electron to the gold electrode), and is then back-reduced by the electron donor **Rel′**.

12.7.4   (Supra)molecular shift register
A shift register is a form of memory which in essence consists of a set of memory cells (*registers*) connected in a line (Fig. 12.21). Each cell stores one bit of information. During each clock cycle, the contents of each cell are *shifted* to the next element to the right, the bit that was in the last register (n) is transferred to (read by) the external circuit, and the first register (1) receives (is written with) a new bit of information. Electronic shift registers are used as circuit elements for a variety of applications.

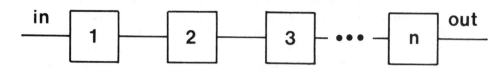

Fig. 12.21 — Scheme of a shift register.

Beratan and coworkers [137,138] have described a shift register memory at the molecular level based on photoinduced electron transfer processes. Their idea is illustrated by Fig. 12.22. From the macroscopic point of view, the register would

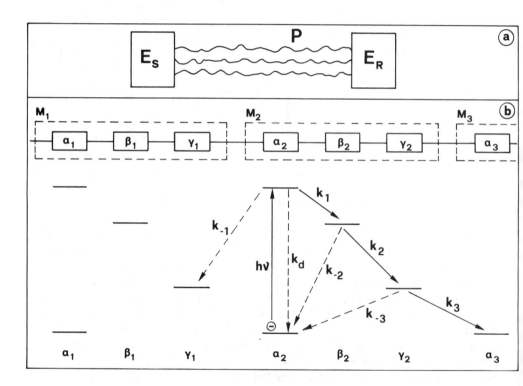

Fig. 12.22 — (Supra)molecular shift register [138].

consist (Fig. 12.22a) of a sending electrode (S) and a receiving electrode (R) connected by many (about 5000, see later) polymer chains (P). Each polymer chain should be made of the same number (for example 600, see later) identical monomer (M) units, schematically shown in the upper part of Fig. 12.22b. Each one of these units constitutes a memory element and should consist of three different molecular components, α, β, and γ as indicated in Fig. 12.22b, where the one-electron energy level of the chain is also schematized. The α elements are supposed to possess a low-lying excited energy level, whereas this should not be the case for β and γ. The bit of information is related to the presence (or absence) of an electron in the lower energy level of α. If, for example, α_2 and α_3 have an electron and α_1 and α_4 do not, the stored bit sequence is 0110.

The shifts to the right of the information bits (that is, of the electrons) need a power supply and a system clock. In principle, both could be provided by periodic pulses of light. Each α-β-γ unit, in fact, constitutes a supramolecular triad as those widely discussed in Chapter 5. Light excitation of the α_2-component, for example, would be followed (solid arrows, Fig. 12.22b) by the successive exoergonic electron-transfer processes to β_2, γ_2 and α_3. The electron originally present in α_3 should have been synchronously shifted to α_4 by the same photochemical mechanism.

The operation of such a system requires that the following conditions exist (for a complete discussion, see [138]):

(1) the light pulse must be sufficiently intense to allow a $> 99.9\%$ probability of excitation of each α^-, and short enough to prevent the occurrence of two transitions to the right in one clock cycle;
(2) the initial transfer from $^*\alpha_2^-$ to β_2 (k_1) must be much faster than the decay of $^*\alpha_2^-$ to its ground state α_2^- (k_d);
(3) the transfer to the right must always be much faster than the transfer to the left $(k_1 \gg k_{-1}; k_2 \gg k_{-2}; k_3 \gg k_{-3})$;
(4) the transfer from $^*\alpha_2^-$ to β_2 must be much faster than from γ_2^- to α_2;
(5) if the polymer repeat unit is about 20 Å long, about 500–600 units are needed to make a micrometre-length chain, the scale needed for incorporation with the rest of the device;
(6) about 5000 polymer chains are required to obtain a signal detectable with sufficient accuracy;
(7) all the polymer chains must have exactly the same length to have a synchronous net signal at the receiving electrode;
(8) the polymer ends should be covalently linked to the electrodes.

Some of the above requirements are quite complex and difficult to meet [138]. Supramolecular structures such as a porphyrin covalently-linked to two different quinones and a $Ru(bpy)_3^{2+}$ complex covalently-linked to two different amines have been suggested as possible monomeric units. Besides being (i) suitable for polymer synthesis, such units should exhibit appropriate (ii) redox levels, (iii) electronic coupling, and (iv) Franck–Condon factors, and should also (v) have rigid connectors, and (vi) be stable towards photodegradation.

In spite of these limitations and of the formidable technical difficulties that must be overcome to fabricate a real device, the idea of an $(\alpha\beta\gamma)_n$ (supra)molecular shift register is most significant and will certainly stimulate important studies in the near future.

12.8 SYNTHESIS AND SELF-ORGANIZATION

Progress in the design and engineering of **PMD**s is, of course, strictly related to progress in chemical synthesis.

Over the last 150 years, chemists have learned how to synthesize simple molecules. After World War II, macromolecular chains and nets made of a very large number of repeating units (polymers) have been developed. In the last 20 years, considerable attention has been directed to the synthesis of large molecules (macrocycles, amphiphiles, etc.) that can interact with other molecules to give rise to organized entities of higher complexity (host–guest systems, liquid crystals, etc.). More recently, the previously-neglected area of covalent supramolecular synthesis has begun to developed. Some important steps in this direction are the following:

(1) molecules which contain desired "pieces of information" (spectroscopic properties, redox properties, etc.) can be linked together in a desired sequence [37,38];
(2) suitable metal complexes (even oligometallic species) can be used as "ligands" to synthesize in a controlled way compounds of higher nuclearity ("complexes as ligands" strategy) [84,139];
(3) simple ways of making apparently complex artificial organic structures have been devised ("structure-directed" synthesis) [140–143];
(4) rigid linear structures which have accurately defined lengths in the 1–10 nm scale can be prepared (construction sets for "nanotechnology") [144-146];
(5) production of polypeptides and polynucleotides with predetermined amino acid sequences [147] is a routine procedure;
(6) a wide range of general synthetic approaches to the design and construction of scaffolding-like materials are available [148];
(7) photochemically-active components can be attached to electrodes [149], and appended to polymers [59,86,120,150,151] and DNA duplexes [152];
(8) it is possible to devise synthetic strategies which enable a systematic molecular morphogenesis from small molecules to macroscopic matter with control of size, shape, topology, flexibility, and surface chemistry (starburst dendrimers) [153].

At the same time, powerful techniques for the study of the electronic topography of surfaces and adsorbed layers with atomic resolution (for example, the scanning tunnelling microscope [154]) have been developed.

Very important for future developments is the fact that chemists are now learning to design molecular species capable of undergoing self-assembly into well-defined structures. Self-assembly is an aspect of *self-organization* of matter, a topic which is currently attracting much attention in several branches of science (chemistry, physics, biology, mathematics, social science) [155]. The most striking example of self-organization of matter is the Belousov–Zhabotinski reaction, that is the oxidation of malonic acid by potassium bromate in acid solution, catalysed by a redox couple like $Ce^{4+/3+}$ and $Fe(phen)_3^{3+/2+}$ [156,157]. Mixing these very simple reactants in appropriate concentrations gives origin to an oscillating process whose amazing properties have markedly changed our view of chemical dynamics. In fact, the mixture of the above "lifeless" reactants gives rise to an "alive" reaction that measures times, propagates information, discriminates between past and future and between left and right, may construct mosaic structures and, if pushed too hard, starts to vacillate and behaves erratically (we could say "irrationally') [155,156]. Furthermore, when the redox catalyst used is $Ru(bpy)_3^{3+/2+}$, the reaction spontaneously generates regular flashes of light (artificial firefly) [158].

The spontaneous onset of self-organized states in chemical reactions shows that matter is not inert; rather, it can have its own ability for internal organization. This holds true also for *molecular assembling*, as demonstrated by living systems [159]. The spontaneous formation of the double helix of nucleic acids represents the self-assembling of a supramolecular structure induced by a pattern of intermolecular interactions. The ordered self-assembling of artificial amphiphilic molecules to give mono- and multilayers, micelles, liquid crystals, etc. has long been known (for some recent papers, see [160–165]). The potentialities of such systems for the exploitation of **PMDs** are enormous [166–172] (see, for example, section 12.7.3), but they have not yet been fully explored.

Even more interesting for these purposes is the self-assembling of a controlled number of molecular components to yield well-defined supramolecular structures. In the last few years several systems have in fact been reported where spontaneous and directed generation of organized supramolecular entities takes place, following an "Aufbau" plan contained in the structural elements of the components. For example:

(1) oligopyridine strands and their derivatives self-assemble around metal ions to give double-helix structures [173–175] (see section 11.2.4); this self-organization process has recently been used to obtain the nucleoside-substituted helicate **12.1** [176];

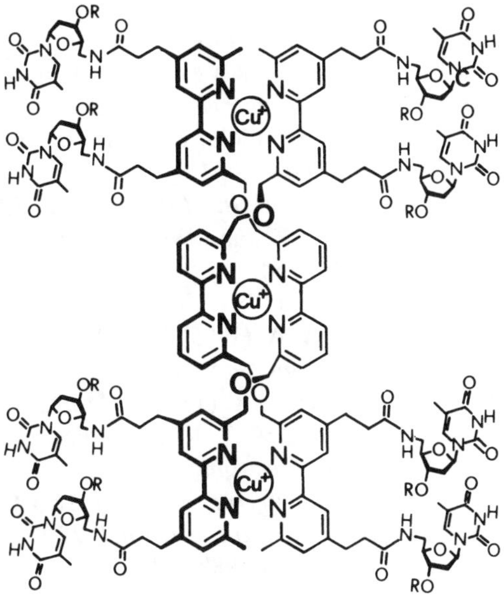

12.1

(2) self-assembling induced by metal-ligand [141,177] or electron donor–acceptor [143,178,179] interactions lies at the basis of the synthesis of catenates, knots, rotaxanes, and related species (section 11.2);

(3) spontaneous self-assembling of enolate-type ligands with four metal ions leads to tetranuclear complexes which have a high degree of symmetry [180];

(4) autocatalysis has been observed in a self-replicating system, a phenomenon that can be regarded as a primitive sign of life [105,181];
(5) molecular recognition and self-assembling of two complementary molecular components can result in the formation of liquid crystalline phases, which constitutes a macroscopic reading and amplification of molecular information [182];
(6) a monolayer of a molecular redox reagent can be spontaneously assembled on a microfabricated array of electrodes [183];
(7) spontaneous formation of superhelical strands have been reported [167,184];
(8) steps have been made on the way towards the synthesis of a self-assembling, cation-conducting channel [185].

These and other recent advances in chemical synthesis make possible the design of a variety of **PMD**s. Outstanding developments in this field are therefore expected for the near future.

REFERENCES

[1] Aviram, A. and Ratner, M. A. (1974) *Chem. Phys. Lett.* **29** 277.
[2] Carter, F. L. (ed.) (1982) *Molecular electronic devices*. Dekker.
[3] Carter, F. L. (1984) *Physica* **10D** 175.
[4] Wrighton, M. S. (1985) *Comments Inorg. Chem.* **4** 269.
[5] Joachim, C. and Launay, J. P. (1984) *Nouv. J. Chim.* **8** 723.
[6] Carter, F. L. (ed.) (1987) *Molecular electronic devices II*. Dekker.
[7] Balzani, V., Moggi, L., and Scandola, F. (1987) In Balzani, V. (ed.) *Supramolecular photochemistry*. Reidel, p. 1.
[8] Lehn, J. M. (1988) *Angew. Chem. Int. Edit. Engl.* **27** 89.
[9] Carter, F. L., Siatkowski, R. E., and Wohltjen, H. (eds) (1988) *Molecular electronic devices*. North-Holland.
[10] Stolarczyk, L. Z. and Piela, L. (1984) *Chem. Phys. Lett.* **85** 451.
[11] Haddon, R. C. and Lamola, A. A. (1985) *Proc. Natl. Acad. Sci. USA* **82** 1874.
[12] Aviram, A. (1988) *J. Am. Chem. Soc.* **110** 5687.
[13] Gilmanshin, R. I. and Lazarev, P. I. (1988) *J. Molec. Electronics* **4** S83.
[14] Metzger, R. M. and Panetta, C. A. (1989) *J. Molec. Electronics* **5** 1.
[15] Hong, F. T. (1989) *J. Molec. Electronics* **5** 163.
[16] Aviram, A. (1989) *Angew. Chem. Int. Ed. Engl.* **28** 520.
[17] Haarer, D. (1989) *Angew. Chem. Int. Ed. Engl.* **28** 1544.
[18] Clarkson, M. A. (1989) *Byte* **May 1989** 268.
[19] Hader, D. P. and Tevini, M. (1987) *General photobiology*. Pergamon.
[20] Breton, J. and Vermeglio, H. (eds) (1988) *The photosynthetic bacterial reaction center. Structure and dynamics*. Plenum.
[21] Deisenhofer, J. and Michel, H. (1989) *Angew. Chem. Int. Ed. Engl.* **28** 829.
[22] Huber, R. (1989) *Angew. Chem. Int. Ed. Engl.* **28** 848.
[23] Boxer, S. G., Goldstein, R. A., Lockhart, D. J., Middendorf, T. R., and Takiff, L. (1989) *J. Phys. Chem.* **93** 8280.
[24] Friesner, R. A. and Won, Y. (1989) *Photochem. Photobiol.* **50** 83.

[25] Feher, G., Allen, J. P., Okamura, M. Y., and Rees, D. C. (1989) *Nature* **339** 111.

[26] Porter, G. (1982) In Coyle, J. D., Hill, R. R., and Roberts, D. R. (eds) *Light, chemical change, and life*. The Open University Press (Milton Keynes), p. 362.

[27] Balzani, V., Juris, A., and Scandola, F. (1986) In Pelizzetti, E. and Serpone, N. (eds) *Homogeneous and heterogeneous photocatalysis*. Reidel, p. 1.

[28] Polymeropoulos, E. E., Moebius, D., and Kuhn, H. (1978) *J. Chem. Phys.* **68** 3918.

[29] Arden, W. and Fromherz, P. (1980) *J. Electrochem. Soc.* **127** 370.

[30] Moebius, D. (1981) *Acc. Chem. Res.* **14** 63.

[31] Van der Auweraer, M., Willig, F., and Charlé, K.P. (1986) *Chem. Phys. Lett.* **128** 214.

[32] Fujihira, M., Nishiyama, K., and Yamada, H. (1985) *Thin Solid Films* **132** 77.

[33] Sato, H., Kawasaki, M., Kasatani, K., Higuchi, Y., Azuma, T., and Nishiyama, Y. (1988) *J. Phys. Chem.* **92** 754.

[34] Nishikata, Y. Morikawa, A., Kakimoto, M., Imai, Y., Hirata, Y., Nishiyama, K., and Fujihira, M. (1989) *J. Chem. Soc., Chem. Commun.* 1772.

[35] Balzani, V. (ed.) (1987) *Supramolecular photochemistry*. Reidel.

[36] Connolly, J. S. and Bolton, J. R. (1988) In Fox, M. A. and Chanon, M. (eds) *Photoinduced electron transfer*. Part D. Elsevier, p. 303.

[37] Gust, D. and Moore, T. A. (eds) (1989) *Tetrahedron Symposium in Print on "Covalently linked donor-acceptor species for mimicry of photosynthetic electron and energy transfer" Tetrahed.* **45** N.15 4699–4912.

[38] Scandola, F., Indelli, M. T., Chiorboli, C., and Bignozzi, C. A. (1990) *Topics Curr. Chem.* **158** 73.

[39] Wasielewski, M. R., Niemczyk, M. P., Svec, W. A., and Pewitt, E. B. (1985) *J. Am. Chem. Soc.* **107** 5562.

[40] Wasielewski, M. R. and Niemczyk, M. P. (1984) *J. Am. Chem. Soc.* **106** 5043.

[41] Gust, D., Moore, T. A., Moore, A. L., Barrett, D., Harding, L. O., Makings, L. R., Liddell, P. A., De Schryver, F. C., Van der Auweraer, M., Bensasson, R. V., and Rougee, M. (1988) *J. Am. Chem. Soc.* **110** 321.

[42] Gust, D., Moore, T. A., Moore, A. L., Seely, G. R., Liddell, P. A., Barrett, D., Harding, L. O., Ma, X. C., Lee, S. J., and Gao, F. (1989) *Tetrahed.* **45** 4867.

[43] Palazzotto, M. C., Sahyun, M. R. V., Serpone, N., and Sharma, D. K. (1989) *J. Chem. Phys.* **90** 3373.

[44] Surridge, N. A., Hupp, J. T., McClanahan, S. F., Gould, S., and Meyer, T. J. (1989) *J. Phys. Chem.* **93** 304.

[45] Shafirovich, V. Y. and Shilov, A. E. (1988) *Isr. J. Chem.* **38** 149.

[46] Zamaraev, K. I., Lymar, S. V., Khramov, M. I., and Parmon, V. N. (1988) *Pure Appl. Chem.* **60** 1039.

[47] Kalyanasundaram, K., Graetzel, M., and Pelizzetti, E. (1986) *Coord. Chem. Rev.* **69** 57.

[48] Pichat, P. and Fox, M. A. (1988) In Fox, M. A. and Chanon, M. (eds) *Photoinduced electron transfer*. Part D. Elsevier, p. 241.

[49] Somorjai, G. A. (1989). In Serpone, N. and Pelizzetti, E. (eds) *Photocatalysis. Fundamentals and applications*. Wiley, p. 251.

[50] Shiragami, T., Pac, C., and Yanagida, S. (1990) *J. Phys. Chem.* **94** 504, and references therein..

[51] Gregg, B. A., Fox, M. A., and Bard, A. J. (1989) *Tetrahed.* **45** 4707.

[52] Vlček, A. (1982) *Coord. Chem. Rev.* **43** 39.

[53] Astruct, D. (1987) *Comments Inorg. Chem.* **6** 61.

[54] Beer, P. D. (1989) *Chem. Soc. Rev.* **18** 409.

[55] Bignozzi, C. A., Paradisi, C., Roffia, S., and Scandola, F. (1988) *Inorg. Chem.* **27** 408.

[56] De Cola, L., Belser, P., Ebmeyer, F., Barigelletti, F., Vögtle, F., von Zelewsky, A., and Balzani, V. (1990) *Inorg. Chem.* **29** 495.

[57] Denti, G., Campagna, S., Sabatino, L., Serroni, S., Ciano, M., and Balzani, V. (1990) *Inorg. Chem.*, in press.

[58] Meyer, T. J. (1989) In Norris, J. R., Jr. and Meisel, D. (eds) *Photochemical energy conversion*. Elsevier, p. 75.

[59] Younathan, J. N., McClanahan, S. F., and Meyer, T. J. (1989) *Macromolecules* **22** 1048.

[60] Ennis, P. M. and Kelly, J. M. (1989) *J. Phys. Chem.* **93** 5735.

[61] George, G. N., Prince, R. C., and Cramer, S. P. (1989) *Science* **243** 789.

[62] Sivaraja, M., Philo, J. S., Lary, J., and Dismukes, G. C. (1989) *J. Am. Chem. Soc.* **111** 3221.

[63] Vincent, J. B., Christmas, C., Chang, H. R., Li, Q., Boyd, P. D. W., Huffman, J. C., Hendrickson, D. N., and Christou, G. (1989) *J. Am. Chem. Soc.* **111** 2086.

[64] Luneva, N. P., Knerelman, E. I., Shafirovich, V. Y., and Shilov, A. E. (1989) *New J. Chem.* **13** 107.

[65] Seta, P., Bienvenue, E., Moore, A. L., Mathis, P., Bensasson, R. V., Liddell, P., Passiki, P. J., Joy, A., Moore, T. A., and Gust, D. (1985) *Nature* **316** 653.

[66] Moore, T. A., Gust, D., Moore, A. L., Bensasson, R. V., Seta, P., and Bienvenue, E. (1987) In Balzani, V. (ed.) *Supramolecular photochemistry*. Reidel, p. 283.

[67] Launay, J. P. (1987) In Carter, F. L. (ed.) *Molecular electronic devices II*. Dekker, p. 39.

[68] Hoffman, B. M. and Ratner, M. A. (1987) *J. Am. Chem. Soc.* **109** 6237.

[69] Aviram, A. (1988) *J. Molec. Electronics* **4** S99.

[70] Joachim, C. and Launay, J. P. (1990) *J. Molec. Electronics* **6** 37.

[71] Aviram, A., Joachim, C., and Pomerantz, M. (1988) *Chem. Phys. Lett.* **146** 490.

[72] Balzani, V. (1989) *Gazz. Chim. Ital.* **119** 311.

[73] Bignozzi, C. A., Indelli, M. T., and Scandola, F. (1989) *J. Am. Chem. Soc.* **111** 5192.

[74] Fromherz, P. and Arden, W. (1980) *J. Am. Chem. Soc.* **102** 6211.

[75] Dzhabiev, T. S., Nadtochenko, V. A., Rubtsov, I. V., and Smirnov, V. A. (1987) *Doklady Phys. Chem.* **294** 598.

[76] Takami, A. and Mataga, N. (1987) *J. Phys. Chem.* **91** 618.

[77] Tamai, N., Yamazaki, T., Yamazaki, I., Mizuma, A., and Mataga, N. (1987) *J. Phys. Chem.* **91** 3503.

[78] Kimizuka, N. and Kunitake, T. (1989) *J. Am. Chem. Soc.* **111** 3758.

[79] Nagamura, T., Kamata, S., Toyozawa, K., and Ogawa, T. (1990) *Ber. Bunsenges. Phys. Chem.* **94** 87.

[80] Yamazaki, I., Tamai, N., and Yamazaki, T. (1990) *J. Phys. Chem.* **94** 516.

[81] Amadelli, R., Argazzi, R., Bignozzi, C. A., and Scandola, F. (1990) *J. Am. Chem. Soc.* **112** 7099.

[82] Campagna, S., Denti, G., Sabatino, L., Serroni, S., Ciano, M., and Balzani, V. (1989) *J. Chem. Soc., Chem. Commun.* 1500.

[83] Murphy, W. R., Brewer, K. J., Gettliffe, G., and Petersen, J. D. (1989) *Inorg. Chem.* **28** 81.

[84] Denti, G., Campagna, S., Sabatino, L., Serroni, S., Ciano, M., and Balzani, V., (1990) *Inorg. Chim. Acta* **176** 175.

[85] Gust, D., Moore, T. A., Moore, A. L., Makings, L. R., Seely, G. R., Ma, X. C., Trier, T. T., and Gao, F. (1988) *J. Am. Chem. Soc.* **110** 7567; Gust, D., Moore, T. A., Moore, A. L., Lee, S. J., Bittersmann, A., Rehms, A., Belford, R. E., Luttrull, D. K., DeGraziano, J., Ma, X. C., Gao, F., and Trier, T. T. (1990) *Abstracts XIII IUPAC Symposium on photochemistry*, Warwick (UK).

[86] Strouse, G. F., Worl, L. A., Younathan, J. N., and Meyer, T. J. (1989) *J. Am. Chem. Soc.* **111** 9101.

[87] Blasse, G. (1987). In Balzani, V. (ed.) *Supramolecular photochemistry*. Reidel, p. 355.

[88] Blasse, G. (1986) *Recl. Trav. Chim. Pays-Bas* **105** 143.

[89] Mahiou, R., Metin, J., Fournier, M. T., Cousseins, J. C., and Jacquier, B. (1989) *J. Luminescence* **43** 51.

[90] Azuel, F., Pecile, D., and Morin, D. (1975) *J. Electrochem. Soc.* **122** 101.

[91] Reisfeld, R. and Jorgensen, C. K. (1977) *Lasers and excited states of rare earths*. Springer-Verlag, Chap. 4.

[92] Cockroft, N. J., Jones, G. D., and Syme, R. W. G. (1989) *J. Luminescence* **43** 275.

[93] Malta, O. L., Antic-Fidancev, E., Lemaitre-Blaise, M., Dexpert-Ghys, J., and Piriou, B. (1986) *Chem. Phys. Lett.* **129** 557.

[94] Rettig, W. (1986) *Angew. Chem. Int. Ed. Engl.* **25** 971.

[95] Lippert, E., Rettig, W., Bonačič-Koutecký V., Heisel, F., and Miehé, J. A. (1987) *Adv. Chem. Phys.* **68** 1.

[96] Rettig, W. (1988) *Appl. Phys. B* **45** 145.

[97] Joachim, C. and Launay, J. P. (1986) *Chem. Phys.* **109** 93.

[98] Launay, J. P., Woitellier, S., Sowinska, M., Tourrel, M., and Joachim, C. (1988) In Carter, F. L., Siatkowski, R. E., and Wohltjen, H. (eds) *Molecular electronic devices*. North-Holland, p. 171.

[99] Shinkai, S. and Manabe, O. (1984) *Topics Curr. Chem.* **121** 67.

[100] Shinkai S., Nakaji, T., Ogawa, T., Shigematsu, K., and Manabe, O. (1981) *J. Am. Chem. Soc.* **103** 111.

[101] Shinkai, S., Honda, Y., Kusano, Y., and Manabe, O. (1982) *J. Chem. Soc., Chem. Commun.* 848.

[102] Losensky, H. W., Spelthann, H., Ehlen, A., Vögtle, F., and Bargon, J. (1988) *Angew. Chem. Int. Ed. Engl.* **27** 1189.

[103] Hosseini, M. W., Lehn, J. M., Jones, K. C., Plute, K. E., Mertes, K. B., and Mertes, M. P. (1989) *J. Am. Chem. Soc.* **111** 6330.

[104] Kelly, T. R., Zhao, C., and Bridger, G. J. (1989) *J. Am. Chem. Soc.* **111** 3744.

[105] Rebek, J., Jr. (1990) *Angew. Chem. Int. Ed. Engl.* **29** 245.

[106] Jorgensen, C. K. (1981) In Connolly, J. S. (ed.) *Photochemical conversion and storage of solar energy.* Academic, p. 79.

[107] Potember, R. S., Hoffman, R. C., Kim, S. H., Speck, K. R., and Stetyick, K. A. (1988) *J. Molec. Electronics* **4** 5.

[108] Emmelius, M., Pawlowski, G., and Vollmann, H. W. (1989) *Angew. Chem. Int. Ed. Engl.* **28** 1445.

[109] Parthenopoulos, D. A. and Rentzepis, P. M. (1989) *Science* **245** 843.

[110] Ebbesen, T. W. (1986) *Appl. Opt.* **25** 2193.

[111] Shimidzu, T., Iyoda, T., and Honda, K. (1988) *Pure Appl. Chem.* **60** 1025.

[112] Dabestani, R., Bard, A. J., Campion, A., Fox, M. A., Mallouk, T. E., Webber, S. E., and White, J. M. (1988) *J. Phys. Chem.* **92** 1872.

[113] Vlachopoulos, N., Liska, P., Augustynski, J., and Graetzel, M. (1988) *J. Am. Chem. Soc.* **110** 1216.

[114] Vrachnou, E., Graetzel, M., and McEvoy, A. J. (1989) *J. Electroanal. Chem.* **258** 193.

[115] Enea, O., Moser, J., and Graetzel, M. (1989) *J. Electroanal. Chem.* **259** 59.

[116] Andrieux, C. P., Haas, O., and Savéant, J. M. (1986) *J. Am. Chem. Soc.* **108** 8175.

[117] Degani, Y. and Heller, A. (1989) *J. Am. Chem. Soc.* **111** 2357.

[118] Licht, S., Cammarata, V., and Wrighton, M. S. (1989) *Science* **243** 1176.

[119] Jernigan, J. C., Surridge, N. A., Zvanut, M. E., Silver, M., and Murray, R. W. (1989) *J. Phys. Chem.* **93** 4620.

[120] Jernigan, J. C. and Murray, R. W. (1990) *J. Am. Chem. Soc.* **112** 1034.

[121] Serpone, N. and Pelizzetti, E. (eds) (1989) *Photocatalysis. Fundamentals and applications.* Wiley.

[122] Graetzel, M. (ed.) (1983) *Energy resources through photochemistry and catalysis.* Academic.

[123] Gerisher, H. and Willig, F. (1976) *Topics Curr. Chem.* **61** 31.

[124] Memming, R. (1984) *Prog. Surface Sci.* **17** 7.

[125] Gerischer, H. (1973) *Ber. Bunsenges. Phys. Chem.* **77** 771.

[126] Desilvestro, J., Graetzel, M., Kavan, L., and Moser, J. (1985) *J. Am. Chem. Soc.* **107** 2988.

[127] Furlong, D. N., Welles, D., and Sasse, W. H. F. (1986) *J. Phys. Chem.* **90** 1107.

[128] Graetzel, M. (1989) In Norris, J. R., Jr. and Meisel, D. (eds) *Photochemical energy conversion.* Elsevier.

[129] Bignozzi, C. A., Chiorboli, C., Indelli, M. T., Rampi, M. A., and Scandola, F. (1990) *Coord. Chem. Rev.* **97** 299.

[130] Bignozzi, C. A., Chiorboli, C., Davila, J., Indelli, M. T., and Scandola, F. (1989) *Inorg. Chem.* **28** 4350.

[131] Sessler, J. L., Johnson, M. R., and Lin, T. Y. (1989) *Tetrahed.* **45** 4767.

[132] Fujihira, M., Nishiyama, K., and Aoki, K. (1988) *Thin Solid Films* **160** 317.

[133] Fujihira, M. and Yamada, H. (1988) *Thin Solid Films* **160** 125.

[134] Kondo, T., Yamada, H., Nishiyama, K., Suga, K., and Fujihira, M. (1989) *Thin Solid Films* **179** 463.

[135] Fujihira, M. and Sakomura, M. (1989) *Thin Solid Films* **179** 471.

[136] Fujihira, M., Sakomura, M., and Kamei, T. (1989) *Thin Solid Films* **180** 43.

[137] Hopfield, J. J., Onuchic, J. N., and Beratan, D. N. (1988) *Science* **241** 817.

[138] Hopfield, J. J., Onuchic, J. N., and Beratan, D. N. (1989) *J. Phys. Chem.* **93** 6350.

[139] Thanyasiri, T. and Sinn, E. (1989) *J. Chem. Soc., Dalton Trans.* 1187.

[140] Sendhoff, N., Weissbarth, K. H., and Vögtle, F. (1987) *Angew. Chem. Int. Ed. Engl.* **26** 777.

[141] Dietrich-Bucheker, C. O. and Sauvage, J. P. (1987) *Chem. Rev.* **87** 795.

[142] Kohnke, F. H., Mathias, J. P., and Stoddart, J. F. (1989) *Angew. Chem. Int. Ed. Engl.* **28** 1103.

[143] Stoddart, J. F. (1990) In Schneider, H. J. and Dürr, H. (eds) *Frontiers in supramolecular organic chemistry and photochemistry.* Verlag Chemie, in press.

[144] Kenny, P. W. and Miller, L. L. (1988) *J. Chem. Soc., Chem. Commun.* 84.

[145] Kaszynski, P. and Michl, J. (1988) *J. Am. Chem. Soc.* **110** 5225.

[146] Murthy, G. S., Hassenruck, K., Lynch, V. M., and Michl, J. (1989) *J. Am. Chem. Soc.* **111** 7262.

[147] Merrifield, R. B. (1985) *Angew. Chem. Int. Ed. Engl.* **24** 799.

[148] Hoskins, B. F. and Robson, R. (1990) *J. Am. Chem. Soc.* **112** 1546.

[149] Hidalgo-Luangdilok, C., Bocarsly, A. B., and Woods, R. E. (1990) *J. Phys. Chem.* **94** 1918.

[150] Deronzier, A. and Essakalli, M. (1990) *J. Chem. Soc., Chem. Commun.* 242.

[151] Surridge, N. A., McClanahan, S. F., Hupp, J. T., Danielson, E., Gould, S., and Meyer, T. J. (1989) *J. Phys. Chem.* **93** 294.

[152] Telser, J., Cruickshank, K. A., Schanze, K. S., and Netzel, T. L. (1989) *J. Am. Chem. Soc.* **111** 7211.

[153] Tomalia, D. A., Naylor, A. M., and Goddard III, W. A. (1990) *Angew. Chem. Int. Ed. Engl.* **29** 138.

[154] Avouris, P. (1990) *J. Phys. Chem.* **94** 2246.

[155] Babloyantz, A. (1986) *Molecules, dynamics, and life.* Wiley.

[156] Field, R. J. and Schneider, F. W. (1989) *J. Chem. Educ.* **66** 195.

[157] Zhabotinskii, A. M. (1964) *Dokl. Akad. Nauk SSSR* **157** 392.

[158] Bolletta, F. and Balzani, V. (1982) *J. Am. Chem. Soc.* **104** 4250.

[159] Pörschke D. and Eigen, M. (1971) *J. Mol. Biol.* **62** 361.

[160] Lipatov, Y. S. (1987) *Doklady Phys. Chem.* **296** 920.

[161] Markowitz, M. A., Janout, V., Castner, D. G., and Regen, S. L. (1989) *J. Am. Chem. Soc.* **111** 8192.

[162] Richards, R. D. C., Hawthorne, W. D., Hill, J. S., White, M. S., Lacey, D., Semlyen, J. A., Gray, G. W., and Kendrick, T. C. (1990) *J. Chem. Soc., Chem. Commun.* 95.

[163] Cometti, G., Dalcanale, E., Du vosel, A., and Levelut, A. M. (1990) *J. Chem. Soc., Chem. Commun.* 163.

[164] Shinkai, S., Nishi, T., Ikeda, A., Matsuda, T., Shimamoto, K., and Manabe, O. (1990) *J. Chem. Soc., Chem. Commun.* 303.

[165] Rong, D., Hong, H. G., Kim, Y. I., Krueger, J. S., Mayer, J. E., and Mallouk, T. E. (1990) *Coord. Chem. Rev.* **97** 237.

[166] Kuhn, H. (1987) In Carter, F. L. (ed.) *Molecular electronic devices II.* Dekker, p. 411.

[167] Fuhrhop, J. H. and Boettcher, C. (1990) *J. Am. Chem. Soc.* **112** 1768.

[168] Ringsdorf, H., Schlarb, B., and Venzmer, J. (1988) *Angew. Chem. Int. Ed. Engl.* **27** 113.

[169] Eidenschink, R. (1989) *Angew. Chem. Int. Ed. Engl. Adv. Mater.* **28** 1424.

[170] Groh, W., Lupo, D., and Sixl, H. (1989) *Angew. Chem. Int. Ed. Engl. Adv. Mater.* **28** 1548.

[171] Laschewsky, A. (1989) *Angew. Chem. Int. Ed. Engl. Adv. Mater.* **28** 1574.

[172] Simon, J., Bassoul, P., and Norvez, S. (1989) *New. J. Chem.* **13** 13.

[173] Lehn, J. M., Rigault, A., Siegel, J., Harrowfield, J., Chevrier, B., and Moras, D. (1987) *Proc. Natl. Acad. Sci. USA* **84** 2565.

[174] Lehn, J. M. and Rigault, A. (1988) *Angew. Chem. Int. Ed. Engl.* **27** 1095.

[175] Constable, E. C. and Ward, M. D. (1990) *J. Am. Chem. Soc.* **112** 1256.

[176] Koert, U., Harding, M. M., and Lehn, J. M. (1989) In *Proc. Workshop on supramolecular organic chemistry and photochemistry, Saarbrücken (FRG) August 27–September 1, 1989,* p. P35.

[177] Dietrich-Bucheker, C. O. and Sauvage, J. P. (1989) *Angew. Chem. Int. Ed. Engl.* **28** 189.

[178] Ashton, P. R., Goodnow, T. T., Kaifer, A. E., Reddington, M. V., Slawin, A. M. Z., Spencer, N., Stoddart, J. F., Vicent, C., and Williams, D. J. (1989) *Angew. Chem. Int. Ed. Engl.* **28** 1396.

[179] Anelli, P. L., Ashton, P. R., Ballardini, R., Balzani, V., Gandolfi, M. T., Goodnow, T. T., Kaifer, A. E., Pietraszkiewicz, M., Prodi, L., Reddington, M. V., Slawin, A. M. Z., Spencer, N., Stoddart, J. F., Vicent, C., and Williams, D. J. *J. Am. Chem. Soc.*, submitted.

[180] Saalfrank, R. W., Stark, A., Bremer, M., and Hummel, H. U. (1990) *Angew. Chem. Int. Ed. Engl.* **29** 311.

[181] Tjivikua, T., Ballester, P., and Rebek, J., Jr. (1990) *J. Am. Chem. Soc.* **112** 1249.

[182] Brienne, M. J., Gabard, J., Lehn, J. M., and Stibor, I. (1989) *J. Chem. Soc., Chem. Commun.* 1868.

[183] Hickman, J. J., Zou, C., Ofer, D., Harvey, P. D., and Wrighton, M. S. (1989) *J. Am. Chem. Soc.* **111** 7271.

[184] Yanagawa, H., Ogawa, Y., Furuta, H., and Tsuno, K. (1989) *J. Am. Chem. Soc.* **111** 4567.

[185] Gokel, G. W., Echegoyen, L., Kim, M., Hernandez, J. C., and De Jesus, M. (1989) *J. Inclusion Phenom.* **7** 73.

Author index

author, page (chapter.reference number)

Aartsma, T. J., 343 (11.139)
Abelt, C. J., 258 (9.127)
Aboul-Enein, A., 299 (10.297), 305 (10.297)
Abruna, H., 125 (5.155)
Abu-Elgheit, M., 345 (11.144)
Adachi, T., 345 (11.147)
Adamson, A. W., 20 (1.90), 25 (2.1), 41 (2.1), 180 (6.95), 186 (6.114), 187 (6.114), 188 (6.114)
Adar, E., 298 (10.255), 299 (10.256)
Addington, J. W., 232 (8.34), 234 (8.34), 235 (8.34)
Agam, G., 334 (11.101, 11.102)
Agbaria, R. A., 289 (10.185), 290 (10.188)
Ahn, B. T., 340 (11.116)
Akabori, S., 205 (7.79)
Akiyama, S., 271 (10.66)
Akkaya, E. U., 271 (10.68)
Al-Joubbeh, M., 268 (10.47), 273 (10.47)
Alak, A. M., 290 (10.194, 10.196)
Alam, I., 267 (10.22)
Albin, M., 326 (11.45, 93 (5.25)
Albini, A., 251 (9.29)
Albrecht-Gary, A. M., 268 (10.47), 273 (10.47), 337 (11.108)
Alder, J. F., 270 (10.59)
Alegria, G., 91 (5.16)
Alexander, J., 19 (1.84)
Alexandrescu, A., 299 (10.270), 304 (10.270)
Alfano, R., 198 (7.10)
Alfimov, M. V., 201 (7.30, 7.31)
Allen, G., 348 (11.158)
Allen, J. P., 90 (5.9), 356 (12.25), 363 (12.25)
Allen, M. M., 188 (6.122), 189 (6.122)
Allen, M. T., 198 (7.11), 199 (7.11)
Allwood, B. L., 278 (10.103, 10.105), 285 (10.105, 10.148, 10.149, 10.150, 10.151), 286 (10.151),
Aloisi, G. G., 257 (9.97)

Alpha, B., 327 (11.47, 11.50), 328 (11.52), 329 (11.49, 11.50, 11.51, 11.52, 11.53, 11.54)
Amadelli, R., 186 (6.112), 370 (12.81), 380 (12.81), 381 (12.81), 382 (12.81)
Amatore, C., 258 (9.125)
Amirav, A., 19 (1.30), 256 (9.78)
Amouyal, E., 212 (7.110)
Anderson, A. G., 295 (10.235)
Anderson, H. L., 285 (10.147)
Andreetti, G. D., 329 (11.61, 11.62)
Andrieux, C. P., 380 (12.116)
Anelli, P. L., 19 (1.79), 279 (10.112), 334 (11.105, 11.106), 336 (11.106), 387 (12.179)
Anner, O., 251 (9.30), 256 (9.30, 9.80)
Anson, F. C., 320 (11.21)
Anthon, D., 128 (5.162), 129 (5.162)
Antic-Fidancev, E., 375 (12.93)
Anzai, J., 216 (7.143, 7.145), 219 (7.179)
Aoki, K., 201 (7.41), 383 (12.132)
Arad-Yellin, R., 292 (10.215)
Arai, Y., 280 (10.133)
Archer, M. D., 117 (5.129), 118 (5.129)
Archer, R. D., 240 (8.60)
Arcioni, A., 217 (7.149)
Arden, W., 361 (12.29), 363 (12.29), 370 (12.74), 383 (12.74)
Arduini, A., 329 (11.60, 11.61, 11.62)
Argazzi, R., 186 (6.112), 370 (12.81), 380 (12.81), 381 (12.81), 382 (12.81)
Armaroli, N., 341 (11.122)
Armitage, B., 48 (2.25), 70 (3.25), 259 (9.137)
Arrhenius, T. S., 19 (1.46), 116 (5.121)
Asahi, T., 258 (9.123)
Ashton, P. R., 19 (1.42, 1.52, 1.79), 285 (10.152), 287 (10.156), 334 (11.104, 11.106), 336 (11.106), 387 (12.178, 12.179)
Ashworth, D. C., 270 (10.59)
Astruct, D., 364 (12.53)
Atwell, G. A., 306 (10.323)

Auburn, P. R., 180 (6.100), 257 (9.90)
Auerbach, R. A., 171 (6.55), 173 (6.55, 6.59, 6.61), 174 (6.61)
Augustynski, J., 380 (12.113), 382 (12.113)
Aviram, A., 355 (12.1, 12.12, 12.16), 367 (12.1, 12.12, 12.69, 12.71), 380 (12.12, 12.16, 12.69, 12.71)
Avouris, P., 386 (12.154)
Axup, A. W., 93 (5.25)
Ayala, N. P., 257 (9.91, 9.93)
Azuel, F., 375 (12.90)
Azuma, T., 361 (12.33), 363 (12.33), 370 (12.33)

Baba, H., 257 (9.98)
Babloyantz, A., 386 (12.155)
Bacciola, D., 219 (7.174)
Backsay, G. B., 95 (5.56)
Baczynskyj, L., 19 (1.12)
Baird, W., 299 (10.300)
Bajaj, A. V., 268 (10.40)
Baker, A. D., 174 (6.74, 174 6.75), 348 (11.160, 11.161)
Bakker, C. J., 253 (9.44)
Bakker, N. A. C., 105 (5.99), 252 (9.37), 253 (9.37), 254 (9.37)
Balaji, V., 102 (5.77)
Baldeshwieler, J. D., 98 (5.67)
Ballardini, R., 19 (1.79), 70 (3.23), 227 (8.13), 232 (8.37, 8.38), 233 (8.37, 8.38, 8.47), 234 (8.37, 8.38), 235 (8.37, 8.38), 237 (8.37, 8.38), 240 (8.56), 260 (9.146), 274 (10.85), 275 (10.85, 10.92, 10.93), 276 (10.92, 10.93, 10.96, 10.97), 277 (10.93, 10.96), 280 (10.85), 289 (10.93), 319 (11.9), 328 (11.52), 329 (11.52), 330 (11.9), 334 (11.106), 336 (11.106), 387 (12.179)
Ballester, P., 388 (12.181)
Balzani, V., 19 (1.69, 1.73, 1.77, 1.79, 1.81), 20 (1.89, 1.99), 25 (2.1), 41 (2.1), 45 (2.12, 2.13), 46 (2.13, 2.19), 47 (2.19, 2.21), 48 (2.13, 2.21), 49 (2.29), 51 (3.1), 57 (3.9), 72 (3.35), 73 (3.9, 3.35), 76 (4.2), 77 (4.2, 4.3), 82 (4.2), 84 (4.22, 4.23), 86 (4.22, 4.38), 95 (5.54), 111 (5.110), 124 (5.145), 144 (5.202) 168 (6.42), 170 (6.50), 171 (6.57), 174 (6.67, 6.69, 6.70, 6.71), 176 (6.70), 177 (6.69, 6.70, 6.71), 178 (6.67, 6.69, 6.70, 6.71, 6.83), 179 (6.67), 181 (6.70, 6.71, 6.105, 6.107, 6.108), 182 (6.70, 6.71, 6.105, 6.107, 6.108), 183 (6.108), 184 (6.70, 6.71, 6.107, 6.108), 186 (6.113), 187 (6.113), 188 (6.113), 202 (7.52), 203 (7.62), 212 (7.107, 7.109), 227 (8.6, 8.12, 8.13), 231 (8.12, 8.20, 8.21, 8.23), 232 (8.20, 8.27, 8.36, 8.37, 8.38), 233 (8.37, 8.38, 8.46, 8.47, 8.48, 8.49), 234 (8.49, 8.36, 8.37, 8.38, 8.48), 235 (8.20, 8.36, 8.37, 8.38, 8.48, 8.49), 237 (8.37, 8.38), 239 (8.6), 240 (8.27, 8.55, 8.56), 241 (8.46, 8.48), 242 (8.74), 243 (8.89), 260 (9.146), 274 (10.82, 10.85, 10.86), 275 (10.85, 10.92, 10.93), 276 (10.92, 10.93, 10.96, 10.97), 277 (10.93, 10.96), 280 (10.85, 10.86, 10.136), 281 (10.136, 10.137, 10.138), 282 (10.137, 10.138), 287 (10.82), 289 (10.86, 10.93), 301 (10.309), 306 (10.309, 10.321, 10.322), 319 (11.7, 11.8, 11.9, 11.10), 320 (11.13), 321 (11.10, 11.26, 11.28), 322 (11.10), 323 (11.26, 11.34, 11.35), 324 (11.13, 11.34), 325 (11.10), 327 (11.48, 11.50), 328 (11.52), 329 (11.50, 11.51,11.52, 11.63), 330 (11.7, 11.8, 11.9, 11.10, 11.65, 11.66), 331 (11.73, 11.74), 332 (11.74, 11.75), 334 (11.106), 336 (11.106), 341 (11.122, 11.125), 347 (11.7), 355 (12.7), 356 (12.7), 357 (12.7), 359 (12.7, 12.27), 360 (12.7), 361 (12.7, 12.35), 365 (12.56, 12.57, 12.66), 366 (12.7), 367 (12.7, 12.72), 368 (12.7), 370 (12.82, 12.84), 371 (12.82), 373 (12.87), 374 (12.7, 12.87), 379 (12.7), 382 (12.84), 386 (12.84, 12.158), 387 (12.179)
Bancroft, D. P., 300 (10.308)
Barbara, P. F., 342 (11.138), 346 (11.138)
Bard, A. J., 217 (7.152), 234 (8.50), 301 (10.311), 364 (12.51), 380 (12.51, 12.112)
Bargon, J., 213 (7.111), 378 (12.102),
Barigelletti, F., 124 (5.145), 170 (6.50), 171 (6.57), 173 (6.60), 174 (6.69), 177 (6.69), 178 (6.69), 212 (7.107, 7.109), 257 (9.99), 301 (10.309), 306 (10.309, 10.322), 319 (11.7), 330 (11.7, 11.65), 331 (11.73, 11.74), 332 (11.74, 11.75), 340 (11.120), 341 (11.120, 11.122), 347 (11.7), 365 (12.56)
Barltrop, J. A., 25 (2.1), 41 (2.1)
Barnett, J. R., 232 (8.43), 241 (8.43)
Barqawi, K. R., 125 (5.150)
Barr, D., 19 (1.26)
Barrett, D., 19 (1.58), 116 (5.126), 145 (5.126), 146 (5.126), 147 (5.212), 148 (5.222, 5.223), 150 (5.222, 5.223), 169 (6.46, 6.47), 363 (12.41, 12.42), 364 (12.41)
Barton, J. K., 19 (1.67), 151 (5.235, 5.236), 299 (10.263, 10.264, 10.266, 10.267, 10.268, 10.269, 10.270, 10.271, 10.272, 10.273, 10.274, 10.275, 10.276, 10.277, 10.278, 10.279, 10.280, 10.281, 10.282), 300 (10.266, 10.276), 301 (10.280), 302 (10.280), 303 (10.271, 10.280), 304 (10.268, 10.269, 10.270, 10.271, 10.274, 10.276, 10.280), 305 (10.268, 10.272, 10.273, 10.279, 10.280, 10.281), 306 (10.275, 10.279)
Basile, L. A., 299 (10.270, 10.277, 10.278), 304 (10.270)
Bassoul, P., 387 (12.172)
Basu, S., 257 (9.108)
Batteas, J. D., 122 (5.143)
Baucom, D. A., 171 (6.52), 172 (6.52)
Bauer, H., 270 (10.53)
Bauer, W. R., 300 (10.304)
Bauerle, P., 165 (6.29), 166 (6.29)
Bauman, J., 168 (6.38)
Baumann, W., 202 (7.48, 7.51, 7.53), 254 (9.54)
Baxter, D. V., 98 (5.67)
Beaudet, R. A., 251 (9.33), 252 (9.33)
Beaumont, P. C., 299 (10.288), 305 (10.288)
Becker, R. S., 201 (7.40), 217 (7.151)
Becker, W. G., 234 (8.50)

Beddard, G. S., 120 (5.136), 122 (5.141), 147
(5.209, 5.214), 274 (10.84)
Beecroft, R. A., 255 (9.72)
Beer, P. D., 204 (7.71, 7.73, 7.75), 267 (10.32,
10.33), 270 (10.54), 364 (12.54)
Beeston, R. F., 332 (11.77)
Beitz, J. V., 48 (2.23), 93 (5.27), 95 (5.27, 5.63),
97 (5.63)
Belford, R. E., 150 (5.224), 370 (12.85)
Bell, T. W., 19 (1.41)
Bell, W. L., 118 (5.133), 254 (9.51)
Bellobono, I. R., 216 (7.144)
Belser, P., 19 (1.50), 83 (4.18), 124 (5.145), 170
(6.50), 171 (6.57), 173 (6.60), 212 (7.107,
7.109), 301 (10.309), 306 (10.309, 10.322), 319
(11.7), 330 (11.7), 331 (11.72, 11.73, 11.74)
332 (11.74, 11.75, 11.80), 347 (11.7), 365
(12.56)
Bencini, A., 279 (10.116, 10.120, 10.131)
Bender, C. J., 248 (9.8)
Bender, M. L., 288 (10.172)
Benedix, R., 230 (8.18)
Bensasson, R. V., 19 (1.56, 1.58), 116 (5.126),
145 (5.126, 5.204, 5.205), 146 (5.126), 147
(5.205, 5.220), 148 (5.222), 150 (5.222), 169
(6.45), 363 (12.41), 364 (12.41), 365 (12.65,
12.66)
Benson, D. R., 287 (10.163)
Beratan, D. N., 95 (5.46, 5.47, 5.48, 5.50), 98
(5.66), 113 (5.118), 114 (5.118), 384 (12.137,
12.138), 385 (12.138)
Berezin, I. V., 197 (7.2)
Bergelson, L. D., 217 (7.148)
Bergkamp, M. A., 118 (5.132)
Bernhard, P., 320 (11.19, 11.21, 11.22, 11.23),
332 (11.81)
Bernhardt, P. V., 279 (10.132)
Bernstein, J. S., 133 (5.184), 134 (5.184), 135
(5.184)
Bertelson, R. C., 197 (7.6), 200 (7.6)
Besnard, M., 250 (9.15)
Bhanumathi, N., 288 (10.173)
Bhattacharyya, K., 202 (7.57, 7.59), 298 (10.249,
10.250)
Bianchi, A., 279 (10.116, 10.120, 10.130, 10.131)
Bidd, I., 178 (6.86)
Bienvenue, E., 147 (5.219, 5.220), 365 (12.65,
12.66)
Bignozzi, C. A., 19 (1.76), 60 (3.12, 3.13, 3.14,
3.15), 79 (4.6), 82 (4.9), 94 (5.37), 132 (5.37),
125 (5.151), 136 (5.186), 137 (5.186), 140
(5.195, 5.196), 142 (5.195, 5.196), 144 (5.195,
5.196), 161 (6.7), 170 (6.7), 184 (6.109,
6.110), 185 (6.109), 186 (6.109, 6.112), 187
(6.117, 6.118), 188 (6.118), 189 (6.123), 231
(8.23), 301 (10.310), 306 (10.310), 348
(11.164, 11.165), 361 (12.38), 365 (12.55), 367
(12.73), 369 (12.73), 370 (12.81), 380 (12.81),
381 (12.38, 12.81, 12.129, 12.130), 382 (12.81)
Billing, R., 230 (8.17, 8.18, 8.19)
Birks, J. B., 39 (2.7), 40 (2.7), 248 (9.3), 251
(9.3, 9.20), 252 (9.20)

Bittersman, E., 150 (5.224)
Bittersmann, A., 370 (12.85)
Bixon, M., 91 (5.13), 95 (5.55)
Bizzarro, G., 256 (9.85)
Blackburn, C., 204 (7.73), 267 (10.33)
Blacker, A. J., 305 (10.316), 306 (10.316)
Blair, H. S., 218 (7.172, 7.173)
Blanc, J., 197 (7.5), 200 (7.5)
Blanchard-Desce, M., 19 (1.46), 116 (5.121)
Blank, M., 213 (7.112)
Blank, N. E., 253 (9.47), 254 (9.47)
Blaskie, M. W., 340 (11.1l7)
Blasse, G., 161 (6.1), 329 (11.53, 11.54), 373
(12.87), 374 (12.87, 12.88),
Blau, W. J., 299 (10.285)
Blondeel, G., 273 (10.81)
Bocarsly, A. B., 386 (12.149)
Bock, C. R., 347 (11.155)
Bockman, T. M., 232 (8.28), 239 (8.28)
Boens, N., 161 (6.6), 257 (9.102)
Boettcher, C., 387 (12.167), 388 (12.167)
Boeyens, J. C. A., 279 (10.113)
Böhmer, V., 267 (10.29), 329 (11.58), 333 (11.90)
Boillot, M. L., 212 (7.110)
Boldaji, M., 180 (6.104)
Bolletta, F., 45 (2.12, 2.13), 46 (2.13), 48 (2.13),
57 (3.9), 73 (3.9), 232 (8.36), 233 (8.49), 234
(8.36, 8.49), 235 (8.36, 8.49), 386 (12.158)
Bolt, J. D., 292 (10.203)
Bolton, J. R., 19 (1.55), 51 (3.2), 94 (5.33), 113
(5.119), 116 (5.33), 117 (5.128, 5.129), 118
(5.128, 5.129, 5.130), 119 (5.134), 120 (5.33),
121 (5.33), 251 (9.17), 252 (9.17), 361 (12.36)
Bonacic-Koutecky, V., 202 (7.45), 203 (7.61),
377 (12.95)
Bonazzi, A., 231 (8.20), 232 (8.20), 235 (8.20),
327 (11.48),
Bond, A. M., 320 (11.16)
Bond, P. J., 300 (10.304)
Boos, H., 202 (7.49)
Borkent, J. H., 253 (9.43), 254 (9.43)
Bortolus, P., 86 (4.40), 200 (7.23, 7.25), 257
(9.99), 297 (10.242), 298 (10.247)
Bos, M., 267 (10.25),
Bottcher, W., 178 (6.89)
Bottino, F., 287 (10.159)
Bouas-Laurent, H., 255 (9.71), 260 (9.145,
9.146), 268 (10.47), 272 (10.76), 273 (10.47,
10.77, 10.78, 10.79), 275 (10.76), 276 (10.76,
10.97)
Bouchard, D. A., 259 (9.134)
Boucher, H. A., 320 (11.16)
Bourson, J., 167 (6.34, 6.35, 6.37), 168 (6.34,
6.35), 269 (10.52), 271 (10.52)
Boxer, S. G., 20 (1.97), 90 (5.12), 356 (12.23)
Boyd, P. D. W., 365 (12.63)
Boyde, S., 191 (6.129)
Boyer, J. H., 289 (10.184)
Bradshaw, J. S., 279 (10.114)
Braterman, P. S., 124 (5.146)
Bäruchle, Chr., 201 (7.33)
Braun, A. M., 127 (5.156)

Braunstein, C. H., 174 (6.74)
Braunstein, C. H., 348 (11.160)
Bremer, M., 387 (12.180)
Brenman, B. A., 257 (9.92)
Brenner, D., 19 (1.27)
Breslow, R., 19 (1.66), 288 (10.170), 289 (10.170)
Breton, J., 20 (1.94), 90 (5.6), 91 (5.15), 356 (12.20), 370 (12.20)
Brewer, K. J., 181 (6.106), 182 (6.106), 370 (12.83)
Briaire, J., 270 (10.53)
Bridger, G. J., 378 (12.104)
Brienne, M. J., 388 (12.182)
Bright, F. V., 295 (10.227)
Brittinger, C., 254 (9.54)
Brown, D. B., 54 (3.5)
Brown, G. H., 197 (7.4, 7.5, 7.6), 200 (7.5, 7.6)
Brown, G. M., 51 (3.3)
Brown, P. A., 170 (6.49)
Brubaker, G. R., 324 (11.43)
Brucker, G. A., 346 (11.152, 11.153)
Brunink, J. A. J., 267 (10.23), 333 (11.97)
Brus, L. E., 342 (11.138), 346 (11.138)
Bryant, J. A., 267 (10.34, 10.35), 287 (10.161)
Bryson, N. J., 19 (1.27)
Buckner, M. T., 340 (11.116)
Bugge, K. E., 267 (10.23), 333 (11.97),
Buhner, M., 285 (10.155)
Bünzli, J. C., 272 (10.75)
Burgi, H. B., 320 (11.23)
Busch, D. H., 267 (10.9)
Busmanny, H. G., 257 (9.109)

Cabrera, I., 217 (7.153, 7.154, 7.155)
Calcaterra, L. T., 98 (5.68, 5.69, 5.70), 99 (5.70), 100 (5.69, 5.70), 101 (5.70)
Caldwell, R. A., 261 (9.149)
Calestani, G., 329 (11.60, 11.62)
Calgari, S., 216 (7.144)
Callahan, R. W., 51 (3.3), 144 (5.210), 145 (5.210), 174 (6.78), 180 (6.78)
Calvert, J. G., 25 (2.1), 41 (2.1)
Calvin, M., 257 (9.111)
Cammarata, V., 380 (12.118)
Campagna, S., 170 (6.50), 174 (6.70, 6.71), 176 (6.70), 177 (6.70, 6.71), 178 (6.70, 6.71, 6.83, 6.86), 181 (6.70, 6.71, 6.105, 6.107, 6.108), 182 (6.70, 6.71, 6.105, 6.107, 6.108), 183 (6.108), 184 (6.70, 6.71, 6.107, 6.108), 301 (10.309), 306 (10.309, 10.321), 319 (11.7), 330 (11.7), 347 (11.7), 365 (12.57), 370 (12.82, 12.84), 371 (12.82), 382 (12.84), 386 (12.84)
Campion, A., 380 (12.112)
Canceill, J., 19 (1.37)
Canceill, J., 287 (10.162), 333 (11.94)
Carassiti, V., 20 (1.89), 25 (2.1), 41 (2.1), 84 (4.22), 86 (4.22), 186 (6.113), 187 (6.113), 188 (6.113), 280 (10.136), 281 (10.136), 320 (11.13), 324 (11.13)
Carmack, R. A., 268 (10.46)
Carter, F. L., 19 (1.74), 93 (5.31, 5.32), 211 (7.103, 7.104, 7.105), 216 (7.135), 217 (7.147),

342 (11.136), 343 (11.136), 344 (11.136), 345 (11.136), 355 (12.2, 12.3, 12.6, 12.9), 367 (12.3, 12.67), 377 (12.98), 380 (12.2, 12.3, 12.6, 12.9, 12.67), 387 (12.166)
Carter, M. T., 301 (10.311)
Casabò, J., 267 (10.37)
Casnati, A., 329 (11.63)
Caspar, J. V., 125 (5.149), 161 (6.3), 170 (6.51), 243 (8.88), 299 (10.260), 330 (11.68) 331 (11.68)
Castella, M., 256 (9.78)
Castellan, A., 260 (9.145), 272 (10.76), 275 (10.76), 276 (10.76)
Castello, M., 279 (10.120)
Castelvetro, V., 86 (4.38), 281 (10.137), 282 (10.137)
Castner, D. G., 387 (12.161)
Catena, G. C., 295 (10.227)
Catlow, B., 250 (9.13)
Cavagnat, R. M., 250 (9.15)
Cave, R. J., 98 (5.67)
Cesario, M., 287 (10.162), 333 (11.94), 339 (11.111)
Chachaty, C., 19 (1.56), 116 (5.126), 145 (5.126, 5.204, 5.205, 5.206), 146 (5.126, 5.206), 147 (5.205)
Chakravorti, S., 204 (7.74)
Chan, C., 151 (5.238)
Chan, C. K., 306 (10.319)
Chance, B., 90 (5.4), 92 (5.20)
Chang, C.-H., 90 (5.8)
Chang, C. H., 299 (10.290), 304 (10.290)
Chang, H. R., 365 (12.63)
Chanon, M., 19 (1.55, 1.59, 1.69), 61 (3.16), 71 (3.27), 94 (5.33, 5.34), 95 (5.34), 116 (5.33), 120 (5.33), 121 (5.33), 144 (5.202), 226 (8.3), 227 (8.12, 8.3), 231 (8.12), 241 (8.3, 8.64), 242 (8.3), 251 (9.16, 9.17, 9.29, 9.30), 252 (9.16, 9.17), 256 (9.30), 361 (12.36), 364 (12.48), 380 (12.48)
Chardon-Noblat, S., 167 (6.32)
Charl, K. P., 361 (12.31), 363 (12.31)
Chattopadhyay, N., 298 (10.249)
Chaudhuri, P., 279 (10.110)
Checchi, L., 188 (6.120), 233 (8.44), 234 (8.44), 235 (8.44), 237 (8.44)
Chen, D., 267 (10.24)
Chen, M. J., 295 (10.228)
Chen P., 125 (5.153, 5.154), 126 (5.153, 5.154), 127 (5.153, 5.154), 128 (5.162, 5.163), 129 (5.162, 5.163)
Cheung, S. T., 257 (9.95)
Chevrier, B., 341 (11.124), 387 (12.173)
Chiorboli, C., 19 (1.76), 60 (3.15), 94 (5.37), 132 (5.37), 136 (5.186), 137 (5.186), 161 (6.7), 170 (6.7), 184 (6.109, 6.110), 185 (6.109), 186 (6.109), 187 (6.118), 188 (6.118), 189 (6.123), 233 (8.44), 234 (8.44), 235 (8.44, 8.51), 237 (8.44), 301 (10.310), 306 (10.310), 361 (12.38), 381 (12.38, 12.129, 12.130)
Choi, J. D., 345 (11.145)
Chou, M., 178 (6.89)

Chou, P., 343 (11.139)
Chowdhury, M., 257 (9.108)
Christensen, J. J., 267 (10.12), 278 (10.98)
Christmas, C., 365 (12.63)
Christodoulou, D., 279 (10.121)
Christou, G., 365 (12.63)
Chung, C. J., 290 (10.200)
Chung, W. S., 295 (10.237)
Ciampolini, M., 279 (10.124)
Ciano, M., 84 (4.23), 174 (6.70, 6.71), 176 (6.70),
    177 (6.70, 6.71), 178 (6.70, 6.71, 6.83), 181
    (6.70, 6.71, 6.105, 6.107, 6.108), 182 (6.70,
    6.71, 6.105, 6.107, 6.108), 183 (6.108), 184
    (6.70, 6.71, 6.107, 6.108), 231 (8.20), 232
    (8.20, 8.27), 235 (8.20), 240 (8.27, 8.55), 275
    (10.93), 276 (10.93), 277 (10.93), 289 (10.93),
    306 (10.321), 321 (11.26), 323 (11.26, 11.34,
    11.35), 324 (11.34), 327 (11.48), 365 (12.57),
    370 (12.82, 12.84), 371 (12.82), 382 (12.84),
    386 (12.84)
Ciardelli, F., 215 (7.121), 218 (7.121), 219 (7.121,
    7.174, 7.175, 7.177, 7.178, 7.180)
Clarkson, M. A., 355 (12.18), 380 (12.18)
Clegg, W., 19 (1.26)
Cline, III J. I., 290 (10.198)
Cline Love, L. J., 80 (4.7), 290 (10.193, 10.197),
    295 (10.229, 10.230)
Closs, G. L., 48 (2.27), 57 (3.10), 70 (3.24), 73
    (3.10, 3.36), 95 (5.49), 98 (5.64, 5.68, 5.69,
    5.70, 5.72, 5.73, 5.74), 99 (5.64, 5.70, 5.72),
    100 (5.64, 5.69, 5.70, 5.74), 101 (5.64, 5.70,
    5.72), 102 (5.73, 5.75, 5.76), 162 (6.8, 6.9,
    6.10, 6.11, 6.12), 163 (6.9)
Cockroft, N. J., 375 (12.92)
Cocks, A. T., 212 (7.108)
Cohen, H., 240 (8.58)
Cohen, M. D., 251 (9.24)
Collart, P., 257 (9.98, 9.106)
Collet, A., 19 (1.37), 287 (10.162), 333 (11.94)
Collman, J. P., 19 (1.20)
Colquhoun, H. M., 19 (1.49), 85 (4.33), 267
    (10.10), 274 (10.10, 10.88, 10.91), 275 (10.10,
    10.94, 10.95), 278 (10.102, 10.103), 280
    (10.10), 289 (10.10)
Comba, P., 324 (11.41, 11.42), 325 (11.41), 326
    (11.41)
Cometti, G., 387 (12.163)
Connolly, J. S., 19 (1.55), 94 (5.33), 116 (5.33,
    5.126), 117 (5.128), 118 (5.128, 5.130, 5.133),
    120 (5.33), 121 (5.33), 145 (5.126), 146
    (5.126), 251 (9.17), 252 (9.17), 254 (9.51), 361
    (12.36), 380 (12.106)
Conrad, D., 348 (11.158)
Constable, E. C., 19 (1.61), 342 (11.126, 11.127),
    387 (12.175)
Contolini, N., 290 (10.196)
Cooley, L. F., 129 (5.164), 130 (5.164)
Corbin, D. R., 299 (10.260)
Cormier, R. A., 118 (5.133), 254 (9.51)
Corvaja, C., 256 (9.85)
Costa, J. C., 240 (8.57), 323 (11.36)

Cotsaris, E., 19 (1.6), 98 (5.65), 102 (5.65, 5.87,
    5.88, 5.91), 103 (5.65), 104 (5.65, 5.88), 105
    (5.65, 5.91), 107 (5.101), 113 (5.101), 162
    (6.13, 6.14), 163 (6.13, 6.15)
Cotton, F. A., 35 (2.6), 83 (4.10), 319 (11.11),
    327 (11.11)
Coucouvanis, D., 279 (10.121)
Courtney, S. H., 198 (7.19), 296 (10.239),
Cousseins, J. C., 374 (12.89)
Cowan, J. A., 147 (5.214)
Cowley, D. J., 202 (7.48)
Cox, G. S., 292 (10.202, 10.205), 295 (10.228)
Coxon, J. M., 20 (1.87)
Coyle, J. D., 25 (2.1), 41 (2.1), 356 (12.26)
Cragg, D. E., 114 (5.120), 115 (5.120)
Crago, K. T., 289 (10.184)
Craig, D. C., 102 (5.95, 5.97), 106 (5.95, 5.97),
    107 (5.95, 5.97)
Cram, D. J., 19 (1.3, 1.12, 1.13, 1.15), 83 (4.20),
    85 (4.20), 204 (7.67), 267 (10.4, 10.19, 10.34,
    10.35), 268 (10.46), 274 (10.89), 278 (10.89),
    287 (10.161), 329 (11.59), 333 (11.87, 11.88,
    11.89)
Cram, J. M., 267 (10.4)
Cram, M. J., 274 (10.89), 278 (10.89)
Cramer, F., 283 (10.143)
Cramer, S. P., 365 (12.61)
Creager, E. E., 147 (5.216)
Creaser, I. I., 132 (5.174, 5.175), 320 (11.14), 321
    (11.25, 11.29, 11.31, 11.32, 11.33), 322 (11.32,
    11.33), 323 (11.33), 324 (11.42)
Creed, D., 261 (9.149)
Creutz, C., 52 (3.4), 54 (3.6), 56 (3.6), 62 (3.6),
    90 (5.2), 132 (5.178, 5.182), 133 (5.182), 143
    (5.198), 144 (5.199), 178 (6.89, 6.90), 227
    (8.11), 232 (8.31), 240 (8.31), 319 (11.3)
Croke, D. T., 299 (10.285)
Crosby, G. A., 42 (2.10), 72 (3.34), 127 (5.160),
    134 (5.160), 319 (11.2)
Cruickshank, K. A., 151 (5.237, 5.238), 306
    (10.318, 10.319, 10.320), 386 (12.152)
Crutchley, R. J., 93 (5.25), 348 (11.159)
Cundliffe, E., 300 (10.303)
Currel, L. J., 299 (10.296), 304 (10.296)
Curry, M., 125 (5.154), 126 (5.154), 127 (5.154)
Curtis, J. C., 133 (5.184), 134 (5.184), 135
    (5.184), 227 (8.5), 231 (8.5), 232 (8.5, 8.30,
    8.39), 239 (8.53), 240 (8.58)
Czarnik, A. W., 271 (10.68)

D'Agostino, J., 198 (7.13), 217 (7.13)
D'Angelantonio, M., 240 (8.56), 242 (8.73)
D'Epenoux, B., 147 (5.219)
D'Souza, F., 254 (9.50)
D'Souza, V. T., 288 (10.172)
Da Silva, J. J. R. F., 272 (10.74)
Dabestani, R., 380 (12.112)
Dahenens, M., 287 (10.158)
Dalcanale, E., 267 (10.38), 287 (10.161), 387
    (12.163)
Dalton, J., 118 (5.132)
Daney, M., 272 (10.76), 275 (10.76), 276 (10.76)

Daney, N., 260 (9.145)
Danielson, E., 125 (5.153), 126 (5.153), 127 (5.153), 128 (5.162), 129 (5.162), 137 (5.188), 147 (5.215), 148 (5.215), 150 (5.227), 180 (6.97), 386 (12.151)
Danishefsky, A. T., 299 (10.268, 10.270), 304 (10.268, 10.270), 305 (10.268)
Dannenberg, J. J., 299 (10.266), 300 (10.266)
Dapporto, P., 279 (10.120)
Darwent, J. R., 241 (8.65)
Das, P. K., 168 (6.44)
Das, S., 217 (7.151)
Daub, J., 214 (7.116)
Davidson, R. S., 248 (9.7), 255 (9.72)
Davila, J., 60 (3.15), 136 (5.186), 137 (5.186), 168 (6.43), 184 (6.109), 185 (6.109), 186 (6.109), 348 (11.165), 381 (12.130)
Davis, J. H., Jr., 19 (1.80)
Davison, A., 19 (1.27)
Day, P., 54 (3.8), 56 (3.8), 71 (3.29), 95 (5.62)
De B. Costa, S. M., 272 (10.74)
De Boer, Th. J., 90 (5.1), 253 (9.42, 9.43, 9.44, 9.45), 254 (9.43)
De Cola, L., 19 (1.50), 83 (4.18), 171 (6.57), 173 (6.60), 174 (6.69), 177 (6.69), 178 (6.69), 306 (10.322), 330 (11.65), 331 (11.72, 11.73, 11.74) 332 (11.74, 11.75), 340 (11.120), 341 (11.120, 11.122), 365 (12.56)
De Haas, M. P., 102 (5.81, 5.88, 5.90, 5.92), 104 (5.88, 5.90, 5.92), 105 (5.81), 163 (6.16)
De Jesus, M., 388 (12.185)
De Jong, B., 102 (5.81), 105 (5.81)
De Jong, F., 267 (10.5)
De Ridder, D. J. A., 178 (6.88)
De Rosa, G., 178 (6.83)
De Santis, G., 279 (10.126)
De Schryver, F. C., 19 (1.58), 116 (5.126), 145 (5.126), 146 (5.126), 148 (5.222), 150 (5.222), 161 (6.6), 226 (8.4), 244 (8.4), 252 (9.38), 255 (9.66, 9.69), 257 (9.96, 9.98, 9.102, 9.106), 260 (9.139), 363 (12.41), 364 (12.41)
De Silva, A. P., 271 (10.67, 10.69)
De Vault, D. C., 90 (5.4), 92 (5.20)
DeArmond, L. K., 124 (5.147)
Deboer, C. D., 85 (4.25)
Deeble, D. J., 299 (10.288), 305 (10.288)
DeFelippis, M. R., 130 (5.166)
Degani, Y., 150 (5.229), 298 (10.255), 380 (12.117)
DeGraff, B. A., 257 (9.91, 9.93), 290 (10.198, 10.201), 291 (10.201)
DeGraziano, J., 150 (5.224), 370 (12.85)
Deisenhofer, J., 20 (1.95), 90 (5.7, 5.10), 91 (5.7), 356 (12.21)
Del Campo, N., 250 (9.15)
Del Paggio, A. A., 340 (11.117)
Delaney, J. K., 121 (5.137, 5.138)
Dellonte S., 233 (8.48), 234 (8.48), 235 (8.48), 241 (8.48), 243 (8.89), 257 (9.99), 327 (11.48)
DeLuccia, F. J., 295 (10.229, 10.230)
Demas, J. N., 79 (4.4, 4.5), 232 (8.34), 234 (8.34), 235 (8.34), 257 (9.91, 9.93), 290 (10.198, 10.201), 291 (10.201), 348 (11.162, 11.163)
Denny, W. A., 306 (10.323)
Denti, G., 174 (6.70, 6.71), 176 (6.70), 177 (6.70, 6.71), 178 (6.70, 6.71, 6.83), 181 (6.70, 6.71, 6.105, 6.107, 6.108), 182 (6.70, 6.71, 6.105, 6.107, 6.108), 183 (6.108), 184 (6.70, 6.71, 6.107, 6.108), 306 (10.321), 365 (12.57), 370 (12.82, 12.84), 371 (12.82), 382 (12.84), 386 (12.84)
Deronzier, A., 242 (8.79), 386 (12.150)
Dervan, P. B., 113 (5.115, 5.116, 5.117), 114 (5.116, 5.117), 299 (10.262), 304 (10.262, 10.313, 10.314, 10.315), 306 (10.262)
Deshayes, K., 214 (7.115)
Desilvestro, J., 381 (12.126)
Desvergne, J. P., 255 (9.71), 260 (9.145, 9.146), 268 (10.47), 272 (10.76), 273 (10.47, 10.77, 10.78, 10.79), 275 (10.76), 276 (10.97, 10.76)
Detellier, C., 243 (8.86)
Detzer, N., 202 (7.51)
Devanathan, S., 296 (10.240)
Dewan, J. C., 19 (1.27)
Dexpert-Ghys, J., 375 (12.93)
Di Casa, M., 279 (10.126)
Di Grazia, M., 287 (10.159)
Diamond, P., 292 (10.206)
Dick, B., 198 (7.17)
Dick, K., 283 (10.144, 10.145), 284 (10.144, 10.145)
Diederich, F., 19 (1.19), 85 (4.35), 267 (10.13, 10.26, 10.30), 283 (10.13, 10.144, 10.145), 284 (10.144, 10.145), 287 (10.30, 10.163), 333 (11.93, 11.95)
Dieter, T., 174 (6.75)
Dietrich-Buchecker, C. O., 19 (1.51, 1.71), 334 (11.100), 337 (11.100, 11.107, 11.108, 11.110), 338 (11.100), 339 (11.111, 11.113, 11.115), 340 (11.118, 11.119), 341 (11.113), 386 (12.141), 387 (12.141, 12.177)
Dietrich, B., 19 (1.17), 83 (4.14), 85 (4.37), 279 (10.127, 10.128), 280 (10.127, 10.128), 281 (10.127, 10.128)
Dijkstra, P. J., 267 (10.23), 333 (11.97)
Dirksen, G. J., 329 (11.53, 11.54)
Dirkx, I. P., 90 (5.1)
Dismukes, G. C., 365 (12.62)
Dix, J. P., 268 (10.48)
Dixon, D. W., 93 (5.24)
Doany, F. E., 198 (7.21)
Dobkowski, J., 202 (7.56)
Dobson, S. M., 279 (10.113)
Dodsworth, E., 180 (6.100)
Doizi, D., 145 (5.205), 147 (5.205)
Dolphin, D., 280 (10.135)
Dose, E. V., 174 (6.64)
Doughty, S. M., 275 (10.95), 278 (10.103)
Drake, J. M., 289 (10.186)
Dressick, W. J., 125 (5.149), 242 (8.78), 290 (10.198)
Drew, M. G. B., 342 (11.126)

Du vosel, A., 387 (12.163)
Dubowchik, G. M., 168 (6.40)
Ducasse, L., 251 (9.18)
Duesing, R., 128 (5.163), 129 (5.163), 135 (5.185), 136 (5.185)
Dung, B., 19 (1.2)
Durante, V. A., 133 (5.183)
Durham, B., 330 (11.68), 331 (11.68)
Dürr, H., 127 (5.156), 227 (8.13), 243 (8.81), 274 (10.85), 275 (10.85), 280 (10.85), 331 (11.69), 386 (12.143), 387 (12.143)
Dutta, R., 298 (10.249)
Dutton, P. L., 91 (5.16), 92 (5.20), 110 (5.106)
Duveneck, G., 255 (9.67, 9.68)
Duveneck, G. L., 296 (10.241), 297 (10.241)
Dvolaitzky, M., 19 (1.46), 116 (5.121)
Dykstra, R. E., 261 (9.152)
Dzhabiev, T. S., 370 (12.75)

Eaton, D. F., 161 (6.3), 289 (10.181, 10.182), 292 (10.181, 10.182, 10.215), 295 (10.181, 10.234, 10.235), 298 (10.182), 299 (10.260)
Ebbesen, T. W., 380 (12.110)
Ebbeson, T. W., 242 (8.75)
Ebmeyer, F., 83 (4.21), 85 (4.21), 171 (6.57), 306 (10.322), 331 (11.73, 11.74), 332 (11.74, 11.75, 11.76, 11.79), 365 (12.56)
Echegoyen, L., 388 (12.185)
Eckert, C., 251 (9.18)
Edwards, A. K., 174 (6.73), 177 (6.73), 181 (6.73)
Effenberger, F., 165 (6.29), 166 (6.29)
Ege, D., 48 (2.24), 259 (9.135, 9.136)
Egorochkin, A. N., 252 (9.34, 9.35), 253 (9.34)
Ehlen, A., 213 (7.111), 378 (12.102),
Ehrenfreund, M., 19 (1.84)
Eich, M., 217 (7.156)
Eichen, Y., 299 (10.258)
Eichner, M., 260 (9.144)
Eichorn, G. L., 66 (3.19), 68 (3.19)
Eidenschink, R., 387 (12.169)
Eigen, M., 387 (12.159)
Eisenberg, R., 232 (8.42)
Eisenthal, K. B., 296 (10.241), 297 (10.241), 298 (10.251)
El-Bayoumi, M. A., 345 (11.144)
El Khalifa, M., 19 (1.7)
Elisei, F., 257 (9.97)
Elliott, C. M., 129 (5.164), 130 (5.164), 147 (5.215), 148 (5.215), 173 (6.59)
Emmelius, M., 380 (12.108)
Emming, C. S., 215 (7.118)
Endicott, J. F., 47 (2.20), 187 (6.119), 190 (6.126), 324 (11.38, 11.43)
Enea, O., 380 (12.115)
Engel, M., 217 (7.153)
Ennis, P. M., 150 (5.228), 365 (12.60), 380 (12.60)
Epp, O., 90 (5.7), 91 (5.7)
Ericson, J. L., 19 (1.13)
Ericson, J. L., 267 (10.35)

Erlanger, B. F., 197 (7.1, 7.3), 204 (7.69), 213 (7.112), 215 (7.120, 7.122)
Ernsting, N. P., 201 (7.42)
Eskola, S. M., 250 (9.14)
Esposito, F., 242 (8.79)
Essakalli, M., 386 (12.150)
Evans, B. R., 208 (7.90)
Evans, S., 120 (5.136)
Evans, T. R., 292 (10.208)
Everly, R. M., 241 (8.62)
Evers, A., 279 (10.113)
Eyring, H., 26 (2.2)

Fabbri, D., 219 (7.178)
Fabbrizzi, L., 19 (1.34), 279 (10.108, 10.124, 10.126)
Fages, F., 255 (9.71), 268 (10.47), 273 (10.47, 10.78, 10.79)
Falcetta, M. F., 95 (5.58), 102 (5.58), 254 (9.48)
Fan, P., 299 (10.299), 304 (10.299)
Faraggi, M., 130 (5.166)
Farid, S., 48 (2.24, 2.25), 70 (3.25), 259 (9.135, 9.136, 9.137, 9.138), 261 (9.148, 9.150), 292 (10.208)
Farrow, S. J., 299 (10.300)
Faus, J., 279 (10.120)
Fayer, M. D., 168 (6.38)
Feeney, M. M., 299 (10.289), 305 (10.289)
Feher, G., 90 (5.9), 356 (12.25), 363 (12.25)
Felker, P. M., 113 (5.116, 5.117), 114 (5.116, 5.117), 198 (7.20)
Femia, R. A., 80 (4.7), 290 (10.197)
Fendler, J. H., 271 (10.72), 274 (10.84), 289 (10.183)
Fenton, D. E., 279 (10.117)
Ferguson, J., 19 (1.28), 251 (9.27), 252 (9.39)
Fernandez, A., 232 (8.40)
Ferraudi, G. J., 20 (1.91), 25 (2.1), 41 (2.1)
Fery-Forgues, S., 271 (10.70, 10.71)
Fessner, W. D., 19 (1.31)
Fiedler, J., 19 (1.84)
Field, R. J., 386 (12.156)
Figard, J. E., 179 (6.92, 6.93, 6.94), 180 (6.93)
Finckh, P., 95 (5.53), 108 (5.103, 5.104, 5.105), 109 (5.103, 5.104, 5.105), 110 (5.103)
Finkele, U., 91 (5.17)
Finklea, H. O., 125 (5.155)
Finocchiaro, P., 287 (10.159), 329 (11.55),
Firestone, A., 19 (1.41)
Fischer, C., 214 (7.116)
Fischer, E., 198 (7.14, 7.15), 201 (7.29)
Fischer, P., 287 (10.160)
Fischer, S.F., 95 (5.57)
Fissi, A., 215 (7.121), 218 (7.121), 219 (7.121, 7.174, 7.175, 7.177, 7.178, 7.180)
Fitzgerald, M. C., 332 (11.77)
Flamigni, L., 298 (10.247)
Fleischauer, P., 319 (11.1)
Fleischauer, P. D., 20 (1.90), 25 (2.1), 41 (2.1), 180 (6.95), 186 (6.114), 187 (6.114), 188 (6.114), 319 (11.1)
Fleisher, M. B., 299 (10.273), 305 (10.273)

Fleming, G. R., 73 (3.36), 91 (5.15), 102 (5.75), 162 (6.8) 198 (7.19), 296 (10.239)
Fonda, H. N., 118 (5.133), 254 (9.51)
Force, R. K., 121 (5.140)
Ford, P. C., 133 (5.183), 180 (6.95), 190 (6.127), 240 (8.58)
Fornasier, R., 288 (10.168)
Forster, L. S., 186 (6.115), 188 (6.115)
Frster, Th., 198 (7.12), 342 (11.128)
Forsyth, G. A., 342 (11.126)
Foster, R., 248 (9.2, 9.6), 251 (9.2), 252 (9.2)
Fournier, M. T., 374 (12.89)
Fox, L. S., 131 (5.171)
Fox, M. A., 19 (1.55, 1.59, 1.69), 61 (3.16), 71 (3.27), 94 (5.33, 5.34), 95 (5.34), 116 (5.33), 120 (5.33), 121 (5.33), 144 (5.202), 202 (7.58), 217 (7.152), 226 (8.3), 227 (8.3, 8.12), 231 (8.12), 241 (8.64, 8.3), 242 (8.3), 251 (9.16, 9.17, 9.29, 9.30), 252 (9.16, 9.17), 256 (9.30), 361 (12.36), 364 (12.48, 12.51), 380 (12.48, 12.51, 12.112)
Franco, C., 131 (5.173)
Frank, A. J., 299 (10.258)
Frank, C. W., 218 (7.166)
Frank, R., 232 (8.35), 233 (8.35), 234 (8.35)
Frankel, D. A., 216 (7.140)
Franken, S., 333 (11.83)
Franzke, D., 215 (7.127)
Frauenfelder, H., 90 (5.4), 92 (5.20)
Freitag, R. A., 173 (6.59)
Friesner, R. A., 20 (1.98), 356 (12.24),
Frhling, J. C., 254 (9.54)
Fromherz, P., 151 (5.234), 306 (10.317), 361 (12.29), 363 (12.29), 370 (12.74), 383 (12.74)
Fronczek, F. R., 287 (10.159)
Fuchs, Y., 174 (6.75)
Fugate, R. D., 345 (11.145)
Fuhrhop, J. H., 387 (12.167), 388 (12.167)
Fujie, M., 288 (10.169)
Fujihira, M., 361 (12.32, 12.34), 363 (12.32, 12.34), 383 (12.132, 12.133, 12.134, 12.135, 12.136)
Fujii, M., 207 (7.86), 260 (9.143)
Fujimoto, A., 346 (11.150)
Fujimoto, B. S., 299 (10.295)
Fujimura, H., 267 (10.36)
Fujita, T., 90 (5.3), 288 (10.166)
Fujiwara, Y., 292 (10.214)
Fultz, W. C., 132 (5.177)
Funabashi, M., 258 (9.124)
Furlong, D. N., 381 (12.127)
Furue, M., 171 (6.53, 6.54), 172 (6.58)
Furuta, H., 388 (12.184)

Gabard, J., 19 (1.37), 388 (12.182)
Gadzepko, V. P. Y., 117 (5.129), 118 (5.129)
Gafney, H. D., 174 (6.74, 6.75), 241 (8.69), 242 (8.69), 348 (11.160, 11.161)
Gahan, L. R., 321 (11.29, 11.31), 324 (11.42)
Gale, E. F., 300 (10.303)
Galili, T., 168 (6.41)
Gamache, R. E. Jr., 340 (11.117)

Gandolfi, M. T., 19 (1.79), 227 (8.13), 232 (8.37, 8.38), 233 (8.37, 8.38, 8.47), 234 (8.37, 8.38), 235 (8.37, 8.38), 237 (8.37, 8.38), 260 (9.146), 274 (10.85), 275 (10.85, 10.92, 10.93), 276 (10.92, 10.93, 10.96, 10.97), 277 (10.93, 10.96), 280 (10.85), 289 (10.93), 334 (11.106), 336 (11.106), 387 (12.179)
Ganesh, K. N., 19 (1.11)
Gao, F., 148 (5.221, 5.223), 149 (5.221), 150 (5.223, 5.224), 169 (6.48), 363 (12.42), 370 (12.85)
Garcia-España, E., 279 (10.116, 10.120, 10.131)
Garret, T. M., 83 (4.17), 333 (11.84)
Gegiou, D., 198 (7.14, 7.15)
Gehrtz, M., 201 (7.33)
Geiger, T., 321 (11.27)
Geller, G. G., 113 (5.115)
Gelroth, J. A., 179 (6.92, 6.93), 180 (6.93)
Geoffroy, G. L., 20 (1.92), 25 (2.1), 41 (2.1)
George, G. N., 365 (12.61)
Gerald, III R., 255 (9.74), 256 (9.74)
Gerischer, H., 381 (12.123, 12.125)
Gerritsma, G. J., 267 (10.25), 268 (10.44), 270 (10.44)
Gettliffe, G., 181 (6.106), 182 (6.106), 370 (12.83)
Getz, D., 163 (6.22)
Geuder, W., 285 (10.155)
Geue, R. J., 320 (11.14), 321 (11.25, 11.31)
Ghidini, E., 267 (10.23), 329 (11.60, 11.62), 333 (11.97)
Ghosh, S., 270 (10.62)
Gill, D., 289 (10.185), 290 (10.188)
Gilmanshin, R. I., 355 (12.13), 380 (12.13)
Giniger, R., 19 (1.30)
Giordano, J., 347 (11.155)
Giusti, M., 279 (10.131)
Glaudemans, C. P. J., 19 (1.36)
Glauser, W. A., 254 (9.60)
Gleiter, R., 19 (1.39)
Gleria, M., 257 (9.99)
Gliozzi, A., 215 (7.120)
Goddard, III W. A., 19 (1.85), 51 (3.1), 98 (5.67), 386 (12.153)
Goedeweeck, R., 255 (9.66), 257 (9.106)
Gokel, G. W., 274 (10.87), 280 (10.87), 388 (12.185)
Goldberg, J. M., 299 (10.268, 10.274), 304 (10.268, 10.274), 305 (10.268)
Goldstein, R. A., 20 (1.97), 90 (5.12), 356 (12.23)
Golovin, M. N., 180 (6.100)
Gonzales, M. C., 119 (5.134), 295 (10.231)
Goodings, E. P., 278 (10.102)
Goodman, J. L., 258 (9.128, 9.129, 9.130), 259 (9.130)
Goodnow, T. T., 19 (1.52, 1.79), 334 (11.104, 11.106), 336 (11.106), 387 (12.178, 12.179)
Goodwin, K. V., 257 (9.88)
Gordon, M., 251 (9.23)
Goren, Z., 298 (10.255), 299 (10.256)
Gormin, D., 343 (11.140, 11.141), 346 (11.140)

Grner, H., 232 (8.40), 244 (8.91), 299 (10.286, 10.296), 301 (10.286), 304 (10.296)
Goswami, K., 243 (8.82)
Goto, T., 208 (7.89)
Goto, Y., 119 (5.135)
Gould, I. R., 48 (2.24, 2.25), 70 (3.25), 259 (9.135, 9.136, 9.137, 9.138)
Gould, S., 363 (12.44), 386 (12.151)
Gourdon, A., 212 (7.110)
Gouterman, M., 111 (5.113), 280 (10.135)
Grabowska, A., 44 (2.11)
Grabowski, Z. R., 44 (2.11), 202 (7.47, 7.48, 7.52)
Graetzel, M., 47 (2.21), 48 (2.21), 321 (11.27), 364 (12.47), 380 (12.47, 12.113, 12.114, 12.115, 12.122), 381 (12.122, 12.126, 12.128), 382 (12.113)
Graham, D. R., 304 (10.312)
Graiver, D., 334 (11.101)
Grammenudi, S., 19 (1.8), 83 (4.16), 331 (11.70, 11.73, 11.74), 332 (11.70, 11.74),
Grassi, G., 102 (5.93, 5.98), 103 (5.93), 104 (5.93, 5.98), 106 (5.98), 107 (5.101), 113 (5.101), 163 (6.17, 6.19), 216 (7.133), 217 (7.153)
Gratton, E., 167 (6.36), 168 (6.36)
Gray, G. W., 387 (12.162)
Gray, H. B., 93 (5.25, 5.26), 131 (5.171, 5.172)
Green, N. D., 162 (6.12)
Green, N. J., 95 (5.49), 98 (5.70, 5.72), 99 (5.70, 5.72), 100 (5.70), 101 (5.70, 5.72)
Greenberg, A., 102 (5.78), 107 (5.78), 113 (5.78), 198 (7.8), 199 (7.8), 202 (7.8), 203 (7.8)
Greene, B. I., 198 (7.21)
Gregg, B. A., 217 (7.152), 364 (12.51), 380 (12.51)
Greiner, S. P., 167 (6.31)
Grellmann, K. H., 202 (7.47)
Griebel, D., 283 (10.145), 284 (10.145)
Gries, W. K., 285 (10.155)
Griffin, R. D., 348 (11.166)
Grimes, R. N., 19 (1.80)
Groh, W., 387 (12.170),
Gronkiewicz, M. K., 256 (9.84)
Gross, M., 279 (10.129)
Gruler, H., 218 (7.171)
Grzybowski, J. J., 332 (11.82)
Guardigli, M., 329 (11.63)
Guarr, T., 93 (5.29)
Guarr, T. F., 178 (6.87)
Gubelmann, M., 123 (5.144), 274 (10.83)
Guerriero, P., 272 (10.75)
Guett, J. P., 271 (10.70, 10.71)
Guglielmetti, G., 267 (10.38)
Guglielmetti, R., 201 (7.34, 7.37, 7.39)
Guilhem, J., 279 (10.115), 287 (10.162), 333 (11.92, 11.94), 339 (11.111, 11.112)
Guinand, G., 260 (9.145), 272 (10.76), 275 (10.76), 276 (10.76)
Gunaratne, H. Q. N., 271 (10.69)
Gunner, M. R., 91 (5.16)
Guo, R. K., 218 (7.167, 7.168)

Gushurst, A. K. I., 340 (11.119)
Gust, D., 19 (1.56, 1.58), 94 (5.35, 5.38), 116 (5.126), 145 (5.38, 5.126, 5.204, 5.205, 5.206, 5.207), 146 (5.126, 5.206), 147 (5.205, 5.207, 5.212, 5.220), 148 (5.38, 5.221, 5.222, 5.223), 149 (5.38, 5.221), 150 (5.38, 5.222, 5.223, 5.224), 169 (6.45, 6.46, 6.47, 6.48), 361 (12.37), 363 (12.41, 12.42), 364 (12.41), 365 (12.65, 12.66), 370 (12.85)
Gutsche, C. D., 19 (1.10), 267 (10.16, 10.22), 329 (11.57)

Haarer, D., 355 (12.17), 380 (12.17)
Haas, O., 380 (12.116)
Haas, Y., 251 (9.30), 256 (9.30, 9.80), 326 (11.44)
Haasnoot, J. G., 174 (6.69), 177 (6.69), 178 (6.69, 6.88)
Habata, Y., 205 (7.79)
Haddon, R. C., 355 (12.11), 380 (12.11)
Hader, D. P., 20 (1.93), 197 (7.7), 198 (7.7), 356 (12.19), 370 (12.19)
Hadjoudis, E., 346 (11.151)
Haenel, M. W., 253 (9.47), 254 (9.47)
Haga, M., 174 (6.72), 176 (6.72), 177 (6.72)
Hage, R., 174 (6.69), 177 (6.69), 178 (6.69, 6.88)
Hagen, K. S., 320 (11.20)
Hahn, F. E., 83 (4.17), 333 (11.84)
Haim, A., 226 (8.1), 230 (8.16), 231 (8.16), 232 (8.16, 8.33), 233 (8.33), 234 (8.33), 237 (8.33), 240 (8.16)
Hall, B., 178 (6.84)
Hall, D. O., 102 (5.93, 5.98), 103 (5.93), 104 (5.93, 5.98), 106 (5.98), 107 (5.101), 113 (5.101), 163 (6.17, 6.19), 216 (7.133), 217 (7.153)
Hallock, J. S., 188 (6.122), 189 (6.122)
Halpern, A. M., 258 (9.112)
Halpern, J., 95 (5.60)
Halton, B., 20 (1.87)
Hamada, F., 210 (7.101), 293 (10.217, 10.219, 10.220, 10.221, 10.223), 294 (10.217), 298 (10.217)
Hamada, N., 218 (7.165, 7.170)
Hamai, S., 290 (10.189, 10.195), 292 (10.195, 10.211)
Hamilton, A. D., 19 (1.24), 168 (6.40)
Hamilton, G. S., 267 (10.21)
Hammershoi, A., 132 (5.175), 321 (11.33), 322 (11.33), 323 (11.33)
Hampton, T., 289 (10.184)
Hanaoka, K., 288 (10.169)
Hanck, K. W., 124 (5.147)
Hancock, R. D., 267 (10.15), 279 (10.113)
Härd, T., 299 (10.299), 304 (10.299)
Harding, L. O., 19 (1.58), 148 (5.222, 5.223), 150 (5.222, 5.223)
Harding, L. O., 363 (12.41, 12.42), 364 (12.41)
Harding, M. M., 387 (12.176)
Harkema, S., 267 (10.23, 10.25), 333 (11.97)
Harriman, A., 121 (5.139), 122 (5.143), 123 (5.144), 168 (6.41, 6.42, 6.43) 241 (8.65), 273 (10.81), 274 (10.83)

Harrison, R. J., 122 (5.141), 147 (5.214)
Harrowfield, J., 341 (11.124), 387 (12.173)
Harrowfield, J. M., 320 (11.14), 321 (11.25), 324 (11.42), 330 (11.64)
Hart, H., 19 (1.22, 1.45)
Harvey, P. D., 388 (12.183)
Hasebe, Y., 216 (7.145)
Hasegawa, M., 267 (10.36)
Hashimoto, S., 292 (10.209, 10.210), 294 (10.226)
Hashimoto, Y., 267 (10.36)
Haspeslagh, L., 282 (10.140)
Hassenruck, K., 19 (1.38)
Hassenruck, K., 386 (12.146)
Hassoon, S., 163 (6.23, 6.24), 164 (6.23, 6.24)
Häubling, C., 216 (7.133)
Hawley, M. D., 61 (3.16)
Hawthorne, W. D., 387 (12.162)
Hayakawa, K., 298 (10.253)
Hayard, R. C., 83 (4.12)
Hayashi, A., 256 (9.81)
Hayashi, K., 201 (7.32), 218 (7.32, 7.160)
Hayashi, R., 218 (7.166, 7.167, 7.168)
Hayashi, T., 255 (9.62, 9.63)
Haymore, B. L., 278 (10.98)
Headford, C. E. L., 129 (5.164), 130 (5.164)
Heath, G. A., 124 (5.146)
Heatherington, A. L., 188 (6.122), 189 (6.122)
Hegelson, R. C., 287 (10.161),
Heiligman-Rim, R., 201 (7.29)
Heisel, F., 202 (7.45), 212 (7.110), 251 (9.18), 377 (12.95)
Heitele, H., 95 (5.53), 108 (5.102, 5.103, 5.104, 5.105), 109 (5.102, 5.103, 5.104, 5.105), 110 (5.103)
Heldt, J., 346 (11.140), 343 (11.140, 11.141)
Helene, C., 299 (10.288), 305 (10.288)
Helgeson, R. C., 19 (1.13), 268 (10.46)
Heller, A., 150 (5.229), 380 (12.117)
Hemmert, C., 341 (11.122)
Hendrickson, D. N., 365 (12.63)
Hendrickx, M., 282 (10.140)
Hennig, H., 230 (8.17, 8.18, 8.19), 240 (8.59, 8.60, 8.61)
Henrick, K., 279 (10.109)
Henry, B. R., 39 (2.7), 40 (2.7)
Heppener, M., 19 (1.6), 98 (5.65), 102 (5.65, 5.87, 5.89) 103 (5.65), 104 (5.65), 105 (5.65)
Herbich, J., 202 (7.56), 346 (11.149)
Here, W. J., 107 (5.100), 113 (5.100)
Heremans, K., 180 (6.103)
Herkstroeter, W. G., 292 (10.208)
Herlt, A. J., 320 (11.14), 321 (11.25)
Hermant, R. M., 105 (5.99), 252 (9.37), 253 (9.37), 254 (9.37)
Hernandez, J. C., 388 (12.185)
Herret, H., 218 (7.171)
Herrmann, U., 271 (10.73)
Hertzberg, R. P., 304 (10.313, 10.314)
Herz, A., 260 (9.144)
Herzberg, G., 26 (2.3), 29 (2.3), 35 (2.3, 2.5)
Hickman, J. J., 388 (12.183)
Hicks, J. M., 298 (10.251)

Hidalgo-Luangdilok, C., 386 (12.149)
Highland, R. G., 127 (5.160), 134 (5.160)
Higuchi, T., 289 (10.177, 10.178)
Higuchi, Y., 257 (9.105), 361 (12.33), 363 (12.33), 370 (12.33)
Hilinski, E. F., 164 (6.25), 258 (9.125, 9.126), 259 (9.131, 9.134)
Hill, C. L., 259 (9.134)
Hill, J. S., 387 (12.162)
Hill, R. R., 356 (12.26)
Himanen, J. P., 290 (10.199)
Hinatu, J., 258 (9.118)
Hino, H., 292 (10.207)
Hino, T., 258 (9.117)
Hino, Y., 290 (10.191), 292 (10.207)
Hinschberger, J., 273 (10.79)
Hintze, R. E., 180 (6.95)
Hiort, C., 299 (10.301), 304 (10.301)
Hirata, Y., 122 (5.142), 361 (12.34), 363 (12.34)
Hirose, K., 271 (10.66)
Hirotsu, K., 289 (10.177, 10.178)
Hirshberg, Y., 201 (7.29)
Ho, T.F., 51 (3.2), 113 (5.119), 118 (5.130)
Hochstrasser, R. M., 198 (7.18, 7.21)
Hodgson, S. M., 19 (1.26)
Hoffman, B. M., 367 (12.68), 380 (12.68)
Hoffman, M. Z., 241 (8.67, 8.70, 8.71), 242 (8.70, 8.71, 8.73, 8.74), 243 (8.67, 8.84)
Hoffman, R. C., 380 (12.107)
Hoffmann, R., 95 (5.42), 107 (5.100), 113 (5.100)
Hofkens, J., 252 (9.38)
Hofstra, U., 147 (5.217)
Hogen-Esch, T. E., 226 (8.2)
Hohlneicher, G., 198 (7.17)
Hollmann, G., 270 (10.55)
Holten, D., 91 (5.18), 92 (5.19), 111 (5.111), 257 (9.89)
Holyle, C. E., 161 (6.2)
Holzapfel, W., 91 (5.17)
Holzwarth, A. R., 116 (5.126), 145 (5.126), 146 (5.126)
Honda, K., 214 (7.117), 380 (12.111)
Honda, Y., 206 (7.82), 207 (7.83), 378 (12.101), 379 (12.101)
Hong, F. T., 355 (12.15), 380 (12.15)
Hong, H. G., 387 (12.165)
Hong, X., 93 (5.24)
Honig, B., 198 (7.10)
Hopfield, J. J., 95 (5.46, 5.47, 5.48), 113 (5.115, 5.116, 5.117), 114 (5.116, 5.117), 384 (12.137, 12.138), 385 (12.138)
Hopkins, A. S., 232 (8.43), 241 (8.43)
Hopkins, R. B., 267 (10.30), 287 (10.30), 333 (11.95)
Horrocks, W. D., 326 (11.45)
Horsman-van den Dool, L. E., 102 (5.81), 105 (5.81)
Hosek, W., 348 (11.161)
Hoskins, B. F., 386 (12.148)
Hosseini, M. W., 83 (4.17), 85 (4.37), 86 (4.38), 279 (10.119, 10.122, 10.125, 10.127, 10.128, 10.129), 280 (10.127, 10.128), 281 (10.127,

10.128, 10.137, 10.138), 282 (10.137, 10.138), 333 (11.84), 378 (12.103)
Houben, J. L., 215 (7.122), 219 (7.174, 7.175, 7.177, 7.180)
Houlding, V., 321 (11.27)
Hrnjez, B., 202 (7.58)
Huang, J., 295 (10.227)
Huber, R., 20 (1.96), 90 (5.7, 5.11), 91 (5.7), 356 (12.22)
Huber, W., 19 (1.84)
Huddleston, R. K., 48 (2.23), 93 (5.27), 95 (5.27)
Huffman, J. C., 259 (9.131), 365 (12.63)
Hugdahl, J., 147 (5.211)
Huizer, A. H., 346 (11.154), 347 (11.154)
Humer, W., 127 (5.158), 128 (5.158), 134 (5.158)
Hummel, H. U., 387 (12.180)
Hünig, S., 285 (10.155)
Hunter, C. A., 120 (5.136), 285 (10.147)
Hunziker, M., 174 (6.63), 180 (6.63)
Hupp, J. T., 363 (12.44), 386 (12.151)
Huppert, D., 342 (11.134), 343 (11.134), 345 (11.134), 346 (11.134)
Hurley, J. K., 117 (5.128), 118 (5.128)
Hurst, J. K., 139 (5.189)
Hush, N. S., 19 (1.6), 46 (2.16), 54 (3.7), 62 (3.7), 66 (3.18), 68 (3.18) 95 (5.54, 5.56, 5.59), 98 (5.65), 102 (5.65, 5.83, 5.84, 5.85, 5.86, 5.87, 5.88, 5.89, 5.90, 5.91), 103 (5.65), 104 (5.65, 5.88, 5.90), 105 (5.65, 5.91), 162 (6.13, 6.14), 163 (6.13, 6.15), 230 (8.15)
Huston, M. E., 271 (10.68)
Hyatt, J. A., 19 (1.29)
Hyla-Kryspin, I., 282 (10.140)

Ichimura, K., 201 (7.41), 213 (7.113), 295 (10.236)
Ichinaga, A. K., 340 (11.118)
Ide, R., 90 (5.3)
Iga, R., 218 (7.164)
Iitaka, Y., 280 (10.134), 283 (10.141)
Ikeda, A., 267 (10.31), 387 (12.164)
Ikeda, N., 257 (9.98), 346 (11.148)
Ikeda, T., 168 (6.39), 215 (7.128), 216 (7.136), 217 (7.157, 7.158)
Ilten, D. F., 257 (9.111)
Imabayashi, S., 255 (9.70)
Imai, A., 294 (10.226)
Imai, Y., 361 (12.34), 363 (12.34)
Imamura, A., 107 (5.100), 113 (5.100)
Imamura, M., 93 (5.28)
Imanishi, Y., 292 (10.209)
Indelli, A., 321 (11.28)
Indelli, M. T., 19 (1.76, 1.82), 60 (3.15), 70 (3.23), 94 (5.37), 125 (5.151), 132 (5.37), 136 (5.186), 137 (5.186), 161 (6.7), 170 (6.7), 184 (6.109, 6.110), 185 (6.109), 186 (6.109), 187 (6.117, 6.118), 188 (6.118, 6.121), 189 (6.123), 235 (8.51), 301 (10.310), 306 (10.310), 342 (11.135), 348 (11.164), 361 (12.38), 367 (12.73), 369 (12.73), 381 (12.38, 12.129, 12.130)

Infelta, P. P., 127 (5.156)
Ingham, K. C., 345 (11.144)
Inoue, Y., 274 (10.87), 280 (10.87)
Interrante, L. V., 347 (11.155)
Inuzuka, K., 204 (7.72), 346 (11.150)
Iqbal, M., 267 (10.22)
Irie, M., 201 (7.32), 215 (7.124), 218 (7.32, 7.124, 7.160, 7.161, 7.162, 7.163, 7.164)
Irvine, M. P., 122 (5.141)
Isaacs, N. S., 19 (1.42)
Ishihara, K., 216 (7.142), 218 (7.165, 7.170)
Ishihara, M., 216 (7.131)
Ishikawa, N., 254 (9.49)
Isied, S. S., 93 (5.21), 130 (5.167, 5.168, 5.169), 131 (5.169)
Itai, A., 283 (10.141)
Ito, S., 168 (6.39)
Itoh, M., 256 (9.76, 9.81), 260 (9.141), 292 (10.214), 345 (11.147)
Iwamoto, K., 216 (7.138)
Iwata, S., 256 (9.83)
Iwayanagi, T., 218 (7.162)
Iyoda, T., 214 (7.117), 347 (11.157), 380 (12.111)
Izaki, K., 347 (11.157)
Izatt, R. M., 267 (10.12), 278 (10.98)

Jackman, D. C., 180 (6.104
Jacquet, L., 180 (6.103)
Jacquier, B., 374 (12.89)
Janata, J., 270 (10.58)
Janout, V., 387 (12.161)
Jansen, M., 287 (10.160)
Jao, T. C., 274 (10.84)
Jazwinski, J., 279 (10.115), 305 (10.316), 306 (10.316), 333 (11.92)
Jernigan, J. C., 150 (5.225, 5.226), 380 (12.119, 12.120), 386 (12.120)
Joachim, C., 116 (5.124), 211 (7.102, 7.104, 7.105), 355 (12.5), 367 (12.5, 12.70, 12.71), 376 (12.97), 377 (12.97, 12.98), 380 (12.5, 12.70, 12.71)
Johansen, O., 190 (6.128)
Johansson, L. B. A., 217 (7.148)
Johnson, C. E., 232 (8.42)
Johnson, D. G., 111 (5.110, 5.112, 5.114), 112 (5.114), 113 (5.114), 114 (5.120), 115 (5.120), 147 (5.217), 251 (9.32), 252 (9.32), 255 (9.74), 256 (9.74), 259 (9.32)
Johnson, M. D., 57 (3.10), 73 (3.10), 98 (5.72, 5.73), 99 (5.72), 101 (5.72), 102 (5.73), 162 (6.9, 6.12), 163 (6.9)
Johnson, M. R., 147 (5.211, 5.213, 5.216, 5.218), 167 (6.33), 383 (12.131)
Johnston, K. P., 202 (7.58)
Jones, A. G., 19 (1.27)
Jones, G. II, 19 (1.59), 226 (8.3) 227 (8.3), 241 (8.3, 8.67, 8.72), 242 (8.3), 243 (8.67, 8.82), 251 (9.16), 252 (9.16),
Jones, G. D., 375 (12.92)
Jones, K. C., 378 (12.103)
Jones, W. E., Jr., 191 (6.129)

Joran, A. D., 113 (5.115, 5.116, 5.117), 114 (5.116, 5.117)
Jordan, K. D., 95 (5.58), 102 (5.58, 5.77), 254 (9.48)
Jorgensen, C. K., 375 (12.91), 380 (12.106)
Jortner, J., 91 (5.13), 94 (5.40), 95 (5.40, 5.55)
Joseph, J., 289 (10.184)
Joy, A., 147 (5.220), 169 (6.45), 365 (12.65)
Juris, A., 124 (5.145), 170 (6.50), 178 (6.86), 212 (7.107, 7.109), 233 (8.46), 241 (8.46), 301 (10.309), 306 (10.309), 319 (11.7), 330 (11.7, 11.66), 347 (11.7), 359 (12.27)

Kaden, T. A., 279 (10.123)
Kadkhodayan, M., 259 (9.134)
Kadoma, Y., 219 (7.179)
Kaganove, S. N., 298 (10.252)
Kahana, N., 204 (7.76)
Kaifer, A. E., 19 (1.52, 1.79), 334 (11.104, 11.106), 336 (11.106), 387 (12.178, 12.179)
Kaiser, W., 91 (5.17)
Kaizu, Y., 254 (9.49)
Kakimoto, M., 361 (12.34), 363 (12.34)
Kalisky, Y., 201 (7.35, 7.36), 218 (7.36),
Kalleymeyn, G. W., 19 (1.12)
Kalyanasundaram, K., 174 (6.68), 176 (6.68), 177 (6.79), 178 (6.68, 6.79), 180 (6.68), 289 (10.180), 290 (10.180), 294 (10.180), 319 (11.4), 330 (11.4), 364 (12.47), 380 (12.47)
Kamata, S., 370 (12.79)
Kamei, T., 383 (12.136)
Kameta, K., 278 (10.99)
Kamitori, S., 289 (10.177, 10.178)
Kanatzidis, M. G., 279 (10.121)
Kanda, Y., 48 (2.26), 122 (5.143)
Kane-Maguire, N. A. P., 188 (6.122), 189 (6.122, 6.124), 324 (11.40)
Kaneda, T., 207 (7.86), 260 (9.143), 268 (10.45), 271 (10.66)
Kaneko, H., 293 (10.221)
Kaneko, M., 150 (5.230)
Kano, K., 216 (7.137), 292 (10.209, 10.210), 294 (10.226)
Karbach, S., 19 (1.12, 1.13)
Karen, A., 118 (5.131), 145 (5.203), 258 (9.120)
Karpiuk, J., 329 (11.55, 11.56)
Kasatani, K., 361 (12.33), 363 (12.33), 370 (12.33)
Kasha, M, 42 (2.9), 342 (11.133, 11.136), 343 (11.136, 11.139, 11.140, 11.141), 344 (11.136), 345 (11.133, 11.136, 11.143), 346 (11.133, 11.140)
Kaszynski, P., 386 (12.145)
Kato, S., 218 (7.165, 7.170)
Katz, N. E., 144 (5.199)
Kauzmann, W., 31 (2.4)
Kavan, L., 381 (12.126)
Kawabata, Y., 216 (7.132)
Kawasaki, M., 361 (12.33), 363 (12.33), 370 (12.33)
Kearns, D. R., 299 (10.299), 304 (10.299)

Kelley, D. F., 129 (5.164), 130 (5.164), 346 (11.152, 11.153)
Kellmann, A., 201 (7.34, 7.37, 7.38, 7.39)
Kelly, C. K., 243 (8.83)
Kelly, J. M., 150 (5.228), 299 (10.283, 10.284, 10.285, 10.287, 10.288, 10.289), 301 (10.284, 10.287), 302 (10.287), 303 (10.287), 304 (10.287), 305 (10.283, 10.284, 10.287, 10.288, 10.289), 365 (12.60), 380 (12.60) Kelly, T. R., 378 (12.104)
Kendrick, T. C., 387 (12.162)
Kennelly, T., 241 (8.69), 242 (8.69)
Kenny, P. W., 19 (1.40), 386 (12.144),
Kern, J. M., 19 (1.51), 337 (11.107), 339 (11.113), 341 (11.113)
Kersey, K. M., 111 (5.112), 114 (5.120), 115 (5.120)
Keskisaari, L., 250 (9.14)
Khemiss, A. K., 337 (11.110)
Khramov, M. I., 363 (12.46)
Kiggen, W., 331 (11.71) 332 (11.71), 333 (11.71, 11.83)
Kihara, N., 267 (10.36)
Kikukawa, K., 208 (7.89)
Kim, M., 388 (12.185)
Kim, S. H., 380 (12.107)
Kim, Y., 132 (5.181)
Kim, Y. E., 19 (1.13)
Kim, Y. H., 19 (1.12)
Kim, Y. I., 387 (12.165)
Kimball, G. E., 26 (2.2)
Kimizuka, N., 370 (12.78)
Kimura, E., 280 (10.134)
Kimura, K., 218 (7.169)
Kinda, H., 215 (7.130)
Kinoshita, J., 216 (7.139), 219 (7.176)
Kinoshita, S., 172 (6.58)
Kira, A., 93 (5.28)
Kirchhoff, J. R., 257 (9.88), 340 (11.117, 11.118)
Kirk, A. D., 324 (11.37, 11.39)
Kirmaier, C., 91 (5.18), 111 (5.111), 257 (9.89)
Kirsch-De Mesmaeker, A., 174 (6.76), 180 (6.76, 6.103), 299 (10.282, 10.284, 10.289), 301 (10.284), 305 (10.284, 10.289)
Kisch, H., 232 (8.40, 8.41), 242 (8.80)
Kissler, B., 19 (1.39)
Kisslinger, J., 178 (6.85), 180 (6.85), 232 (8.25, 8.32), 239 (8.25, 8.32)
Kitamura, N., 255 (9.70), 257 (9.105)
Klapper, M. H., 130 (5.166)
Klpffer, W., 342 (11.132)
Knerelman, E. I., 365 (12.64)
Knibbe, H., 254 (9.55, 9.56)
Knobler, C. B., 19 (1.3, 1.13)
Knobler, C. B., 267 (10.34), 274 (10.90), 278 (10.90), 287 (10.161), 333 (11.89)
Knöchel, T., 214 (7.116)
Knops, P., 287 (10.160)
Kobashi, H., 258 (9.124)
Kobayashi, H., 254 (9.49)
Kobayashi, N., 292 (10.207)
Kober, E. M., 125 (5.149), 170 (6.51)

Koch, M., 285 (10.155)
Kochi, J. K., 232 (8.28), 239 (8.28), 251 (9.31), 252 (9.31), 258 (9.125, 9.126), 259 (9.131, 9.132, 9.133)
Koert, U., 387 (12.176)
Koester, V. J., 127 (5.159), 134 (5.159)
Koga, K., 283 (10.141, 10.142)
Koga, N., 110 (5.107)
Kohnke, F. H., 19 (1.9, 1.42), 275 (10.92, 10.93), 276 (10.92, 10.93), 277 (10.93), 278 (10.103), 285 (10.148, 10.149), 289 (10.93), 386 (12.142)
Kojima, M., 254 (9.53)
Kokubu, T., 295 (10.236)
Kolle, V., 321 (11.27)
Kominato, T., 216 (7.138)
Komiya, H., 90 (5.9)
Komizu, H., 216 (7.132)
Komorowski, S. J., 346 (11.149)
Kondo, T., 383 (12.134)
Kong, J. L. Y., 110 (5.106)
Konijnenberg, J., 346 (11.154), 347 (11.154)
Koo Tze Mew, P., 271 (10.73)
Koper, N. W., 102 (5.82), 254 (9.61), 255 (9.61)
Korpela, T., 290 (10.199)
Kosower, E. M., 342 (11.134), 343 (11.134), 345 (11.134), 346 (11.134)
Kotzyba-Hilbert, F., 19 (1.37), 268 (10.47), 273 (10.47)
Kounot, T., 205 (7.78)
Koutecky, J., 203 (7.61)
Kowase, K., 298 (10.253)
Kozankiewicz, B., 256 (9.84)
Kozik, M., 131 (5.171)
Koziolawa, A., 345 (11.146)
Krakowiak, K. E., 279 (10.114)
Kramer, R., 267 (10.28)
Krause, R. A., 319 (11.6), 330 (11.6)
Krause, S. A., 271 (10.65)
Kress, N., 348 (11.159)
Krijnen, B., 105 (5.99), 252 (9.37), 253 (9.37), 254 (9.37)
Krishnan, V., 254 (9.50), 273 (10.80), 279 (10.106, 10.107)
Kroger, P., 132 (5.182), 133 (5.182), 232 (8.31), 240 (8.31)
Krongauz, V., 217 (7.154, 7.155)
Kroon, J., 102 (5.93, 5.94, 5.95, 5.97, 5.98), 103 (5.93), 104 (5.93, 5.98), 106 (5.95, 5.97, 5.98), 107 (5.95, 5.97), 163 (6.17, 6.19)
Kroto, H. W., 19 (1.21)
Krueger, J. S., 387 (12.165)
Kruger, C., 253 (9.47), 254 (9.47)
Krysanov, S. A., 201 (7.30, 7.31)
Kryukov, A. Y., 260 (9.142)
Kubota, M., 90 (5.3)
Kuczynski, J. P., 241 (8.68), 243 (8.68)
Kuhn, H., 93 (5.31), 361 (12.28), 363 (12.28), 387 (12.166)
Kühnle, W., 255 (9.67, 9.68), 257 (9.109)
Kumagai, T., 205 (7.79),
Kumar, C. V., 151 (5.235, 5.236), 299 (10.271, 10.274, 10.275, 10.279, 10.280), 301 (10.280), 302 (10.280), 303 (10.271, 10.280), 304 (10.271, 10.274, 10.280), 305 (10.279, 10.280), 306 (10.275, 10.279)
Kumar, G. S., 200 (7.28), 215 (7.28), 218 (7.28)
Kundu, T., 202 (7.57)
Kunitake, T., 216 (7.137), 370 (12.78)
Kunkely, H., 19 (1.70), 127 (5.161), 140 (5.191, 5.192, 5.193, 5.194), 214 (7.116), 227 (8.10, 8.7, 8.8, 8.9), 231 (8.7), 232 (8.10, 8.26, 8.29), 239 (8.26, 8.54), 257 (9.87)
Kunst, A. G. M., 102 (5.93), 103 (5.93), 104 (5.93), 163 (6.17)
Kurata, N., 118 (5.131), 145 (5.203)
Kurihara, S., 168 (6.39), 215 (7.128), 217 (7.157, 7.158)
Kuroda, H., 253 (9.44)
Kuroda, N., 171 (6.53, 6.54)
Kuroda, Y., 292 (10.203)
Kusano, Y., 205 (7.78), 206 (7.82), 207 (7.85), 208 (7.89, 7.92), 378 (12.101), 379 (12.101)
Kushida, T., 172 (6.58)
Kusumoto, Y., 289 (10.187)
Kuzmin, M. G., 251 (9.25)

Lacey, D., 387 (12.162)
Lacombe, L., 287 (10.158, 10.162), 333 (11.94)
Lahner, S., 232 (8.41)
Lam, J. D., 278 (10.98)
Lambeck, P. V., 268 (10.44), 270 (10.44)
Lamming, G. R., 19 (1.26)
Lamola, A. A., 72 (3.31), 85 (4.24), 355 (12.11), 380 (12.11)
Lamos, M. L., 299 (10.294)
Lamparski, H., 216 (7.140)
Land, E. J., 19 (1.56), 130 (5.165), 145 (5.204, 5.205), 147 (5.205), 169 (6.45)
Langkilde, F. W., 198 (7.22)
Lapouyade, R., 251 (9.18)
Lappin, A. G., 241 (8.68), 243 (8.68), 279 (10.111)
Larkindale, J. P., 250 (9.11)
Larson, J. M., 270 (10.60, 10.61)
Larson, J. R., 255 (9.74), 256 (9.74)
Larson, R., 232 (8.24)
Larson, S. L., 332 (11.77)
Larsson, S., 95 (5.44, 5.51, 5.52), 116 (5.123)
Lary, J., 365 (12.62)
Laschewsky, A., 387 (12.171)
Lascombe, J., 250 (9.15)
Lattanzi, G., 233 (8.48), 234 (8.48), 235 (8.48), 241 (8.48), 243 (8.89)
Launay, J. P., 116 (5.125), 132 (5.125), 211 (7.102, 7.103, 7.104, 7.105, 7.106), 212 (7.110), 355 (12.5), 367 (12.5, 12.67, 12.70), 376 (12.97), 377 (12.97), 377 (12.98), 380 (12.5, 12.67, 12.70)
Launikonis, A., 132 (5.175), 321 (11.30, 11.31, 11.33), 322 (11.33, 11.33)
Laurence, G. S., 45 (2.12), 233 (8.49), 234 (8.49), 235 (8.49)
Law, K. Y., 202 (7.55)
Law, T. K., 218 (7.173)

Lawrance, G. A., 320 (11.16), 324 (11.42)
Lawrence, D. S., 274 (10.90), 278 (10.90)
Lawrence, G., 279 (10.132)
Lawson, J. M., 102 (5.97), 106 (5.97), 107 (5.97)
Lay, P. A., 320 (11.16, 11.20), 321 (11.29, 11.30, 11.31), 347 (11.156)
Lazarev, P. I., 355 (12.13), 380 (12.13)
Le Bozec, H., 125 (5.152), 126 (5.152), 127 (5.152)
Le Bris, M. T., 271 (10.70, 10.71)
Le Doan, T., 299 (10.288), 305 (10.288)
Le Noble, W. J., 295 (10.237)
Lecomte, J. P., 299 (10.289), 305 (10.289)
Ledwith, A., 232 (8.43), 241 (8.43)
Lee, B., 168 (6.39)
Lee, E. D., 292 (10.204)
Lee, J., 348 (11.166)
Lee, L., 179 (6.93), 180 (6.93)
Lee, S. J., 148 (5.223), 150 (5.223, 5.224), 363 (12.42), 370 (12.85)
Lees, A. J., 178 (6.80, 6.81, 6.82)
Lehman, W. R., 19 (1.56), 145 (5.204)
Lehn, J. M., 19 (1.5, 1.17, 1.37, 1.46, 1.60), 51 (3.1), 83 (4.14, 4.19), 85 (4.19, 4.37), 86 (4.38), 116 (5.121, 5.122), 123 (5.144), 178 (6.86), 205 (7.80), 267 (10.3, 10.20), 268 (10.20, 10.47), 273 (10.47), 274 (10.82, 10.83), 279 (10.20, 10.115, 10.119, 10.122, 10.125, 10.127, 10.128, 10.129), 280 (10.20, 10.127, 10.128), 281 (10.127, 10.128, 10.137, 10.138), 282 (10.137, 10.138), 284 (10.3) 287 (10.82, 10.158), 305 (10.316), 306 (10.316), 327 (11.46, 11.47, 11.50), 328 (11.52), 329 (11.49, 11.50, 11.51, 11.52, 11.53, 11.54, 11.56), 333 (11.91, 11.92), 341 (11.123, 11.124, 11.125), 355 (12.8), 378 (12.103), 387 (12.173, 12.174, 12.176), 388 (12.182)
Lehnmann, J., 19 (1.36)
Lei, Y., 187 (6.119), 190 (6.126)
Leigh, J. S., 92 (5.20), 110 (5.106)
Leighton, P., 122 (5.141)
Leland, B. A., 113 (5.115, 5.116, 5.117), 114 (5.116, 5.117)
Lemaitre-Blaise, M., 375 (12.93)
Lengel, R. K., 340 (11.117)
Lenoble, C., 201 (7.40), 217 (7.151)
Leppkes, R., 19 (1.43)
Lerman, L. S., 299 (10.261)
Lersch, W., 91 (5.13)
Lessard, R. B., 190 (6.126)
Leupin, W., 306 (10.323)
Levanon, H., 168 (6.41, 6.44)
Levelut, A. M., 387 (12.163)
Lever, A. B. P., 180 (6.100, 6.101, 6.102), 240 (8.61), 257 (9.90), 348 (11.159)
Levine, A., 259 (9.132)
Lewis, F. D., 261 (9.147, 9.152)
Lewis, N. A., 132 (5.179, 5.180)
Leydier, L., 212 (7.110)
Li, Q., 365 (12.63)
Li, X., 292 (10.202)
Liang, N., 98 (5.74), 100 (5.74)

Liauw, S., 51 (3.2), 113 (5.119)
Libienbaum, D., 333 (11.92)
Licht, S., 380 (12.118)
Liddell, P. A., 19 (1.56, 1.58), 116 (5.126), 145 (5.126, 5.204, 5.205, 5.206, 5.207), 146 (5.126, 5.206), 147 (5.205, 5.207, 5.212, 5.220), 148 (5.222, 5.223), 150 (5.222, 5.223), 169 (6.46, 6.47), 363 (12.41), 363 (12.42), 364 (12.41), 365 (12.65)
Lieber, C. M., 132 (5.181)
Liebman, J. F., 102 (5.78), 107 (5.78), 113 (5.78), 198 (7.8), 199 (7.8), 202 (7.8), 203 (7.8)
Lilie, J., 239 (8.52), 319 (11.12), 320 (11.12)
Lilienbaum, D., 279 (10.115)
Lim, B. T., 256 (9.77), 257 (9.77)
Lim, E. C., 251 (9.19, 9.26), 256 (9.77), 257 (9.26, 9.77, 9.86)
Liman, U., 216 (7.140)
Lin, C. T., 178 (6.89)
Lin, J. X., 255 (9.73)
Lin, R., 178 (6.87)
Lin, R. J., 267 (10.36)
Lin, T. Y., 147 (5.216, 5.218), 167 (6.33), 383 (12.131)
Lindoy, L. F., 267 (10.14)
Lindqvist, L., 201 (7.34, 7.37, 7.39)
Lindsey, J. S., 121 (5.137, 5.138), 170 (6.49)
Linschitz, H., 121 (5.137)
Lipatov, Y. S., 387 (12.160)
Lippard, S. J., 19 (1.68), 300 (10.304, 10.305, 10.306, 10.307, 10.308)
Lippert, E., 202 (7.45, 7.49), 377 (12.95)
Liska, P., 380 (12.113), 382 (12.113)
Lister-James, J., 19 (1.27)
Liu, J. H., 298 (10.254)
Liu, J. Y., 117 (5.129), 118 (5.129)
Llobet, A., 125 (5.150)
Loach, P. A., 110 (5.106)
Lobenstine, E. W., 299 (10.294)
Locke, R. J., 257 (9.86)
Lockhart, D. J., 20 (1.97), 90 (5.12) 356 (12.23)
Loder, J. W., 243 (8.85)
Loeb, B. L., 137 (5.188), 180 (6.97)
Loeschen, R., 164 (6.26)
Lofters, S., 174 (6.75)
Lhör, H. G., 19 (1.53), 80 (4.8), 268 (10.41, 10.50), 269 (10.41, 10.51), 270 (10.41)
Long, E. C., 299 (10.281), 305 (10.281)
Loock, B., 147 (5.219)
Lopatin, M. A., 252 (9.34, 9.35), 253 (9.34)
Lopes, S., 219 (7.177)
Lora, S., 257 (9.99)
Losensky, H. W., 213 (7.111), 284 (10.146), 378 (12.102)
Lozach, B., 287 (10.162), 333 (11.94)
Lüddecke, E., 86 (4.39), 200 (7.26), 297 (10.243)
Luder, W., 202 (7.49)
Ludi, A., 174 (6.63), 180 (6.63)
Ludmer, Z., 251 (9.24)
Ludwig, R., 19 (1.41)
Luneva, N. P., 365 (12.64)
Lupo, D., 387 (12.170)

Lustig, H., 163 (6.23), 164 (6.23)
Lutter, H. D., 267 (10.26)
Luttrull, D. K., 150 (5.224), 370 (12.85)
Lydon, J. D., 321 (11.31)
Lymar, S. V., 363 (12.46)
Lynch, V. M., 19 (1.38)
Lynch, V. M., 386 (12.146)

Ma, X. C., 148 (5.221, 5.223), 149 (5.221), 150
  (5.223, 5.224), 169 (6.48), 363 (12.42), 370
  (12.85)
Maass, G., 271 (10.73)
MacInnis, J. M., 73 (3.36), 102 (5.75)
Mack, K. B., 174 (6.65, 6.78), 180 (6.78)
MacQueen, D. B., 180 (6.96)
Madge, D., 299 (10.299), 304 (10.299)
Maestri, M., 232 (8.36), 234 (8.36), 235 (8.36),
  240 (8.56)
Magnuson, R. H., 130 (5.169), 131 (5.169)
Maguire, M., 93 (5.29)
Mahiou, R., 374 (12.89)
Maier, S., 165 (6.29), 166 (6.29)
Maitland, G. C., 248 (9.9)
Majenz, W., 203 (7.64)
Maki, A. H., 164 (6.25), 270 (10.62)
Makings, L. R., 19 (1.58), 116 (5.126), 145
  (5.126, 5.207), 146 (5.126), 147 (5.207, 5.212),
  148 (5.221, 5.222), 149 (5.221), 150 (5.222),
  169 (6.47, 6.48), 363 (12.41), 364 (12.41), 370
  (12.85)
Mal'tsev, E. I., 260 (9.142)
Malba, V., 241 (8.67), 243 (8.67)
Malin, J. M., 140 (5.190)
Malini, R., 279 (10.106, 10.107)
Mallouk, T. E., 380 (12.112), 387 (12.165)
Malta, O. L., 375 (12.93)
Malthete, J., 19 (1.46), 116 (5.121)
Mamo, A., 287 (10.159), 329 (11.55)
Manabe, O., 204 (7.70, 7.72), 205 (7.70, 7.77,
  7.78), 206 (7.82), 207 (7.83, 7.85), 208 (7.87,
  7.70, 7.88, 7.89, 7.91, 7.92), 209 (7.93), 215
  (7.129, 7.130), 216 (7.131), 267 (10.31), 268
  (10.43), 377 (12.99, 12.100), 378 (12.100,
  12.101), 379 (12.101), 387 (12.164)
Manda, E., 216 (7.132)
Mandler, D., 299 (10.257)
Manfrin, M. F., 45 (2.12), 86 (4.38), 233 (8.49),
  234 (8.49), 235 (8.49), 281 (10.137, 10.138),
  282 (10.137, 10.138)
Mangani, S., 279 (10.130, 10.131)
Mangiafico, T., 267 (10.22)
Mangini, A., 202 (7.49)
Mann, K. R., 333 (11.98)
Manring, L. E., 242 (8.75)
Marcandalli, B., 216 (7.144)
Marconi, A., 125 (5.151), 348 (11.164)
Marcus, R. A., 46 (2.15, 2.18), 47 (2.15, 2.18), 66
  (3.17, 3.20), 68 (3.17, 3.20), 69 (3.17), 90
  (5.4), 91 (5.14), 92 (5.20), 93 (5.22), 230
  (8.14)
Mariani, M., 279 (10.126)
Markowitz, M. A., 387 (12.161)

Marnot, P. A., 340 (11.118)
Marsau, P., 260 (9.145), 268 (10.47), 272 (10.76),
  273 (10.47, 10.79), 275 (10.76), 276 (10.76)
Marshall, J. L., 125 (5.149), 131 (5.172)
Marshall, L. E., 304 (10.312)
Martell, A. E., 267 (10.15, 10.24), 279 (10.119)
Martelli, A., 298 (10.247)
Marti, K., 19 (1.12)
Martic, P. A., 292 (10.208)
Martin, J. L., 91 (5.15)
Martin, L. L., 320 (11.24), 324 (11.42)
Martin, R. H., 19 (1.23)
Martin, R. L., 320 (11.24)
Martinek, K., 197 (7.2)
Maruyama, K., 166 (6.30)
Masaki, S., 90 (5.3)
Masetti, F., 257 (9.97)
Masnovi, J. M., 258 (9.125, 9.126), 259 (9.131,
  9.132, 9.133)
Masschelein, A., 174 (6.76), 180 (6.76, 6.103),
  299 (10.284), 301 (10.284), 305 (10.284)
Masuhara, H., 257 (9.98, 9.101, 9.102), 258
  (9.114, 9.115, 9.116, 9.117, 9.118)
Mataga, N., 48 (2.22, 2.26), 90 (5.3), 116 (5.127),
  118 (5.131), 122 (5.142, 5.143), 145 (5.203),
  202 (7.51, 7.53), 203 (7.63), 248 (9.6), 254
  (9.54), 255 (9.62, 9.63, 9.64, 9.65, 9.69), 257
  (9.98, 9.101, 9.102), 258 (9.114, 9.115, 9.116,
  9.117, 9.118, 9.119, 9.120, 9.121, 9.122, 9.123,
  9.124), 260 (9.139), 346 (11.148), 370 (12.76,
  12.77)
Mathias, J. P., 19 (1.9), 386 (12.142)
Mathis, G., 329 (11.49)
Mathis, P., 19 (1.56), 145 (5.204, 5.205), 147
  (5.205, 5.220), 167 (6.32), 365 (12.65)
Matsubara, T., 132 (5.182), 133 (5.182), 232
  (8.31), 240 (8.31)
Matsuda, T., 208 (7.89), 267 (10.31), 387
  (12.164)
Matsui, Y., 288 (10.166, 10.169)
Matsumoto, H., 292 (10.209, 10.210)
Matsumoto, M., 216 (7.132)
Matsuo, K., 215 (7.129)
Matsuo, T., 299 (10.259)
Matsuzawa, H., 254 (9.49)
Mattay, J., 251 (9.28)
Mattes, S. L., 48 (2.24), 259 (9.135), 261 (9.148,
  9.150)
Matthews, P., 174 (6.73), 177 (6.73), 181 (6.73)
Matzanke, B. F., 85 (4.31)
Mau, A. W. H., 132 (5.174, 5.175), 190 (6.128),
  243 (8.85), 321 (11.29, 11.30, 11.31, 11.32,
  11.33), 322 (11.32, 11.33), 323 (11.33), 324
  (11.41, 11.42), 325 (11.41), 326 (11.41)
Maud, J. M., 275 (10.95), 278 (10.102)
Mauerall, D., 233 (8.45)
Mauzerall, D. C., 121 (5.137, 5.138)
Maverick, E. F., 19 (1.13), 333 (11.89)
Maxwell, B., 164 (6.26)
Mayer, J. E., 387 (12.165)
Mayo, S. L., 93 (5.25)

Mayoh, B., 71 (3.29)
McAleer, J. F., 204 (7.73), 267 (10.33)
McAuley, A., 279 (10.111)
McCarthy, M. G., 321 (11.31)
McClanahan, S. F., 150 (5.227, 5.231), 363 (12.44), 365 (12.59), 370 (12.59), 380 (12.59), 386 (12.59, 12.151)
McConnell, D. J., 299 (10.283, 10.284, 10.285, 10.288), 301 (10.284), 305 (10.283, 10.284, 10.288)
McConnell, H. M., 71 (3.26), 95 (5.61), 97 (5.61)
McCullough, J. J., 261 (9.151),
McEvoy, A. J., 380 (12.114)
McHale, J. L., 250 (9.10, 9.12)
McInnis, J. M., 162 (6.8)
McIntosh, A. R., 117 (5.128), 118 (5.128, 5.130), 119 (5.134)
McKervey, M. A., 267 (10.29), 333 (11.90)
McLendon, G., 93 (5.23, 5.29), 131 (5.173)
McMahon, R. J., 121 (5.140)
McMillin, D. R., 241 (8.62, 8.63), 257 (9.88, 9.89, 9.94), 299 (10.298), 340 (11.116, 11.117, 11.118, 11.119), 341 (11.121)
McMorrow, D., 343 (11.139), 345 (11.143)
McMurry, J. E., 95 (5.58), 102 (5.58), 254 (9.48)
McMurry, T. J., 83 (4.17), 333 (11.84, 11.85)
Meade, T. J., 93 (5.26), 267 (10.9)
Meares, C. F., 299 (10.290, 10.293), 304 (10.290), 306 (10.293)
Mecati, A., 329 (11.63)
Megarity, E. D., 85 (4.24), 198 (7.13), 217 (7.13)
Megehee, E. G., 232 (8.42)
Mehta, G., 19 (1.33)
Mei, H. Y., 299 (10.272), 305 (10.272)
Meisel, D., 48 (2.22), 91 (5.16), 102 (5.76), 114 (5.120), 115 (5.120), 116 (5.127), 142 (5.197), 145 (5.197), 151 (5.233), 161 (6.3), 162 (6.10), 251 (9.32), 252 (9.32), 258 (9.121), 259 (9.32), 365 (12.58), 370 (12.58), 380 (12.58), 381 (12.128)
Memming, R., 381 (12.124)
Menif, R., 267 (10.24)
Menju, A., 201 (7.32), 218 (7.32)
Merkert, J. W., 147 (5.215), 148 (5.215), 173 (6.59)
Merriam, M. J., 250 (9.10)
Merrifield, R. B., 386 (12.147)
Mertes, K. B., 378 (12.103)
Mertes, M. P., 378 (12.103)
Mertesdorf, C., 216 (7.133)
Mes, G. F., 102 (5.80, 5.81, 5.82), 105 (5.81), 254 (9.61), 255 (9.61)
Meshoyer, R., 299 (10.280), 301 (10.280), 302 (10.280), 303 (10.280), 304 (10.280), 305 (10.280)
Metin, J., 374 (12.89)
Metts, L., 198 (7.13), 217 (7.13)
Metzger, R. M., 355 (12.14), 367 (12.14), 380 (12.14)
Meyer, M., 202 (7.60)
Meyer, T. J., 51 (3.3), 94 (5.36), 124 (5.148), 125 (5.149, 5.150, 5.152, 5.153, 5.154, 5.155), 126

(5.152, 5.153, 5.154), 127 (5.152, 5.153, 5.154), 128 (5.162, 5.163), 129 (5.162, 5.163), 133 (5.184), 134 (5.184), 135 (5.184, 5.185), 136 (5.185), 137 (5.187, 5.188), 144 (5.210), 145 (5.210), 147 (5.215, 5.36), 148 (5.215), 150 (5.227, 5.231, 5.232), 151 (5.232, 5.233), 170 (6.51), 180 (6.97), 186 (6.111), 191 (6.129, 6.130), 227 (8.5), 231 (8.5), 232 (8.5, 8.30, 8.39), 239 (8.53), 242 (8.78), 243 (8.88), 319 (11.5), 330 (11.5, 11.68), 331 (11.68), 348 (11.158), 363 (12.44), 365 (12.58, 12.59), 370 (12.58, 12.59, 12.86), 372 (12.86), 380 (12.58, 12.59), 386 (12.59, 12.86, 12.151)
Mialocq, J. C., 145 (5.205), 147 (5.205), 202 (7.60)
Michel-Beyerle, M. E., 90 (5.5), 91 (5.13), 95 (5.53, 5.55), 108 (5.102, 5.103, 5.104, 5.105), 109 (5.102, 5.103, 5.104, 5.105), 110 (5.103)
Michel, H., 20 (1.95), 90 (5.7, 5.10), 91 (5.7), 356 (12.21)
Micheloni, M., 279 (10.116, 10.118, 10.120, 10.130, 10.131)
Michl, J., 19 (1.38), 203 (7.61), 386 (12.145, 12.146)
Middendorf, T. R., 20 (1.97), 90 (5.12), 356 (12.23)
Miehé, J. A., 202 (7.45), 212 (7.110), 251 (9.18), 377 (12.95)
Miki, K., 90 (5.7), 91 (5.7)
Milgron, L. R., 168 (6.43)
Millard, R. R., 198 (7.21)
Miller, D. B., 189 (6.124), 324 (11.40)
Miller, J. R., 48 (2.23, 2.27), 57 (3.10), 70 (3.24), 71 (3.28), 73 (3.10) 93 (5.27), 95 (5.27, 5.63), 97 (5.63), 98 (5.64, 5.68, 5.69, 5.70, 5.71, 5.72, 5.73, 5.74), 99 (5.64, 5.70, 5.72), 100 (5.64, 5.69, 5.70, 5.74), 101 (5.64, 5.70, 5.72), 102 (5.73, 5.76, 5.91), 105 (5.91), 162 (6.9, 6.10, 6.11, 6.12), 163 (6.9, 6.15)
Miller, L. L., 19 (1.40), 386 (12.144)
Miller, P. K., 189 (6.124)
Milosavljevic, B. H., 93 (5.30), 241 (8.68), 243 (8.68)
Milton, S. V., 164 (6.25)
Minami, T., 207 (7.83, 7.85), 208 (7.91, 7.92)
Minsek, D. W., 111 (5.112, 5.114), 112 (5.114), 113 (5.114), 114 (5.120), 115 (5.120), 251 (9.32), 252 (9.32), 255 (9.74), 256 (9.74), 259 (9.32)
Minto, F., 257 (9.99)
Miravitlles, C., 267 (10.37)
Misumi, S., 90 (5.3), 118 (5.131), 119 (5.135), 145 (5.203), 207 (7.86), 255 (9.62, 9.63), 258 (9.120), 260 (9.143), 268 (10.45), 271 (10.66)
Miyasaka, H., 258 (9.122), 346 (11.148)
Miyazaki, K., 209 (7.93)
Miyazaki, O., 204 (7.72)
Mizuma, A., 370 (12.77)
Mobius, K., 95 (5.55), 167 (6.31)
Möbius, D., 216 (7.134), 361 (12.28, 12.30), 363 (12.28, 12.30)
Mock, W. L., 19 (1.18)

Moggi, L., 19 (1.73), 45 (2.12), 51 (3.1), 77 (4.3), 84 (4.23), 86 (4.38), 232 (8.36), 233 (8.49), 234 (8.36, 8.49), 235 (8.36, 8.49), 240 (8.55), 281 (10.137, 10.138), 282 (10.137, 10.138), 321 (11.26), 323 (11.26, 11.35), 355 (12.7), 356 (12.7), 357 (12.7), 359 (12.7), 360 (12.7), 361 (12.7), 366 (12.7), 367 (12.7), 368 (12.7), 374 (12.7), 379 (12.7)

Mohammad, T., 299 (10.300)

Mohri, Y., 216 (7.138)

Molotkowsky, J. G., 217 (7.148)

Momenteau, M., 147 (5.219)

Montagnoli, G., 197 (7.3), 204 (7.69), 215 (7.120, 7.122)

Montanari, F., 279 (10.112)

Monti, S., 86 (4.40), 200 (7.23, 7.24, 7.25), 201 (7.34, 7.37, 7.38, 7.43), 297 (10.242), 298 (10.247), 341 (11.125)

Moody, R., 259 (9.138)

Moore, A. L., 19 (1.56, 1.58), 116 (5.126), 145 (5.126, 5.204, 5.205, 5.206, 5.207), 146 (5.126, 5.206), 147 (5.205, 5.207, 5.220), 148 (5.221, 5.222, 5.223), 149 (5.221), 150 (5.222, 5.223, 5.224), 169 (6.45, 6.48), 363 (12.41, 12.42), 364 (12.41), 365 (12.65, 12.66), 370 (12.85)

Moore, K. J., 179 (6.93), 180 (6.93)

Moore, T. A., 19 (1.56, 1.58), 94 (5.35, 5.38), 116 (5.126), 145 (5.38, 5.126, 5.204, 5.205, 5.206, 5.207), 146 (5.126, 5.206), 147 (5.205, 5.207, 5.212, 5.220), 148 (5.38, 5.221, 5.222, 5.223), 149 (5.38, 5.221), 150 (5.38, 5.222, 5.223, 5.224), 169 (6.45, 6.46, 6.47, 6.48), 361 (12.37), 363 (12.41, 12.42), 364 (12.41), 365 (12.65, 12.66), 370 (12.85)

Moras, D., 341 (11.124), 387 (12.173)

Moret, E., 272 (10.75)

Morgan, L., 174 (6.73), 177 (6.73), 180 (6.99), 181 (6.73)

Morgan, R., 174 (6.75)

Morikawa, A., 361 (12.34), 363 (12.34)

Morin, D., 375 (12.90)

Morita, T., 270 (10.63)

Moriwaki, F., 210 (7.101), 293 (10.217, 10.219, 10.220, 10.221), 294 (10.217), 295 (10.233), 298 (10.217)

Morokuma, K., 95 (5.49), 248 (9.5)

Morrison, H., 164 (6.26, 6.27, 6.28), 165 (6.28), 299 (10.300)

Morrison, L. E., 151 (5.238), 306 (10.319, 10.320)

Moser, C. C., 91 (5.16)

Moser, J., 380 (12.115), 381 (12.126)

Moser, J. E., 48 (2.25), 70 (3.25), 259 (9.136, 9.137)

Moss, R. E., 270 (10.59)

Motekaitis, R. J., 279 (10.119)

Moustakali-Mavridis, I., 346 (11.151)

Moya, L., 244 (8.90)

Mugnier, J., 167 (6.34, 6.35, 6.36, 6.37), 168 (6.34, 6.35, 6.36), 211 (7.106)

Mulazzani, Q. G., 232 (8.27), 240 (8.27, 8.56, 8.57), 241 (8.70), 242 (8.70, 8.73), 323 (11.34, 11.35, 11.36), 324 (11.34)

Müllen, K., 19 (1.84)

Müller, G., 85 (4.31)

Muller, J. G., 332 (11.82),

Müller, W. M., 205 (7.81), 267 (10.27), 284 (10.146), 287 (10.27), 289 (10.175, 10.176), 333 (11.96)

Mulliken, R. S., 248 (9.1), 251 (9.1), 252 (9.1)

Mulvey, R. E., 19 (1.26)

Murai, K., 210 (7.101), 293 (10.217, 10.219, 10.220, 10.221, 10.223), 294 (10.217), 298 (10.217)

Murase, I., 279 (10.119)

Murphy, W. R., Jr., 174 (6.77), 177 (6.77), 181 (6.106), 182 (6.106), 370 (12.83)

Murray, K. S., 320 (11.24)

Murray, R. W., 125 (5.152), 126 (5.152), 127 (5.152), 150 (5.225, 5.226), 380 (12.119, 12.120), 386 (12.120)

Murthy, G. S., 19 (1.38), 386 (12.146)

Muszkat, K. A., 198 (7.14, 7.15)

Mutai, K., 260 (9.140)

Muthuramu, K., 164 (6.26)

Nadtochenko, V. A., 370 (12.75)

Nag, A., 202 (7.57, 7.59), 298 (10.249, 10.250)

Nagakura, S., 256 (9.83), 260 (9.140)

Nagamura, T., 242 (8.76, 8.77), 243 (8.87), 370 (12.79)

Nagaoka, Y., 271 (10.66)

Nagle, J. K., 125 (5.155), 257 (9.92), 330 (11.68), 331 (11.68)

Nakagaki, R., 260 (9.140)

Nakaji, T., 205 (7.77), 208 (7.87, 7.88), 377 (12.100), 378 (12.100)

Nakamichi, M., 289 (10.179)

Nakamura, H., 150 (5.230), 299 (10.259)

Nakamura, T., 216 (7.132)

Nakashima, K., 271 (10.66)

Nakashima, S., 119 (5.135)

Nakatani, K., 255 (9.69), 260 (9.139)

Nakatsuji, S., 271 (10.66)

Nam, K. C., 267 (10.22)

Narayanaswamy, R., 270 (10.56, 10.57, 10.59)

Narushima, R., 280 (10.133)

Nasielski-Hinkens, R., 174 (6.76), 180 (6.76)

Nasielski, J., 299 (10.284), 301 (10.284), 305 (10.284)

Nath, D. N., 257 (9.108)

Naylor, A. M., 19 (1.85), 51 (3.1), 386 (12.153)

Nazeeruddin, M. K., 174 (6.68), 176 (6.68), 177 (6.79), 178 (6.68, 6.79), 180 (6.68)

Neckers, D. C., 200 (7.28), 215 (7.28), 218 (7.28), 295 (10.232)

Negri, F., 198 (7.22)

Nelson, G., 290 (10.190, 10.192)

Nemeth, G.A., 19 (1.56), 116 (5.126), 145 (5.126, 5.204, 5.205, 5.206, 5.207), 146 (5.126, 5.206), 147 (5.205, 5.207)

Netto-Ferreira, J. C., 293 (10.216)

Netzel, T. L., 118 (5.132), 127 (5.157, 5.158), 128

(5.157, 5.158), 132 (5.182), 133 (5.182), 134 (5.158), 143 (5.198), 151 (5.237, 5.238), 232 (8.31, 8.33), 233 (8.33), 234 (8.33), 237 (8.33), 240 (8.31), 306 (10.318, 10.319, 10.320), 386 (12.152)

Neuberger, K. R., 198 (7.13), 217 (7.13)

Neveux, P. E., 128 (5.162), 129 (5.162)

Neyhart, G. A., 137 (5.187, 5.188), 180 (6.97)

Ng, L., 102 (5.77)

Ngwenya, M. P., 279 (10.113)

Nieger, M., 205 (7.81), 267 (10.27), 287 (10.27), 333 (11.96)

Niemczyk, M. P., 19 (1.57), 48 (2.28), 110 (5.108, 5.109), 111 (5.113, 5.114), 112 (5.114), 113 (5.114), 147 (5.208), 361 (12.39), 362 (12.39, 12.40)

Nieuwenhuis, H. A., 178 (6.88)

Niino, H., 216 (7.132)

Nishi, T., 267 (10.31), 387 (12.164)

Nishida, Y., 205 (7.77)

Nishikata, Y., 361 (12.34), 363 (12.34)

Nishioka, T., 288 (10.166)

Nishitani, S., 118 (5.131), 145 (5.203), 258 (9.120)

Nishiyama, K., 361 (12.32, 12.34), 363 (12.32, 12.34), 383 (12.132, 12.134)

Nishiyama, Y., 361 (12.33), 363 (12.33), 370 (12.33)

Nishizawa, M., 190 (6.127)

Nordén, B., 299 (10.301), 304 (10.301)

Norris, J., 90 (5.8)

Norris, J.R., Jr., 48 (2.22), 91 (5.16), 102 (5.76), 114 (5.120), 115 (5.120), 116 (5.127), 142 (5.197), 145 (5.197), 151 (5.233), 161 (6.3), 162 (6.10), 251 (9.32), 252 (9.32), 258 (9.121), 259 (9.32), 365 (12.58), 370 (12.58), 380 (12.58), 381 (12.128)

Norton, K. A., 139 (5.189)

Norvez, S., 387 (12.172)

Nowacki, J., 202 (7.56)

Nowak, A. K., 122 (5.143)

Nozakura, S., 171 (6.54)

Nuñez, A., 217 (7.150)

Nüsslein, F., 242 (8.80)

O'Brien, D. F., 216 (7.140)

O'Halloran, T. V., 140 (5.190)

Obeng, Y. S., 132 (5.180)

Odashima, K., 283 (10.141, 10.142)

Odell, B., 19 (1.4), 285 (10.154), 286 (10.154), 287 (10.156)

Odgen, M. I., 330 (11.64)

Oesterhelt, D., 91 (5.17)

Oevering, H., 19 (1.6), 98 (5.65), 102 (5.65, 5.87, 5.88, 5.89, 5.90, 5.92, 5.93, 5.96), 103 (5.65, 5.93), 104 (5.65, 5.88, 5.90, 5.92, 5.93), 105 (5.65, 5.96), 162 (6.13, 6.14), 163 (6.13, 6.16, 6.17, 6.18), 252 (9.36), 253 (9.36), 254 (9.36)

Ofer, D., 388 (12.183)

Ogana, T., 216 (7.137)

Ogawa, S., 280 (10.133)

Ogawa, T., 205 (7.77), 208 (7.87, 7.88, 7.89,

7.91), 294 (10.226), 370 (12.79), 377 (12.100), 378 (12.100)

Ogawa, Y., 388 (12.184)

Ogilby, P. R., 254 (9.52)

Ogino, H., 19 (1.78), 334 (11.103)

Ohata, K., 19 (1.78), 334 (11.103)

Ohno, O., 254 (9.49)

Ohta, K., 95 (5.49)

OhUigin, C., 299 (10.283, 10.284, 10.285, 10.288), 301 (10.284), 305 (10.283, 10.284, 10.288)

Ojima, S., 258 (9.122)

Okada, T., 90 (5.3), 118 (5.131), 145 (5.203), 202 (7.51, 7.53), 203 (7.63), 254 (9.54), 255 (9.65, 9.69), 258 (9.120, 9.124), 260 (9.139), 346 (11.148)

Okado, T., 48 (2.26)

Okahata, Y., 216 (7.137, 7.141)

Okajima, S., 256 (9.77), 257 (9.77)

Okamura, M. Y., 356 (12.25), 363 (12.25)

Okubo, T., 290 (10.200), 292 (10.213)

Oliver, A. M., 19 (1.6), 98 (5.65), 102 (5.65, 5.90, 5.91, 5.92, 5.94, 5.95, 5.98), 103 (5.65), 104 (5.65, 5.90, 5.92, 5.98), 105 (5.65, 5.91, 5.96), 106 (5.95, 5.98), 107 (5.95, 5.101), 113 (5.101), 163 (6.15, 6.16, 6.19)

Olmsted, III J., 150 (5.227)

Onuchic, J. N., 95 (5.47, 5.48), 98 (5.66), 384 (12.137, 12.138), 385 (12.138)

Orellana, G., 299 (10.282)

Orgel, L. E., 95 (5.60)

Orgodnik, A., 91 (5.13)

Orioli, P., 279 (10.130, 10.131)

Orlandi, G., 198 (7.16, 7.22), 200 (7.24), 201 (7.38, 7.43)

Orlowski, T. E., 201 (7.35)

Ortholand, J. Y., 19 (1.83), 287 (10.157)

Osa, T., 210 (7.96, 7.97, 7.98, 7.99, 7.100, 7.101), 216 (7.143, 7.145), 219 (7.179), 290 (10.191), 292 (10.207, 10.212), 293 (10.212, 10.217, 10.218, 10.219, 10.220, 10.221, 10.222, 10.223, 10.224, 10.225), 294 (10.217, 10.218, 10.224), 295 (10.233), 298 (10.217, 10.244, 10.245, 10.246)

Osman, A. H., 132 (5.176), 140 (5.191, 5.193), 232 (8.26), 239 (8.26)

Ostrowicki, A., 287 (10.160)

Osuka, A., 166 (6.30)

Otsubo, T., 207 (7.86), 260 (9.143)

Ottolenghi, M., 248 (9.6), 251 (9.22)

Overbeek, J. M., 243 (8.85)

Owens, M., 267 (10.29), 333 (11.90)

Pac, C., 364 (12.50), 380 (12.50)

Paczkowski, J., 295 (10.232)

Paddon-Row, M. N., 19 (1.6), 95 (5.43, 5.58), 98 (5.65), 102 (5.43, 5.58, 5.65, 5.77, 5.78, 5.87, 5.88, 5.89, 5.90, 5.91, 5.92, 5.93, 5.94, 5.95, 5.96, 5.97, 5.98), 103 (5.65, 5.93), 104 (5.65, 5.88, 5.90, 5.92, 5.93, 5.98), 105 (5.65, 5.91, 5.96), 106 (5.95, 5.97, 5.98), 107 (5.78, 5.95, 5.97, 5.101), 113 (5.78, 5.101), 162 (6.13,

6.14), 163 (6.13, 6.15, 6.16, 6.17, 6.18, 6.19), 252 (9.36), 253 (9.36), 254 (9.36, 9.48)
Paek, K., 333 (11.89)
Pague, H. I., 218 (7.172)
Palazzotto, M. C., 363 (12.43)
Pallavicini, P. S., 19 (1.34)
Pallmer, M., 164 (6.26)
Palmans, J. P., 257 (9.96)
Palmer, C. E. A., 257 (9.89)
Palmieri, P., 200 (7.24)
Pandey, B., 164 (6.26)
Panetta, C. A., 355 (12.14), 367 (12.14), 380 (12.14)
Paoletti, P., 279 (10.116, 10.120, 10.130, 10.131)
Pappalardo, S., 287 (10.159), 329 (11.55)
Paradisi, C., 60 (3.14), 140 (5.196), 142 (5.196), 144 (5.196), 365 (12.55)
Park, C. H., 19 (1.25)
Parmon, V. N., 363 (12.46)
Parson, W. W., 91 (5.18), 92 (5.19)
Parsons, B. J., 299 (10.288), 305 (10.288)
Parthenopoulos, D. A., 380 (12.109)
Pascard, C., 279 (10.115), 287 (10.162), 333 (11.92, 11.94), 339 (11.111, 11.112)
Pasimeni, L., 256 (9.85)
Pasman, P., 102 (5.79, 5.82), 253 (9.45, 9.46), 254 (9.61), 255 (9.61)
Passiki, P. J., 365 (12.65)
Patney, H. K., 102 (5.77)
Patonay, G., 290 (10.192), 292 (10.206)
Patterson, H. H., 121 (5.140)
Pawlowski, G., 380 (12.108)
Pecile, D., 375 (12.90)
Pedersen C.J., 19 (1.16), 83 (4.11, 4.13), 85 (4.13), 204 (7.65, 7.66), 267 (10.1, 10.18)
Pelizzetti, E., 174 (6.71), 177 (6.71), 178 (6.71), 181 (6.71), 182 (6.71), 184 (6.71) 231 (8.23), 233 (8.46), 241 (8.46), 330 (11.66), 359 (12.27), 364 (12.47, 12.49), 380 (12.47, 12.49, 12.121)
Penfield, K. W., 98 (5.70), 99 (5.70), 100 (5.70), 101 (5.70), 102 (5.91), 105 (5.91), 163 (6.15)
Pepe, I. M., 215 (7.120)
Perathoner, S., 233 (8.48), 234 (8.48), 235 (8.48), 241 (8.48), 243 (8.89), 327 (11.50), 328 (11.52), 329 (11.50, 11.51, 11.52, 11.53, 11.54)
Perkins, T. A., 127 (5.157, 5.158), 128 (5.157, 5.158), 134 (5.158)
Perly, B., 202 (7.60)
Perotti, A., 19 (1.34), 279 (10.124)
Person, W. B., 248 (9.1), 251 (9.1), 252 (9.1)
Pessiki, P. J., 116 (5.126), 145 (5.126), 146 (5.126), 147 (5.212, 5.220), 169 (6.47)
Peter, F., 279 (10.129)
Peter, R., 242 (8.80)
Peters, K. S., 242 (8.75), 258 (9.128, 9.129, 9.130), 259 (9.130)
Petersen, J. D., 174 (6.66, 6.67, 6.77), 177 (6.66, 6.77), 178 (6.67), 179 (6.67, 6.92, 6.93, 6.94), 180 (6.93, 6.95, 6.96), 181 (6.106), 182 (6.106), 370 (12.83)

Peterson, S. H., 79 (4.4, 4.5), 348 (11.162, 11.163)
Ptillon, F. Y., 19 (1.7)
Petrich, J. W., 255 (9.74), 256 (9.74)
Petrin, M., 270 (10.62),
Pewitt, E. B., 19 (1.57), 110 (5.109), 147 (5.208), 361 (12.39), 362 (12.39)
Philo, J. S., 365 (12.62)
Pichat, P., 364 (12.48), 380 (12.48)
Piela, L., 355 (12.10), 380 (12.10)
Pieroni, O., 215 (7.121), 218 (7.121), 219 (7.121, 7.174, 7.175, 7.177, 7.178, 7.180)
Pietraszkiewicz, M., 19 (1.79), 329 (11.55, 11.56), 334 (11.106), 336 (11.106), 387 (12.179)
Pilette, Y. P., 258 (9.113)
Pimentel, G. C., 19 (1.47)
Pina, F., 84 (4.23), 232 (8.27), 240 (8.27, 8.55, 8.56, 8.57), 281 (10.138), 282 (10.138), 321 (11.26), 323 (11.26, 11.34, 11.35, 11.36), 324 (11.34)
Piotrowiak, P., 57 (3.10), 73 (3.10, 3.36), 98 (5.73), 102 (5.73, 5.75, 5.76), 162 (6.8, 6.9, 6.10), 163 (6.9)
Piriou, B., 375 (12.93)
Pitts, J. N., 25 (2.1), 41 (2.1)
Piuzzi, F., 256 (9.78, 9.79)
Plancherel, D., 327 (11.47)
Plato, M., 95 (5.55)
Plute, K. E., 378 (12.103)
Pochini, A., 329 (11.58, 11.61, 11.62, 11.63)
Poggi, A., 19 (1.34), 279 (10.124)
Poli, P., 279 (10.120)
Politzer, I. R., 289 (10.184)
Poll, T., 285 (10.155)
Pollinger, F., 108 (5.105), 109 (5.105)
Polymeropoulos, E. E., 216 (7.134), 361 (12.28), 363 (12.28)
Pomerantz, M., 367 (12.71), 380 (12.71)
Poonia, N. S., 268 (10.40)
Popma, T. J. A., 268 (10.44), 270 (10.44)
Port, H., 165 (6.29), 166 (6.29)
Porter, G., 121 (5.139), 273 (10.81), 356 (12.26)
Posey, M. R., 118 (5.133), 254 (9.51)
Potember, R. S., 380 (12.107)
Pouget, J., 167 (6.35, 6.37), 168 (6.35), 211 (7.106)
Poulous, A. T., 243 (8.83)
Pourreau, D. B., 127 (5.157), 128 (5.157)
Powers, M. J., 144 (5.210), 145 (5.210)
Prasad, D. R., 241 (8.67, 8.70, 8.71), 242 (8.70, 8.71, 8.74), 243 (8.67, 8.84)
Prince, R. C., 92 (5.20), 365 (12.61)
Prinzbach, H., 19 (1.31)
Prochorow, J., 256 (9.84)
Prodi, L., 19 (1.79), 227 (8.13), 260 (9.146), 274 (10.85), 275 (10.85, 10.93), 276 (10.93, 10.96, 10.97), 277 (10.93, 10.96), 280 (10.85), 289 (10.93), 334 (11.106), 336 (11.106), 387 (12.179)
Prutz, W. A., 130 (5.165)
Puff, H., 333 (11.83)

Purugganan, M. D., 151 (5.236), 299 (10.279), 305 (10.279), 306 (10.279)
Put, J., 161 (6.6)
Puza, M., 252 (9.39)
Pyle, A. M., 299 (10.280, 10.281), 301 (10.280), 302 (10.280), 303 (10.280), 304 (10.280), 305 (10.280, 10.281)

Queimado, M. M., 272 (10.74)
Quici, S., 279 (10.112)

Rabek, J. F., 257 (9.103)
Raber, D. J., 254 (9.60)
Räder, H. J., 19 (1.84)
Ramamurthy, V., 289 (10.182), 292 (10.182), 296 (10.240), 298 (10.182, 10.248), 299 (10.260),
Ramasami, T., 324 (11.43)
Rampi, M. A., 184 (6.110), 188 (6.120), 233 (8.44), 234 (8.44), 235 (8.44, 8.51), 237 (8.44), 381 (12.129)
Rao, B. N., 298 (10.248)
Rao, R. R., 288 (10.173)
Rao, V. P., 295 (10.238)
Raphael, A. L., 299 (10.266, 10.269, 10.278), 300 (10.266), 304 (10.269)
Rapp, K. M., 214 (7.116)
Raselli, A., 320 (11.23)
Ratner, M. A., 355 (12.1), 367 (12.1, 12.68), 380 (12.68)
Rau, H., 86 (4.39), 200 (7.26, 7.27), 232 (8.35), 233 (8.35), 234 (8.35), 297 (10.243)
Raymond, K. N., 83 (4.17), 85 (4.31), 333 (11.84, 11.85, 11.86)
Razuvaev, G. A., 252 (9.34, 9.35), 253 (9.34)
Rebek, J. Jr., 19 (1.14), 85 (4.34, 4.36), 204 (7.74), 267 (10.11), 267 (10.17), 378 (12.105), 388 (12.105, 12.181)
Reck, B., 217 (7.156)
Reddington, M. V., 19 (1.4, 1.52, 1.79), 285 (10.154), 286 (10.154), 287 (10.156), 334 (11.104, 11.105, 11.106), 336 (11.106), 387 (12.178, 12.179)
Reedijk, J., 174 (6.69), 177 (6.69), 178 (6.69, 6.88)
Rees, D. C., 90 (5.9), 356 (12.25), 363 (12.25)
Regen, S. L., 387 (12.161)
Regev, A., 168 (6.41, 6.44)
Rehm, D., 46 (2.14), 48 (2.14), 70 (3.22), 254 (9.56)
Rehmann, J. P., 299 (10.280), 301 (10.280), 302 (10.280), 303 (10.280), 304 (10.280), 305 (10.280)
Rehms, A. A., 150 (5.224), 370 (12.85)
Rehorek, D., 230 (8.17, 8.19), 240 (8.60, 8.61)
Reich, K. A., 304 (10.312)
Reid, C., 256 (9.82)
Reimers, J. R., 95 (5.59)
Reinhoudt, D. N., 267 (10.5, 10.23, 10.25, 10.39), 268 (10.44), 270 (10.44), 333 (11.97)
Reisfeld, R., 375 (12.91)
Reitz, G. A., 290 (10.201), 291 (10.201)
Remers, W. A., 299 (10.265)

Rendell, A. P. L., 95 (5.56)
Reniero, F., 288 (10.168),
Rentzepis, P. M., 111 (5.113), 164 (6.25), 258 (9.125, 9.126, 9.127), 259 (9.131), 380 (12.109)
Rettig, W., 198 (7.8), 199 (7.8), 202 (7.8, 7.44, 7.45, 7.46), 203 (7.8, 7.44, 7.62, 7.63, 7.64), 377 (12.94, 12.95, 12.96)
Reuter, H., 287 (10.160)
Reverberi, S., 329 (11.61)
Reyes, Z. E., 83 (4.17), 333 (11.84)
Reynders, P., 255 (9.68)
Reynolds, P. E., 300 (10.303)
Riccieri, P., 189 (6.125)
Rice-Evans, C., 299 (10.288), 305 (10.288)
Richards, R. D. C., 387 (12.162)
Richardson, D. E., 71 (3.30), 95 (5.45)
Richmond, M. H., 300 (10.303)
Rieger, B., 151 (5.234), 306 (10.317)
Riffaud, M. H., 260 (9.145), 272 (10.76), 275 (10.76), 276 (10.76)
Rigault, A., 19 (1.60), 341 (11.123, 11.124), 387 (12.173, 12.174)
Rigby, M., 248 (9.9)
Rihs, G., 19 (1.31)
Rillema, D. P., 171 (6.52), 172 (6.52), 174 (6.65, 6.73, 6.78), 177 (6.73), 179 (6.91), 180 (6.78, 6.98, 6.99, 6.104), 181 (6.73), 348 (11.158)
Ringsdorf, H., 19 (1.72), 51 (3.1), 161 (6.5), 216 (7.133), 217 (7.153, 7.154, 7.155, 7.156, 7.159), 387 (12.168)
Riordan, J. E., 218 (7.172)
Rius, J., 267 (10.37)
Roach, K. J., 118 (5.130)
Rob, F., 102 (5.79), 253 (9.46)
Robbins, R. J., 19 (1.28), 252 (9.39)
Roberts, D. R., 356 (12.26)
Robin, M. B., 54 (3.8), 56 (3.8)
Robinson, G. W., 348 (11.166)
Robson, R., 386 (12.148)
Rodger, A., 299 (10.301), 304 (10.301)
Rodgers, M. A. J., 241 (8.70), 242 (8.70, 8.73)
Rodriguez-Ubis, J. C., 327 (11.47)
Rodriguez, M., 301 (10.311)
Rodriguez, R., 250 (9.10)
Roffia, S., 60 (3.13, 3.14, 3.15), 82 (4.9), 136 (5.186), 137 (5.186), 140 (5.195, 5.196), 142 (5.195, 5.196), 144 (5.195, 5.196), 184 (6.109), 185 (6.109), 186 (6.109), 365 (12.55)
Rogers, J., 267 (10.22)
Rllig, K., 254 (9.55)
Ron, A., 163 (6.22)
Rondelez, F., 218 (7.171)
Rong, D., 387 (12.165)
Ropartz, S., 299 (10.298)
Rosato, N., 215 (7.122), 219 (7.174, 7.177)
Ross, D. L., 197 (7.5), 200 (7.5)
Roth, H. D., 258 (9.127)
Rothin, A. S., 270 (10.54)
Rotkiewicz, K., 202 (7.47, 7.48)
Rotlin, A. S., 204 (7.75)
Rouge, M., 19 (1.58), 116 (5.126), 145 (5.126),

146 (5.126), 148 (5.222), 150 (5.222), 363 (12.41), 364 (12.41)
Rubin, M. B., 163 (6.21, 6.22, 6.23, 6.24), 164 (6.23, 6.24)
Rubtsov, I. V., 370 (12.75)
Rullire, C., 202 (7.56)
Ruminski, R. R., 178 (6.84)
Ryan, D. A., 140 (5.190)
Rybak, W., 232 (8.33), 233 (8.33), 234 (8.33), 237 (8.33)
Ryu, C. K., 174 (6.62), 190 (6.126)

Saad, Z., 337 (11.108)
Saalfrank, R. W., 387 (12.180)
Sabatino, L., 170 (6.50), 174 (6.70, 6.71), 176 (6.70), 177 (6.70, 6.71), 178 (6.70, 6.71, 6.83), 181 (6.70, 6.71, 6.105, 6.107, 6.108), 182 (6.70, 6.71, 6.105, 6.107, 6.108), 183 (6.108), 184 (6.70, 6.71, 6.107, 6.108), 306 (10.321), 365 (12.57), 370 (12.82, 12.84), 371 (12.82), 382 (12.84), 386 (12.84)
Sabbatini, N., 19 (1.81), 76 (4.2), 77 (4.2), 82 (4.2). 227 (8.6), 231 (8.20, 8.21), 232 (8.20), 233 (8.48), 234 (8.48), 235 (8.20, 8.48), 239 (8.6), 241 (8.48), 243 (8.89), 274 (10.86), 280 (10.86), 289 (10.86), 319 (11.10), 321 (11.10), 322 (11.10), 325 (11.10), 327 (11.48, 11.50), 328 (11.52), 329 (11.50, 11.51, 11.52, 11.53, 11.54, 11.63), 330 (11.10)
Saenger, W., 19 (1.65), 210 (7.94), 283 (10.143), 288 (10.164)
Sahai, R., 171 (6.52), 172 (6.52), 174 (6.73), 177 (6.73), 179 (6.91), 180 (6.98, 6.99, 6.104), 181 (6.73)
Sahyun, M. R. V., 363 (12.43)
Saigo, K., 267 (10.36)
Saillard, J. Y., 19 (1.7)
Saito, R., 292 (10.207)
Saito, T., 122 (5.142)
Saka, R., 210 (7.96, 7.97), 298 (10.245)
Sakai, K., 242 (8.76, 8.77), 243 (8.87)
Sakata, Y, 90 (5.3), 118 (5.131), 119 (5.135), 145 (5.203), 255 (9.62, 9.63), 258 (9.120)
Sakomura, M., 383 (12.135, 12.136)
Sakuragi, H., 254 (9.53)
Sakuragi, M., 201 (7.41)
Salemme, F. R., 90 (5.4)
Salmon, D. J., 125 (5.155), 144 (5.210), 145 (5.210)
Saltbeck, J., 214 (7.116)
Saltiel, J., 85 (4.24), 198 (7.13), 217 (7.13)
Samorí, B., 217 (7.149)
Sampson, R. M., 19 (1.12)
Sandanayake, K. R. A. S., 271 (10.67)
Sanders, G. M., 147 (5.210, 5.217)
Sanders, J. K. M., 19 (1.11), 120 (5.136), 122 (5.141), 147 (5.214), 285 (10.147)
Sangster, D. F., 320 (11.16)
Sano, S., 171 (6.53)
Santi, F., 167 (6.37)
Santos, P. S., 180 (6.102)

Sargeson, A. M., 19 (1.35), 83 (4.15), 84 (4.15), 132 (5.174, 5.175), 320 (11.14, 11.15, 11.16, 11.17, 11.18, 11.19, 11.20, 11.21, 11.22, 11.23, 11.24), 321 (11.25, 11.29, 11.30, 11.31, 11.32, 11.33), 322 (11.32, 11.33), 323 (11.17, 11.33), 324 (11.41, 11.42), 325 (11.41), 326 (11.41), 332 (11.81)
Sasaki, H., 216 (7.143)
Sasse, W. H. F., 132 (5.174, 5.175), 190 (6.128), 243 (8.85), 321 (11.29, 11.30, 11.31, 11.32, 11.33), 322 (11.32, 11.33), 323 (11.33), 324 (11.42), 347 (11.156), 381 (12.127)
Sato, H., 122 (5.142), 361 (12.33), 363 (12.33), 370 (12.33)
Sato, M., 215 (7.129), 216 (7.139), 219 (7.176)
Sato, S., 205 (7.79)
Sattur, P. B., 288 (10.173)
Sauer, K., 131 (5.170)
Sauerer, W., 267 (10.30), 287 (10.30), 333 (11.95)
Sauvage, J. P., 19 (1.17, 1.51, 1.71), 83 (4.14), 167 (6.32), 327 (11.46), 334 (11.100), 337 (11.100, 11.107, 11.108, 11.109, 11.110), 338 (11.100), 339 (11.111, 11.112, 11.113, 11.114, 11.115), 340 (11.118, 11.119), 341 (11.113, 11.122), 386 (12.141), 387 (12.141, 12.177)
Savant, J. M., 380 (12.116)
Scaiano, J. C., 293 (10.216),
Scandola, F., 19 (1.69, 1.73, 1.76, 1.81, 1.82), 45 (2.13), 46 (2.13, 2.19), 47 (2.19, 2.21), 48 (2.13, 2.21), 49 (2.29), 51 (3.1), 57 (3.9), 60 (3.12, 3.13, 3.14, 3.15), 70 (3.23), 72 (3.35), 73 (3.9, 3.35), 76 (4.2), 77 (4.2, 4.3), 79 (4.6), 82 (4.2, 4.9), 94 (5.37), 125 (5.151), 132 (5.37), 136 (5.186), 137 (5.186), 140 (5.195, 5.196), 142 (5.195, 5.196, 5.197), 144 (5.195, 5.196, 5.202), 145 (5.197), 161 (6.7), 170 (6.7), 184 (6.109, 6.110), 185 (6.109), 186 (6.109, 6.112), 187 (6.117, 6.118), 188 (6.118, 6.120, 6.121), 189 (6.123), 227 (8.6, 8.12), 231 (8.12, 8.23), 233 (8.44, 8.46, 8.47), 234 (8.44), 235 (8.44, 8.51), 237 (8.44), 239 (8.6), 241 (8.46), 274 (10.86), 280 (10.86), 289 (10.86), 301 (10.310), 306 (10.310), 319 (11.10), 321 (11.10, 11.28), 322 (11.10), 325 (11.10), 330 (11.10, 11.66), 342 (11.135), 348 (11.164, 11.165), 355 (12.7), 356 (12.7), 357 (12.7), 359 (12.7, 12.27), 360 (12.7), 361 (12.7, 12.38), 365 (12.55), 366 (12.7), 367 (12.7, 12.73), 368 (12.7), 369 (12.73), 370 (12.81), 374 (12.7), 379 (12.7), 380 (12.81), 381 (12.38, 12.81, 12.129, 12.130), 382 (12.81)
Scandola, M., 321 (11.28)
Scarpa, A., 110 (5.106)
Schaafsma, T. J., 147 (5.217)
Schfer, F. P., 254 (9.55)
Schanck, A., 226 (8.4), 244 (8.4)
Schanze, K. S., 127 (5.157, 5.158), 128 (5.157, 5.158, 5.162), 129 (5.162), 131 (5.170), 134 (5.158), 137 (5.187), 151 (5.237), 306 (10.318), 386 (12.152)
Scheer, H., 91 (5.17)
Scherer, P. O. J., 95 (5.57)

Scherer, T., 105 (5.99), 252 (9.37), 253 (9.37), 254 (9.37)
Scheuring, M., 19 (1.36)
Schiavello, M., 174 (6.71), 177 (6.71), 178 (6.71), 181 (6.71), 182 (6.71), 184 (6.71)
Schiffer, M., 90 (5.8)
Schill, G., 333 (11.99)
Schlaepfer, C. W., 281 (10.139)
Schlarb, B., 19 (1.72), 51 (3.1), 161 (6.5), 217 (7.159), 387 (12.168)
Schlessinger, R. H., 85 (4.25)
Schmehl, R. H., 171 (6.55, 6.56), 173 (6.55, 6.59, 6.61), 174 (6.61, 6.62)
Schmidt, F., 215 (7.126)
Schmidt, J. A., 117 (5.128, 5.129), 118 (5.128, 5.129), 259 (9.134)
Schnabel, W., 218 (7.163)
Schneider, F. W., 386 (12.156)
Schneider, H. J., 227 (8.13), 243 (8.81), 267 (10.28), 274 (10.85), 275 (10.85), 280 (10.85), 386 (12.143), 387 (12.143)
Schrieffer, J. R., 90 (5.4), 92 (5.20)
Schroff, L. G., 253 (9.42, 9.44)
Schulte-Frohlinde, D., 299 (10.286, 10.296, 10.297), 301 (10.286), 304 (10.296), 305 (10.297)
Schulten, H. R., 267 (10.29), 333 (11.90)
Schurr, J. M., 299 (10.295)
Schwager, B., 202 (7.51)
Schwartz, H. A., 130 (5.169), 131 (5.169)
Sclosser, H., 165 (6.29), 166 (6.29)
Scott, A. J., 19 (1.26)
Scrimin, P., 288 (10.168)
Scurlock, R. D., 254 (9.52)
Scypinski, S., 289 (10.186), 290 (10.193)
Seddon, K. R., 212 (7.108)
Sedelmeier, G., 19 (1.31)
See, K. A., 267 (10.22)
Seely, G. R., 148 (5.221, 5.223), 149 (5.221), 150 (5.223), 169 (6.48), 363 (12.42), 370 (12.85)
Seghi, B., 19 (1.34), 279 (10.124, 10.126)
Seitz, G., 132 (5.179
Seitz, W. R., 292 (10.204)
Selli, E., 216 (7.144)
Semlyen, J. A., 387 (12.162)
Sendhoff, N., 386 (12.140)
Serizawa, M., 278 (10.101)
Serpone, N., 231 (8.23), 233 (8.46), 241 (8.46, 8.64), 330 (11.66), 359 (12.27), 363 (12.43), 364 (12.49), 380 (12.49), 380 (12.121)
Serroni, S., 170 (6.50), 174 (6.70, 6.71), 176 (6.70), 177 (6.70, 6.71), 178 (6.70, 6.71), 181 (6.70, 6.71, 6.105, 6.107, 6.108), 182 (6.70, 6.71, 6.105, 6.107, 6.108), 183 (6.108), 184 (6.70, 6.71, 6.107, 6.108), 306 (10.321), 365 (12.57), 370 (12.82, 12.84), 371 (12.82), 382 (12.84), 386 (12.84)
Sessions, R. B., 85 (4.37), 279 (10.127, 10.128), 280 (10.127, 10.128), 281 (10.127, 10.128),
Sessler, J. L., 123 (5.144), 147 (5.211, 5.213, 5.216, 5.218), 167 (6.33), 274 (10.83), 383 (12.131)

Seta, P., 147 (5.219, 5.220), 365 (12.65), 365 (12.66)
Sevilla, F., 270 (10.56, 10.57)
Seward, E. M., 267 (10.30), 287 (10.30)
Seward, E. M., 333 (11.95)
Sexton, D. A., 240 (8.58)
Seymour, P., 257 (9.90)
Shafirovich, V. Y., 363 (12.45), 365 (12.64)
Shah, M., 289 (10.184)
Shahriari-Zavareh, H., 275 (10.93), 276 (10.93), 277 (10.93), 278 (10.105), 285 (10.105, 10.150, 10.151), 286 (10.151), 289 (10.93)
Shapira, A., 292 (10.206)
Sharma, D. K., 363 (12.43)
Sharpe, S., 251 (9.33), 252 (9.33)
Shaver, R. J., 174 (6.73), 177 (6.73), 180 (6.104), 181 (6.73)
Shaw, J. R., 171 (6.56)
Sherman, J. C., 287 (10.161), 333 (11.88)
Sherman, S. E., 19 (1.68), 300 (10.307)
Shi, W., 174 (6.75)
Shiga, M., 207 (7.84)
Shigematsu, K., 208 (7.88, 7.91), 377 (12.100), 378 (12.100)
Shih, N. Y., 19 (1.18)
Shilov, A. E., 363 (12.45), 365 (12.64)
Shimada, M., 258 (9.115)
Shimamoto, K., 267 (10.31), 387 (12.164)
Shimidzu, T., 214 (7.114, 7.117), 347 (11.157), 380 (12.111)
Shimomura, M., 216 (7.137)
Shimura, Y., 289 (10.179)
Shin, D. M., 252 (9.40)
Shinkai, S., 85 (4.32), 204 (7.69, 7.70, 7.72), 205 (7.70, 7.77, 7.78), 206 (7.82), 207 (7.83, 7.85), 208 (7.70, 7.87, 7.88, 7.89, 7.91, 7.92), 209 (7.93), 215 (7.129, 7.130), 216 (7.131), 267 (10.31), 268 (10.43), 377 (12.99, 12.100), 378 (12.100, 12.101), 379 (12.101), 387 (12.164)
Shinohara, I., 216 (7.142), 218 (7.165, 7.170)
Shinohara, N., 239 (8.52), 319 (11.12), 320 (11.12)
Shinozaki, T., 278 (10.99)
Shionoya, M., 280 (10.134)
Shiragami, T., 364 (12.50), 380 (12.50)
Shirai, M., 271 (10.64)
Shisuka, H., 258 (9.124), 270 (10.63), 278 (10.99, 10.100, 10.101)
Shoup, M., 178 (6.84)
Siatkowski, R. E., 19 (1.74), 93 (5.32), 211 (7.104, 7.105), 216 (7.135), 217 (7.147), 342 (11.136), 343 (11.136), 344 (11.136), 345 (11.136), 355 (12.9), 380 (12.9), 377 (12.98)
Siebrand, W., 39 (2.7), 40 (2.7), 198 (7.16)
Siegel, J., 341 (11.124), 387 (12.173)
Sieger, H., 289 (10.176)
Siemiarczuk, A., 51 (3.2), 113 (5.119), 118 (5.130), 202 (7.48, 7.54)
Siesel, D. A., 170 (6.49)
Sigman, D. S., 304 (10.312)
Sigwart, C., 19 (1.39)
Sikanyika, H., 204 (7.73)

Sikanyta, H., 267 (10.33,
Silver, J., 295 (10.237)
Silver, M., 150 (5.225), 380 (12.119)
Simic, M. G., 239 (8.52), 319 (11.12), 320 (11.12)
Simkin, D. J., 250 (9.11)
Simmons, H. E., 19 (1.25)
Simon, J., 387 (12.172)
Singer, L. A., 202 (7.50)
Sinn, E., 19 (1.80), 386 (12.139)
Sishta, B., 93 (5.24)
Sisido, M., 292 (10.209)
Sitzmann, E. V., 296 (10.241), 297 (10.241), 298
    (10.251)
Sivaraja, M., 365 (12.62)
Sixl, H., 387 (12.170)
Skelton, B. W., 279 (10.132)
Skobeleva, S. E., 252 (9.35)
Slawin, A. M. Z., 19 (1.4, 1.52, 1.79, 1.83), 278
    (10.103), 285 (10.148, 10.152, 10.153, 10.154),
    286 (10.154), 287 (10.156, 10.157), 334
    (11.104, 11.106), 336 (11.106), 387 (12.178,
    12.179),
Sloper, R. W., 130 (5.165)
Smets, G., 215 (7.123), 218 (7.123)
Smirnov, V. A., 370 (12.75)
Smit, K. J., 102 (5.92), 104 (5.92), 163 (6.16)
Smith, E. B., 248 (9.9)
Smith, M. L., 250 (9.12)
Smith, P. H., 333 (11.85)
Smith, U., 90 (5.8)
Snaith, R., 19 (1.26)
Snow, M. R., 320 (11.14), 321 (11.25), 324
    (11.42)
Soboleva, I. V., 251 (9.25)
Soga, T., 283 (10.142)
Somorjai, G. A., 364 (12.49), 380 (12.49)
Soncini, P., 267 (10.38)
Sone, T., 215 (7.129)
Song, P. S., 345 (11.145)
Soo, L. M., 213 (7.112)
Sotomayor, J., 240 (8.57), 323 (11.36)
Soumillion, J. Ph., 226 (8.4), 244 (8.4)
Sousa, L. R., 270 (10.60, 10.61, 10.62)
Sowinska, M., 211 (7.105, 7.106), 212 (7.110),
    377 (12.98)
Spangler, C. W., 116 (5.125), 132 (5.125)
Spatz, H. Ch., 283 (10.143)
Speck, K. R., 380 (12.107)
Speiser, S., 163 (6.21, 6.22, 6.23, 6.24), 164
    (6.23, 6.24)
Spelthann, H., 213 (7.111), 378 (12.102)
Spencer, N., 19 (1.4, 1.52, 1.79, 1.83), 275
    (10.93), 276 (10.93, 10.96), 277 (10.93, 10.96),
    285 (10.150, 10.151, 10.152, 10.153, 10.154),
    286 (10.151, 10.154), 287 (10.157), 289
    (10.93), 334 (11.104, 11.105, 11.106), 336
    (11.106), 387 (12.178, 12.179)
Springborg, J., 320 (11.14), 321 (11.25)
Sprintschnik, H., 125 (5.155)
Spurr, P. R., 19 (1.31)
Srinivasan, T. N., 288 (10.173)

Staalman, D. J. H., 253 (9.42)
Staab, H. A., 270 (10.53)
Stacy, E. M., 241 (8.63), 257 (9.94), 341 (11.121)
Staerk, H., 257 (9.109)
Stagno D'Alcontres, G., 19 (1.42)
Stark, A., 387 (12.180)
Stein, C. A., 132 (5.179)
Stein, G., 326 (11.44)
Steinberger, B., 299 (10.256)
Steiner, U. E., 257 (9.107)
Stenman, F., 250 (9.14)
Stephens, E. M., 290 (10.201), 291 (10.201)
Stetyick, K. A., 380 (12.107)
Stevens, B., 251 (9.21), 254 (9.57, 9.58, 9.59,
    9.60)
Stevens, R. V., 274 (10.90), 278 (10.90)
Stevetson, B. R., 298 (10.252)
Steyert, D. W., 258 (9.127)
Stibor, I., 388 (12.182)
Stillman, M. J., 118 (5.130)
Stilz, H. U., 91 (5.17)
Stinson, A. J., 179 (6.93), 180 (6.93)
Stobart, S. R., 131 (5.172)
Stoddart, J. F., 19 (1.1, 1.4, 1.9, 1.42, 1.48, 1.49,
    1.52, 1.79, 1.83), 85 (4.33) 267 (10.10), 274
    (10.10, 10.87, 10.88, 10.91), 275 (10.10, 10.92,
    10.93, 10.94, 10.95), 276 (10.92, 10.93, 10.96),
    277 (10.93, 10.96), 278 (10.102, 10.103,
    10.104, 10.105), 280 (10.10, 10.87), 285
    (10.105, 10.148, 10.149, 10.150, 10.151,
    10.152, 10.153, 10.154), 286 (10.151, 10.154),
    287 (10.156, 10.157), 289 (10.10, 10.93,
    10.174), 334 (11.104, 11.105, 11.106), 336
    (11.106), 386 (12.142, 12.143), 387 (12.143,
    12.178, 12.179)
Stolarczyk, L. Z., 355 (12.10), 380 (12.10)
Stolwijk, T. B., 267 (10.39)
Stradowski, C., 299 (10.296, 10.286), 301
    (10.286), 304 (10.296)
Straub, K. D., 111 (5.113)
Street, K. W. Jr., 271 (10.65)
Strekas, T. C., 174 (6.74, 6.75), 241 (8.69), 242
    (8.69), 348 (11.160, 11.161)
Striker, G., 255 (9.68)
Strouse, G. F., 150 (5.232), 151 (5.232), 191
    (6.129, 6.130), 370 (12.86), 372 (12.86), 386
    (12.86)
Stutte, P., 331 (11.71) 332 (11.71), 333 (11.71)
Subrahmanyam, D., 19 (1.33)
Subramanian, R., 299 (10.293), 306 (10.293)
Suck, T. A., 215 (7.118)
Sudhlter, E. J. R., 267 (10.39), 268 (10.44), 270
    (10.44)
Suga, K., 383 (12.134)
Sugi, M., 93 (5.32), 216 (7.135)
Sullivan, B. P., 125 (5.149, 5.155), 137 (5.188),
    170 (6.51), 180 (6.97), 232 (8.39), 242 (8.78)
Sullivan, J. C., 320 (11.16)
Sulpizio, A., 251 (9.29)
Sunamoto, J., 216 (7.138)
Sundquist, W. I., 300 (10.308)

Surridge, N. A., 150 (5.225), 363 (12.44), 380 (12.119), 386 (12.151)
Sutherland, I. O., 270 (10.59)
Sutin, N., 46 (2.17, 2.18), 47 (2.17, 2.18), 59 (3.11), 66 (3.11, 3.19, 3.20), 67 (3.11), 68 (3.11, 3.19, 3.20), 70 (3.11), 90 (5.4), 92 (5.20), 93 (5.22), 95 (5.41), 132 (5.182), 133 (5.182), 143 (5.198), 144 (5.199), 178 (6.89, 6.90), 232 (8.31, 8.33), 233 (8.33), 234 (8.33), 237 (8.33), 240 (8.31), 319 (11.3)
Suzuki, I., 292 (10.212), 293 (10.212, 10.222, 10.223, 10.224, 10.225), 294 (10.224)
Suzuki, T., 218 (7.169), 255 (9.62)
Suzuki, Y., 267 (10.36)
Svec, W. A., 19 (1.57), 110 (5.109), 111 (5.110, 5.112, 5.114), 112 (5.114), 113 (5.114), 114 (5.120), 115 (5.120), 147 (5.208), 361 (12.39), 362 (12.39)
Swain, C. S., 278 (10.98)
Swinnen, A. M., 255 (9.69), 257 (9.96, 9.106)
Syage, J. A., 198 (7.20)
Syamala, M. S., 296 (10.240)
Sykes, A., 333 (11.98)
Syme, R. W. G., 375 (12.92)
Symons, M. C. R., 107 (5.101), 113 (5.101)
Szabo, A. G., 243 (8.86)
Szejtli, J., 85 (4.26), 210 (7.95), 288 (10.165)

Tabushi, I., 85 (4.27), 110 (5.107), 288 (10.167, 10.171), 292 (10.203)
Tachibana, H., 216 (7.132)
Takada, K., 270 (10.63)
Takagi, M., 207 (7.84), 268 (10.42), 269 (10.42)
Takahashi, K., 210 (7.98, 7.99), 290 (10.191), 298 (10.246)
Takahashi, T., 280 (10.134)
Takami, A., 370 (12.76)
Takashima, M., 260 (9.141)
Takeuchi, K. J., 332 (11.82)
Takiff, L., 20 (1.97), 90 (5.12), 356 (12.23)
Takizawa, A., 216 (7.139), 219 (7.176)
Talarmin, J., 19 (1.7)
Tam, S. W., 267 (10.30), 287 (10.30), 333 (11.95)
Tam, W., 161 (6.3), 295 (10.235)
Tamai, N., 161 (6.4), 166 (6.30), 370 (12.77, 12.80)
Tamai, Na., 257 (9.101, 9.102, 9.105)
Tamai, No., 257, (9.105)
Tamaki, T., 201 (7.41), 213 (7.113), 295 (10.236)
Tamilarasan, R., 299 (10.298)
Tanaka, H., 218 (7.161)
Tanaka, J., 256 (9.83)
Tanaka, M., 271 (10.64)
Tanaka, Y., 216 (7.137)
Tang, J., 90 (5.8)
Tanigawa, I., 271 (10.66)
Taniguchi, Y., 218 (7.162)
Tanimoto, Y., 260 (9.141)
Taoda, H., 298 (10.253)
Tapolsky, G., 128 (5.163), 129 (5.163), 135 (5.185), 136 (5.185)
Tarroni, R., 217 (7.149)

Tashiro, M., 207 (7.83)
Tasker, P.A., 279 (10.109)
Tatemitsu, H., 119 (5.135)
Taube, H., 52 (3.4), 71 (3.30), 90 (5.2), 95 (5.45), 144 (5.200)
Tazuke, S., 168 (6.39), 215 (7.128), 216 (7.136), 217 (7.157, 7.158), 218 (7.166, 7.167, 7.168), 255 (9.70), 257 (9.100, 9.104, 9.105)
Telser, J., 151 (5.237, 5.238), 306 (10.318, 10.319, 10.320), 386 (12.152)
Tevini, M., 20 (1.93), 197 (7.7), 198 (7.7), 356 (12.19), 370 (12.19)
Texidor, F., 267 (10.37)
Tfibel, F., 201 (7.34, 7.37, 7.38)
Thanabal, V., 273 (10.80)
Thanyasiri, T., 386 (12.139)
Theis, I., 267 (10.28)
Thiery, U., 127 (5.156)
Thomas, J. K., 93 (5.30), 241 (8.68), 243 (8.68)
Thomas, Ph., 240 (8.59)
Thorn, D. L., 132 (5.177)
Thornber, J. P., 92 (5.19)
Thulstrup, E. W., 217 (7.149)
Tiede, D.M., 90 (5.8), 92 (5.20)
Tien, H. T., 216 (7.146), 217 (7.147)
Tifbel, F., 201 (7.39)
Tjivikua, T., 388 (12.181)
Tkachev, V. A., 260 (9.142)
Tocher, D. A., 19 (1.61), 342 (11.127)
Toda, F., 19 (1.45)
Toffoletti, A., 256 (9.85)
Tokumaru, K., 254 (9.53), 345 (11.147)
Tom, G. M., 144 (5.200)
Tom, R., 169 (6.45)
Toma, H. E., 180 (6.100, 6.101, 6.102), 231 (8.22), 232 (8.22)
Tomalia, D. A., 19 (1.85), 51 (3.1), 386 (12.153)
Tomita, Y., 210 (7.100), 293 (10.218), 294 (10.218), 295 (10.233), 298 (10.244)
Tonellato, U., 288 (10.168)
Torkelson, J. M., 161 (6.2)
Tossi, A. B., 299 (10.283, 10.284, 10.286, 10.287, 10.288, 10.289), 301 (10.284, 10.286, 10.287), 302 (10.287), 303 (10.287), 304 (10.287), 305 (10.283, 10.284, 10.287, 10.288, 10.289)
Tourrel, M., 211 (7.105), 377 (12.98)
Toyozawa, K., 370 (12.79)
Tramer, A., 256 (9.78, 9.79)
Tran, C. D., 289 (10.183)
Treanor, R. L., 217 (7.150)
Trend, J. E., 204 (7.74)
Trier, T. T., 148 (5.221), 149 (5.221), 150 (5.224), 169 (6.48), 370 (12.85)
Trierweiler, H. P., 331 (11.69)
Trueblood, K. N., 19 (1.3), 274 (10.90), 278 (10.90)
Truesdell, K. A., 127 (5.160), 134 (5.160)
Tsay, Y. H., 253 (9.47), 254 (9.47)
Tseng, J. C. C., 202 (7.50)
Tsujino, N., 258 (9.115, 9.116)
Tsujita, Y., 216 (7.139), 219 (7.176)
Tsuno, K., 388 (12.184)

Tucker, J. A., 19 (1.3), 287 (10.161)
Tummler, B., 271 (10.73)
Tundo, P., 271 (10.72), 274 (10.84)
Tunstad, L. M., 287 (10.161)
Turner, D. H., 299 (10.294)
Turro, N. J., 20 (1.86), 25 (2.1), 41 (2.1), 72 (3.31, 3.32), 76 (4.1), 80 (4.1), 85 (4.1), 151 (5.235, 5.236), 163 (6.20), 166 (6.20) 255 (9.75), 290 (10.200), 292 (10.202, 10.203, 10.205, 10.213), 295 (10.228, 10.237, 10.238), 296 (10.241), 297 (10.241), 298 (10.248), 299 (10.271, 10.273, 10.274, 10.275, 10.280, 10.282), 301 (10.280), 302 (10.280), 303 (10.271, 10.280), 304 (10.271, 10.274, 10.280), 305 (10.273, 10.280), 306 (10.275, 10.279), 345 (11.142)
Twigden, S. J., 306 (10.323)
Tysoe, S. A., 348 (11.161)

Ueda, K., 207 (7.83)
Ueno, A., 210 (7.96, 7.97, 7.98, 7.99, 7.100, 7.101), 216 (7.143, 7.145), 219 (7.179), 290 (10.191), 292 (10.207, 10.212), 293 (10.212, 10.217, 10.218, 10.219, 10.220, 10.221, 10.222, 10.223, 10.224, 10.225), 294 (10.217, 10.218, 10.224), 295 (10.233), 298 (10.217, 10.244, 10.245, 10.246)
Ueno, K., 207 (7.84), 268 (10.42), 269 (10.42)
Ugozzoli, F., 267 (10.23), 329 (11.60, 11.61, 11.62), 333 (11.97)
Ulrich, T., 257 (9.107)
Ulstrup, J., 66 (3.21), 70 (3.21), 94 (5.39)
Umemoto, T., 255 (9.63)
Ungaro, R., 267 (10.23), 329 (11.58, 11.60, 11.61, 11.62, 11.63), 333 (11.97)
Uzan, B., 290 (10.188)

Valentekovich, R., 287 (10.163)
Valeur, B., 167 (6.34, 6.35, 6.36, 6.37), 168 (6.34, 6.35, 6.36), 211 (7.106), 269 (10.52), 271 (10.52, 10.70, 10.71)
Van Damme, M., 252 (9.38)
Van der Auweraer, M., 19 (1.58), 116 (5.126), 145 (5.126), 146 (5.126), 148 (5.222), 150 (5.222), 226 (8.4), 244 (8.4), 255 (9.66, 9.69), 257 (9.96, 9.106), 260 (9.139), 361 (12.31), 363 (12.31, 12.41), 364 (12.41)
Van der Plas, H. C., 147 (5.210, 5.217)
Van der Putten, W. J. M., 299 (10.285)
Van der Voort, D., 329 (11.53)
Van der Weerdt, A. J. A., 253 (9.42)
Van Dijk, M., 147 (5.210, 5.217)
Van Dyke, M. W., 304 (10.314, 10.315)
Van Engen, D., 19 (1.24)
Van Gent, J., 268 (10.44), 270 (10.44)
Van Houten, J., 330 (11.67)
Van Ramesdonk, H. J., 102 (5.80, 5.81), 105 (5.81)
Van Staveren, C. J., 267 (10.25)
Van Veggel, F. C. J. M., 267 (10.25)
Van Veldhuizen, A., 147 (5.210)
Vandendriessche, J., 257 (9.102, 9.106)

Vander Donckt, E., 342 (11.131)
Vandereecken, P., 226 (8.4), 244 (8.4)
Vandersall, M.T., 298 (10.251)
Vanhecke, F., 180 (6.103)
Vannikov, A. V., 260 (9.142)
Vanquickenborne, L. G., 41 (2.8), 282 (10.140)
Vanzyl, C. M., 267 (10.21)
Varma, C. A. G. O., 346 (11.154), 347 (11.154)
Vassilian, A., 130 (5.167, 5.168, 5.169), 131 (5.169)
Vaught, J. M., 188 (6.122), 189 (6.122)
Venturi, M., 232 (8.27), 240 (8.27), 242 (8.73), 323 (11.34, 11.35), 324 (11.34)
Venzmer, J., 19 (1.72), 51 (3.1), 161 (6.5), 217 (7.159), 387 (12.168)
Verboom, W., 267 (10.25)
Verhoeven, C., 174 (6.76), 180 (6.76)
Verhoeven, J. W., 19 (1.6), 90 (5.1), 98 (5.65), 102 (5.65, 5.79, 5.80, 5.82, 5.87, 5.88, 5.89, 5.90, 5.92, 5.93, 5.94, 5.95, 5.96, 5.97, 5.98), 103 (5.65, 5.93), 104 (5.65, 5.88, 5.90, 5.92, 5.93, 5.98), 105 (5.65, 5.96, 5.99), 106 (5.95, 5.97, 5.98), 107 (5.95, 5.97, 5.101), 113 (5.101), 162 (6.13, 6.14), 163 (6.13, 6.16, 6.17, 6.18, 6.19), 252 (9.36, 9.37), 253 (9.36, 9.37, 9.42, 9.43, 9.44, 9.45, 9.46), 254 (9.36), 254 (9.37, 9.43, 9.61), 255 (9.61)
Vermeglio, H., 20 (1.94), 90 (5.6), 356 (12.20), 370 (12.20)
Vicens, J., 329 (11.58)
Vicent, C., 19 (1.52, 1.79), 334 (11.104, 11.105, 11.106), 336 (11.106), 387 (12.178, 12.179)
Vigato, P. A., 272 (10.75), 279 (10.117)
Vigneron, J. P., 287 (10.158)
Vilanove, R., 218 (7.171)
Villemure, G., 243 (8.86)
Vinas, C., 267 (10.37)
Vincent, J. B., 365 (12.63)
Vincenti, M., 267 (10.38)
Vining, W. J., 243 (8.88)
Vinod, T. K., 19 (1.22)
Vittorakis, M., 346 (11.151)
Vlachopoulos, N., 380 (12.113), 382 (12.113)
Vlcek, A., 364 (12.52)
Vo-Dinh, T., 290 (10.194, 10.196)
Vogler, A., 19 (1.70), 127 (5.161), 132 (5.176), 140 (5.191, 5.192, 5.193, 5.194), 174 (6.66), 177 (6.66), 178 (6.85), 180 (6.85), 227 (8.10,8.7, 8.8, 8.9), 231 (8.7), 232 (8.10, 8.25, 8.26, 8.29, 8.32), 233 (8.44), 234 (8.44), 235 (8.44), 237 (8.44), 239 (8.25, 8.26, 8.32, 8.54), 257 (9.87), 348 (11.164)
Vogt, W., 267 (10.29), 333 (11.90)
Vögtle, F., 19 (1.2, 1.8, 1.32, 1.43, 1.53, 1.62, 1.63, 1.64), 51 (3.1), 80 (4.8), 83 (4.16, 4.21), 85 (4.21, 4.28, 4.29, 4.30), 171 (6.57), 204 (7.68), 205 (7.81), 213 (7.111), 267 (10.2, 10.27, 10.6, 10.7, 10.8), 268 (10.41, 10.48, 10.48, 10.50), 269 (10.41, 10.51), 270 (10.41, 10.55), 271 (10.73), 284 (10.146), 287 (10.160, 10.27), 289 (10.175, 10.176), 306 (10.322), 331 (11.70, 11.71, 11.73, 11.74), 332 (11.70, 11.71,

11.74, 11.75, 11.76, 11.78, 11.79), 333 (11.71, 11.83, 11.96), 365 (12.56), 378 (12.102), 386 (12.140)
Voitlnder, J., 201 (7.33)
Volk, M., 108 (5.104), 109 (5.104)
Vollmann, H. W., 380 (12.108)
Volosov, A., 95 (5.51, 5.52)
Von Bünau, G., 215 (7.118, 7.119, 7.125, 7.126)
Von Kameke, A., 144 (5.200)
Von Maltzan, B., 167 (6.31)
Von Zelewski, A., 19 (1.50), 83 (4.18), 124 (5.145), 170 (6.50), 171 (6.57), 212 (7.107, 7.109), 281 (10.139), 301 (10.309), 306 (10.309, 10.322), 319 (11.7), 330 (11.7), 331 (11.72, 11.73, 11.74) 332 (11.74, 11.75, 11.80), 347 (11.7), 365 (12.56)
Vos, J. G., 174 (6.69), 177 (6.69), 178 (6.69, 6.88)
Vrachnou, E., 380 (12.114)

Wacholtz, W. F., 171 (6.55), 173 (6.55, 6.59, 6.61), 174 (6.61)
Wada, S., 280 (10.134)
Wade, P. W., 279 (10.113)
Wainwright, K. P., 279 (10.113)
Wakatsuki, Y., 232 (8.41)
Wakeham, W. A., 174 (6.77), 177 (6.77), 248 (9.9)
Wallace, K. C., 324 (11.40)
Wallendael, S. V., 180 (6.104)
Wallon, A., 205 (7.81), 267 (10.27), 287 (10.27), 333 (11.96)
Walsh, P. K., 342 (11.138), 346 (11.138)
Walter, J., 26 (2.2)
Walther, D., 240 (8.59)
Waluk, J., 346 (11.149)
Wan, C. S. K., 51 (3.2), 113 (5.119)
Wang, Y., 295 (10.234, 10.235)
Ward, D. L., 19 (1.45)
Ward, M. D., 19 (1.61), 342 (11.126, 11.127), 387 (12.175)
Ward, R. L., 257 (9.110)
Ware, W. R., 202 (7.54), 251 (9.23), 257 (9.95)
Waring, M., 300 (10.303)
Warman, J. M., 102 (5.81, 5.88, 5.90, 5.92, 5.96), 104 (5.88, 5.90, 5.92), 105 (5.81, 5.96), 163 (6.16, 6.18), 252 (9.36), 253 (9.36), 254 (9.36)
Warner, I. M., 290 (10.190, 10.192), 292 (10.206)
Warshawsky, A., 204 (7.76)
Warshel, A., 198 (7.9)
Wasgestian, H. F., 187 (6.116)
Wasielewski, M. R., 19 (1.57), 48 (2.28), 71 (3.27), 94 (5.34), 95 (5.34), 110 (5.108, 5.109), 111 (5.110, 5.112, 5.113, 5.114), 112 (5.114), 113 (5.114), 114 (5.120), 115 (5.120), 117 (5.128), 118 (5.128), 147 (5.208, 5.217), 169 (6.46), 251 (9.32), 252 (9.32), 255 (9.74), 256 (9.74), 259 (9.32), 361 (12.39), 362 (12.39, 12.40)
Wassermann, N. H., 213 (7.112)
Waterman, K. C., 299 (10.273), 305 (10.273)
Wattley, R. V., 204 (7.74)

Watts, R. J., 330 (11.67)
Wayne, R. P., 20 (1.88), 25 (2.1), 41 (2.1)
Webb, R. T., 171 (6.56)
Webber, S. E., 380 (12.112)
Weber, E., 19 (1.32, 1.54, 1.64), 85 (4.30), 267 (10.8)
Weed, G. C., 292 (10.213)
Weedon, A. C. 51 (3.2), 113 (5.119), 117 (5.128), 118 (5.128, 5.130), 119 (5.134), 295 (10.231)
Weeren, S., 108 (5.105), 109 (5.105)
Weers, J. G., 164 (6.25)
Weiser, J., 287 (10.161)
Weiss, J., 337 (11.109), 339 (11.112)
Weiss, K., 258 (9.112, 9.113)
Weiss, R. G., 217 (7.150), 298 (10.254)
Weissbarth, K. H., 386 (12.140)
Weller, A., 46 (2.14), 48 (2.14), 70 (3.22), 254 (9.55, 9.56), 257 (9.109), 342 (11.129, 11.130)
Welles, D., 381 (12.127)
Wendorff, J. H., 217 (7.156)
Wenska, G., 252 (9.41), 255 (9.41), 260 (9.41)
Werner, T. C., 292 (10.204)
Werner, U., 205 (7.81), 267 (10.27), 284 (10.146), 287 (10.27), 333 (11.96)
Wertz, D. W., 124 (5.147)
West, M. A., 241 (8.65)
Westmoreland, T. D., 125 (5.152), 126 (5.152), 127 (5.152), 128 (5.162), 129 (5.162)
White, A. H., 279 (10.132), 330 (11.64)
White, B. G., 241 (8.66)
White, H. S., 234 (8.50)
White, J. M., 380 (12.112)
White, M. S., 387 (12.162)
Whitten, D. G., 198 (7.11), 199 (7.11), 252 (9.40)
Wieghardt, K., 279 (10.110)
Wienk, M. M., 267 (10.39)
Wientges, H., 253 (9.47), 254 (9.47)
Wierenga, W., 208 (7.90)
Wilbrandt, R., 198 (7.22)
Wilhelm, F. X., 305 (10.316), 306 (10.316)
Wilkinson, G., 83 (4.10), 319 (11.11), 327 (11.11)
Williams, D. J., 19 (1.4, 1.49, 1.52, 1.79, 1.83), 85 (4.33), 201 (7.35, 7.36), 218 (7.36), 267 (10.10), 274 (10.10, 10.91), 275 (10.10, 10.94, 10.95), 278 (10.102, 10.103, 10.105), 280 (10.10), 285 (10.149, 10.105, 10.148, 10.150, 10.151, 10.152, 10.153, 10.154), 286 (10.151, 10.154), 287 (10.156, 10.157), 289 (10.10), 334 (11.104, 11.106), 336 (11.106), 387 (12.178, 12.179)
Williams, R. F. X., 347 (11.155)
Williamson, M. M., 259 (9.134)
Willig, F., 361 (12.31), 363 (12.31), 381 (12.123)
Willner, B., 243 (8.81)
Willner, I., 243 (8.81), 298 (10.255), 299 (10.256, 10.257, 10.258)
Wilner, F. R., 330 (11.64)
Wilowska, A., 121 (5.139), 273 (10.81)
Wilson, G. J., 19 (1.28), 252 (9.39)
Wilson, L. J., 174 (6.64)
Windsor, M. W., 92 (5.19)
Winkler, J. D., 214 (7.115)

Winkler, J. R., 93 (5.26), 131 (5.171), 143 (5.198)
Winscom, C. J., 167 (6.31)
Winzenburg, J., 167 (6.31)
Witt, H., 72 (3.33)
Wittenbeck, P., 215 (7.127)
Wittig, C., 251 (9.33), 252 (9.33)
Woehler, S. E., 93 (5.24)
Wohlers, H. D., 179 (6.93), 180 (6.93)
Wohltjen, H., 19 (1.74), 93 (5.32), 211 (7.104, 7.105), 216 (7.135), 217 (7.147), 342 (11.136), 343 (11.136), 344 (11.136), 345 (11.136), 355 (12.9), 380 (12.9), 377 (12.98)
Woitellier, S., 116 (5.125), 132 (5.125), 211 (7.105), 377 (12.98)
Wokaun, A., 215 (7.127)
Wolf, H. C., 165 (6.29), 166 (6.29)
Wolff, T., 215 (7.118, 7.119, 7.125, 7.126)
Wolstenholme, J. B., 275 (10.94, 10.95), 278 (10.102)
Woltersom, J. A., 208 (7.90)
Won, Y., 20 (1.98), 356 (12.24)
Wong, S. S., 107 (5.101), 113 (5.101)
Woods, R. E., 386 (12.149),
Worl, L. A., 137 (5.188), 150 (5.232), 151 (5.232), 180 (6.97), 191 (6.130), 370 (12.86), 372 (12.86), 386 (12.86)
Wright, D. S., 19 (1.26)
Wright, J. D., 253 (9.44)
Wright, R., 212 (7.108)
Wrighton, M., 198 (7.13), 217 (7.13)
Wrighton, M. S., 20 (1.92), 25 (2.1), 41 (2.1), 121 (5.140), 347 (11.155), 355 (12.4), 364 (12.4), 380 (12.4, 12.118), 388 (12.183)
Wu, K. C., 300 (10.304)
Wu, W., 19 (1.44)
Wu, Z., 164 (6.27, 6.28), 165 (6.28)
Wubbels, G. G., 298 (10.252)

Xu, H. J., 255 (9.73)

Yabe, A., 216 (7.132)
Yakhot, V., 251 (9.24)
Yamada, H., 361 (12.32), 363 (12.32), 383 (12.133, 12.134)
Yamagishi, A., 299 (10.291, 10.292), 304 (10.292)
Yamaguchi, H., 215 (7.128), 216 (7.136)
Yamakita, H., 298 (10.253)
Yamamoto, M., 168 (6.39)
Yamanari, K., 289 (10.179)
Yamashita, I., 207 (7.86), 260 (9.143)
Yamazaki, I., 161 (6.4), 166 (6.30), 257 (9.105), 370 (12.77, 12.80)
Yamazaki, T., 161 (6.4), 370 (12.77, 12.80)
Yanagawa, H., 388 (12.184)
Yanagida, S., 364 (12.50), 380 (12.50)
Yanagita, M., 110 (5.107)

Yang, N. C., 251 (9.32), 252 (9.32), 255 (9.74), 256 (9.74), 259 (9.32), 295 (10.228)
Yao, H., 203 (7.63), 255 (9.65)
Yarwood, J., 248 (9.4), 250 (9.13), 251 (9.4)
Yazdi, P. T., 202 (7.58)
Yeates, T. O., 90 (5.9)
Yellowlees, L. J., 124 (5.146)
Yersin, H., 132 (5.176), 174 (6.66), 177 (6.66), 233 (8.44), 234 (8.44), 235 (8.44), 237 (8.44), 348 (11.164)
Yokoyama, M., 218 (7.169)
Yonemura, H., 299 (10.259)
Yoshida, F., 258 (9.118)
Yoshida, T., 122 (5.142)
Yoshikawa, M., 214 (7.114)
Yoshimura, H., 210 (7.96), 298 (10.245)
Yoshimura, Y., 292 (10.210)
Younathan, J. N., 150 (5.227, 5.231, 5.232), 151 (5.232), 191 (6.130), 365 (12.59), 370 (12.59, 12.86), 372 (12.86), 380 (12.59), 386 (12.59, 12.86)
Yu, Q., 255 (9.73)
Yuan, H. L., 257 (9.100)
Yumoto, T., 298 (10.253)

Zachariasse, K. A., 255 (9.67, 9.68)
Zafiriou, O. C., 85 (4.24), 198 (7.13), 217 (7.13)
Zamaraev, K. I., 363 (12.46)
Zamecka-Krakowiak, D. J., 279 (10.114)
Zannoni, C., 217 (7.149)
Zanobini, F., 279 (10.124)
Zappi, T., 277 (10.96), 278 (10.96)
Zarzycki, R., 274 (10.87), 275 (10.94), 278 (10.103), 280 (10.87), 289 (10.174)
Zengerle, K., 331 (11.69)
Zerbetto, F., 201 (7.43)
Zewail, A. H., 113 (5.116, 5.117), 114 (5.116, 5.117), 198 (7.20)
Zhabotinskii, A. M., 386 (12.157)
Zhao, C., 378 (12.104)
Zhou, M. Q., 267 (10.28)
Zhou, Q. F., 255 (9.73)
Ziessel, R., 178 (6.86), 279 (10.115), 333 (11.92)
Zilkha, A., 334 (11.101, 11.102)
Zimmerman, H. W., 300 (10.302)
Zimmerman, S. C., 19 (1.44), 267 (10.21)
Zinato, E., 186 (6.114), 187 (6.114), 188 (6.114), 189 (6.125)
Zinth, W., 91 (5.17)
Zisk, M. B., 241 (8.72)
Zom, R. L. J., 253 (9.44)
Zou, C., 388 (12.183)
Zulu, M. M., 178 (6.80, 6.81, 6.82)
Zvanut, M. E., 150 (5.225), 380 (12.119)

Subject index

Single supramolecular systems are not indexed because in many cases compact names and formulae are not available. A page number in *italics* indicates the start of a chapter or a section.

activation barrier, *see also* reorganizational energy, 46–48, 68
adducts, 86, *273, 280, 299*
adiabatic electron transfer, 46, 70
antenna effect, 168, 169, 184, 187, 329, *370,* 380–383
antenna-sensitizer complex, 186, 381, 382
anti-Stokes luminescence, 374
assemblies of molecular components, *see* molecular devices *and* photochemical molecular devices
autocatalysis, 388
aza macrocycles, *279*
    adducts with metal complexes, 86, *280*
    complexation of metal cations, *280*

back electron transfer, definition, 63
Belousov–Zhabotinski reaction, 386
bimolecular processes of excited states, *43*
    kinetic aspects, *44*
    thermodynamic aspects, *44*
binuclear metal complexes, 54, *132,* 170–180, 184–190, 272, 333, 341, 342, 367–369, 377
blocker, 85, 86
Born–Oppenheimer approximation, 26, 28, 29
boxes, 285
butterfly, 208

cage cavity, tuning, 331, 332
caged metal ions, *83, 319*
    chromium complexes, *324*
    cobalt complexes, 84, 239, *319*
    lanthanide complexes, 243, *326,* 380
    ruthenium complexes, *330*
calixarenes, 329, 330
capped systems, electron transfer, *120*
carcerands, 333
catalysis, multielectron, 365
catalytic activity, photocontrol, 210, 298
catenands, *338*
catenanes, 285, *334,* 387

catenates, *338*
cation-lock effects, 207, 273
cation-steered interactions, 269, 270
cavity size, photocontrol, 213, *377*
channels, 388
charge injection on semiconductors, *381*
charge recombination, *see* back electron transfer *and* charge separation
charge separation, 89
    biomimetic, 94
    dyads, *98, 102, 107, 110, 125, 132,* 358, 359
    natural photosynthesis, 90
    tetrads, *145,* 363, 364
    transmembrane, 147, *365*
    triads, *145,* 358, 382, 383
charge shift, in covalently linked systems, 89
    electron transfer, 98, 163
    hole transfer, 98, 163
charge-transfer complexes, *see* electron donor–acceptor complexes
charge-transfer emission, 63, 105, 249, *254*
chromoionophores, *268, 280*
concave molecular species, *see* host–guest systems
Condon approximation, 35
connectors, *52,* 89, 97, *161, 197,* 356, 358, 367, *375*
constraints to nuclear motions, *82*
    covalent links, *83*
    host–guest complexation, *85*
coordination ability
    control by acid–base processes, 204
    control by metal coordination, 204
    control by redox processes, 204
    control by temperature, 204
    cooperative, photocontrol, 208, 209
    photocontrol, *204,* 208, 210
covalently linked systems, electron transfer, *89*
    aromatic bridges, *107*
    as models, 93
    bicyclooctane bridges, *113*

capped porphyrins, *120*
  donor–acceptor (CLDA), *89*
  flexible bridges, *116, 121*
  metal complexes, *123*
  norbornylogous bridges, *102*
  oligopeptide bridges, 130
  polyene bridges, *114*
  porphyrin–host, *122*
  porphyrin–quencher, *110*
  porphyrin–quinone, *119*
  porphyrin–viologen, *121*
  steroid-type bridges, *98*
  trypticene bridges, *110*
covalently linked systems, energy transfer, *161*
  conformationally restricted bridges, *161*
  flexible bridges, *167*
  metal complexes, *170*
  metal-containing fragments bridged by
    cyanide, *184*
  metal-containing fragments bridged by
    polyimine ligands, *174*
  norbornylogous bridges, 162
  organic molecular components, 161
  steroid-type bridges, 162
  rigid bridges, 161
Crosby rule, 43
crown ethers, *268*
  adducts with metal complexes, *274*
  anionic crown-ether dyes, 268
  aza crown-ether dyes, 268
  complexation of metal cations, *268*
  cryptands, 268
  energy transfer, 272
  excimer formation, 272
  exciplex formation, 273
  in photoflexible supramolecular systems, *204*
  phenolic crown-ether dyes, 268, 270
  photochemical reactions, 272
  photoinduced electron transfer, 123, 274
  photoinduced proton transfer, 278
  photoresponsive, *204*
  spherands, 268
cryptands, 268, 273, 284, 327, 331
cryptates, 243, 244, 273, 279, *326*, 331
cyclodextrin inclusion compounds, 283, *288*
  co-inclusion effects, 290
  colloidal particles, 299
  energy transfer, 295
  heavy atom phosphorescence enhancement,
    80, 290
  intermolecular excimers, 292
  intermolecular exciplexes, 294
  intramolecular excimers, 292, 293
  intramolecular exciplexes, 294
  intramolecular photodimerization, 293
  metal complexes, 290–292
  photochemistry, *295*
  photoinduced electron transfer, 119, 295
  photoinduced proton transfer, 289
  photophysics, *289*
  protection from back electron transfer, 299
  protection from quenchers, 292

cyclophane hosts, *283*
cylinders, *205*

derivatized electrodes, 380–386
dyads, *see also* charge separation *and* covalently
  linked systems, 19, 94, 358, 367
ditopic receptors, 205, 279, 333
DNA adducts, *299*
  conformation-specific cleavage, 304
  conformation-specific recognition, 304
  intercalation binding, *299*
  ion pairs, 299
  photochemistry, *304*
  spectroscopy and photophysics, *301*
double helical complexes, *see* helicates

efficiency, excited-state processes, 42
electric field effects, 252
electron donor–acceptor complexes, *89, 248, 268,*
  276, 334, 387
  absorption spectra, *251*
  charge-transfer character, 254
  cyclophane-type systems, 253, 254
  dipole moment, 254
  intersystem crossing, 256
  luminescence, *254*
  photoinduced processes, 257
  solvent effects, 258
  supersonic nozzle beams, 256
electronic configurations, states, and transitions,
  30
electronic coupling
  distance dependence, 97, 162
  electron transfer, 54, 62, 67, 70, 71, 94, 97,
    163, 252
  exchange energy transfer, 162, 163
electronic transmission coefficient, 46, 70, 94
electron store, 357, 360, *363*
electron-transfer kinetics, *45, 65,* 94
  classical model, 66, 230
  distance effect, 101, 104, 109, 111, 114, 116,
    139
  effect of driving force, 98, 104, 110
  effects of stereochemistry, 101, 106, 113, *119*
  empirical free-energy relationship, 48
  Marcus free-energy relationship, 47, 68, 230
  Marcus inverted region, 47, 69, 99, 100, 259
  quantum mechanical models, 66, 94
  solvent effects, 99, 104, 109, 118
  superexchange mechanism, 95, 96
  temperature effects, 109
  vs energy transfer, *73,* 162, 163
electron-transfer photosensitization, remote, *365*
electron-transfer processes
  covalently linked systems, *89, 358*
  electron–donor acceptor complexes, 257
  host–guest systems, 123, 273
  ion pairs, *231*
  monolayer assemblies, 93, *382*
  photoinduced, *see also* excited-state processes,
    65
  polymers, 93

proteins, 93
rigid glasses, 93
thermal, *63*
electrostatic work terms, 62
encapsulated metal ions, *see* caged metal ions
encapsulation–deencapsulation, photocontrol, 333
energy-gap law, 39, 95
energy-transfer kinetics, 43, *71, 73*
  coulombic mechanism, 72, 166, 168, 172, 174, 184
  distance dependence, 162
  exchange mechanisms, 45, 72, 162, 174, 184, 187, 189, 190, 370, 371
  vs electron transfer, *73*, 162, 163
energy-transfer photosensitization, remote, *371*
energy-transfer processes
  covalently linked systems, *161, 367*
  crown-ether complexes, 272
  ordered phases, 161
  organzed molecular assemblies, 161
  polymers, 161
  polynuclear metal complexes, *170, 367*
  sensitizer–antenna devices, *381*
  solid-state materials, 161, 373–375
  vectorial, 372, 374
energy up-converter, 357, 374
excimers, 250
  intermolecular, 292
  intramolecular, 255, 292, 293
  photochemical reactions, 261
  transition-metal complexes, 257
  triplet, 256
exciplexes, *248,* 250
  intermolecular, 294
  intramolecular, 294
  luminescence, *254*
  photochemical reactions, 261
  polychromophoric systems, 255, 256
  polymers, 257
  solvent role, 255
  triple, 255
  triplet, 256
excited-state processes
  bimolecular, *43*
  chemical reaction, *40*
  electron transfer, *43, 65, 89, 231, 257, 304, 358*
  energy transfer, *43, 71, 161, 367*
  quenching, *43*
  radiationless deactivation, *38*
  radiative deactivation, *38*
  tautomerization, *345*
  unimolecular, *37*
  vibrational relaxation, *37*
excited-state properties, 59
  control and tuning, *76, 243*
  decay kinetics, *41, 45*
  distortion, 37, 38
  lifetime, 42, 45
  perturbation, *77, 80, 82*
  thermodynamic aspects, 44

face-face interactions, 254
Fermi Golden Rule, 39, 94, 162
fluorescence, *see* light emission
fluorescent concentrators, 380
fluoroionophores, 270
footprinting techniques, 304
fractal semiconductor electrodes, 381
Franck–Condon principle, 36, 37, 46, 67
Franck–Condon factor, 37–40, 94

gates, 378
giant dipole, 104

helicates, *341*, 387
hemicaged complexes, 332
heptanuclear metal complexes, 181–183, 370, 382
holders, 357, 377, 378
hole transfer, *see also* electron transfer, 95
host–guest systems, *85*, 122, *267*
  aza macrocycles, *279*
  crown ethers, *268*
  cyclophanes, *283*
  cyclodextrin inclusion compounds, *288*
  DNA adducts, 299

intercalation, *299*
intercomponent processes, *see also* electron-transfer processes *and* energy-transfer processes, *60*
interface toward light, 358, 367
interfacing terminals, 380
internal conversion, 39
intersystem crossing, 39
intervalence transfer, 55, 56, 61, *132, 140, 211,* 227
intrinsic nuclear barrier, *see* reorganizational energy
ionophores, 268
ion pairs, *226*, 299
  charge-transfer transitions, 227–232
  classification, 226
  contact, 226, 254, 258, 259
  electron-transfer rates, *236*
  energy levels, 227
  geminate, 226, 258, 259
  luminescence, *243*
  photochemical reactions, *237*
  photoinduced electron transfer, *231*
  solvent-separated, 254, 258
  solvent-shared, 258
  static quenching, *233*

Jablonski diagram, 32, 34

Kasha rule, 40, 42
knots, 339, 340, 387

Langmuir–Blodgett films, 363, *382*
  photoswitching, 216
lifetime, definition, 42, 45
light absorption, *34*

as an electron pump, 358
as an electron switch, 358
selection rules, 35
light emission, *38*
  spin-allowed (fluorescence), 38
  spin-forbidden (phosphorescence), 38
  spontaneous, 38
light energy conversion
  into chemical energy, 358, *359*
  into electrical energy, *359*
  up-, *374*
light, nature, *25*
liquid crystals, 21, 387
  photoinduced phase transition, 217
  photoresponsive, *217*
luminescence shifters, 367
luminescent labels, 329, 380
luminophores, 319, 357, 367–373

macrobicyclic hosts, 284, 285
macroscopic devices, 355
magnetic field effects, 257, 260, 299
Marcus inverted region, 47, 69, 99, 110, 259
membranes
  bilayer, photodestabilization, 216
  bilayer, photoinduced current, 216, 365
  liquid, photocontrol of ion transport, 208, 214,
    377, 380
  permeability, photocontrol, 216, 378
  photoinduced aminoacid transport, 216
  photoinduced potential change, 216
  properties, photoregulation, *215*, 216, 380
micellar catalysis, photocontrol, 215, 380
micelles, 387
  photoeffects, *215*
  photoinduced formation, 215
microelectronics, 355
microfabricated array of electrodes, 388
mixed valence systems, *see also* intervalance
  transfer, 54, 377
molecular assemblies, 370, 386
molecular components, 20, 51
  active, 52, 89, *197, 356*
  blocker, *85*, 86
  connecting, *see* connectors
  electron store, 357, 360, *363*
  electron-transfer photosensitizer, 322, 357,
    358, 381–385
  electron transfer relay, 321, 357, *358*, 383–385
  energy up-converter, 357, 374
  heavy-atom perturber, 80
  holder, 357, 377, 378
  intercomponent processes, *60*
  luminophore, 319, 357, 367–373
  perturbing, *76*, 356
  photoisomerizable, *197*, 297, 357, *375*
  properties, *58*
  quencher, *80*, 86
  tuner, *77*, 86
molecular computing elements, 380, 384
molecular devices

artificial vs natural, *356*
electronic, 345, 367, 380
photochemical, *355*
molecular information, reading and
  amplification, 388
molecular layers, 370, 387, 388
molecular photochemistry, principles, 25
molecular probes, 300, 380
molecular recognition, 205, 267, 271, 300, 304,
  388
molecular rectification, 366
molecular scaffolds
  DNA duplexes, 151, 306
  polymers, 150
molecular shift register, 366, *383*
molecular switch, 366
molecular transportation, 267
molecular wires, 116
Mulliken theory, 248

nanotechnology, construction sets, 386
nonadiabatic electron transfer, 48, 70, 95
nuclear frequency, 46
nuclear tunnelling, 95

optical electron transfer, *61*
  covalently linked systems, 89, 104, 140
  ion pairs, 229
  mixed-valence compounds, *see* intervalance
    transfer
optical modulators, 380
optical reading devices, 268
optical signals, remote generation, 372
optical storage memories, 216, 380
organization of molecules
  as requisite for function, 93, 356
  spontaneous, 342, *385*
oxygen indicators, 292

Pauli principle, 31
pentad, 150, 168, 370
phosphorescence, *see* light emission
phosphors, 374
photochemical molecular devices, *355*
  active components, 356
  connecting components, 357
  electronic energy transfer, *367*
  interfacing with the macroscopic world, *380*
  machinery, *356*
  perturbing components, 356
  photoinduced electron transfer, *358*
  photoinduced structural changes, *375*
photochromic systems, *197*
  dual-mode memory, 215
photoeleastic structures, 205
photoflexible supramolecular systems, *197*
photoinduced coreceptor catalysis, *378*
photoinduced electron collection, *365*
photoinduced electron transfer, *65, 89*
  covalently linked systems, *89*
  DNA derivatives, 306

efficiency, 65
macromolecular systems, 150
molecular shift register, *383*
photochemical molecular devices, *358*
rate constant, 65
vectoral, 355, 361
photoinduced proton transfer, *342*
metal complexes, *347*
tautomerization reactions, *345*
photoinduced structural changes, *197*
liquid crystals, *217*
membranes, *215*
micelles, *215*
other systems, *212*
polymers, *217*
supramolecular systems, *204*
systems involving crown ethers, *204*
systems involving cyclodextrins, *210*
systems involving metal complexes, *211*
photoinduced switch, *366, 376*
electric signals, 214, *366, 376*
receptor ability, 205–209, 355, *377*
photoisomerizable component, *197*, 297, 357, *375*
photoisomerizations
azo-type compounds, *199*
effects on supramolecular systems, *204*
olefin-type compounds, *198*
restrictions, 85, 86, 297
spiropyrans, 200
photomechanic effects, 218, 380
photoreactivity
control, 282, 295, 333, 380
effect of cavity dimension, 298
effect of cavity polarity, 298
probe for adduct structure, 87, 284
protection against, 81–86, 169, 187, 276, 370
photoresponsive CDs, *210*
azobenzene appended, 210
azobenzene capped, 210, 298
photoresponsive crown ethers, *204, 377, 378*
butterfly and capped, *208*
cyclindrical and phane-type, *205*
with intramolecular bridges, *204*
photoresponsive
inophores, *204, 205, 208*, 214, 377, 380
ion pump, 213
polymers, *217*
polypeptides, 218
photosensitizers, 319, 330, 331
electron transfer, 322, 357, *358*, 381–385
energy transfer, 357, *367, 381*
photosynthesis, 90–92, 168, 370
artificial, 359
bacterial reaction centre, 90, 91, 356
simulation, 383
phototransistors, 366, 377
phototweezers, 208, 377
photoviscosity effects, 215, 216, 218, 380
photovoltaic cells, molecule-based, 361
podands, 171, 269
polymers, 21, 363, 370, 386
functionalized, 372, 386

photoeffects, 218, 380
photoresponsive, *217*
soluble, 372
polynuclear metal complexes, *see also* bi-, tri-,
tetra-, and heptanuclear metal complexes,
140, 170, 364, 382
electron transfer, *140*
potential energy curves (surfaces), *26*, 36, 39, 47,
55, 61–71, 77–82, 199, 203, 229, 249, 322, 325,
331, 343
potential ligands, *see* holders
proton transfer lasers, 343
pulse radiolysis, 64, 98, 105

quantum yield, 42
quencher, 81, 86
quenching processes, 43, 45, 46, 72
dynamic mechanism, 45, *233*, 238
protection from, 292
static mechanism, *233*, 238

radiationless deactivation, *38*
radiative deactivation, *38*
relays
electron transfer, 321, 357, *358*, 383–385
energy transfer, 357, *367*, 377
reorganizational energy, 46–48, 54, 62, 68, 95
rotaxanes, *334*, 387

scaffolding-like materials, 306, 386
scanning tunneling microscope, 386
Schroedinger equation, 27
second-harmonic generation, 295
second-sphere coordination compounds, 273,
280, 289, 290
self-organization of matter, *385*
semiconductors, 21, 363, 370, *381*
electrodes, 370, *381*
powders, 380
sensors, 270
sepulchrands, 320
sepulchrates, 320, 323
singlet–triplet switch, 165
solar energy conversion, 356, *359*, 381
solvent repolarization, 46
spacers, 331, 332
spectral sensitization, 187, *367*
semiconductor electrodes, *381*
spherands, 268, 333
spin–orbit coupling, 35
starburst dendrimers, 386
Stern–Volmer equation, 45
Stokes shift, 38, 44
supercomplexes, *see also* second-sphere
coordination compounds, 274, 280
superexchange, 95, 96, 101, 109, 111–113, 142
superhelical strands, 388
supermolecules, 19, *51*
as localized multicomponent electronic
systems, *54*
vs large molecules, 51–53
supramolecular

chemistry, 19
photochemistry, 20
properties, *51*
shift register, *383*
synthesis of supramolecular systems, *385*
  complexes-as-ligands strategy, 181, 386
  structure directed synthesis, 386
  template, based on donor–acceptor
    interactions, 334
  template, based on metal coordination, 337

tetrads, 19, 93, *145*, 363, 370, 371
tetranuclear metal complexes, 171–177, *180*, 370,
  371, 387
third-sphere ligands, 289
through-bond interactions, 95–116, 125, 140, 162,
  163, 254

through-space interactions, 95–98, 121, 254
time-resolved microwave conductivity, 104
triads, *see also* charge separation, 19, 93, *145*
trinuclear metal complexes, 140, 144, 171–177,
  180, 184, 381, 382
triplet perturbers, 270
tuner, 78, 86
twisted-intramolecular-charge-transfer species,
  63, *202*, 211, 218, 251, 298, 377
  fluorescence probes in polymers, 218
  ligands in metal complexes, 211, 377

vesicular systems, 363
vibrational relaxation, *37*
vibronic coupling, 36
vision, 197

BC